CHAPMAN & HALL/CRC
Monographs and Surveys in
Pure and Applied Mathematics   134

AN INTRODUCTION

TO SEMIFLOWS

T0187819

CHAPMAN & HALL/CRC
Monographs and Surveys in
Pure and Applied Mathematics   134

AN INTRODUCTION

TO SEMIFLOWS

Albert J. Milani

Norbert J. Koksch

CRC Press
Taylor & Francis Group
Boca Raton London New York

CRC Press is an imprint of the
Taylor & Francis Group, an **informa** business

A CHAPMAN & HALL BOOK

First published 2005 by Chapman & Hall

Published 2019 by CRC Press
Taylor & Francis Group
6000 Broken Sound Parkway NW, Suite 300
Boca Raton, FL 33487-2742

© 2005 by Taylor & Francis Group, LLC
CRC Press is an imprint of Taylor & Francis Group, an Informa business

First issued in paperback 2019

No claim to original U.S. Government works

ISBN 13: 978-0-367-45428-9 (pbk)
ISBN 13: 978-1-58488-458-3 (hbk)

Visit the Taylor & Francis Web site at
http://www.taylorandfrancis.com

and the CRC Press Web site at
http://www.crcpress.com

Library of Congress Cataloging-in-Publication Data

To Martina, Andrea, Daniel, Stephan and Claudia

Contents

Preface

1. In these notes we present some introductory material on a particular class of dynamical systems, called SEMIFLOWS. This class includes, but is not restricted to, systems that are defined, or modelled, by certain types of differential equations of evolution (DEEs in short). Our purpose is to introduce, in a relatively self-contained manner, the basic results of the theory of dynamical systems that can be naturally extended and applied to study the asymptotic behavior of the solutions of the DEEs we consider. Equations of evolution include ordinary differential equations (ODEs in short), partial differential equations of evolution (PDEEs in short), and other types of equations, such as, for instance, stochastic or difference equations. As such, they provide natural examples of dynamical systems, since one of the independent variables (usually called "time") plays a different role than the other variables (which in some situations may be called "space" variables). Thus, in this context, the heat and wave equations are considered as prototypical examples of PDEEs, while elliptic equations such as Laplace's equation are not considered as evolution equations, because in these equations all the variables have the same role. Here, we make the further distinction that "time" evolves continuously; thus, we do not consider stochastic equations, nor, except for some introductory examples, discrete systems (where "time" varies along a sequence).

2. One of the major goals of the theory of dynamical systems is the study of the evolution of a system, with the purpose of predicting, as accurately as possible, the behavior of the system as time becomes large. A quite general feature of the systems we consider, which is shared with other systems, is a property called DISSIPATIVITY. Loosely speaking, this property can be described by the fact that all solutions of these systems eventually enter, and remain, in a bounded set, called ABSORBING SET. Thus, the evolution of the solutions of the system can be studied in this set; as a result, the long time behavior of the system can be described by means of certain subsets of the absorbing set. Among these, we shall consider three types of sets, called respectively ATTRACTORS, EXPONENTIAL ATTRACTORS, and INERTIAL MANIFOLDS. (Exponential attractors are sometimes also known as INERTIAL SETS.) We will present the fundamental properties of these sets, and then proceed to show the existence of some of these sets for a number of dynamical systems, generated by fairly well known physical models. In particular, we shall consider in full detail two particular PDEEs of evolution: a semilinear version of the heat equation, and a corresponding version of the dissipative wave equation. These examples allow us to illustrate the most important features of the theory of semiflows, and to provide a sort of "template" that can then be applied, in a more or less straightforward fashion, to the analysis of other models, with the help of the many specialized references that exist in the literature.

3. Even a quick survey of much of the existing literature on dynamical systems, both at the introductory and the specialized level, reveals that the notion of "dynamical system" is used with many different meanings, according to the specific point of view of the authors. At the opposite extreme, this notion may well be not defined at all. In these notes, we do not attempt to give a general definition of dynamical system; rather, we confine ourselves to a special class of systems, properly known as CONTINUOUS, SEMI-DYNAMICAL SYSTEMS, or CONTINUOUS SEMIFLOWS. Here, the term "continuous" is used to distinguish these systems from DISCRETE ones, where only a sequence of successive time values are considered, and "semi-" refers to the fact that time evolves, i.e. we only consider nonnegative values of the time variable. For brevity, we shall refer to these systems as SEMIFLOWS (their precise definition is given in section 2.2). In the introductory chapter 1, we consider more general TWO-PARAMETER SEMIFLOWS or DYNAMICAL PROCESSES, which allows us to include some nonautonomous difference or differential equations as generators of dynamical systems. However, when our presentation can proceed in a more discursive way, and rigor is not an issue, we conform to the common use and adopt the general term "dynamical system".

4. In general, we say that an ODE defines a semiflow if the corresponding CAUCHY PROBLEM is globally well posed, in the sense we define in section 1.2.1. We can extend this definition to semiflows defined by PDEEs, by interpreting the PDEE as an abstract ODE in a suitable Banach space $\mathcal{X}$ (see remark 3.2 in chapter 3). This is a generalization of the usual interpretation of a system of ODEs as a single differential equation in the Banach space $\mathcal{X} = \mathbb{R}^n$, or in more general finite dimensional vector spaces, and explains the qualification of the systems generated by PDEEs as "infinite dimensional" ones, since in this case $\mathcal{X}$ is in general no longer a finite dimensional space. Examples of PDEEs that can be put in such abstract form are: the Navier-Stokes equations, the Kuramoto-Sivashinski equations, the "original" Burger's equation, the Chafee-Infante and Cahn-Hilliard reaction-diffusion equations, the Korteweg-de Vries and the Maxwell equations (see chapter 6). Indeed, many basic notions and results in the theory of the asymptotic behavior of infinite dimensional dissipative dynamical systems trace their origin in the study of the Navier-Stokes equations of fluid dynamics, and have been inspired by a detailed analysis of both the qualitative properties of their solutions, and their behavior with respect to numerical computations.

5. Not surprisingly, much of the general terminology in the theory of dynamical systems, as well as the general spirit of its qualitative results, borrows directly from the qualitative theory of ODEs in $\mathbb{R}^n$. For example, we shall need to recall some basic results on stability, equilibrium points, periodic orbits, ω-limit sets, etc. On the other hand, in an effort to keep these notes within a reasonable length, we shall be forced to not discuss many other important topics. In particular, we regretfully do not include any result on bifurcation theory. Among the many excellent and fairly complete references on the qualitative theory of ODEs, including ODEs as dynamical systems, we refer for example to Hirsch and Smale, [HS93], Jordan and Smith, [JS87], Perko, [Per91], Amann, [Ama90], and Verhulst, [Ver90]. A few other references, specifically on dynamical systems, are listed in the bibliography. Since so many articles and books are continually being published, it is almost impossible to compile an exhaustive list of references; on the other hand, an internet search can provide all necessary updated references on any particular topic.

6. These notes have their origin in a series of graduate seminars we held at the Universities of Dresden, Wisconsin-Milwaukee and Tsukuba. Most of the material we cover is relatively well known, although some of the results we present, in particular on the existence of an exponential attractor and of an inertial manifold for semilinear dissipative wave equations, even if not entirely new, do not seem to enjoy the recognition we feel they deserve. In part, our intention in writing these notes is to be of some help to "beginners" in the area of infinite dimensional dynamical systems; that is, anyone who, having a solid background in the classical theory of ODEs and some knowledge of functional analysis in Sobolev spaces, wishes to proceed to the study of examples of semiflows arising from DEEs, but may need some "smoothing into" the subject, before turning to more general introductory texts, such as those of Temam, [Tem88], the cycle of lectures by Oleinik, [Ole96], or, most recently, Sell-You, [SY02], and Robinson, [Rob01]. We also hope that these notes may serve as a ready reference to researchers in more applied fields, who feel the need for a clear presentation of the background material and results that are necessary for the study of the specific systems they are interested in. To this end, we have tried to "build up" the material in as careful and gradual progression as possible, with the goal of presenting the main topics (in particular, the construction of the exponential attractor and the inertial manifold), with a larger degree of detail than generally found in most sources in the literature. If successful, our effort should put the reader in a better position to refer to more specific texts on global attractors, exponential attractors, and inertial manifolds, such as, respectively, the books by Babin and Vishik, [BV92], Eden, Foias, Nicolaenko and Temam, [EFNT94], and Constantin, Foias, Nicolaenko and Temam, [CFNT89].

7. These notes are organized as follows. As an introduction to the main ideas of the abstract theory of semiflows, in chapter 1 we present some well known and well studied examples of finite dimensional dynamical systems, generated by such celebrated ODEs as Duffing's equations and Lorenz' equations. In chapter 2 we introduce the general definitions of SEMIFLOWS and their GLOBAL ATTRACTORS, and we present two sufficient conditions that guarantee the existence of the attractor under different assumption on the asymptotic properties of the semiflow. We also describe an alternate construction of the attractor, based on the idea of α-contracting maps. In chapter 3 we apply these results to show that the semiflows generated by two types of semilinear dissipative evolution PDEEs (one parabolic and the other hyperbolic) admit a global attractor in a suitable space of weak solutions. In chapter 4 we briefly develop the theory of EXPONENTIAL ATTRACTORS, and apply this theory to the models of PDEEs considered in chapter 3. In chapter 5 we present Hadamard's GRAPH TRANSFORMATION METHOD for the construction of an INERTIAL MANIFOLD, and apply this method to a one-dimensional version of the PDEEs considered in chapter 3. In chapter 6, we consider a number of other dynamical systems, generated by PDEEs that model various mathematical physics problems, and briefly show how the methods developed in the previous chapters can be applied. In chapter 7 we present a result, due to Mora and Solà-Morales, on the nonexistence of inertial manifolds for the semiflow generated by a one-dimensional version of the hyperbolic model of PDEE considered in chapter 3. Finally, in the Appendix we collect, for the reader's convenience, a list of various definitions and results from the classical theory of ODEs and PDEs, functional and nonlinear analysis, semigroup theory and Lebesgue-Sobolev spaces, that we use in these notes, and provide at least one reference for each of the definitions and theorems we state.

Acknowledgements. Both authors have been partially supported by the Alexander von Humboldt Stiftung. The second author also had partial support from the Japanese Society for the Promotion of Sciences and the Institute of Mathematics of Fudan University in Shanghai, and is grateful to his colleagues at the Institut für Analysis of the TU Dresden for their warm hospitality. We are also greatly indebted to Professor Songmu Zheng of Fudan University for very kind and stimulating discussion, and to the anonymous referees of this book for various helpful suggestions and corrections.

Chapter 1

Dynamical Processes

In this chapter we introduce the definition of DYNAMICAL PROCESS, and the main
ideas of the theory of dynamical systems that we want to investigate. We illustrate
these ideas by examining some simple examples of dynamical processes generated
by finite systems of ODEs and by iterated maps.

1.1 Introduction

1. Roughly speaking, the theory of dynamical systems studies mathematical models of physical "systems" which evolve in time from a "state" which is known at
an initial moment; more specifically, how the evolution of a system depends, or is
influenced by, its initial state. The changing density of a population from a known
number of individuals (e.g., sharks in a regional sea; bacteria in an infected organism;
prey-predator models); the changing of weather patterns in a particular region; the
spreading of a rumor; the vapor trail in the wake of an airplane; the propagation of a
fire: all these would be examples of dynamical systems.

To study the evolution of a system, we assume that its state at each time t can be
described generally by means of a function $t \mapsto u(t)$, where the independent "time"
variable t is measured in a parameter set $\mathcal{T} \subset \mathbb{R}$, and the corresponding dependent
variable is in a set $\mathcal{X}$, called STATE SPACE. We also assume that the state $u(t)$ of the
system at any given time t depends not only on the value of t, but also on its initial
configuration, i.e. on the value u_0 of the system at a previous time t_0, with u_0 and t_0
given or known. A natural goal of the theory is then to study the dependence of the
state $u \in \mathcal{X}$ on the time $t \in \mathcal{T}$ and the INITIAL VALUES $u_0 \in \mathcal{X}$, $t_0 \in \mathcal{T}$. In particular,
we can think of a dynamical system as a way of transforming an initial state u_0 into
a family of subsequent states $u(t)$, parametrized by $t \in \mathcal{T}$. We shall indeed assume
that there is a specified functional dependence of $u \in \mathcal{X}$ from $u_0 \in \mathcal{X}$ and t, $t_0 \in \mathcal{T}$,
described by a map

$$(t, t_0, u_0) \mapsto u(t, t_0, u_0). \tag{1.1}$$

By specifying certain properties of this map, we come to a definition of a special
kind of dynamical systems.

DEFINITION 1.1   *Let $\mathcal{X}$ be an arbitrary set, and $\mathcal{T}$ be one of the sets $\mathbb{N}$, $\mathbb{Z}$, $\mathbb{R}_{\geq 0}$ or $\mathbb{R}$, where $\mathbb{R}_{\geq 0} := [0, +\infty[$. Set*

$$\mathcal{T}_*^2 := \{(t, \tau) \in \mathcal{T} \times \mathcal{T} : t \geq \tau\}.$$

A TWO-PARAMETER SEMIFLOW, *or* DYNAMICAL PROCESS *in $\mathcal{X}$ is a family $S = (S(t, \tau))_{(t,\tau) \in \mathcal{T}_*^2}$ of maps $S(t, \tau) \colon \mathcal{X} \to \mathcal{X}$, which satisfies the following conditions:*

$$\forall t \in \mathcal{T}: \quad S(t,t) = I_{\mathcal{X}} \tag{1.2}$$

(the identity in $\mathcal{X}$), and

$$\forall t_1, t_2, t_3 \in \mathcal{T}: \quad S(t_1, t_2)S(t_2, t_3) = S(t_1, t_3). \tag{1.3}$$

The following are familiar examples of dynamical processes.

Example 1.2
Let $\mathcal{X} = \mathbb{R}$ and $\mathcal{T} = \mathbb{R}$. Let f be a continuous function on $\mathbb{R}$, and $S = (S(t, \tau))_{(t,\tau) \in \mathcal{T}_*^2}$ be the family of maps $S(t, \tau) \colon \mathbb{R} \to \mathbb{R}$ defined by

$$S(t, \tau)x := \left(\exp \left(\int_\tau^t f(s) \, \mathrm{d}s \right) \right) x, \quad x \in \mathbb{R}. \tag{1.4}$$

Then, S is a dynamical process in $\mathbb{R}$. Indeed, verification of (1.2) and (1.3) is immediate.   ∎

Example 1.3
Let $\mathcal{X} = \mathbb{R}^n$, and A be an $n \times n$ matrix. Then, the family $T = (\mathrm{e}^{tA})_{t \in \mathbb{R}}$ of the exponentials of the matrices tA is a linear semigroup in $\mathcal{X}$ (see section A.3). Consequently, the family S defined by

$$S(t, \tau) := \mathrm{e}^{(t-\tau)A}, \quad (t, \tau) \in \mathbb{R}^2,$$

is a dynamical process.   ∎

Note that, in these examples, each map $S(t, \tau)$ is linear; as we shall see, this needs not be the case in general.

According to definition 1.1, a dynamical process S on a set $\mathcal{X}$ consists of a family of transformations of $\mathcal{X}$ into itself, each defined by the map (1.1), that is,

$$\mathcal{X} \ni u_0 \mapsto u(t, \tau, u_0) =: S(t, \tau)u_0 \in \mathcal{X}. \tag{1.5}$$

We are then mainly interested in the dependence of the map $t \mapsto S(t, t_0)u_0$ on the "initial values" t_0 and u_0 or, sometimes, on u_0 only, for fixed t_0. Of course, this requires $\mathcal{X}$ to have some kind of topological structure, and we shall remove the provisional nature of definition 1.1, supplementing it by a number of continuity conditions on the maps $S(t, \tau)$ on $\mathcal{X}$, and of the map $(t, \tau) \mapsto S(t, \tau)$. In particular, as the examples we cited above indicate, we are often interested in being able to describe, or determine, the evolution of a given system "in the future". This question can be clearly related to the asymptotic properties, as $t \to +\infty$ (in $\mathcal{T}$), of the map defined in (1.1). Because of (1.5), we are then naturally led to relate the asymptotic behavior of the function u to some suitable properties of the corresponding dynamical process S, defined by (1.5). For example, a possible question would be to determine all the values $(u_0, t_0) \in \mathcal{X} \times \mathcal{T}$ such that the limit

$$\lim_{t \to +\infty} S(t, t_0) u_0 =: L(u_0) \tag{1.6}$$

exists, for a fixed t_0. As an illustration, if S is the dynamical process defined in (1.4), the limit in (1.6) exists for all $u_0 \in \mathbb{R}$ if f is bounded above by a negative constant. Note that, since in this case $L(u_0) = 0$ for all $u_0 \in \mathbb{R}$, this limit is actually independent of the initial value u_0. Another, related question would be to study the properties of the map $u_0 \mapsto L(u_0)$ defined by (1.6).

2. In these notes, we assume that the state space $\mathcal{X}$ is at least a Banach space (on $\mathbb{R}$), and the underlying time parameter set $\mathcal{T}$ will be either $\mathbb{N}$ or $\mathbb{Z}$, in which case we call the system DISCRETE, or $\mathbb{R}_{\geq 0}$ or $\mathbb{R}$, in which case we call the system CONTINUOUS. In this chapter we propose to give a first idea of the nature of the questions, related to the long time behavior of dynamical processes, that we want to investigate. To do so, we consider some introductory examples of discrete dynamical processes, generated by iterated maps, and of continuous dynamical processes, generated by finite systems of ODEs. In these cases, the Banach space $\mathcal{X}$ has finite dimension, and the corresponding dynamical process is also called FINITE DIMENSIONAL. In chapter 3 we will instead consider INFINITE DIMENSIONAL dynamical processes, generated by PDEs of evolution. In this case, the space $\mathcal{X}$ is infinite dimensional; specifically, a space of functions of some "space" variables, defined on a domain of $\mathbb{R}^n$.

One can find a large amount of examples of this type of systems in specialized texts, such as Jordan-Smith, [JS87], Marsden-McCracken, [MM76], Guckenheimer-Holmes, [GH83], Moon, [Moo92], Alligood-Sauer-Yorke, [ASY96], and many others. Among the most studied examples, we recall the models known as Duffing's equation, the logistic equation, the Lorenz system, and Hénon's horseshoe map. Most of these also illustrate another major goal of the theory of the dynamical systems, which, regretfully, we cannot pursue because of the introductory character of these notes. Namely, all these systems depend on various numerical parameters, and the influence of these parameters on the long time behavior of the system exhibits some striking phenomena, and unexpected similarities among these systems. In particular, even if the parameters are allowed to vary in a continuous fashion, and even if for a certain range of the parameters the evolution of the system seems to be quite "regular", for other parameter ranges a number of other, totally new qualitative phenomena unexpectedly appear. Examples of such phenomena are BIFURCATIONS (see e.g. Marsden-McCracken, [MM76]), FEIGENBAUM CASCADES (e.g. for the logistic map described in section 1.4.4; see e.g. Moon, [Moo92], or Feigenbaum, [Fei78]), and HORSESHOE MAPS (e.g. for Hénon's map, whose iterations converge to a set of so-called FRACTAL type; see Hénon-Pomeau, [HP76]).

1.2 Ordinary Differential Equations

1.   As a first example of dynamical processes, we consider continuous systems generated by an evolution equation of the form

$$\dot{u} = F(t, u), \tag{1.7}$$

where $F \colon \mathbb{R} \times \mathcal{X} \to \mathcal{X}$ is a continuous function on a Banach space $\mathcal{X}$. In this case, we take $\mathcal{T} = \mathbb{R}$ or $\mathcal{T} = \mathbb{R}_{\geq 0}$. If $\mathcal{X}$ is finite dimensional, (1.7) is equivalent to a system of ODEs in $\mathbb{R}^n$, where n is the dimension of $\mathcal{X}$. An example is the system of m coupled pendulums on the same vertical plane: In this case, if $\theta_1, \ldots, \theta_m$ denote the angles of each pendulum with respect to the vertical, then $u = (\theta_1, \dot{\theta}_1, \ldots, \theta_m, \dot{\theta}_m)$ and $\mathcal{X} = \mathbb{R}^{2m}$. We shall, however, be more interested in the case when the dimension of $\mathcal{X}$ is infinite, and (1.7) represents a PDEE, interpreted as an abstract evolution equation in $\mathcal{X}$. An example is the semilinear heat equation

$$u_t = \Delta u + f(u) \tag{1.8}$$

in a domain $\Omega \subset \mathbb{R}^n$, with appropriate boundary conditions. In this case, the space $\mathcal{X}$ is a space of functions defined on Ω; for example, we can consider the Lebesgue space $\mathrm{L}^2(\Omega)$, or the Sobolev space $\mathrm{H}_0^1(\Omega)$, or the Hölder space $\mathrm{C}^{0,\alpha}(\overline{\Omega})$. We can then interpret PDEEs like (1.8) as abstract ODEs in $\mathcal{X}$ by means of the following natural identification. If u is a solution of (1.8), we define a function $t \mapsto \tilde{u}(t) \in \mathcal{X}$ by

$$(\tilde{u}(t))(x) := u(t, x), \qquad x \in \Omega;$$

that is, we consider for each t the image $\tilde{u}(t) \in \mathcal{X}$ as a function of the space variable x. It is common practice to identify u and $\tilde{u}$, introducing the notation

$$u(t, \cdot) := \tilde{u}(t),$$

which we shall often adopt.

2.   We assume that, in accord with the classical (Newtonian) theory, equation (1.7) is deterministic, in the sense that the knowledge of the initial values (t_0, u_0) (and, of course, of F) uniquely determines a solution u, defined for all "future times", of the Cauchy problem corresponding to (1.7), that is

$$\begin{cases} \dot{u} = F(t, u), \\ u(t_0) = u_0. \end{cases} \tag{1.9}$$

More precisely, we assume that under sufficient assumptions on the function F, there is a unique function $u \in C([t_0, +\infty[; \mathcal{X})$, which satisfies the Cauchy problem (1.9), either in the classical sense (if e.g. u is also in $C^1([t_0, +\infty[; \mathcal{X}))$, or in a generalized sense (e.g. almost everywhere in t, or in distributional sense). This solution is typically determined at first only locally in time, that is, on a neighborhood $]t_0 - \alpha, t_0 + \beta[$ of t_0, and then extended uniquely to a function, which is defined at least on the unbounded interval $]t_0 - \alpha, +\infty[$, and solves problem (1.9) on the whole interval $[t_0, +\infty[$. We usually denote this extended function again by u. Of course, in some cases the local solution u could also be extended to the left of $t_0 - \alpha$; however, since in the context of evolution problems we are mostly interested in what happens in "the future", we will generally not be too concerned about the possibility of extending u to the left of t_0. (We also note in passing that, when trying to do so, we sometimes meet additional problems, such as the lack of backward uniqueness.) Thus, when in the sequel we use the term "global solution", we always refer to solutions that are defined globally at least to the *right* of t_0, i.e. for all $t \geq t_0$.

Clearly, the possibility of extending a local solution to a global one must in general be proven for each specific problem. This can be done in different ways; a common one is to show that any local solution satisfies a number of so-called A PRIORI ES-TIMATES. These are bounds on the solution which are independent of the particular time interval where the solution is defined, and therefore allow us to extend any local solution uniquely to a global one, by means of a repeated application of the local existence result.

3.   Having thus established a unique solution u of the Cauchy problem (1.9) for all choices of initial values (t_0, u_0), we are then interested in the asymptotic behavior of $u(t)$ as $t \to +\infty$. More specifically, we would like to understand how this behavior is determined (if at all) by the initial values u_0 and t_0 (or, in some cases, by u_0 only). To this end, it is convenient to introduce more proper notations. To emphasize that the solution u depends not only on t, but also on the initial values (t_0, u_0), we consider u as a function defined on $\mathbb{R} \times \mathbb{R} \times \mathcal{X}$, with values in $\mathcal{X}$, and write $u(t, t_0, u_0)$ to indicate the image of the point (t, t_0, u_0) by u. Next, we realize that the solution of the Cauchy problem (1.9) defines a family

$$S := (S(t, t_0))_{(t, t_0) \in \Theta}, \quad \Theta := \{(t, s) : t \geq s\}, \tag{1.10}$$

of operators $S(t, t_0) : \mathcal{X} \to \mathcal{X}$, parametrized by the pair (t, t_0) in the half-plane Θ. Each operator $S(t, t_0)$ is defined by

$$S(t, t_0)u_0 := u(t, t_0, u_0), \quad u_0 \in \mathcal{X}. \tag{1.11}$$

This family S is called the family of SOLUTION OPERATORS associated to (or, defined by) equation (1.7). Standard uniqueness theorems on solutions of the Cauchy problem (1.9) can then be used to verify that S satisfies conditions (1.2) and (1.3) of definition 1.1; hence, S is a dynamical process on $\mathcal{X}$. We say that S is GENERATED by problem (1.9). We also say that the map $t \mapsto S(t, t_0)u_0$ defined in (1.11) is a MO-TION of the dynamical process S, corresponding to the initial values (t_0, u_0), and the image of this motion is an ORBIT of the system (a more precise definition of motions and orbits will be given in section 1.2.4).

Example 1.4
The Cauchy problem

$$
\begin{cases}
\dot{y} = f(t)y, \\
y(t_0) = y_0
\end{cases}
\tag{1.12}
$$

generates the dynamical process S defined in (1.4). Indeed, (1.12) has the unique solution

$$
y(t) = S(t, t_0) y_0 .
$$

$\square$

1.2.1 Well-Posedness

From now on, we shall consider the value of t_0 in the Cauchy problem (1.9) as fixed; in fact, unless otherwise specified, we shall always choose $t_0 = 0$. We are then interested in the question of the dependence of solutions of (1.9) on the other initial value u_0. This question is naturally related to the WELL-POSEDNESS of the Cauchy problem (1.9). This means that solutions of (1.7) should not only be uniquely determined by the choice of the initial value u_0, but they should also depend continuously on u_0, in a specified topology.

Since we are interested in the long-time behavior of the solutions, a crucial distinction has to be made between the notion of well-posedness on arbitrary, but bounded, time intervals $[0, T]$, and that of well-posedness in the whole interval $[0, +\infty[$. Explicitly, we explain the first of these notions in

DEFINITION 1.5 *The Cauchy problem (1.9) is* WELL POSED IN THE LARGE *if for all $u_0 \in \mathcal{X}$, and all T, $\varepsilon > 0$, there exists $\delta > 0$ such that for all $v_0 \in \mathcal{X}$ and all $t \in [0, T]$,*

$$
\|u_0 - v_0\| < \delta \quad \Longrightarrow \quad \|u(t) - v(t)\| < \varepsilon ,
\tag{1.13}
$$

where u and v are the unique solutions of (1.7) with $u(0) = u_0$ and $v(0) = v_0$, and $\|\cdot\|$ denotes the norm of $\mathcal{X}$.

We remark that, in the theory of finite dimensional dynamical systems, definition 1.5 is often referred to as "continuity with respect to time and initial conditions". Note that, in (1.13), δ depends not only on the initial value u_0, but, in general, also on T. That is, we can define a function $(\varepsilon, T) \mapsto \delta(\varepsilon, T)$ (this function may often be defined only implicitly). If δ can be chosen independently of T, the solutions of (1.7) depend continuously on the initial data on all of $[0, +\infty[$; this corresponds to the Lyapunov stability of the solutions of (1.7) (see definition A.6). In contrast, it is well known that well-posedness in the large is not sufficient to guarantee stability, since the dependence of δ on T may be "bad", in the sense that

$$\liminf_{T \to +\infty} \delta(\varepsilon, T) = 0.$$

To show this, it is sufficient to consider the following elementary example.

Example 1.6
Consider the Cauchy problems for the ODEs

$$\dot{u} = -u, \qquad\qquad (1.14)$$

$$\dot{u} = u, \qquad\qquad (1.15)$$

with initial data at $t = 0$. Both problems have globally defined unique solutions for each choice of initial values, but only the first is globally well posed for $t \geq 0$. In fact, when checking (1.13) we can take $\delta = \varepsilon$ for (1.14), but for (1.15) we are forced to take $\delta = \varepsilon e^{-T}$, so in this case $\delta \to 0$ as $T \to +\infty$. We can interpret this in another way, realizing that the effect of any error in the initial value for equation (1.14) becomes negligible, up to arbitrary tolerance, if sufficient time is allowed to pass; on the contrary, even if two solutions of equation (1.15) are initially very close, after sufficient time they will be arbitrarily apart. Indeed, for (1.14), given any M and $\varepsilon > 0$, even if initially $|u_0 - v_0| \geq M$, it will be $|u(t) - v(t)| \leq \varepsilon$ for all $t \geq \ln(M/\varepsilon)$, while for (1.15), given again any M and $\varepsilon > 0$, even if initially $|u_0 - v_0| \leq \varepsilon$, it will be $|u(t) - v(t)| \geq M$ for all $t \geq \ln(M/\varepsilon)$. For instance, if we approximate $u_0 = \pi$ by $v_0 = 3.141$, the initial error is less than 10^{-3}, but for the corresponding solutions of (1.15) we have $|u(t) - v(t)| \geq 10^3$ for all $t \geq \ln(10^3/(\pi - 3.141)) \approx 14.5087$. This phenomenon is illustrated in figures 1.1 and 1.2. In terms of Lyapunov stability, the

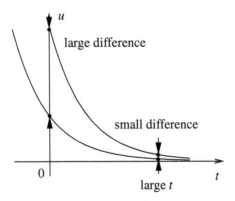

Figure 1.1: Exponential stability for $\dot{u} = -u$: A large difference in initial values still results in a small difference of the solutions after sufficient time.

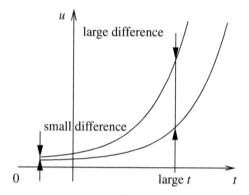

Figure 1.2: Exponential loss of information for $\dot{u} = u$: Even a small difference in initial values is drastically amplified after sufficient time.

point $u = 0$ in the phase space $\mathcal{X} = \mathbb{R}$, which corresponds to the solution $u(t) \equiv 0$ of both equations, is (uniformly) stable only for system (1.14), while system (1.15) is highly unstable under arbitrarily small perturbations of the initial value $u_0 = 0$. In fact, if in (1.15) $\pm u_0 > 0$, then as $t \to +\infty$, $u(t) \to \pm\infty$ (exponentially, of course), even if $|u_0| < \varepsilon$. Loosely speaking, this means that all control on the solution is lost if sufficient time is allowed to elapse.   ☐

1.2.2 Regular and Chaotic Systems

As we have mentioned, the theory of dynamical systems is largely concerned with the behavior of the orbits $t \mapsto u(t)$ as $t \to +\infty$, and, more specifically, with how such behavior is influenced by the choice of the initial value u_0. This explains the use of notations like (1.19) below, which emphasize the dependence of the solution, at each time t, on its initial value u_0.

With a great degree of simplification, we distinguish between two kinds of situations, which we call REGULAR and CHAOTIC. This choice of terms is rather arbitrary, and by no means universal; indeed, we find many different definitions of regularity and chaos in the literature, and even among those definitions that are mathematically rigorous, no one is universally accepted. Rather, different definitions are preferred for different applications.

Roughly speaking, regular systems are those for which perturbations in the initial values will influence the orbits only for a short period of time (called TRANSIENT). After this time, different orbits would have the same qualitative behavior, and in particular the same asymptotic behavior. This type of situation is usually described by theorems like those on the asymptotic stability of a system, or the existence of limit cycles. In some cases, the asymptotic behavior is even *independent* of the initial values, in the sense that two orbits, even if starting from two points that are arbitrarily apart, after sufficient time (i.e. the transient, whose length depends on how far apart the initial values are) they will be *and remain* arbitrarily close to each other, and so exhibit the same qualitative asymptotic behavior.

Chaotic systems are instead those for which a sort of opposite situation holds. That is, these systems are extremely sensitive to even small variations of the initial values, in the sense that "close" initial conditions eventually move arbitrarily apart. The evolution of this type of system will be "regular" for a short time only (this is in general a consequence of some result analogous to the well-posedness of ODEs on *compact* time intervals). However, if observed for sufficiently long time periods, these systems not only do not exhibit any indication of convergence towards any sort of stable or periodic configuration, but their evolution seems to be totally unpredictable. More precisely, we give the following

DEFINITION 1.7   *Let* $S = (S(t,t_0))_{(t,t_0) \in T_*^2}$ *be a dynamical process on* $\mathcal{X}$. *S is said to* DEPEND SENSITIVELY ON ITS INITIAL CONDITIONS *if there is* $R > 0$ *such that for all* $t_0 \in T$, $x_0 \in \mathcal{X}$, *and all* $\delta > 0$, *there are* $y_0 \in \mathcal{X}$ *and* $t_1 \in T$, $t_1 \geq t_0$, *such that*

$$0 < \|x_0 - y_0\| \leq \delta \quad \text{and} \quad \|S(t_1,t_0)x_0 - S(t_1,t_0)y_0\| \geq R. \tag{1.16}$$

We remark that the notion of sensitive dependence of a dynamical process on its initial conditions is a natural generalization of that of uniform Lyapunov instability for ODEs (see section A.1).

Example 1.8
Consider the dynamical processes S_1 and S_2 in $\mathbb{R}$ generated, respectively, by the ODEs (1.14) and (1.15) of example 1.6. Then S_1 is regular, while S_2 depends sensitively on its initial conditions. In fact, given ε and R such that $0 < \varepsilon < R$, any two solutions x and y of (1.14) which initially differ by R will be such that $|x(t) - y(t)| \leq \varepsilon$ for all $t \geq t_0 := \ln \frac{R}{\varepsilon}$. That is, t_0 is the transient after which these solutions will always differ by at most ε. In contrast, any two solutions x and y of (1.15) which differ initially by ε will be such that $|x(t) - y(t)| \geq R$ for all $t \geq t_0 := \ln \frac{R}{\varepsilon}$ (compare to (1.16)). $\square$

Example 1.8 shows that there are dynamical processes for which *no matter how close* two initial values may be, if sufficient time is allowed to pass the corresponding orbits will be arbitrarily apart. That is, the asymptotic behavior of these systems, which is still completely and uniquely determined by their initial values (the systems *are* deterministic), may be drastically different. To put this in another way, in this type of system all relevant information carried by the initial data is rapidly lost, and, consequently, it becomes impossible to maintain any reasonable control on the evolution of the system. Examples of this kind of situation are the smoke of a cigarette, the dynamics of large populations, of traffic patterns, economic cycles, etc. Probably, the most familiar example is that of the various meteorological models for the evolution of weather, whose prediction is in general relatively accurate only in a short time range (and the shorter the time interval, the better the prediction), but after sufficient time all predictions lose any practical value.

In section 1.4 we shall see some other simple examples of systems that exhibit chaotic behavior, as described by their being sensitive to their initial conditions. Before proceeding, we mention another possible way of describing chaotic systems, whereby "an orbit can begin roughly anywhere and end up roughly anywhere". More precisely, given any two open subsets $\mathcal{U}$ and $\mathcal{V}$ of $\mathcal{X}$, there is $x_0 \in \mathcal{U}$ such that the corresponding orbit intersect $\mathcal{V}$. For an exhaustive discussion of these, and other, possible descriptions of the chaotic behavior of a dynamical system, we refer e.g. to Robinson, [Rob99], and to Alligood, Sauer and Yorke, [ASY96].

Of course, the possibility of determining whether a given system is regular or chaotic (we should rather say, whether the system may exhibit chaotic features or is guaranteed not to) is of extreme importance in applications, for at least two reasons. First, actual initial values depend on physical measurements, and are therefore never "exact" (this is not just a problem of the "real world": Even in a simple numerical exercise in ODEs, initial values like $\sqrt{2}$ can only be introduced within approximations). Second, because in practice we cannot afford to observe the evolution of a system for very long time periods (deadlines have to be met, computer simulation time is expensive ...). Moreover, even if we could, we are still bound to observations in *finite* time intervals, and there is no guarantee that any such period of time, in which we may see "irregular" behavior, is still not part of a very long transient, after which the system may yet settle into a regular evolution.

In these notes, we are concerned with a sort of intermediate situation between the two extremes described above. There are in fact examples of systems, whose evolution may appear to be chaotic, and yet after sufficient time their solutions seem to settle into a pattern that preserves a certain degree of order, which allows for some control of the disturbances typical of a chaotic regime. This type of behavior is usually better seen in the state space $\mathcal{X}$, to which the solution curves $(u(t))_{t\geq 0}$ belong. More precisely, these systems are characterized by the existence of some subsets of $\mathcal{X}$, to which the solution curves appear to converge (in the topology of the phase space), as $t \to +\infty$. These subsets are therefore called ATTRACTING SETS, and can be thought of as a generalization of the sets, such as stationary points or limit cycles, that are known to be attracting for regular systems of ODEs. Thus, for example, if a bounded attractor exists, two solutions which started at close initial values may still be quite apart at arbitrary later times (indication of chaos), but their distance cannot be arbitrarily large, since they both converge to the same attractor. In this sense the system is still controllable. Thus, even if we cannot decide whether a given system is regular, it is clearly desirable that we be able to determine if it at least possesses an attractor. Indeed, if this is the case, we would then know that, even if the system may possibly evolve chaotically, it will nevertheless settle into some type of controlled behavior. This is of course of fundamental importance in applications.

1.2.3 Dependence on Parameters

In many physical examples, the equation (1.7) which models the evolution of a dynamical process may also depend on various numerical parameters, such as, for instance, the dielectric and permeability constants in Maxwell's equations, or the viscosity coefficient in Navier-Stokes' equations of fluid dynamics. In this case, equation (1.7) takes the more general form

$$\dot{u} = F(\lambda, t, u), \qquad \lambda \in \Lambda \subset \mathbb{R}^m, \tag{1.17}$$

and the corresponding solution operator also depends on the parameters λ. In applications, it is of course of great importance to have a good knowledge of how the evolution of a system is influenced not only by (small) variations of the initial value u_0, but also by (small) variations of these parameters. For example, if the arm of a robot has the task of repeatedly moving an object from one position to another, and its motion is governed by a differential equation like (1.17), we are interested in the choice of parameters that make such motion as smooth as possible, and to avoid those that may make it irregular or, worse, chaotic.

We will not present any theoretical results on the dependence of dynamical systems, in particular infinite dimensional ones, on numerical parameters, since this topic is too extensive and specialized, and a large quantity of the available insights and results are most often obtained by means of extensive and robust numerical simulation. Indeed, an experimental analysis of the equations modelling many physical examples indicates that various kinds of bifurcation phenomena typically occur at different, increasing values of λ. We refer to Temam, [Tem88, ch. 1], for a very general outline of various scenarios that are possible.

1.2.4   Autonomous Equations

1.   Most classical results on the theory of the asymptotic behavior of dynamical systems involve systems generated by evolution equations (1.7) that are AUTONOMOUS. These systems, which occur quite frequently in applications, correspond to the case when the function F in (1.7) is independent of t, that is, when (1.7) has the form

$$\dot{u} = F(u), \tag{1.18}$$

with $F: \mathcal{X} \to \mathcal{X}$ continuous. For example, the heat equation (1.8) is autonomous. In this case, we can always reduce ourselves, by a shift of the time coordinate, to a fixed choice of t_0. This means that the operators of S have the form

$$S(t, \tau) = S(t - \tau, t_0) =: \tilde{S}(t - \tau),$$

where now $\tilde{S} = (\tilde{S}(t))_{t \geq t_0}$ is a one-parameter family of operators, i.e. a SEMIFLOW, on $\mathcal{X}$. In particular, we choose $t_0 = 0$ for simplicity. We use again the letter S to denote this one-parameter family; that is, we write $S = (S(t))_{t \geq 0}$, and (1.11) reads

$$S(t)u_0 = u(t, 0, u_0). \tag{1.19}$$

In particular, conditions (1.2) and (1.3) of definition 1.1 are satisfied if S is a SEMI-GROUP of (not necessarily linear) operators on $\mathcal{X}$, i.e. if

$$S(0) = I_{\mathcal{X}} \tag{1.20}$$

(the identity in $\mathcal{X}$), and for all $t, s \geq 0$,

$$S(t+s) = S(t)S(s) \tag{1.21}$$

(fig. 1.3). Indeed, if S is the solution operator defined by the autonomous equation

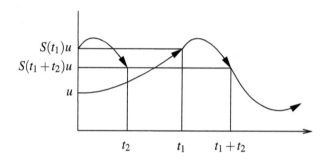

Figure 1.3: The action of the semigroup.

(1.18), (1.20) holds by virtue of the initial condition

$$S(0)u_0 = u_0 \qquad \text{for all } u_0 \in \mathcal{X}.$$

To show (1.21), we note that for all $t, s \geq 0$,

$$S(t)S(s)u_0 = v(t),$$

where v is the solution of the Cauchy problem

$$\begin{cases} \dot{v} = F(v), \\ v(0) = S(s)u_0 = u(s). \end{cases}$$

Thus, setting $w(t) := v(t-s)$, we have

$$\dot{w}(t) = \dot{v}(t-s) = F(v(t-s)) = F(w(t))$$

and

$$w(s) = v(0) = u(s).$$

By the assumed uniqueness of solutions of the differential equation, we conclude that $w(t) = u(t)$ for all $t \geq 0$. In particular,

$$u(t+s) = w(t+s) = v(t); \tag{1.22}$$

and since $u(t+s) = S(t+s)u_0$ and

$$v(t) = S(t)v(0) = S(t)S(s)u_0,$$

(1.22) means that (1.21) holds.

Clearly, this argument may fail if the differential system is not autonomous, since then

$$\dot{v}(t-s) = F(t-s, v(t-s)),$$

and in general

$$F(t-s, w) \neq F(t, w).$$

Example 1.9

The first order autonomous system

$$\begin{cases} \dot{u} = & v \\ \dot{v} = & -u \end{cases} \tag{1.23}$$

generates the dynamical system S in $\mathbb{R}^2$, defined by

$$S(t)(x, y) := A(t)(x, y)^\top, \qquad t \in \mathbb{R},$$

where $A(t)$ is the 2×2 matrix defined by

$$A(t) := \begin{pmatrix} \cos t & \sin t \\ -\sin t & \cos t \end{pmatrix},$$

and $\top$ denotes transposition. Indeed, it is immediate to verify that for all $(x, y) \in \mathbb{R}^2$, the vector function $t \mapsto U(t) := A(t)(x, y)^\top$ solves system (1.23) with initial values $U(0) = (x, y)^\top$. Furthermore, $A(0) = I$, and for all t and $s \in \mathbb{R}$,

$$
\begin{aligned}
A(t+s) &= \begin{pmatrix} \cos(t+s) & \sin(t+s) \\ -\sin(t+s) & \cos(t+s) \end{pmatrix} \\
&= \begin{pmatrix} \cos t \cos s - \sin t \sin s & \sin t \cos s + \cos t \sin s \\ -\sin t \cos s - \cos t \sin s & \cos t \cos s - \sin t \sin s \end{pmatrix} \\
&= \begin{pmatrix} \cos t & \sin t \\ -\sin t & \cos t \end{pmatrix} \begin{pmatrix} \cos s & \sin s \\ -\sin s & \cos s \end{pmatrix} \\
&= A(t)A(s).
\end{aligned}
$$

$\square$

Example 1.10

The solution operator defined by the ODE

$$\dot{u} = \cos t$$

is not a semigroup. Indeed, for arbitrary t and $s \in \mathbb{R}$ we have

$$u(t) = u_0 + \sin t,$$
$$S(t+s)u_0 = u_0 + \sin(t+s),$$
$$S(t)S(s)u_0 = u_0 + \sin s + \sin t.$$

On the other hand, the solution operator defined by the autonomous ODE

$$\dot{u} = 1 - u$$

is indeed a semigroup. In fact, for arbitrary t and $s \in \mathbb{R}$ we have

$$u(t) = (u_0 - 1)e^{-t} + 1,$$
$$S(t+s)u_0 = (u_0 - 1)e^{-(t+s)} + 1,$$
$$S(t)S(s)u_0 = (S(s)u_0 - 1)e^{-t} + 1$$
$$= \big((u_0 - 1)e^{-s} + 1 - 1\big)e^{-t} + 1$$
$$= (u_0 - 1)e^{-s-t} + 1.$$

$\Box$

Except for some elementary introductory examples, in these lectures we shall only consider autonomous systems. For an extensive account on the nonautonomous case, see e.g. Haraux, [Har91].

2. When a system is autonomous, we call the corresponding family S of solution operators a SEMIFLOW on $\mathcal{X}$, and the space $\mathcal{X}$ is often called the PHASE SPACE of the dynamical system. The map $u \colon [0, +\infty[\to \mathcal{X}$ defined by

$$u(t) := S(t)u_0, \qquad t \geq 0,$$

is called a MOTION, and the image of u in $\mathcal{X}$, i.e. the subset (or curve)

$$\gamma_{u_0} := \bigcup_{t \geq 0} u(t) \subset \mathcal{X},$$

is called the ORBIT of the motion u, starting at u_0. (When the system is not autonomous, we would need to consider the product $\mathbb{R} \times \mathcal{X}$ as an extended phase space.) Then, the asymptotic behavior of solutions of (1.18) is related to the evolution of the corresponding orbits, as subsets of $\mathcal{X}$. Indeed, the recourse to the notion of orbits in the phase space (as opposed to that of solution of the differential equation) quite naturally allows us to introduce, together with the appropriate instruments from analysis in metric spaces to determine limiting behaviors etc., a more geometric approach, in which we study and exploit the topological properties of the orbits, seen in their own right as subsets of the phase space $\mathcal{X}$. The example of definition of stability in the theory of ODEs is a familiar one; another example in two dimensions of space, i.e. for $\mathcal{X} = \mathbb{R}^2$, is the Poincaré-Bendixon theorem (see e.g. theorem A.32), which describes conditions under which the orbits of an autonomous system of two ODEs converge, in a suitable sense, to a limit cycle.

3.   In conclusion, we have seen in what sense an autonomous differential equation
(1.18) generates a continuous semiflow S, by means of the solution operator defined
in (1.19). If F is sufficiently regular, S is also differentiable. It is worth to point
out that the converse is also true; that is, a differentiable semiflow $S = (S(t))_{t \geq 0}$ is
always generated by an autonomous ODE.

PROPOSITION 1.11
Let S be a semiflow defined on $\mathcal{X}$, and assume that for all $x_0 \in \mathcal{X}$, the map

$$[0, +\infty[\ni t \mapsto S(t)x_0 \in \mathcal{X}$$

is differentiable at $t = 0$. Let $F : \mathcal{X} \to \mathcal{X}$ be defined by

$$F(x) := \frac{d}{dt}(S(t)x)\Big|_{t=0}, \quad x \in \mathcal{X},$$

*and, for $x_0 \in \mathcal{X}$ and $t \geq 0$, set $x(t) := S(t)x_0$. Then x is differentiable in $[0, +\infty[$, and
satisfies the autonomous Cauchy problem*

$$\begin{cases} \dot{x} = F(x), \\ x(0) = x_0. \end{cases} \tag{1.24}$$

PROOF   Fix $t_0 \geq 0$. For $t \geq t_0$, we compute that

$$\frac{x(t) - x(t_0)}{t - t_0} = \frac{S(t)x_0 - S(t_0)x_0}{t - t_0} = \frac{S(t - t_0 + t_0)x_0 - S(t_0)x_0}{t - t_0}$$
$$= \frac{S(t - t_0)S(t_0)x_0 - S(t_0)x_0}{t - t_0}. \tag{1.25}$$

Let $y_0 := S(t_0)x_0$ and $\theta := t - t_0$. Then from (1.25)

$$\frac{x(t) - x(t_0)}{t - t_0} = \frac{S(\theta)y_0 - y_0}{\theta}.$$

Since the map $t \mapsto S(t)y_0$ is differentiable at $t = 0$, we have that, as $\theta \to 0$

$$\frac{x(t) - x(t_0)}{t - t_0} \longrightarrow \frac{d}{d\theta}(S(\theta)y_0)\Big|_{\theta=0} = F(y_0) = F(S(t_0)x_0).$$

This proves that x is differentiable from the right, and

$$x'_+(t) = F(x(t)).$$

If instead $0 < t < t_0$, we compute that

$$x'_-(t_0) = \lim_{t \to t_0^-} \frac{x(t) - x(t_0)}{t - t_0} = \lim_{s \to t_0^+} \frac{x(2t_0 - s) - x(t_0)}{t_0 - s}$$

$$= \lim_{s \to t_0^+} \frac{S(t_0 - s)y_0 - y_0}{t_0 - s} = \lim_{\theta \to 0^+} \frac{S(\theta)y_0 - y_0}{\theta}$$
$$= F(y_0) = F(x(t_0)).$$

This proves that x is differentiable also from the left, and

$$x'_-(t) = F(x(t)).$$

Hence, x is differentiable, and satisfies the equation of (1.24). The initial value of (1.24) is obviously taken.   □

1.3 Attracting Sets

We have mentioned that in some cases, even if the evolution of a system appears to be chaotic, a certain degree of order seems to be preserved, in the sense that the orbits of the system appear to settle into a somewhat regular pattern, described by the fact that they converge, or at least remain "close", to some subset of $\mathcal{X}$. We can often describe this situation in terms of subsets that are ATTRACTING, or at least ABSORBING, in the following sense.

1. Absorbing Sets.   In the theory of ODEs, a first step in the study of the asymptotic behavior of the solution of a given system is to recognize that these solutions are bounded as $t \to +\infty$. Analogously, given a dynamical system S on a Banach space $\mathcal{X}$, it may be possible, in some cases, to recognize the existence of a subset $\mathcal{B} \subset \mathcal{X}$ into which all orbits, or at least those starting from some subset $\mathcal{U} \subseteq \mathcal{X}$ containing $\mathcal{B}$, enter and, after possibly leaving $\mathcal{B}$ a finite number of times, eventually remain in $\mathcal{B}$ for ever. This set $\mathcal{B}$ is thus called an ABSORBING SET. If a *bounded* absorbing set exists, this is taken as an expression of a specific property of the system, generically called DISSIPATIVITY.

2. Attracting Sets.   When an absorbing set exists, it is sometimes possible to also recognize the existence of a smaller subset $\mathcal{A} \subset \mathcal{B}$, to which all orbits starting from $\mathcal{U}$ converge as $t \to +\infty$ after having entered $\mathcal{B}$; see fig. 1.4.

(The precise definition of convergence of an orbit to a set of $\mathcal{X}$ is given in section 2.1 of chapter 2.) Such sets $\mathcal{A}$ are generally called ATTRACTING SETS. We will see that if a dynamical system admits an attractor, it necessarily has an absorbing set as well. Attracting sets may have a quite complicated geometric or topological structure (they may be self-similar sets, or FRACTALS), and the convergence of the orbits to these sets may be quite slow. However, these sets often possess some important properties, that may allow for a better understanding of the evolution of the system (in particular, if the system appears to be chaotic). For example, the set $\mathcal{A}$ may be *compact*, and (often but not always) it may have a *finite fractal dimension*

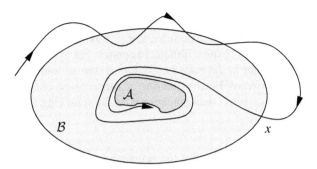

Figure 1.4: Absorbing and attracting sets. After entering the absorbing set $\mathcal{B}$ for the last time at x, the orbit remains in $\mathcal{B}$, and then converges to $\mathcal{A}$.

(the definition of which we recall in section 2.8 of chapter 2). The set $\mathcal{A}$ may also be *invariant*, which means that

$$S(t)\mathcal{A} = \mathcal{A} \qquad \text{for all } t \geq 0. \tag{1.26}$$

That is, if $u_0 \in \mathcal{A}$ then $u(t) = S(t)u_0 \in \mathcal{A}$ for all $t \geq 0$ and, conversely, every $u_0 \in \mathcal{A}$ is on some orbit starting from some point in $\mathcal{A}$.

3. Attractors.   Bounded, positively invariant attracting sets are generally called ATTRACTORS. Of particular importance are attractors that are finite dimensional, because the corresponding dynamics is also finite dimensional. Indeed, the invariance of the attractor implies, by (1.26), that orbits which originate in the attractor remain in the attractor for all future times; consequently, the evolution of a system on a finite dimensional attractor would essentially be governed by a finite system of ODEs. In fact, a celebrated theorem of Mañé, [Mañ81], states that if a dynamical system possesses a finite dimensional attractor, this attractor can be generated by (or, as it is sometimes said, is "imbedded into") a finite system of ODEs. This result allows us to reduce, at least in principle, the study of the long time behavior of orbits which converge to a finite dimensional attractor to that of the solutions of a finite dimensional system of ODEs on $\mathcal{A}$.

This question, together with the description of the corresponding ODEs, is one of the most challenging problems in the theory of dynamical systems. Moreover, in most cases the reduction of the study of the evolution of the system on the attractor cannot be pursued in practice, because of several difficulties, which partially motivate the search for "friendlier" sets, such as the inertial manifolds discussed below. For example, we have mentioned the generally nonsmooth geometrical or topological structure of the attractor, which may cause the corresponding ODEs to only admit generalized solutions. Another problem, of special importance in applications, is that in many cases the available estimates on the dimension of the attractor, and therefore on the dimension of the system of ODEs, are simply too large for computational feasibility. For instance, in meteorology it is not uncommon to have estimates of the order of 10^m, $m \geq 20$. Also, attractors are in most cases not sufficiently stable under perturbations of the data, so that their numerical approximations, and the consequent propagation of errors, may be quite difficult to control. For example, approximations of attractors with respect to the Hausdorff distance (see section 2.1) are in general only upper semicontinuous. Finally, the rate of convergence of the orbits to the attractor may really be no better than polynomial, as the following example shows.

Example 1.12
Consider the semiflow S generated by the autonomous ODE

$$\dot{u} = f(u) := -u^3 . \tag{1.27}$$

The attractor of S is the set $\mathcal{A} = \{0\}$, but the convergence of the orbits to $\mathcal{A}$ is at most polynomial, as we see from the explicit solution of the Cauchy problem relative to (1.27) with initial value $u(0) = u_0$, that is,

$$u(t) = \frac{u_0}{\sqrt{1 + 2u_0^2 t}} .$$

$\square$

4. Inertial Manifolds.   On the other hand, there are systems whose attractors do not present this type of difficulties, since they are imbedded into a finite dimensional Lipschitz manifold $\mathcal{M}$ of $\mathcal{X}$, and the orbits converge to this manifold with a uniform exponential rate. Such a set $\mathcal{M}$ is called an INERTIAL MANIFOLD of the system (fig. 1.5). When an inertial manifold exists, the evolution of the semiflow on the

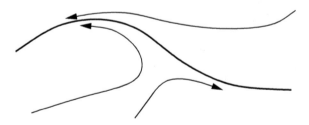

Figure 1.5: Inertial Manifolds.

manifold is governed by a finite system of ODEs, called the INERTIAL FORM of the semiflow. This finite system of ODEs will in general admit solutions with a certain degree of smoothness, depending on the smoothness of the manifold. Since orbits converge to the inertial manifold with a uniform exponential rate, we see that, in turn, the dynamics on the manifold will be a good description of the long time behavior of solutions of equation (1.7). Clearly, the possibility of imbedding the attractor into an inertial manifold provides an indirect way to obtain the above mentioned desired system of ODEs. Moreover, the uniformity of the rate of convergence of the orbits to the manifold makes these systems extremely stable under perturbations and numerical approximations. Unfortunately, there are not many examples of systems which are known to admit an inertial manifold; among these, we mention the semiflows generated by a number of reaction-diffusion equations of "parabolic" type, and by the corresponding hyperbolic (small) perturbations of these equations. A typical model is that of the so-called Chafee-Infante equations, which we present in chapter 5.

5. Exponential Attractors.   An intermediate situation occurs when a system admits a so-called EXPONENTIAL ATTRACTOR. These sets, which are also sometimes called INERTIAL SETS in the literature, are somehow intermediate between attractors and inertial manifolds, in the sense that while they do not necessarily have a smooth structure, they can still be imbedded into a finite system of ODEs. In addition, these sets retain at least three of the features of the inertial manifolds that attractors do not necessarily have: the finite dimensionality, the exponential convergence of the orbits, and a high degree of stability with respect to approximations (for example, continuity with respect to the Hausdorff distance). This means that when an exponential attractor exists, after an "exponentially short" transient the dynamics of the system are essentially governed by a finite system of ODEs (the classical image is that of an airplane, landing at a "fast" speed and then "slowly" taxiing to the arrival gate).

The following is a simple example of a regular system, whose solutions converge exponentially to its attractor.

Example 1.13
Consider the function $f\colon [0,1] \to [0,1]$ defined by $f(x) = (1+x)^{-1}$, and the corresponding discrete system $(S^n)_{n\in\mathbb{N}}$ defined by the iterated sequence (1.30). This system has an attractor, which is the set $\mathcal{A} = \{\ell\}$, with $\ell := (\sqrt{5}-1)/2$. We now show that $\mathcal{A}$ is also an exponential attractor; that is, there is $\alpha > 0$ such that, for all initial values $x_0 \in [0,1]$,

$$|S^n x_0 - \ell| \le e^{-\alpha n}. \tag{1.28}$$

Indeed, setting

$$S^n x_0 = f(x_n) =: x_{n+1},$$

we see that this sequence converges to the positive solution of the equation $x = f(x)$, which is precisely ℓ. Since $\ell = f(\ell)$, we compute that

$$\ell - x_{n+1} = f(\ell) - f(x_n) = \frac{x_n - \ell}{(1+\ell)(1+x_n)}. \tag{1.29}$$

Since $1 + \ell > \frac{3}{2}$ and $1 + x_n \ge 1$ for each n, we deduce from (1.29) that

$$|x_{n+1} - \ell| \le \tfrac{2}{3}|x_n - \ell|,$$

from which we conclude that

$$|x_n - \ell| \le (\tfrac{2}{3})^n |x_0 - \ell| \le (\tfrac{2}{3})^n.$$

This shows that (1.28) holds, with e.g. $\alpha = \ln \tfrac{3}{2} > 0$. We explicitly note that α is independent of the initial values: this ensures that the iterates $S^n x_0$ converge to $\mathcal{A}$ with a uniform rate. $\square$

Of course, not all dynamical systems will possess attractors, exponential attractors or inertial manifolds. In the sequel, we shall try to present a theory, by now quite well established, that provides a number of sufficient conditions on the system for at least some of these sets to exist. In particular, since attractors will contain stationary and periodic solutions of (1.17), this theory is really a natural extension of the classical theory of stability for ODEs.

1.4   Iterated Sequences

Not surprisingly, many of the ideas (and difficulties) in the theory of continuous dynamical systems already surface in the context of discrete dynamical systems generated by ITERATED SEQUENCES. These are sequences $(u_n)_{n \in \mathbb{N}} \subset \mathcal{X}$, of the form

$$u_{n+1} = f(u_n), \tag{1.30}$$

where f is a map of $\mathcal{X}$ into itself. Thus, each sequence is completely determined by its initial value u_0, assigned separately. Iterated sequences generate a DISCRETE dynamical system $S := (S^n)_{n \in \mathbb{N}}$ on $\mathcal{X}$, defined by

$$S^0 = I, \quad S^{n+1} = S \circ S^n,$$

where S^n is the n-th iterate of S, and $\circ$ denotes the composition of maps in $\mathcal{X}$. Thus, $\mathcal{T} = \mathbb{N}$, and the orbits of S are the sequences $(S^n u_0)_{n \in \mathbb{N}}$. We are interested in how the behavior of each such sequence, as $n \to +\infty$, depends on its initial term u_0.

In this section we present some well known examples of discrete systems in $\mathbb{R}^n$, $n \le 3$, each defined by a sequence like (1.30).

For future reference, we recall the following

DEFINITION 1.14   *Let $F \colon \mathcal{X} \to \mathcal{X}$ be a map (not necessarily linear), and $x \in \mathcal{X}$.*

1. x is a FIXED POINT of F if $x = F(x)$.

2. A fixed point x of F is said to be STABLE if, given any neighborhood $\mathcal{U}$ of x there is another neighborhood $\tilde{\mathcal{U}} \subset \mathcal{U}$ of x such that for all x_0 in $\tilde{\mathcal{U}}$, the corresponding recursive sequence $(x_n)_{n \in \mathbb{N}}$, starting at x_0 and defined by $x_{n+1} = F(x_n)$, is contained in $\mathcal{U}$. Otherwise, x is said to be UNSTABLE.

3. *A fixed point x of F is said to be* ATTRACTIVE *if for all x_0 in a neighborhood of x, the above defined recursive sequence $(x_n)_{n \in \mathbb{N}}$ converges to x.*

4. *A stable and attractive fixed-point is called* ASYMPTOTICALLY STABLE.

5. *A point x is said to be p*-PERIODIC *($p \in \mathbb{N}$) if $F^p(x) = x$.*

Note that not all stable fixed points are attractive, as we see by taking $F(x) = x$. For this map, each point x is a stable, but not attractive, fixed point. On the other hand, we have the following

THEOREM 1.15
Let $X = \mathbb{R}$, and x_0 be a fixed point of a C^1 function F. Then x_0 is asymptotically stable if $|F'(x_0)| < 1$, while if $|F'(x_0)| > 1$, x_0 is unstable.

PROOF Without loss of generality, we can confine ourselves to symmetric neighborhoods of x_0.
 1) Assume first that $|F'(x_0)| < 1$. There exists then a number $\varepsilon \in]|F'(x_0)|, 1[$, and, correspondingly, a number $\delta > 0$ such that if $|x - x_0| \leq \delta$, then

$$|F(x) - F(x_0)| \leq \varepsilon|x - x_0|. \tag{1.31}$$

Since $\varepsilon < 1$ and $F(x_0) = x_0$, (1.31) implies that

$$|F(x) - x_0| \leq |x - x_0| \leq \delta.$$

Consequently, we can repeat estimate (1.31), and obtain that for all $n \in \mathbb{N}_{\geq 1}$,

$$|F^n(x) - x_0| \leq \varepsilon^n|x - x_0|. \tag{1.32}$$

From this, it follows that x_0 is asymptotically stable: Indeed, given any neighborhood $\mathcal{U} :=]x_0 - \rho, x_0 + \rho[$ of x_0, let $\delta \in]0, \rho]$, and set $\tilde{\mathcal{U}} :=]x_0 - \delta, x_0 + \delta[$. Then, (1.32) implies that if $x \in \tilde{\mathcal{U}}$, each iterate $F^n(x)$ is in $\mathcal{U}$, because

$$|F^n(x) - x_0| \leq \varepsilon^n|x - x_0| \leq \delta \leq \rho.$$

Thus, x_0 is stable; clearly, (1.32) also implies that x_0 is also attractive.
 2) Conversely, assume that $|F'(x_0)| > 1$. Then, as before, given any $a \in]1, |F'(x_0)|[$, we can determine $\gamma > 0$ such that if $|x - x_0| \leq \gamma$, then

$$|F(x) - x_0| \geq a|x - x_0|. \tag{1.33}$$

We wish to prove that there is $\bar{\rho} > 0$ such that for all $\delta \in]0, \bar{\rho}]$, there are $\bar{x}$ and $\bar{n}$ such that

$$|\bar{x} - x_0| \leq \delta \quad \text{and} \quad |F^n(\bar{x}) - x_0| \geq \bar{\rho}.$$

Arguing by contradiction, taking $\rho = \gamma$, we can determine $\delta \in \,]0, \gamma]$ such that if $|x - x_0| \leq \delta$, then for all $n \in \mathbb{N}_{>0}$,

$$|F^n(x) - x_0| \leq \rho = \gamma. \tag{1.34}$$

Now, (1.34) and (1.33) imply that for all n,

$$|F^n(x) - x_0| \geq a^{n-1} |F(x) - x_0| ; \tag{1.35}$$

but since $|x - x_0| \leq \delta \leq \gamma$, (1.33) implies that, in fact,

$$|F^n(x) - x_0| \geq a^n |x - x_0| \tag{1.36}$$

for all n. Choose then, for example, $x = x_0 + \frac{1}{2}\delta$. Then, (1.36) implies that

$$\gamma \geq |F^n(x) - x_0| \geq \tfrac{1}{2}\delta a^n . \tag{1.37}$$

Since $a > 1$, letting $n \to +\infty$ in (1.37) we achieve the desired contradiction. $\square$

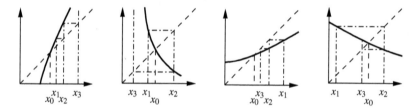

Figure 1.6: The four possibilities: $F'(x_0) > 1$, $F'(x_0) < -1$, $0 < F'(x_0) < 1$, $-1 < F'(x_0) < 0$.

We remark that when $|F'(x_0)| = 1$, x_0 can be either attractive, or unstable. This is easily seen by considering a function F which changes concavity at x_0. For example, if $F'(x_0) = 1$, and F changes from convex to concave at x_0, then x_0 is attractive, while if F changes from concave to convex at x_0, then x_0 is unstable.

1.4.1 Poincaré Maps

Given a continuous dynamical system, it is in many cases possible to construct a discrete one, whose asymptotic behavior is essentially the same as that of the continuous system. One way to do so is to choose a sequence $(t_n)_{n \in \mathbb{N}}$ of equidistant values $t_n \to \infty$ and, given a solution of the continuous autonomous system (1.18), to consider the corresponding sequence $(u_n)_{n \in \mathbb{N}}$ of points $u_n := u(t_n)$ in the phase space $\mathcal{X}$. Clearly, each of these points lies on the orbit starting at u_0. This choice defines a map $\Phi \colon \mathcal{X} \to \mathcal{X}$, by

$$u_{n+1} = \Phi(u_n) . \tag{1.38}$$

Maps constructed in this way are called STROBOSCOPIC MAPS. For example, the choice $t_n = n+1$ in (1.38) yields the sequence $(u_n)_{n \in \mathbb{N}}$, defined by

$$S := S(1), \quad u_{n+1} = S^n u_0 \quad \text{for } n \in \mathbb{N}.$$

We can visualize a stroboscopic map by considering the graph of u in the product space $[0, +\infty[\times \mathcal{X}$; that is, the set

$$\text{graph } u := \{(t, u(t)) : t \geq 0\}. \tag{1.39}$$

Then, the sequence in $\mathcal{X}$ defined by the stroboscopic map (1.38) is obtained by projecting on $\mathcal{X}$ the points $(t_n, u(t_n))$.

In the case of finite dimensional systems, a remarkable construction is that of the so-called POINCARÉ MAPS. These maps are constructed by fixing a hyperplane $\Sigma \subset \mathbb{R}^n$, called a POINCARÉ SECTION, and considering on Σ the sequence of points P_n defined by the "first returns" of the (graph of the) solution on Σ, i.e. by the successive intersections of the semiorbit $\{u(t) : t \geq 0\}$ with Σ (figs. 1.7 and 1.8). Indeed,

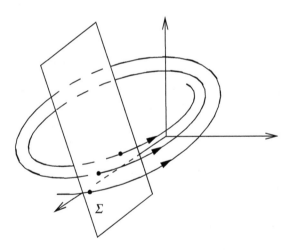

Figure 1.7: The Poincaré section.

Poincaré maps are sometimes also known as "first return" maps. More precisely, we consider again the intersection of the graph (1.39) with $\mathbb{R} \times \Sigma$ (both as subsets of $\mathbb{R} \times \mathbb{R}^n$), and construct the sequence of points $(u(t_n))_{n \in \mathbb{N}} \subseteq \Sigma$, as ordered by the first argument t_n; that is, by the time of the n-th intersection of the orbit with the hyperplane Σ. Set $u_n := u(t_n)$. The sequence $(u_n)_{n \in \mathbb{N}}$ can then be considered as a recursive sequence on Σ, defined by a map

$$u_{n+1} = \Phi_\Sigma(u_n).$$

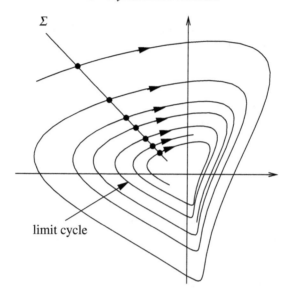

Figure 1.8: In the plane, the Poincaré section is a line.

The map Φ_Σ is called the POINCARÉ MAP associated to the semiflow defined by (1.18). Poincaré maps can thus be used to study the asymptotic behavior of a continuous semiflow, by reducing it to a discrete one. For example, if (1.7) has a periodic solution with period T, the Poincaré map with sampling synchronized with the period, i.e. with $t_n = nT$, will have a fixed point (fig. 1.9). Of course, for a given ODE, or system of ODEs, even autonomous ones, it may not be clear how to find suitable sampling sequences $(t_n)_{n \in \mathbb{N}}$, and extensive numerical experimentation may well be required.

Finally, we mention that the notion of Poincaré maps can be generalized to infinite dimensional continuous dynamical systems (see e.g. Marsden-McCracken, [MM76]).

1.4.2   Bernoulli's Sequences

We start with an example that illustrates the phenomenon of the loss of information from the initial data after sufficient time is allowed to pass.

The so-called BERNOULLI'S SEQUENCE is the recursive sequence $x_{n+1} = f(x_n)$ generated by the function $f: [0,1] \to [0,1]$ defined by

$$f(x) := 2x - \lfloor 2x \rfloor ,$$

where $\lfloor x \rfloor$ denotes the integer part of x (that is, the largest integer less than or equal to x). Note that f is not continuous at $x = \frac{1}{2}$ (fig. 1.10); however, f can be nicely described as a so-called "circle-doubling" map, if we identify the endpoints of the domain interval $[0,1]$ with each other. More precisely, if we define $g: [0,1] \to \mathbb{R}^2$ by

$$g(x) := (\cos(2\pi f(x)), \sin(2\pi f(x))) ,$$

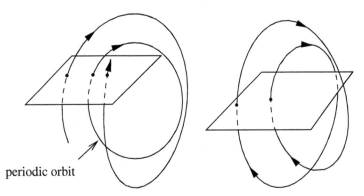

periodic orbit

Figure 1.9: Periodic and 2-periodic orbits produce fixed points in a Poincaré section.

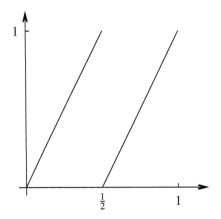

Figure 1.10: $f(x) = 2x - \lfloor 2x \rfloor$.

then g is continuous also at $x = \frac{1}{2}$. In fact, the point $g(x)$ runs twice around the unit circle, once as $x \in [0, \frac{1}{2}[$, and another as $x \in [\frac{1}{2}, 1[$.

It is easy to study the stability of Bernoulli's sequences: $x = 0$ is the only stationary point of f, and if $x_0 = \frac{1}{2}$ or $x_0 = 1$, then $x_1 = 0$, so $x_n = 0$ for all $n \geq 1$. Consider then any other initial value x_0 different from 0, $\frac{1}{2}$ and 1, and let $\varepsilon > 0$ be such that $y_0 := x_0 + \varepsilon$ is in the same half interval $]0, \frac{1}{2}[$ or $]\frac{1}{2}, 1[$ which contains x_0. Then

$$y_1 - x_1 = f(x_0 + \varepsilon) - f(x_0) = 2\varepsilon.$$

Next, if y_1 and x_1 are both still in the same half interval, we proceed to compute in the same way that

$$y_2 - x_2 = f(y_1) - f(x_1) = 2(y_1 - x_1) = 4\varepsilon.$$

Proceeding in this fashion, we see that $y_n - x_n = 2^n \varepsilon$, as long as y_{n-1} and x_{n-1} are in the same half interval. This computation shows that the distance between orbits grows exponentially; this has the consequence that at a certain point the orbits "must separate", no matter how close they were initially. In fact, we have that $y_n - x_n \geq \frac{1}{2}$ as soon as $n \geq \log_2 \frac{1}{2\varepsilon}$, and after this point the difference $y_{n+1} - x_{n+1}$ is no longer controllable.

One way to interpret this situation is that all information deriving from the knowledge of x_0 is eventually lost. For example, if x_0 represents the "true" initial value in an experiment, and $x_0 \pm \varepsilon$ is its actual measurement, after a number of steps equal to $\log_2 \frac{1}{2\varepsilon}$ no meaningful control of the error between the true and the approximated initial values is maintained. Figure 1.11 illustrates this phenomenon, by comparing

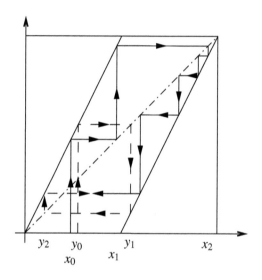

$$y_2 \qquad y_0 \qquad\qquad y_1 \qquad\qquad\qquad x_2$$
$$x_0 \qquad\qquad x_1$$

Figure 1.11: Two Bernoulli sequences.

the evolution of the sequences corresponding to different approximations x_0 and y_0 of $\frac{\pi}{4}$. After the first iteration, the points x_1 and y_1 are still close, but at the second iteration, $x_2 \approx 1$, while $y_2 \approx 0$.

In fact, the evolution of Bernoulli's sequence is chaotic, in the sense that it is sensitive to its initial conditions (see definition 1.7). We actually show more than (1.16); namely, we show that there is $R > 0$ such that for all $\delta \in \,]0, 1[$ and all x_0, $y_0 \in [0, 1]$ such that $|x_0 - y_0| \leq \delta$, there is $k \in \mathbb{N}$ such that $|x_k - y_k| \geq R$. To see this, we proceed by contradiction. Thus, we assume that for all $r > 0$ there are $\delta_r \in \,]0, 1[$ and $x_0, y_0 \in [0, 1]$ such that $|x_0 - y_0| \leq \delta_r$, and for all $k \in \mathbb{N}$,

$$|x_k - y_k| \leq r. \tag{1.40}$$

We fix $r = \frac{1}{4}$, and determine $\delta_{1/4}$, x_0 and y_0 accordingly. Our previous discussion implies that there are infinitely many indices $m \in \mathbb{N}$ such that x_m and y_m are in different half-intervals. In fact, if instead there were m_0 such that the sequences $(x_m)_{m \geq m_0}$ and $(y_m)_{m \geq m_0}$ are in the same half-interval, then we would deduce that

$$\tfrac{1}{2} \geq |x_m - y_m| = 2^{m-m_0}|x_{m_0} - y_{m_0}|,$$

which is a contradiction as $m \to +\infty$. It follows that there is at least one pair of subsequences $(x_{m_p})_{p \in \mathbb{N}}$ and $(y_{m_p})_{p \in \mathbb{N}}$ such that for all $p \in \mathbb{N}$, either

$$x_{m_p} < \tfrac{1}{2} \leq y_{m_p} \qquad \text{or} \qquad y_{m_p} < \tfrac{1}{2} \leq x_{m_p}. \tag{1.41}$$

If e.g. the first of (1.41) holds, recalling that $|x_{m_p} - y_{m_p}| \leq \tfrac{1}{4}$ we easily see that $x_{m_p} \geq y_{m_p}$; therefore, recalling (1.40), we deduce the contradiction

$$\tfrac{1}{4} \geq |x_{m_p+1} - y_{m_p+1}| = x_{m_p+1} - y_{m_p+1} = 1 - 2\left(y_{m_p} - x_{m_p}\right) \geq \tfrac{1}{2}.$$

With a totally analogous computation, we find the same result if the second of (1.41) holds. Hence, we conclude that Bernoulli's sequence is sensitive to its initial conditions.

The loss of information characteristic of Bernoulli's sequence can be described explicitly. Indeed, let x_0 be represented in the binary system by the series

$$x_0 = \sum_{n=1}^{\infty} \frac{\alpha_n}{2^n}, \qquad \alpha_n \in \{0, 1\}.$$

Then

$$x_1 = 2x_0 - \lfloor 2x_0 \rfloor = \sum_{n=1}^{\infty} \frac{\alpha_n}{2^{n-1}} - \left\lfloor \sum_{n=1}^{\infty} \frac{\alpha_n}{2^{n-1}} \right\rfloor = \alpha_1 + \sum_{n=2}^{\infty} \frac{\alpha_n}{2^{n-1}} - \alpha_1 = \sum_{n=1}^{\infty} \frac{\alpha_{n+1}}{2^n}.$$

This means that f moves the digits of the fractional part of each number x_n one position to the left, and subtracts the unit that may so result. For example, if

$$x_0 = 0.1101001 = \tfrac{1}{2} + \tfrac{1}{4} + \tfrac{1}{16} + \tfrac{1}{128},$$

then

$$x_1 = (1 + \tfrac{1}{2} + \tfrac{1}{8} + \tfrac{1}{64}) - 1 = 0.101001.$$

Now, in any numerical approximation, the initial value x_0 is known only up to a finite number of digits of its fractional part. If m is this number, after m iterations of Bernoulli's map we obtain $x_m = 0$; that is, we reach the fixed point of the map. Thus, all information from x_0 is lost in a finite number of steps.

1.4.3   Tent Maps

Another example of the phenomenon of the strong dependence of a system on its initial data, which numerically translates into a drastic loss of significant information, is provided by the iterated sequence (1.30), corresponding to the family of functions $f_\lambda : [0,1] \to [0,1]$ defined for $\lambda > 0$ by

$$f_\lambda(x) = \begin{cases} 2\lambda x & \text{for } 0 \le x \le \frac{1}{2}, \\ 2\lambda(1-x) & \text{for } \frac{1}{2} < x \le 1. \end{cases} \tag{1.42}$$

Each f_λ is an example of a so-called TENT MAP (fig. 1.12). Note that if $\lambda = \frac{1}{2}$, the

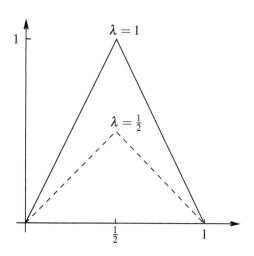

Figure 1.12: Tent maps for $\lambda = \frac{1}{2}$ and $\lambda = 1$.

sequence is constant at least after its second term; otherwise, it is possible to show (see e.g. Moon, [Moo92]), that the evolution of the corresponding dynamical system is regular if $\lambda < \frac{1}{2}$, and chaotic (in the sense that it is sensitive to its initial conditions, see definition 1.7) if $\lambda > \frac{1}{2}$. In particular, we show this for $\lambda = 1$.

Example 1.16
Let S be the dynamical system in $\mathcal{X} = [0,1]$ defined by the function f_1 of (1.42), which can be written as

$$f_1(x) = 1 - |1 - 2x| =: f(x). \tag{1.43}$$

Then, S is sensitive to its initial conditions.

To show this, given $x_0 \in [0,1]$, we consider the corresponding recursive sequence defined by $x_{n+1} = f(x_n)$ (i.e., the orbit of S starting at x_0). As was the case for Bernoulli's sequences, there are infinitely many terms of these sequences that fall in each of the subintervals $L := [0, \frac{1}{2}]$ and $R := [\frac{1}{2}, 1]$. In fact, for each $k \in \mathbb{N}_{>0}$, it is possible to keep track of the half interval in which the term x_k will fall, by means of the following device, which is adapted from Alligood, Sauer and Yorke, [ASY96, ch. 1.8]. Given k choices $S_1, \ldots, S_k$ of the letters L or R, we define a subinterval $S_1 \ldots S_k$ of $[0,1]$, by

$$x_0 \in S_1 \ldots S_k \iff x_j \in S_{j+1} \quad \text{for } j = 0, \ldots, k-1 \tag{1.44}$$

(thus, S_1 is either L or R). There are exactly 2^k such subintervals, each having length $\frac{1}{2^k}$; we can order these in a family $\mathcal{I} := \{\mathcal{I}_1, \ldots, \mathcal{I}_{2^k}\}$. Note that

$$\bigcup_{1 \le j \le 2^k} \mathcal{I}_j = [0,1], \tag{1.45}$$

and that, if $x_0 \in S_1 \ldots S_k$, then

$$x_1 \in S_2 \ldots S_k, \quad x_2 \in S_3 \ldots S_k, \quad \ldots, \quad x_{k-1} \in S_k. \tag{1.46}$$

The way the family $\mathcal{I}$ is actually ordered is of no importance here, except for the case $k = 2$, in which we have the ordering

$$[0,1] = LL \cup LR \cup RR \cup RL = [0, \tfrac{1}{4}] \cup [\tfrac{1}{4}, \tfrac{1}{2}] \cup [\tfrac{1}{2}, \tfrac{3}{4}] \cup [\tfrac{3}{4}, 1], \tag{1.47}$$

which is easily verified. By way of illustration, we consider the case $k = 3$, where the 8 subintervals are

$$LLL = [0, \tfrac{1}{8}], \quad LLR = [\tfrac{1}{8}, \tfrac{1}{4}], \quad LRR = [\tfrac{1}{4}, \tfrac{3}{8}], \quad LRL = [\tfrac{3}{8}, \tfrac{1}{2}],$$
$$RRL = [\tfrac{1}{2}, \tfrac{5}{8}], \quad RRR = [\tfrac{5}{8}, \tfrac{3}{4}], \quad RLR = [\tfrac{3}{4}, \tfrac{7}{8}], \quad RLL = [\tfrac{7}{8}, 1]$$

(in accord with (1.45)). For instance, suppose that $x_0 \in LRL$. Then, by definition (1.44),

$$x_0 \in L = [0, \tfrac{1}{2}], \quad x_1 \in R = [\tfrac{1}{2}, 1], \quad x_2 \in L = [0, \tfrac{1}{2}], \tag{1.48}$$

and this is in accord with (1.46). By (1.43), the first of (1.48) implies that $x_1 = 2x_0$. The second of (1.48) implies then that $\frac{1}{4} \le x_0 \le \frac{1}{2}$; i.e., $x_0 \in LR$. Again by (1.43), the second of (1.48) also implies that $x_2 = 2 - 2x_1 = 2 - 4x_0$; then, the third of (1.48) finally implies that $\frac{3}{8} \le x_0 \le \frac{1}{2}$, as claimed.

We are now ready to show the sensitivity of the semiflow S to its initial conditions. We claim that, given $\delta \in {]0,1[}$ and $x_0 \in [0,1]$, there are $y_0 \in [0,1]$, with $0 < |x_0 - y_0| \le \delta$, and $k \in \mathbb{N}_{>0}$, such that

$$|x_k - y_k| \ge \tfrac{1}{4} \tag{1.49}$$

(compare to (1.16)).

To show this, fix $\delta \in {]0,1[}$ and $x_0 \in [0,1]$. Let $k \in \mathbb{N}_{>0}$ be such that $\frac{1}{2^k} < \delta$. Then the interval $]x_0 - \delta, x_0 + \delta[\cap [0,1]$ contains at least one subinterval $S_1 \ldots S_k S_{k+1} S_{k+2}$, with $x_0 \in S_1 \ldots S_k S_{k+1} S_{k+2}$. Fix one of these subintervals, for which there are the four possibilities

$$S_0 \ldots S_k LL, \quad S_0 \ldots S_k LR, \quad S_0 \ldots S_k RR, \quad S_0 \ldots S_k RL.$$

We choose then y_0 in the subinterval where x_0 is, but with the last letters L and R reversed; that is, respectively, in

$$S_0 \ldots S_k RR , \quad S_0 \ldots S_k RL , \quad S_0 \ldots S_k LL , \quad S_0 \ldots S_k LR .$$

Then, since both x_0 and y_0 are in the larger subinterval $S_1 \ldots S_k$, whose length is $\frac{1}{2^k} < \delta$, we have that $|x_0 - y_0| < \delta$, and (1.49) holds. Indeed, suppose e.g. that

$$x_0 \in S_0 \ldots S_k LL , \quad y_0 \in S_0 \ldots S_k RR .$$

Then, by (1.46), $x_k \in LL$ and $y_k \in RR$, which, recalling (1.47), means that

$$0 \leq x_k \leq \tfrac{1}{4} , \quad \tfrac{1}{2} \leq y_k \leq \tfrac{3}{4} .$$

Thus,

$$|x_k - y_k| = y_k - x_k \geq \tfrac{1}{2} - \tfrac{1}{4} = \tfrac{1}{4} .$$

An analogous argument shows that the same inequality holds in each of the remaining three possibilities. Thus, (1.49) follows, proving the sensitivity of the semiflow S defined by (1.43) to its initial conditions. □

From these two examples, we could surmise that the chaotic behavior of a discrete system may be a consequence of the fact that the functions that define the sequence (1.30) are not regular (for Bernoulli's sequences, f is not continuous; for the tent maps, f is not differentiable). The next example shows that we can in fact have chaotic behavior even with C^∞ maps.

1.4.4 Logistic Maps

A regularized version of the tent maps is provided by the family of functions $f_\lambda : \mathbb{R} \to \mathbb{R}$ defined by

$$f_\lambda (x) = \lambda x (1 - x) ,$$

see fig. 1.13. When $0 \leq \lambda \leq 4$, each f_λ maps the interval $[0, 1]$ into itself; the corresponding iterated sequence (1.30) is called a LOGISTIC SEQUENCE. These sequences are a normalized version of the sequence $(x_n)_{n \in \mathbb{N}}$, defined by

$$x_{n+1} = a x_n - b x_n^2 ,$$

which describes a model of population growth; the coefficient a represents a constant growth rate, and b measures an external inhibiting factor. In absence of the latter, i.e. when $b = 0$, the terms of the sequence reduce to

$$x_{n+1} = a^{n+1} x_0 .$$

In this case, the fixed point $x = 0$ is stable if $|a| < 1$, and unstable if $|a| > 1$. If $a = 1$ we have a 1-periodic orbit, and if $a = -1$, a 2-periodic orbit.

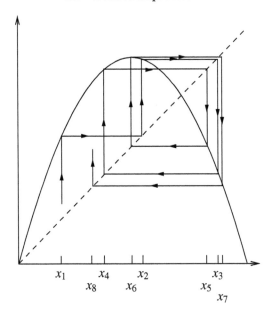

Figure 1.13: A sequence generated by the logistic map.

To find the stationary points of the logistic sequence, we consider the equation $f(x) = x$ (having dropped the index λ for convenience), i.e.

$$x = \lambda x(1-x).$$

This equation has the two solutions $x = 0$ and $x = s_\lambda := 1 - \frac{1}{\lambda}$. Since $s_\lambda \in [0,1]$ if and only if $\lambda \geq 1$, we conclude that if $\lambda < 1$, $x = 0$ is the only stationary point. Since $s_1 = 0$, the same is true for $\lambda = 1$. If $x_0 = 1$, then $x_n = 0$ for all $n \geq 1$. Since $f'(0) = \lambda$, recalling theorem 1.15 we see that the stationary point $x = 0$ is stable if $\lambda < 1$, and unstable if $\lambda > 1$; similarly, since $f'(s_\lambda) = 2 - \lambda$, s_λ is stable if $1 < \lambda < 3$, unstable if $\lambda > 3$. We also see directly that $x = s_1 = 0$ and $x = s_3$ are stable also if, respectively, $\lambda = 1$ and $\lambda = 3$.

As we know, stationary points of the sequence correspond to 1-periodic orbits. To find 2-periodic orbits, we look for the stationary points of the second iterate of f, i.e. for solutions of the equation

$$x = f(f(x)) =: f^2(x) = \lambda^2 x(1-x)(\lambda x^2 - \lambda x + 1),$$

(fig. 1.14). Of course, $f^2(0) = 0$ and $f^2(s_\lambda) = s_\lambda$, since a fixed point of f is also a fixed point of any of its iterates. Other fixed points of f^2 are found by solving the equivalent equation

$$\frac{x - f^2(x)}{x(x - s_\lambda)} =: Q(x) = 0.$$

We compute that

$$Q(x) = \lambda^2 \left(\lambda x^2 - (1+\lambda)x + 1 + \lambda^{-1} \right);$$

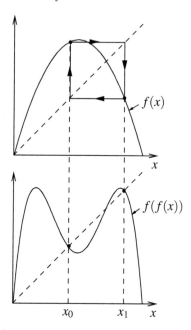

Figure 1.14: 2-periodic orbits: x_0 and x_1 are fixed points of $f(f(x))$.

since the discriminant of Q equals $\Delta_Q := (\lambda + 1)(\lambda - 3)$, for $\lambda > 3$ there are two more fixed points of f^2; these produce two 2-periodic orbits for the sequence. For $\lambda = 3$, $\Delta_Q = 0$, $s_3 = \frac{2}{3}$, and

$$Q(x) = -9(3x^2 - 4x + \tfrac{4}{3}) = -27(x - s_3)^2 \,;$$

thus, there still is only one 1-periodic orbit. At this point, we should proceed with the study of the stability of these fixed points, and then look for fixed points of further iterates of f, and so on. However, the analytical complexity of these computations is such that it is far more effective to resort to numerical experimentation and geometric or topological arguments. As can be seen in Moon, [Moo92], the numerical evidence shows that, as λ increases to 4, the corresponding orbits first exhibit period doubling, then turn to chaotic behavior. In fact, proceeding as in example 1.16, we can easily show that the logistic sequence corresponding to $\lambda = 4$ (in which case the range of f_4 is all of $[0, 1]$) is sensitive to its initial conditions.

1.5 Lorenz' Equations

In this and the next section, we present two examples of continuous, finite dimensional dynamical systems, which admit an attractor for certain values of their parameters. This values are found by numerical experiment; the existence of the attractor can be confirmed by the methods described in the next chapter.

1.5.1   The Differential System

We begin with the so-called (and quite famous) LORENZ' EQUATIONS , which are the system of the three autonomous differential equations

$$\begin{cases} \dot{x} = -\sigma x + \sigma y \\ \dot{y} = rx - y - xz \\ \dot{z} = -bz + xy, \end{cases} \qquad (1.50)$$

with σ, r and $b > 0$. This system was proposed by Lorenz in [Lor63] as an approximation, with the three degrees of freedom σ, r, and b, of the Boussinesq equations modelling the convective motion of a stratified bidimensional fluid heated by convection from below, such as air over the earth's surface. As such, it provides a model, admittedly oversimplified, of an atmospheric phenomenon of interest in meteorology. Our goal is to study the behavior of solutions to (1.50), in relation to the parameter r (the Rayleigh's number), keeping σ and b fixed. In the next chapter we shall show that for all values of σ, b and r, and for all initial values $(x_0, y_0, z_0) \in \mathcal{X} = \mathbb{R}^3$, system (1.50) has a unique global solution; that this system generates a semiflow S on $\mathbb{R}^3$; that S admits a bounded absorbing set in $\mathbb{R}^3$; that, as a consequence, system (1.50) has a compact attractor $\mathcal{A}$.

For certain values of r, the structure of this attractor is relatively well understood. Although most detailed information can be obtained by means of extensive numerical experimentation, we present here some results that can be established by simple analytical techniques. For a more extensive study of Lorenz' equations, we refer e.g. Sparrow, [Spa82], Marsden-McCracken, [MM76], and Guckenheimer-Holmes, [GH83].

1.5.2   Equilibrium Points

The stationary points of (1.50) are obtained by solving the system

$$\begin{cases} -\sigma(x - y) = 0 \\ rx - y - xz = 0 \\ -bz + xy = 0. \end{cases}$$

We obtain that if $r \leq 1$, the origin is the only equilibrium point, while if $r > 1$ there are exactly three equilibrium points: The origin $O = (0, 0, 0)$ and the two other points

$$C_{\pm} := (\pm\sqrt{b(r-1)}, \pm\sqrt{b(r-1)}, r - 1).$$

To study the stability of these three points, recalling theorem A.7 we linearize system (1.50) at each point, and consider the sign of the real part of the eigenvalues of the

Jacobian matrix

$$J(x,y,z) = \begin{pmatrix} -\sigma & \sigma & 0 \\ r-z & -1 & -x \\ y & x & -b \end{pmatrix}$$

at each equilibrium point. At the origin the characteristic polynomial is

$$\det[J(O) - \lambda I] = \det \begin{pmatrix} -(\sigma+\lambda) & \sigma & 0 \\ r & -(1+\lambda) & 0 \\ 0 & 0 & -(b+\lambda) \end{pmatrix}$$

$$= -(b+\lambda)\underbrace{(\lambda^2 + (\sigma+1)\lambda + \sigma(1-r))}_{:= P(\lambda)}.$$

Thus, $J(O)$ always has at least the real negative eigenvalue $\lambda_1 = -b$. The discriminant Δ_P of the polynomial P equals

$$\Delta_P = (\sigma-1)^2 + 4\sigma r;$$

since $\Delta_P > 0$, $J(O)$ has in fact three real eigenvalues. If $r < 1$, all coefficients of P are positive, so the eigenvalues are all negative, the unique equilibrium point O is a stable node, and is in fact the attractor of the system (i.e., $\mathcal{A} = \{O\}$). If $r > 1$, one eigenvalue of $J(O)$ is positive, so the origin is an unstable saddle, with a 2-dimensional stable manifold $\mathcal{M}^s(O)$ attracted by O, and a one-dimensional unstable manifold $\mathcal{M}^u(O)$ repelled by it; see definition 2.22 in chapter 2. At the points $C_\pm$, $r > 1$ and the characteristic polynomial is

$$\det[J(C_\pm) - \lambda I] = \det \begin{pmatrix} -(\sigma+\lambda) & \sigma & 0 \\ 1 & -(1+\lambda) & \mp\sqrt{b(r-1)} \\ \pm\sqrt{b(r-1)} & \pm\sqrt{b(r-1)} & -(b+\lambda) \end{pmatrix}$$

$$= -(\lambda^3 + (\sigma+b+1)\lambda^2 + b(\sigma+r)\lambda + 2b\sigma(r-1))$$

$$=: -P_1(\lambda).$$

Again, at least one eigenvalue λ_1 is real. The other two eigenvalues λ_2 and λ_3 are either both real, or complex conjugate. If they are real, elementary calculus shows that, since $r > 1$, both are negative; hence, C_+ and C_- are stable nodes. If instead λ_2 and λ_3 are complex nonreal, to study the sign of their real part we set $\lambda_2 = \zeta = u + iv$ and $\lambda_3 = \bar{\zeta} = u - iv$, and proceed as follows. Writing the characteristic equation as

$$P_1(\lambda) = (\lambda - \lambda_1)(\lambda - \zeta)(\lambda - \bar{\zeta}) = 0,$$

we obtain the equation

$$\lambda^3 - (2u + \lambda_1)\lambda^2 + (|\zeta|^2 + 2\lambda_1 u)\lambda - \lambda_1|\zeta|^2 = 0.$$

We easily verify that, since $v \neq 0$, $u = \mathfrak{Re}\,\zeta = 0$ if and only if the product of the coefficients of λ^2 and λ equals the constant term. In terms of the original form of $P_1(\lambda)$, this translates into the condition

$$b(\sigma+b+1)(\sigma+r) = 2b\sigma(r-1). \tag{1.51}$$

As an equation in r, if $\sigma \neq b+1$, (1.51) has the solution

$$r_* = \frac{\sigma(\sigma+b+3)}{\sigma-b-1}.$$

It can then be shown (see e.g. Sparrow, [Spa82]) that $\mathfrak{Re}\,\lambda < 0$ if $r < r_*$, while $\mathfrak{Re}\,\lambda > 0$ if $r > r_*$. It follows that if $1 < r < r_*$, the stationary points $C_\pm$ are both stable, and every orbit converges to one of these points. Thus, there is an attractor $\mathcal{A}$, consisting of the points C_-, C_+, and the unstable manifold $\mathcal{M}^u(O)$ connecting C_- to C_+. More precisely, there is a value $r_1 \in\,]1, r_*[$ such that:

1. If $1 < r < r_1$, all three eigenvalues of $J(C_\pm)$ are real negative;

2. If $r_1 < r < r_*$, there are two complex conjugate eigenvalues with negative real part. In this range, $\mathcal{M}^u(O)$ circles around C_- and C_+;

3. If $r > r_*$, the two complex conjugate eigenvalues have positive real part, so the stationary points O, C_+ and C_- are all unstable (fig. 1.15).

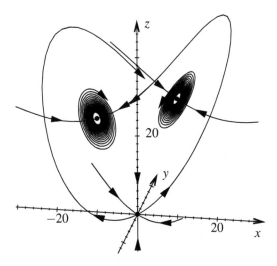

Figure 1.15: Behavior of the orbits near the equilibrium points (unstable case).

The numerical evidence confirms the existence of an attractor. Near $C_\pm$, orbits arrive along the stable manifolds $\mathcal{M}^s(C_\pm)$ (corresponding to the real negative eigenvalue of $J(C_\pm)$), and spiral out along the two-dimensional surface $\mathcal{M}^u(C_\pm)$. This behavior was first observed by Lorenz, in whose original computations the parameter values are $\sigma = 10$, $b = 8/3$; corresponding to these values, $r_* = 470/19 \approx 24.74$, and $r_1 \approx 24.06$. Lorenz' so-called BUTTERFLY ATTRACTOR is observed at $r = 28$ (fig. 1.16).

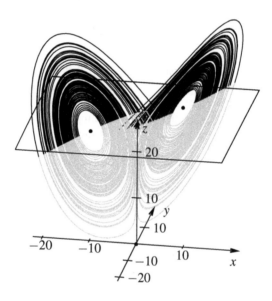

Figure 1.16: The "butterfly" attractor.

1.6 Duffing's Equation

1.6.1 The General Model

The second example we consider is that of the so-called DUFFING EQUATION, which describes the motion of a vibrating spring subject to a nonlinear restoring term. The corresponding ODE model is determined in accord to Hooke's law. Taking into account the dissipation effects due to friction (as measured by a numerical coefficient k), and assuming the presence of a periodic forcing term, the evolution of this system is governed by the nonlinear second order ODE

$$\ddot{x} + k\dot{x} + x^3 - x = \lambda \cos \omega t, \qquad (1.52)$$

where k, λ and $\omega > 0$. Equation (1.52) is equivalent to the first order system

$$\begin{cases} \dot{x} = y \\ \dot{y} = -ky + x - x^3 + \lambda \cos \omega t, \end{cases} \qquad (1.53)$$

which is of type (1.17), with $u = (x, y) \in \mathcal{X} = \mathbb{R}^2$. It is not difficult to verify that (1.53) has, for each $\lambda \in \mathbb{R}$ and $u_0 \in \mathbb{R}^2$, a unique global solution $u(\cdot, u_0, \lambda) \in C^1([0, +\infty[; \mathbb{R}^2)$.

As was the case for the logistic equation, the asymptotic behavior of the solutions of system (1.53) is sharply influenced by the values of the parameter λ. When $\lambda = 0$, system (1.53) is autonomous, and the asymptotic behavior of its solutions can be studied with elementary techniques. In this case, (1.53) becomes

$$\begin{cases} \dot{x} = y \\ \dot{y} = x - x^3 - ky. \end{cases} \qquad (1.54)$$

The stationary points of this system are the origin O, and the points $C_\pm := (\pm 1, 0)$. To study the stability of these stationary points, we refer again to theorem A.7, and consider the characteristic polynomial of the linearized system, i.e.

$$P(x, y; \mu) := \det[J(x, y) - \mu I] = \mu(\mu + k) + 3x^2 - 1.$$

At the origin, $P(0, 0; \mu) = \mu^2 + k\mu - 1$, which has two real distinct roots with opposite sign; thus, $(0, 0)$ is a saddle point. At $C_\pm$, $P(\pm 1, 0; \mu) = \mu^2 + k\mu + 2$. Thus, if $k > 2\sqrt{2}$, there are two real, distinct, negative eigenvalues, and $(\pm 1, 0)$ are stable nodes. If instead $k < 2\sqrt{2}$, there are two complex conjugate eigenvalues. Since these have negative real part, both C_+ and C_- are stable sinks (fig. 1.17). The unstable

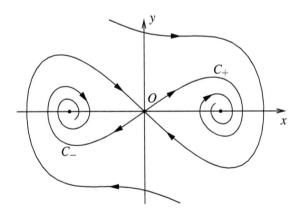

Figure 1.17: The origin is an unstable saddle point; C_+ and C_- are stable sinks.

manifolds of O converge to the fixed points $C_\pm$; correspondingly, solutions of (1.54) with initial values on these manifolds converge to the stationary solutions $(\pm 1, 0)$.

When $\lambda \neq 0$, the analogous of (1.54), i.e. system (1.53), is not autonomous, and will not have stationary solutions. However, since the system has a periodic forcing term, it may have periodic solutions, at least for some values of λ. In this case, we can carry out an analysis of the asymptotic behavior of the system by means of Poincaré maps. Namely, for each integer m we consider the sequence $(u_n)_{n \in \mathbb{N}}$, with $u_n := u(2mn\pi\omega^{-1}, u_0, \lambda)$. Each choice of m defines a stroboscopic map, and we can try to find the fixed points (if any) of these maps, and determine their stability. These fixed points would then correspond to periodic solutions of the system. For example, figure 1.18 refers to the stroboscopic map defined by system (1.53), with

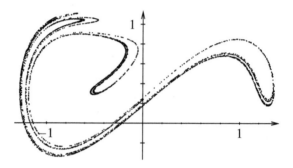

Figure 1.18: Attractor for Duffing's equation.

$\lambda = 0.5$, $k = 0.3$ and $\omega = 1$, corresponding to the case $m = 1$. Recall that this map is obtained by recording the projections of the points $(u(2n\pi, u_0, 0.5))_{n \in \mathbb{N}}$. This stroboscopic map shows evidence of the existence of an attractor. We mention explicitly that the particular shape of this attractor depends on the choice of the sampling points t_n; for example, the sequence $(u(2\pi(n+1)\omega^{-1}, u_0, 0.5))_{n \in \mathbb{N}}$ defines a different stroboscopic map, and produces an attractor with a different geometrical shape. More generally, numerical experiments show that for small values of λ there are at first periodic solutions with successive doublings of the period; then, at $\lambda = \frac{1}{2}$, the evolution of the system appears to be chaotic. We refer to Jordan-Smith, [JS87], and Guckenheimer-Holmes, [GH83], for a thorough examination of these cases.

1.6.2   A Linearized Model

To better illustrate the procedure we have outlined for the nonautonomous case, we consider a simpler, linearized version of Duffing's equation (1.52). More precisely, we consider the equation

$$\ddot{x} + \dot{x} - 2x = -5(\sin t - \cos t),$$

which we transform as usual into the system

$$\begin{cases} \dot{x} = y \\ \dot{y} = 2x - y - 5(\sin t - \cos t). \end{cases} \tag{1.55}$$

As proposed, we consider the Poincaré sequence on the phase space $\mathcal{X} = \mathbb{R}^2$

$$x_n = x(2n\pi), \quad y_n = y(2n\pi), \tag{1.56}$$

and look for fixed points of this sequence. The initial value problem for (1.55) with initial conditions $x(0) = x_0$, $y(0) = y_0$ has the explicit solution

$$\begin{cases} x = \frac{1}{3}(2x_0 + y_0)e^t + \frac{1}{3}(x_0 - y_0 + 3)e^{-2t} + 2\sin t - \cos t, \\ y = \frac{1}{3}(2x_0 + y_0)e^t - \frac{2}{3}(x_0 - y_0 + 3)e^{-2t} + 2\cos t + \sin t \ (= \dot{x}). \end{cases} \tag{1.57}$$

Correspondingly, the sequence (1.56) becomes

$$\begin{cases} x_n = \frac{1}{3}(2x_0 + y_0)e^{2n\pi} + \frac{1}{3}(x_0 - y_0 + 3)e^{-4n\pi} - 1, \\ y_n = \frac{1}{3}(2x_0 + y_0)e^{2n\pi} - \frac{2}{3}(x_0 - y_0 + 3)e^{-4n\pi} + 2. \end{cases} \tag{1.58}$$

Let $P = (-1, 2)$. We immediately see that P is a fixed point for the sequence (1.58) if and only if the initial conditions (x_0, y_0) are such that

$$2x_0 + y_0 = 0, \quad x_0 - y_0 + 3 = 0,$$

i.e. if and only if $(x_0, y_0) = P$. Corresponding to this choice of initial values, we see from (1.57) that (1.55) has the periodic solution

$$x = 2\sin t - \cos t, \quad y = 2\cos t + \sin t.$$

To determine the stability of this solution, we study the stability of P as fixed point of the sequence (1.58). Setting

$$\mathcal{M}^s(P) = \{(x,y): 2x + y = 0\},$$
$$\mathcal{M}^u(P) = \{(x,y): x - y + 3 = 0\}$$

(these are two straight lines, intersecting at P), we easily verify that $\mathcal{M}^s(P)$ and $\mathcal{M}^u(P)$ are the stable and unstable manifolds of P (see definition 2.22 of section 2.3.3 below). Indeed, from (1.58) we have that

$$(x_0, y_0) \in \mathcal{M}^s(P) \implies (x_n, y_n) \to (-1, 2) \quad \text{as } n \to +\infty,$$

while

$$(x_0, y_0) \in \mathcal{M}^u(P) \implies (x_n, y_n) \to (-1, 2) \quad \text{as } n \to -\infty.$$

Equivalently, if $(x_0, y_0) \in \mathcal{M}^s(P) \cup \mathcal{M}^u(P)$, then (x_0, y_0) belongs to a motion $t \mapsto (x(t), y(t))_{t \in \mathbb{R}}$ of (1.57), with

$$(x(t), y(t)) \to (-1, 2)$$

as $t \to +\infty$ if $(x_0, y_0) \in \mathcal{M}^s(P)$, or as $t \to -\infty$ if $(x_0, y_0) \in \mathcal{M}^u(P)$. If instead $(x_0, y_0) \notin \mathcal{M}^s(P) \cup \mathcal{M}^u(P)$, the sequence $((x_n, y_n))_{n \in \mathbb{N}}$ will not have a limit point, regardless of how close (x_0, y_0) is to $(-1, 2)$. Thus, the fixed point $(-1, 2)$ is unstable.

1.7 Summary

We conclude this chapter with a brief summary of the main ideas of this introduction. We consider a DYNAMICAL SYSTEM, defined by an abstract autonomous
ODE

$$\frac{d}{dt}u = F(u), \qquad u(t) \in \mathcal{X}, \qquad (1.59)$$

where $\mathcal{X}$ is a Banach space and $F \in C(\mathcal{X}; \mathcal{X})$. System (1.59) is FINITE DIMEN-
SIONAL if $\dim \mathcal{X} < +\infty$. Since the dynamical system (1.59) is autonomous, if the
corresponding Cauchy problem is well posed in the large the differential equation
(1.59) defines a CONTINUOUS SEMIGROUP S, also called a SEMIFLOW. We are interested in determining how the asymptotic properties of this semiflow, i.e. its behavior
as $t \to +\infty$, depend on the initial values we attach to (1.59). If the system contains
some numerical parameters, we are also interested in how these may influence the
asymptotic behavior of the system.

In many situations, it is possible to describe the long-time behavior of a dynamical
system by means of a bounded ATTRACTOR, to which all the orbits converge as $t \to$
$+\infty$, independently of where they originate. This attractor may be compact, and also
have finite dimension. We would like to have criteria that allow us to recognize the
existence of such an attractor, even if its structure may in general be quite complex,
and very little may be known of its geometric or differential properties (although for
many types of important physical examples there may be reasonable estimates on
its dimension). On the other hand, there are favorable examples where we can in
fact determine that the attractor is contained into a finite dimensional, exponentially
attracting manifold of $\mathcal{X}$ (the so-called INERTIAL MANIFOLD), or at least into a
compact, finite dimensional EXPONENTIAL ATTRACTOR. In these cases, since orbits
converge exponentially fast to these sets, after a relatively short transient time the
dynamics of the system are essentially governed by a finite system of ODEs. The
number of the equations in this system is in general determined by the dimension of
these sets. We would therefore like to identify some criteria that allow us to deduce
the existence of such sets and, possibly, meaningful estimates on their dimension.

Chapter 2

Attractors of Semiflows

In this chapter we introduce the definitions of the SEMIFLOW associated to a dissipative autonomous dynamical system in a Banach space, and of the attractor of this semiflow. We discuss some of the most relevant properties of semiflows and their attractors. As we shall see, the ideas and results presented in this chapter are a natural generalization to the infinite dimensional case of many well known notions of the qualitative theory of ODEs (where the dimension of the space is finite). Throughout this chapter, $\mathcal{X}$ is a Banach space, with norm $\|\cdot\|$ and induced distance d; however, most of the results we establish also hold in general complete metric spaces.

In the sequel, we will adopt the following notation. If $\mathcal{E} \subseteq \mathbb{R}$ and $a \in \mathcal{E}$, we set

$$\mathcal{E}_{\geq a} := \{x \in \mathcal{E} : x \geq a\}.$$

For example, we have already referred to the set

$$\mathbb{R}_{\geq 0} = [0, +\infty[.$$

Analogous definitions hold for the sets $\mathcal{E}_{>a}$, $\mathcal{E}_{\leq a}$ and $\mathcal{E}_{<a}$.

2.1 Distance and Semidistance

Since attractors are sets to which orbits converge, we need to recall the following facts on the distance and semidistance of two subsets of a metric space $\mathcal{X}$, with distance d.

DEFINITION 2.1 *Given two subsets $A, B \subseteq \mathcal{X}$ and $x \in \mathcal{X}$, we set*

$$d(x, B) := \inf_{b \in B} d(x, b), \tag{2.1}$$

$$\partial(A, B) := \sup_{a \in A} d(a, B) = \sup_{a \in A} \inf_{b \in B} d(a, b), \tag{2.2}$$

$$\text{dist}(A, B) := \max\{\partial(A, B), \partial(B, A)\}. \tag{2.3}$$

We remark that the map

$$(\mathcal{A}, \mathcal{B}) \mapsto \partial(\mathcal{A}, \mathcal{B})$$

is a SEMIDISTANCE; that is, ∂ is not symmetric, and the equality $\partial(\mathcal{A}, \mathcal{B}) = 0$ doesn't necessarily imply that $\mathcal{A} = \mathcal{B}$ (take e.g. $\mathcal{A} \subset \mathcal{B}$). Moreover, we can even have $\partial(\mathcal{A}, \mathcal{B}) = 0$ with $\mathcal{A} \supset \mathcal{B}$. This is for example the case when $\mathcal{B}$ is open, and $\mathcal{A} = \bar{\mathcal{B}}$. However, we have

PROPOSITION 2.2

Let $\mathcal{A}$ and $\mathcal{B}$ be subsets of $\mathcal{X}$, such that $\partial(\mathcal{A}, \mathcal{B}) = 0$. Then $\mathcal{A} \subseteq \bar{\mathcal{B}}$. In particular, if $\mathcal{B}$ is closed, $\mathcal{A} \subseteq \mathcal{B}$.

PROOF   By (2.2), the condition $\partial(\mathcal{A}, \mathcal{B}) = 0$ implies that

$$\inf_{b \in \mathcal{B}} d(a, b) = 0$$

for all $a \in \mathcal{A}$. Thus, for each $a \in \mathcal{A}$ we can find a sequence $(b_n)_{n \in \mathbb{N}} \subset \mathcal{B}$, such that $d(a, b_n) \to 0$. This means that $b_n \to a$, so $a \in \bar{\mathcal{B}}$. □

From the last part of proposition 2.2, we immediately deduce

COROLLARY 2.3

The restriction of the map $(\mathcal{A}, \mathcal{B}) \mapsto \mathrm{dist}(\mathcal{A}, \mathcal{B})$ to the family of closed subsets of $\mathcal{X}$ is a metric.

PROOF   We only have to show that $\mathrm{dist}(\mathcal{A}, \mathcal{B}) = 0$ implies $\mathcal{A} = \mathcal{B}$ for closed $\mathcal{A}$, $\mathcal{B} \subseteq \mathcal{X}$. Indeed, (2.3) implies that both $\partial(\mathcal{A}, \mathcal{B}) = 0$ and $\partial(\mathcal{B}, \mathcal{A}) = 0$. The conclusion then follows from proposition 2.2. □

The distance defined in corollary 2.3 is known as the HAUSDORFF DISTANCE of closed sets in $\mathcal{X}$.

2.2 Discrete and Continuous Semiflows

2.2.1 Types of Semiflows

We start by defining various types of flows and semiflows, as follows.

DEFINITION 2.4   *Let T be one of the sets $\mathbb{R}$, $\mathbb{R}_{\geq 0}$, $\mathbb{N}$, or $\mathbb{Z}$. A SEMIFLOW on $\mathcal{X}$ is a family $S := (S(t))_{t \in T}$ of continuous (but not necessarily linear) maps in $\mathcal{X}$, i.e.*

such that for all $t \in T$,

$$S(t) \in C(\mathcal{X}, \mathcal{X}), \tag{2.4}$$

which satisfies the so-called SEMIGROUP *conditions*

$$S(0) = I, \tag{2.5}$$
$$S(t + t') = S(t)S(t'), \tag{2.6}$$

for all t, $t' \in T$, and the additional continuity condition

$$S(\cdot)x \in C(T; \mathcal{X}) \tag{2.7}$$

for all $x \in \mathcal{X}$. Furthermore:

1. *If T is either $\mathbb{R}$ or $\mathbb{Z}$, the semiflow is called a* FLOW.

2. *If T is either $\mathbb{R}$ or $\mathbb{R}_{\geq 0}$, the flow (respectively, the semiflow) is called* CONTINUOUS.

3. *If T is either $\mathbb{Z}$ or $\mathbb{N}$, the flow (respectively, the semiflow) is called* DISCRETE.

Conditions (2.5) and (2.6) were already introduced in (1.20) and (1.21), and refer to the semigroup or group properties of S. S becomes a semiflow if it satisfies the additional requirements of the continuity of the maps

$$x \mapsto S(t)x \qquad \text{and} \qquad t \mapsto S(t)x,$$

respectively for all fixed $t \in T$ and all fixed $x \in \mathcal{X}$, as required in conditions (2.4) and (2.7). Note that the continuity in t, that is (2.7), is trivially satisfied for discrete flows and semiflows. Also, if S is a flow, S is in particular a GROUP of continuous operators on $\mathcal{X}$.

Alternatively, S may be called:

1. A CONTINUOUS SEMI-DYNAMICAL SYSTEM if $T = \mathbb{R}_{\geq 0}$;

2. A CONTINUOUS DYNAMICAL SYSTEM if $T = \mathbb{R}$;

3. A DISCRETE SEMI-DYNAMICAL SYSTEM if $T = \mathbb{N}$;

4. A DISCRETE DYNAMICAL SYSTEM if $T = \mathbb{Z}$.

The term "continuous" therefore distinguishes these systems from "discrete" ones, where the "time" set T is discrete, such as those considered in section 1.4. Likewise, the prefix "semi-" refers to the fact that we only consider nonnegative values of the time variable (continuous or discrete).

In particular, a discrete semiflow S can be identified with the continuous mapping $\tilde{S} \colon \mathcal{X} \to \mathcal{X}$, defined by $\tilde{S} := S(1)$. In fact, $S(n) = \tilde{S}^n$, for all $n \in T$.

If T is either $\mathbb{R}_{\geq 0}$ or $\mathbb{N}$, in definition 2.4 we do not require that $S(t)$ be invertible for any $t \in T$. However, if $S(t)$ is invertible for each $t \in T$, we can extend S to the parameter set

$$T_- := \{-t : t \in T\}, \tag{2.8}$$

and therefore to a flow, by setting

$$S(-t) := S(t)^{-1}$$

for each $t \in T$. This definition extends (2.6) in a natural way; more precisely, we have

PROPOSITION 2.5

1. *Let T be either $\mathbb{Z}$ or $\mathbb{R}$, and $S = (S(t))_{t \in T}$ be a flow on $\mathcal{X}$. Then for all $t \in T$, $S(t)$ is invertible, and $S(t)^{-1} = S(-t)$.*

2. *Let T be either $\mathbb{N}$ or $\mathbb{R}_{\geq 0}$, and $S = (S(t))_{t \in T}$ be a semiflow on $\mathcal{X}$, such that $S(t)$ is invertible for each $t \in T$. Define T_- as in (2.8), and set*

$$T_* := T \cup T_- .$$

For $t \in T_*$, define

$$\tilde{S}(t) := \begin{cases} S(t) & \text{if } t \in T, \\ S(-t)^{-1} & \text{if } t \in T_- . \end{cases}$$

Then the family $\tilde{S} := (\tilde{S}(t))_{t \in T_*}$ is a flow in $\mathcal{X}$.

PROOF   The first claim is immediate, since for all $t \in T$

$$S(t)S(-t) = S(t-t) = S(0) = I .$$

To prove that $\tilde{S}$ is a flow, it is sufficient to show the semigroup property (2.6). This is immediate if both $t, t' \in T$. Assume then that $t \in T$ and $t' \in T_-$. We first show that for all $x \in \mathcal{X}$,

$$\tilde{S}(t')\tilde{S}(t)x = \tilde{S}(t'+t)x . \tag{2.9}$$

We must distinguish two cases, according to whether $t' + t \in T$ or not. If $t' + t \in T$, (2.9) reads

$$S(-t')^{-1}S(t)x = S(t'+t)x , \tag{2.10}$$

and this is established by applying the operator $S(-t')$ to both sides of (2.10), recalling that this operator is injective. If instead $t' + t \in T_-$, (2.9) reads

$$S(-t')^{-1}S(t)x = S(-t'-t)^{-1}x . \tag{2.11}$$

Let $y \in \mathcal{X}$ be such that $S(-t'-t)y = x$. Then, since $-t' \in \mathcal{T}$,

$$S(t)S(-t'-t)y = S(t)S(-t-t')y = S(-t')y.$$

This means that $y = S(-t')^{-1}S(t)x$; thus, (2.11) holds. We now show that also

$$\tilde{S}(t)\tilde{S}(t')x = \tilde{S}(t+t')x \qquad (2.12)$$

for all $x \in \mathcal{X}$. Again, if $t'+t \in \mathcal{T}$, (2.12) reads

$$S(t)S(-t')^{-1}x = S(t+t')x. \qquad (2.13)$$

Let $z \in \mathcal{X}$ be such that $S(-t')z = x$. Then, since $t \in \mathcal{T}$,

$$S(t+t')x = S(t+t')S(-t')z = S(t)z,$$

which means that (2.13) holds. If instead $t+t' \in \mathcal{T}_-$, (2.12) reads

$$S(t)S(-t')^{-1}x = S(-t-t')^{-1}x, \qquad (2.14)$$

and this is established by applying the operator $S(-t-t')$, which is injective, to both sides of (2.14).

To conclude the proof of proposition 2.5, we must still consider the case when both $t, t' \in \mathcal{T}_-$; that is, we must show that for all $x \in \mathcal{X}$,

$$S(-t)^{-1}S(-t')^{-1}x = S(-t-t')^{-1}x = S(-t')^{-1}S(-t)^{-1}x. \qquad (2.15)$$

To prove the first of these identities, let $y := S(-t-t')^{-1}x$, and $z := S(-t)^{-1}x$. Then, since both $-t$ and $-t' \in \mathcal{T}$,

$$x = S(-t)z = S(-t-t')y = S(-t)S(-t')y. \qquad (2.16)$$

Since $S(-t)$ is injective, (2.16) implies that $z = S(-t')y$; thus, since $S(-t')$ is also invertible,

$$y = S(-t')^{-1}z = S(-t')^{-1}S(-t)^{-1}x.$$

This means that the first of (2.15) holds. The second identity is proven in the same way, exchanging the role of t and t'. $\square$

In particular, if S is a semiflow generated by an evolution equation, the operators $S(-t)$, $t \geq 0$, will be defined whenever the equation can be uniquely solved "in the past", i.e. for $t < 0$ as well. In particular, this requires that for each $t > 0$ the operator $S(t)$ be bijective (see proposition 2.13 below).

In the sequel, we shall adopt the following

CONVENTION 2.6   *The underlying time-parameter set for a semiflow or flow S is always denoted by $\mathcal{T}$, i.e., $S = (S(t))_{t \in \mathcal{T}}$. Moreover, in order to avoid unnecessary complications in formulas and sentences, we agree that if S is a semiflow, then all*

*time variables like t, τ, θ, s run only in this time set T, or in the set $T_- := \{-t \in T\}$
if they are negative, or in the set $T_* := T \cup T_-$ if they can be positive or negative.*

For example, if $T = \mathbb{N}$ and $(\Omega_s)_{s \in T}$ is a family of subsets of $\mathcal{X}$, the intersection

$$\bigcap_{s \geq \tau} \Omega_s$$

has to be understood as an abbreviation of

$$\bigcap_{\substack{s, \tau \in \mathbb{N} \\ s \geq \tau}} \Omega_s \,.$$

Similarly, when we write $s \leq -t$ with $t \geq 0$, we understand that the inequality $-s \geq t$
holds in the set T, where both t and $-s$ are. Finally, we recall that if $T = \mathbb{R}_{\geq 0}$, then
$T_- = \mathbb{R}_{\leq 0}$ and $T_* = \mathbb{R}$.

2.2.2 Example: Lorenz' Equations

As we have seen in chapter 1, if the Cauchy problem for an autonomous evolution
equation is well posed on $\mathbb{R}_{\geq 0}$, it generates a continuous semiflow S on $\mathcal{X}$, defined by
the identification of the function $t \mapsto u(t, u_0) =: S(t)u_0$ as the solution to the Cauchy
problem

$$\begin{cases} \dot{u} = F(u) \\ u(0) = u_0 \,. \end{cases}$$

As an example, we verify that Lorenz' equations (1.50) define a semiflow on $\mathbb{R}^3$.

PROPOSITION 2.7
*For all values of σ, r and b (not necessarily positive), the system of Lorenz' equations
(1.50) defines a continuous semiflow in $\mathcal{X} = \mathbb{R}^3$.*

PROOF   1. System (1.50) has at least a local solution, determined by standard
existence and uniqueness results (see e.g. theorem A.1). That is, for each $u_0 :=
(x_0, y_0, z_0) \in \mathbb{R}^3$ there exist $T(u_0) > 0$ and a unique, maximally defined solution

$$u(\cdot, u_0) \colon [0, T(u_0)[\, \to \mathbb{R}^3$$

of the Lorenz' equations (1.50), with $u(0, u_0) = u_0$. This defines $S(t)$ at least for
$t \in [0, T(u_0)[$, by

$$S(t)u_0 := u(t, u_0) \,.$$

2. We now prove that each such local solution can be extended to a global one. We
achieve this, by establishing an A PRIORI ESTIMATE on each local solution $u(\cdot, u_0)$,
which shows that if $T(u_0) < +\infty$, then each function $t \mapsto u(t, u_0)$ is bounded in

the interval $[0, T(u_0)[$. This would allow us to continue the solution beyond $T(u_0)$, contradicting the fact that $T(u_0)$ is finite. Thus, the a priori estimate yields that $T(u_0) = +\infty$ for all $u_0 \in \mathbb{R}^3$. Setting

$$u(t, u_0) = (x(t, u_0), y(t, u_0), z(t, u_0)),$$

dropping the arguments t and u_0, and using the differential equations (1.50), we obtain

$$\frac{d}{dt}|u|^2 = 2x\dot{x} + 2y\dot{y} + 2z\dot{z} = -2\sigma x^2 + 2(r + \sigma)xy - 2y^2 - 2bz^2 = 2u\,M u^\top, \quad (2.17)$$

where $\top$ denotes transposition, $|\cdot|$ denotes the Euclidean norm in $\mathbb{R}^3$, and

$$M := \begin{pmatrix} -\sigma & \frac{1}{2}(r + \sigma) & 0 \\ \frac{1}{2}(r + \sigma) & -1 & 0 \\ 0 & 0 & -b \end{pmatrix}.$$

Let Λ be the largest eigenvalue of M, i.e.

$$\Lambda := \max\left\{ \frac{1}{2}\left(\sqrt{(\sigma + r)^2 + (\sigma - 1)^2} - (\sigma + 1) \right), -b \right\}.$$

Expressing u with respect to the basis of the eigenvectors of M, which are orthogonal, and can therefore be chosen to be orthonormal, we obtain from (2.17) that

$$\frac{d}{dt}|u(t, u_0)|^2 \leq 2\Lambda|u(t, u_0)|^2.$$

From this we conclude that for all $t \in [0, T(u_0)[$,

$$|u(t, u_0)| \leq |u_0|e^{\Lambda t} \leq |u_0|\max\{1, e^{\Lambda T(u_0)}\}. \quad (2.18)$$

This estimate shows that $u(\cdot, u_0)$ is bounded in $[0, T(u_0)[$, as claimed. As we have discussed, from this it follows that the operators $S(t)$ are defined for all $t \geq 0$. The semigroup properties of S now follow from the fact that system (1.50) is autonomous, and condition (2.7) follows from the fact that the function $t \mapsto u(t, u_0)$ is continuously differentiable. We also mention that later on (in proposition 2.65) we shall show that if σ and b are positive, we can obtain a better estimate than (2.18), that is, a bound on $|u(t, u_0)|$ independent of t. This estimate would clearly allow us to show global existence at once.

3. We proceed then to prove the global well-posedness and the continuity of each operator $S(t)$. In fact, we show that for each $t > 0$, $S(t)$ is locally Lipschitz continuous, in the sense that for all $t > 0$ and all bounded subsets $\mathcal{G} \subset \mathbb{R}^3$, there exists $L > 0$ such that for all $u_0, \bar{u}_0 \in \mathcal{G}$,

$$|S(t)u_0 - S(t)\bar{u}_0| \leq L|u_0 - \bar{u}_0|, \quad (2.19)$$

with L a continuous, increasing function of t and $\sigma_{\mathcal{G}} := \sup_{g \in \mathcal{G}} |g|$. To show this, set

$$(x(t), y(t), z(t)) := S(t)u_0, \quad (\bar{x}(t), \bar{y}(t), \bar{z}(t)) := S(t)\bar{u}_0.$$

Then, the difference $S(t)u_0 - S(t)\bar{u}_0 =: (\xi(t), \eta(t), \chi(t))$ solves the system

$$\begin{cases} \dot{\xi} = \sigma\eta - \sigma\xi \\ \dot{\eta} = r\xi - \eta - xz + \bar{x}\bar{z} \\ \dot{\chi} = -b\chi + xy - \bar{x}\bar{y}. \end{cases}$$

As before, dropping the argument t and using Schwarz' inequality, we obtain

$$\begin{aligned} \frac{d}{dt}|S(t)u_0 - S(t)\bar{u}_0|^2 &= 2\xi\dot{\xi} + 2\eta\dot{\eta} + 2\chi\dot{\chi} \\ &= 2\left(-\sigma\xi^2 + (r+\sigma-z)\xi\eta - \eta^2 + y\xi\chi - b\chi^2\right) \\ &\leq 2\left(-\sigma\xi^2 + (r+\sigma)\xi\eta - \eta^2 - b\chi^2\right) - 2z\xi\eta + 2y\xi\chi \\ &\leq 2\left(C + |S(t)u_0|\right)|S(t)u_0 - S(t)\bar{u}_0|^2, \end{aligned}$$

where $C := \max\{|\sigma|, |r+\sigma|, |b|\}$. Thus, recalling (2.18),

$$\frac{d}{dt}|S(t)u_0 - S(t)\bar{u}_0|^2 \leq \begin{cases} 2(C+|u_0|) \cdot |S(t)u_0 - S(t)\bar{u}_0|^2 & \text{if } \Lambda \leq 0, \\ 2(C+|u_0|e^{\Lambda t}) \cdot |S(t)u_0 - S(t)\bar{u}_0|^2 & \text{if } \Lambda > 0. \end{cases}$$

Thus, after integration we obtain that if $\Lambda \leq 0$,

$$|S(t)u_0 - S(t)\bar{u}_0| \leq |u_0 - \bar{u}_0| \cdot \underbrace{e^{(C+|u_0|)t}}_{:= \varphi_1(t, |u_0|)}$$

while, if $\Lambda > 0$,

$$|S(t)u_0 - S(t)\bar{u}_0| \leq |u_0 - \bar{u}_0| \underbrace{\exp\left(Ct + |u_0|\Lambda^{-1}(e^{\Lambda t} - 1)\right)}_{:= \varphi_2(t, |u_0|)}.$$

This estimate shows that (2.19) holds, with

$$L := \max\{\varphi_1(t, \sigma_{\mathcal{G}}), \varphi_2(t, \sigma_{\mathcal{G}})\}.$$

$\square$

We remark that the solution operator associated to system (1.53), i.e. to Duffing's equation, is not a semiflow, but a so-called PROCESS, or TWO-PARAMETER SEMI-FLOW. This is because the semigroup properties fail, due to the fact that the system is not autonomous. Nevertheless, it is possible to establish analogous a priori estimates, which would allow us to show that for all values of k, λ and ω (not necessarily positive), the solutions to the Cauchy problem for Duffing's equation, corresponding to arbitrary initial data, are globally and uniquely defined, and depend continuously on the initial data. Moreover, Duffing's equation defines a discrete semiflow in $\mathbb{R}^2$, by means of the stroboscopic maps described in section 1.4.1.

We conclude this section by remarking that our whole discussion on the asymptotic behavior of the solutions to Lorenz' and Duffing's equations obviously requires that these solution be globally defined. For Lorenz' equations, we established this fact by resorting to a continuation argument, which in turn was based on the possibility of obtaining a global bound on any local solution of the equations (see (2.18)). This procedure is fairly general, and estimates of this type are usually known as A PRIORI estimates. In section 2.9.1 at the end of this chapter we present some well known methods of obtaining a priori estimates.

2.3 Invariant Sets

In this section we introduce a number of sets, which are invariant with respect to a semiflow or flow $S = (S(t))_{t \in T}$, with T as in definition 2.4. Such sets include the orbits, the ω-limit sets, and the stable and unstable manifolds of the semiflow. Since the "time-set" T can be either continuous or discrete, the definitions and properties we present hold for both continuous and discrete semiflows or flows.

In the sequel, we often adopt the following notation. If $B \subseteq X$ and $T \in T$, we set

$$\mathcal{B}_T := \bigcup_{t \geq T} S(t)\mathcal{B} \qquad (2.20)$$

(recall our convention 2.6).

DEFINITION 2.8 *Let $S = (S(t))_{t \in T}$ be a semiflow on a Banach space X. A subset $Y \subseteq X$ is* POSITIVELY INVARIANT *(respectively,* NEGATIVELY INVARIANT*) for S if for all $t \geq 0$,*

$$S(t)\mathcal{Y} \subseteq \mathcal{Y} \quad (\text{respectively,} \quad S(t)\mathcal{Y} \supseteq \mathcal{Y});$$

$\mathcal{Y}$ is INVARIANT *if for all $t \in T$,*

$$S(t)\mathcal{Y} = \mathcal{Y}.$$

We immediately have

PROPOSITION 2.9
Let S be a flow, and $\mathcal{Y} \subseteq X$. Then $\mathcal{Y}$ is negatively invariant if and only if $S(t)\mathcal{Y} \subseteq \mathcal{Y}$ for all $t \leq 0$.

PROOF Assume first that $\mathcal{Y}$ is negatively invariant. Let $y \in \mathcal{Y}$ and $t \leq 0$, and set $\theta = -t$. Then $\theta \geq 0$, so $y \in S(\theta)\mathcal{Y}$. Thus, there is $x \in \mathcal{Y}$ such that $y = S(\theta)x$.

Since $S(t)y = S(t)S(\theta)x = S(t+\theta)x = S(0)x = x$, we deduce that $S(t)y \in \mathcal{Y}$. Thus, $S(t)\mathcal{Y} \subseteq \mathcal{Y}$. Conversely, assume this holds, and let $y \in \mathcal{Y}$, $\theta \geq 0$. Then, $t = -\theta \leq 0$, so $x := S(t)y \in \mathcal{Y}$. Then, $y = (S(t))^{-1}x = S(-t)x = S(\theta)x \in S(\theta)\mathcal{Y}$. This shows that $\mathcal{Y}$ is negatively invariant.   □

We know from ODE theory that simple examples of invariant sets are stationary points and periodic orbits. If S is a flow, other examples are given by complete orbits, and the stable and unstable manifolds of a stationary point.

2.3.1 Orbits

DEFINITION 2.10   *Let $S = (S(t))_{t \in \mathcal{T}}$ be a semiflow on $\mathcal{X}$, and $x \in \mathcal{X}$.*

1. *The* FORWARD ORBIT *originating (or starting) at x is the set*

$$\gamma_+(x) := \bigcup_{t \geq 0}\{S(t)x\} = \bigcup_{t \geq 0} S(t)\{x\}.$$

2. *A* BACKWARD ORBIT *ending at x is the image $\mathrm{im}(u)$ of a function $u\colon \mathcal{T}_- \to \mathcal{X}$ such that $u(0) = x$, and for all $t \geq 0$ and $s \leq -t$,*

$$u(t+s) = S(t)u(s). \tag{2.21}$$

If the backward orbit ending at x is unique, we denote it by $\gamma_-(x)$.

3. *A* COMPLETE ORBIT *through x is the image $\mathrm{im}(u)$ of a function $u\colon \mathcal{T}_* \to \mathcal{X}$ such that $u(0) = x$, and (2.21) holds for all $t \geq 0$ and $s \in \mathcal{T}_*$. If the complete orbit through x is unique, we denote it by $\gamma(x)$.*

REMARK 2.11   1. The forward orbit $\gamma_+(x)$ is the image $\mathrm{im}(u)$ of the function $u \in C(\mathcal{T}_+, \mathcal{X})$ defined by

$$u(t) := S(t)x, \qquad t \in \mathcal{T}_+,$$

where $\mathcal{T}_+ := \mathcal{T} \cap [0, +\infty[$.

2. Let $\beta = \mathrm{im}(u)$, with $u\colon \mathcal{T}_- \to \mathcal{X}$ satisfying (2.21), be a backward orbit ending at x, and let $y \in \beta$. Then x belongs to the forward orbit $\gamma_+(y)$, starting at y. Indeed, there is $\tau \geq 0$ with $y = u(-\tau)$. By (2.21) we have

$$x = u(0) = u(\tau - \tau) = S(\tau)u(-\tau) = S(\tau)y;$$

hence, $x \in \gamma_+(y)$. Moreover, (2.21) implies the continuity of u, since $S(\cdot)u(s)$ is continuous for $s \in \mathcal{T}_-$.

3. An analogous statement holds for complete orbits.   □

Now, we can rephrase the definition of invariance of a set by saying that a set $\mathcal{Y}$ is positively invariant if every forward orbit starting from a point in $\mathcal{Y}$ is contained in

$\mathcal{Y}$, and that it is invariant if for every point y in $\mathcal{Y}$ there is a complete orbit through y which is contained in $\mathcal{Y}$.

When there is no risk of confusion, it is common to refer to forward orbits simply as orbits; we shall indeed often do so. Note that in definition 2.10, the uniqueness of the backward orbit, or of the backward part of a complete orbit, is not implied.

The following example illustrates the possible lack of uniqueness of backward orbits.

Example 2.12
Consider the semiflow on $\mathcal{X} = [0, 1]$ generated by the logistic map of section 1.4.4, with $\lambda = 4$, that is, by the function $S: [0, 1] \to [0, 1]$ defined by $S(x) = 4x(1 - x)$ (whose range is all of $[0, 1]$). We know that the point $s_4 = \frac{3}{4}$ is a fixed point of S; hence, the singleton $\gamma_1 := \{\frac{3}{4}\}$ is, trivially, a complete orbit through s_4. On the other hand, we easily verify that the sequence $\gamma_2 := (x_n)_{n \in \mathbb{Z}}$ defined recursively by

$$x_n := \tfrac{3}{4}, \quad x_{-n-1} := \tfrac{1}{2}(1 - \sqrt{1 - x_{-n}}) \qquad \text{for } n \in \mathbb{N}$$

is such that $x_0 = s_4$ and $S(x_n) = x_{n+1}$ for each $n \in \mathbb{Z}$. Hence, this sequence is another complete orbit through s_4. Obviously, $\gamma_2 \neq \gamma_1$; however, as expected, the forward parts of these two orbits coincide. $\qquad\qquad$ ▯

The following result shows that the bijectivity of a semiflow is a sufficient condition for the uniqueness of backward and complete orbits.

PROPOSITION 2.13
Assume that S is a semiflow on $\mathcal{X}$, such that $S(t)$ is invertible for all $t \in \mathcal{T}$. Then backward and complete orbits through any $x \in \mathcal{X}$ are unique, and

$$\gamma_-(x) = \bigcup_{t \geq 0} S(t)^{-1}\{x\}, \quad \gamma(x) = \gamma_-(x) \cup \gamma_+(x). \tag{2.22}$$

PROOF 1. Suppose β_1 and β_2 are two backward orbits ending at x. They are then the images of two functions $u_1, u_2 \in C(\mathcal{T}_-; \mathcal{X})$; therefore, given $z \in \beta_1$, there is $s \in \mathcal{T}_-$ such that $z = u_1(s)$. Then, $t := -s \in \mathcal{T}$, and from (2.21)

$$S(t)z = S(t)u_1(s) = u_1(t + s) = u_1(0) = x,$$

as well as

$$S(t)u_2(s) = u_2(t + s) = u_2(0) = x.$$

Since $S(t)$ is invertible, $z = u_2(s)$; hence, $z \in \beta_2$. This shows that $\beta_1 \subseteq \beta_2$. Changing the role of β_1 and β_2 proves that $\beta_1 = \beta_2$. Thus, there is a unique backward orbit $\gamma_-(x)$ ending at x.
2. Next, let $w: \mathcal{T}_- \to \mathcal{X}$ be defined by

$$w(s) := S(-s)^{-1}x.$$

Then $w \in C(\mathcal{T}_-, \mathcal{X})$, and $w(0) = x$. Moreover, for all $t \geq 0$ and $s \leq -t$, since $w(t+s) = S(-t-s)^{-1}x$ we have that

$$x = S(-t-s)w(t+s),$$

and therefore

$$S(t)x = S(-s)w(t+s).$$

But since also $x = S(-s)w(s)$, we deduce that

$$S(t)x = S(t)S(-s)w(s) = S(-s)S(t)w(s)$$

as well. Thus, (2.21) holds for w, because of the invertibility of $S(-s)$. It follows that the image of w is a backward orbit ending at x; since we have proven that backward orbits are unique, the first of (2.22) follows.

3. The uniqueness of complete orbits and the second of (2.22) are proven similarly.

□

2.3.2   Limit Sets

As we know from the theory of ODEs, a natural way to study the limit behavior of orbits as $t \to \pm\infty$ in the phase space $\mathcal{X}$ is to study the topological properties of limit sets. Essentially, these are sets constructed from the union of forward or backward orbits (when these are unique), starting either from single points (limit sets of a point), or from (bounded) subsets of $\mathcal{X}$. To study these sets, for $s \geq 0$ we define the set $\mathcal{Y}_s$ as in (2.20), and

$$\mathcal{A}_{-s} := \bigcup_{t \geq s} S(t)^{-1}\mathcal{Y},$$

where for each $t \in \mathcal{T}$ we have set

$$S(t)^{-1}\mathcal{Y} := \{x \in \mathcal{X} : S(t)x \in \mathcal{Y}\} \tag{2.23}$$

(that is, the preimage of $\mathcal{Y}$ under $S(t)$). Note that if $S(t)$ is invertible for all $t \in \mathcal{T}$, then, by the first of (2.22),

$$\mathcal{A}_{-s} \subseteq \bigcup_{y \in \mathcal{Y}} \gamma_-(y).$$

Note, also, that if $s \geq s'$,

$$\mathcal{Y}_s \subseteq \mathcal{Y}_{s'} \quad \text{and} \quad \mathcal{A}_{-s} \subseteq \mathcal{A}_{-s'}.$$

DEFINITION 2.14   *Let S be a semiflow on $\mathcal{X}$, and $\mathcal{Y} \subseteq \mathcal{X}$. The set*

$$\omega(\mathcal{Y}) := \bigcap_{s \geq 0} \overline{\mathcal{Y}_s} = \bigcap_{s \geq 0} \overline{\bigcup_{t \geq s} S(t)\mathcal{Y}}$$

*is called the ω-*LIMIT SET *of* $\mathcal{Y}$ *(again, recall convention 2.6). The set*

$$\alpha(\mathcal{Y}) := \bigcap_{s \geq 0} \overline{\mathcal{A}_{-s}} = \bigcap_{s \geq 0} \overline{\bigcup_{t \geq s} S(t)^{-1} \mathcal{Y}}$$

(recall (2.23)) is called the α-LIMIT SET *of* $\mathcal{Y}$ *(fig. 2.1).*

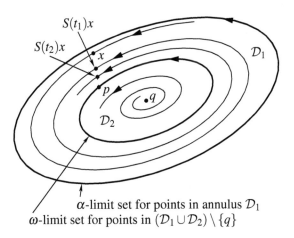

α-limit set for points in annulus $\mathcal{D}_1$
ω-limit set for points in $(\mathcal{D}_1 \cup \mathcal{D}_2) \setminus \{q\}$

Figure 2.1: α- and ω-sets.

Note that ω- and α-limit sets are closed. If $\mathcal{Y} = \{u_0\}$, we simply write $\omega(u_0)$ to denote the limit set of the point u_0, and similarly for $\alpha(u_0)$. We note explicitly that if S is a discrete (semi)flow, then

$$\omega(\mathcal{Y}) = \bigcap_{n \geq 0} \overline{\bigcup_{m \geq n} S^m(\mathcal{Y})}, \quad S := S(1).$$

We now prove a result which we will often use in the sequel; namely, that each point of an ω-limit set can be approximated by points that are on forward orbits starting in their defining sets $\mathcal{Y}$.

PROPOSITION 2.15
Let S be a semiflow on $\mathcal{X}$, *let* $\mathcal{Y} \subset \mathcal{X}$, *and* $z \in \mathcal{X}$. *Then* $z \in \omega(\mathcal{Y})$ *if and only if there exist sequences* $(y_m)_{m \in \mathbb{N}} \subset \mathcal{Y}$ *and* $(t_m)_{m \in \mathbb{N}} \subseteq \mathcal{T}$, *with* $t_m \to +\infty$, *such that*

$$z = \lim_{m \to \infty} S(t_m) y_m.$$

PROOF   Assume first that $z \in \omega(\mathcal{Y})$. Then $z \in \overline{\mathcal{Y}_s}$ for all $s \geq 0$; thus, there exist sequences $(z_k^s)_{k \in \mathbb{N}}$ from each $\mathcal{Y}_s$, converging to z as $k \to +\infty$. This means that for all $\varepsilon > 0$ and all $s \geq 0$ there exists $K_{\varepsilon,s} \in \mathbb{N}$ such that for all $k \geq K_{\varepsilon,s}$, $d(z, z_k^s) < \varepsilon$. Choosing $s = m \in \mathbb{N}_{>0}$, $\varepsilon = \frac{1}{m}$ and $k = K_{\frac{1}{m},m} =: k_m$, we deduce that for all $m \in \mathbb{N}_{>0}$ there exists $z_m \in \mathcal{Y}_m$ such that $d(z_m, z) < \frac{1}{m}$ (take $z_m = z_{k_m}^m$). Then, $z = \lim z_m$. Now, for each $m \in \mathbb{N}_{>0}$,

$$z_m \in \mathcal{Y}_m = \bigcup_{t \geq m} S(t)\mathcal{Y}$$

(recall convention 2.6). Thus, there are $t_m \geq m$ and $y_m \in \mathcal{Y}$ such that $z_m = S(t_m)y_m$. Consequently, $S(t_m)y_m$ converges to z, and $t_m \geq m \to +\infty$. Conversely, assume there exist sequences $(y_m)_{m \in \mathbb{N}} \subset \mathcal{Y}$ and $(t_m)_{m \in \mathbb{N}} \subset \mathcal{T}$, such that $t_m \to +\infty$ and $z = \lim S(t_m)y_m$. Then for all $s \geq 0$ there exists m such that $t_m \geq s$. Therefore,

$$S(t_m)y_m \in S(t_m)\mathcal{Y} \subseteq \mathcal{Y}_s,$$

and $z \in \overline{\mathcal{Y}_s}$. This is true for all $s \geq 0$, so

$$z \in \bigcap_{s \geq 0} \overline{\mathcal{Y}_s} = \omega(\mathcal{Y}).$$

This ends the proof of proposition 2.15.   ▯

For future reference, we note explicitly the following "discrete" version of proposition 2.15.

PROPOSITION 2.16
Let S be a discrete semiflow on $\mathcal{X}$, let $\mathcal{Y} \subset \mathcal{X}$, and $z \in \mathcal{X}$. Then $z \in \omega(\mathcal{Y})$ if and only if there exist sequences $(y_m)_{m \in \mathbb{N}} \subset \mathcal{Y}$ and $(n_m)_{m \in \mathbb{N}} \subseteq \mathcal{T}$, with $n_m \to +\infty$, such that

$$z = \lim_{m \to \infty} S^{n_m} y_m, \tag{2.24}$$

where, we recall, $S := S(1)$.

A result analogous to proposition 2.15 holds for α-limit sets; that is, points in an α-limit set can be approximated by points which have the property that the orbits starting from these points all intersect $\mathcal{Y}$. More precisely,

PROPOSITION 2.17
Let S be a semiflow on $\mathcal{X}$, $\mathcal{Y} \subset \mathcal{X}$, and $z \in \mathcal{X}$. Then $z \in \alpha(\mathcal{Y})$ if and only if there exist sequences $(z_n)_{n \in \mathbb{N}} \subset \mathcal{X}$ and $(t_n)_{n \in \mathbb{N}} \subseteq \mathcal{T}$ such that $t_n \to +\infty$, $z = \lim_{n \to \infty} z_n$ and $y_n := S(t_n)z_n \in \mathcal{Y}$. In particular, if S is a flow then $z \in \alpha(\mathcal{Y})$ if and only if there exist sequences $(y_n)_{n \in \mathbb{N}} \subset \mathcal{Y}$ and $(t_n)_{n \in \mathbb{N}} \subseteq \mathcal{T}$ such that $t_n \to +\infty$ and $z = \lim_{n \to \infty} S(-t_n)y_n$.

PROOF   Assume first that $z \in \alpha(\mathcal{Y})$. Then $z \in \overline{\mathcal{A}_{-s}}$ for all $s \in \mathcal{T}$, with $s \geq 0$. As in the proof of proposition 2.15, this implies that for all $n \in \mathbb{N}_{>0}$ there is $z_n \in \mathcal{A}_{-n}$ such that $d(z_n, z) < \frac{1}{n}$. Then, $z = \lim_{n \to \infty} z_n$. Now, for each $n \in \mathbb{N}$,

$$z_n \in A_{-n} = \bigcup_{\tau \geq n} S(\tau)^{-1} \mathcal{Y}.$$

Thus, $y_n := S(t_n)z_n \in \mathcal{Y}$ for some $t_n \geq n$. Conversely, assume that $z = \lim_{n \to \infty} z_n$ and $t_n \to +\infty$, with $y_n = S(t_n)z_n \in \mathcal{Y}$. Then for all $s \geq 0$ there exists n such that $t_n \geq s$. Since

$$z_n \in S(t_n)^{-1}\{y_n\} \subseteq S(t_n)^{-1}\mathcal{Y} \subseteq A_{-s},$$

it follows that $z = \lim_{n \to \infty} z_n \in \overline{A_{-s}}$. This is true for all $s \geq 0$, and therefore

$$z \in \bigcap_{s \geq 0} \overline{A_{-s}} = \alpha(\mathcal{Y}).$$

This ends the proof of proposition 2.17. □

Propositions 2.15 and 2.17 can be applied to the case when $\mathcal{Y}$ contains a single point:

COROLLARY 2.18
Let S be a semiflow and $x \in \mathcal{X}$. Then

$$\omega(x) = \bigcap_{s \geq 0} \overline{\bigcup_{t \geq s}\{S(t)x\}} = \{z \in \mathcal{X} : \exists t_n \to +\infty \text{ with } S(t_n)x \to z\}.$$

If S is a flow, then

$$\alpha(x) = \bigcap_{s \geq 0} \overline{\bigcup_{t \geq s}\{S(t)^{-1}x\}} = \{z \in \mathcal{X} : \exists t_n \to +\infty \text{ with } S(t_n)^{-1}x \to z\}.$$

Another consequence of propositions 2.15 and 2.17 is the following useful property of ω- and α-limit sets, whose proof is immediate:

COROLLARY 2.19
If $\mathcal{Y} \subset \mathcal{Z} \subset \mathcal{X}$, then $\omega(\mathcal{Y}) \subset \omega(\mathcal{Z})$ and $\alpha(\mathcal{Y}) \subset \alpha(\mathcal{Z})$.

2.3.3 Stability of Stationary Points

1. In analogy to the theory of ODEs (compare to definition A.6), we adopt the following "natural" terminology.

DEFINITION 2.20 *Let $S = (S(t))_{t \in T}$ be a semiflow on $\mathcal{X}$. A point $x \in \mathcal{X}$ is a* STATIONARY POINT *for S if*

$$S(t)x = x \quad \text{for all } t \in T.$$

A stationary point x of S is said to be:

1. STABLE, *if for any neighborhood $\mathcal{U}$ of x there is a neighborhood $\mathcal{V} \subset \mathcal{U}$ of x such that any motion starting in $\mathcal{V}$ is defined and contained in $\mathcal{U}$ for all $t \geq 0$;*

2. UNSTABLE, *if it is not stable;*

3. ASYMPTOTICALLY STABLE, *if x is stable and there is a neighborhood $\mathcal{V}$ of x such that any motion starting in $\mathcal{V}$ converges to x, i.e. if for all $x_0 \in \mathcal{V}$,*

$$\lim_{t \to +\infty} S(t)x_0 = x$$

(see fig. 2.2, and compare to definition 1.14). Finally, if a flow S in $\mathbb{R}^n$ is defined

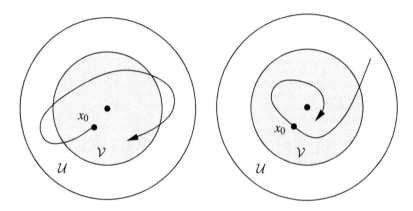

Figure 2.2: Stability and Asymptotic stability.

by an autonomous ODE

$$\dot{u} = F(u), \tag{2.25}$$

a stationary point of S is also called an EQUILIBRIUM POINT.

Thus, equilibrium points are solutions of the algebraic equation $F(x) = 0$. We know from classical stability theory for ODEs (see theorem A.7) that, if $F \colon \mathbb{R}^n \to \mathbb{R}^n$ is of class C^1, and if all eigenvalues of the matrix $A = F'(x)$ have nonzero real part, then the stability of equilibrium points x of (2.25) is related to the sign of the real part of the eigenvalues of A. Indeed, in this case the Hartman-Grobman theorem (see theorem A.8) states that, in a neighborhood of x, the qualitative behavior of the flow generated by (2.25) is the same as that of the flow $S = (e^{tA})_{t \in \mathbb{R}}$ generated by the linearized equation

$$\dot{u} = Au.$$

This motivates the following

DEFINITION 2.21   *1. Let S be the flow in $\mathbb{R}^n$ generated by the ODE (2.25). An equilibrium point x is* HYPERBOLIC *if all the eigenvalues of the matrix $A = F'(x)$ have nonzero real part.*

2. More generally, let $\mathcal{X}$ be a Banach space, and $F: \mathcal{X} \to \mathcal{X}$ be of class C^1. A fixed point x of F is HYPERBOLIC *if the spectrum of the linear operator $A = F'(x)$ does not intersect the unit circle $\{|z| = 1\}$ of $\mathbb{C}$.*

To see how these two notions of hyperbolicity of fixed points are related, consider a matrix A and, for each fixed $t \in \mathbb{R}$, the linear operator $F = e^{tA}$ in $\mathcal{X} = \mathbb{R}^n$. Then, $F'(x) = e^{tA}$ for all $x \in \mathcal{X}$, and the number Λ is an eigenvalue of e^{tA} iff $\Lambda = e^{\lambda t}$, with λ an eigenvalue of A. Let $\lambda = a + ib$. Then,

$$|\Lambda| = e^{at} = e^{(\mathfrak{Re}\,\lambda)t}.$$

Consequently, $|\Lambda| \neq 1$ iff $\mathfrak{Re}\,\lambda \neq 0$.

In particular, by theorem A.7, hyperbolic equilibrium points x of (2.25) are either unstable (if $\mathfrak{Re}\,\lambda > 0$ for at least one eigenvalue λ of $F'(x)$), or asymptotically stable (if $\mathfrak{Re}\,\lambda < 0$ for all eigenvalues λ of $F'(x)$). We refer e.g. to Hirsch-Smale, [HS93], for more details.

2.   We now recall the definitions of stable and unstable manifolds of a stationary point.

DEFINITION 2.22   *Let S be a semiflow, and $x \in \mathcal{X}$ be a stationary point for S. The* STABLE MANIFOLD *and the* UNSTABLE MANIFOLD *of x are the sets $\mathcal{M}^s(x)$ and $\mathcal{M}^u(x)$ defined, respectively, by*

$$\mathcal{M}^s(x) := \left\{ z \in \mathcal{X} : \lim_{t \to +\infty} S(t)z = x \right\} \tag{2.26}$$

and

$$\mathcal{M}^u(x) := \left\{ z \in \mathcal{X} : S(t)^{-1}z \text{ is defined for all } t \geq 0, \text{ and } \lim_{t \to +\infty} S(t)^{-1}z = x \right\}.$$

In other words, $\mathcal{M}^s(x)$ consists of the points of $\mathcal{X}$ which are origins of forward orbits converging to x as $t \to +\infty$, while $\mathcal{M}^u(x)$ consists of the points of $\mathcal{X}$ which are end points of backward orbits, converging to x as $t \to -\infty$. The sets $\mathcal{M}^s(x)$ and $\mathcal{M}^u(x)$ are nonempty, since $x \in \mathcal{M}^s(x) \cap \mathcal{M}^u(x)$ (since x is a fixed point, trivially $S(t)^{-1}x = x$ for all $t \geq 0$). Moreover, we have

PROPOSITION 2.23

Let x be a stationary point of a flow S. Then

$$\mathcal{M}^u(x) = \left\{ z \in \mathcal{X} : \lim_{t \to -\infty} S(t)z = x \right\}. \tag{2.27}$$

If x is stable, $\mathcal{M}^u(x) = \{x\}$.

PROOF   The first claim is immediate. As for the second, assume the contrary. There is then $z \in \mathcal{M}^u(x)$, with $z \neq x$. Thus, $\varepsilon := \frac{1}{2}\|x - z\| > 0$. Let $\mathcal{U} = B(x, \varepsilon)$, and determine a neighborhood $\mathcal{V} = B(x, \delta)$ of x, in accord with definition 2.20, such that for all $y \in B(x, \delta)$ and all $t \geq 0$, $S(t)y \in B(x, \varepsilon)$. Since

$$x = \lim_{\theta \to -\infty} S(\theta)z = \lim_{t \to +\infty} S(-t)z = \lim_{t \to +\infty} S(t)^{-1}z,$$

there is $t_0 \geq 0$ such that $S^{-1}(t)z \in B(x, \delta)$ for all $t \geq t_0$. Let $y := S(t_0)^{-1}z$. Then, $y \in B(x, \delta)$, so $S(t_0)y \in B(x, \varepsilon)$. Since $S(t_0)y = z$, we reach a contradiction. $\qquad\Box$

The definitions of stability and instability of a stationary point, and of their stable and unstable manifolds, come together in the following definition.

DEFINITION 2.24   *Let γ be a complete orbit of a flow S on $\mathcal{X}$. Let x_s, x_u be two distinct stationary points of S, respectively stable and unstable. γ is a HETEROCLINIC ORBIT joining x_s to x_u if*

$$\gamma \subseteq \mathcal{M}^u(x_u) \cap \mathcal{M}^s(x_s).$$

This terminology is justified by (2.22), (2.26) and (2.27), which imply that for any $x \in \gamma$,

$$\lim_{t \to -\infty} S(t)x = x_u, \qquad \lim_{t \to +\infty} S(t)x = x_s.$$

For example, figure 2.3 shows the two heteroclinic orbits γ_+ and γ_- joining the

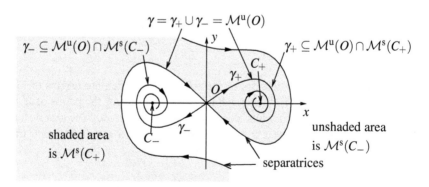

Figure 2.3: Heteroclinic Orbits for Duffing's Equation
γ_+ is the heteroclinic orbit joining O to C_+,
γ_- is the heteroclinic orbit joining O to C_-.

unstable stationary point O of the flow generated by Duffing's equations (1.54) to its asymptotically stable stationary points C_+ and C_-.

2.3.4   Invariance of Orbits and ω-Limit Sets

As in the finite-dimensional case (i.e., for systems of ODEs), the stable and unstable manifolds of a stationary point are examples of invariant sets. More precisely:

PROPOSITION 2.25
Let S be a semiflow on $\mathcal{X}$, and x a fixed point of S. The stable manifold $\mathcal{M}^s(x)$ is positively invariant. If $S(t)$ is invertible for all $t > 0$, the unstable manifold $\mathcal{M}^u(x)$ is negatively invariant.

PROOF   1. Assume first that $z \in S(t)\mathcal{M}^s(x)$ for some $t \geq 0$ in the time set $\mathcal{T}$. Then, $z = S(t)y$ for some $y \in \mathcal{M}^s(x)$. Since

$$\lim_{\theta \to +\infty} S(\theta)y = x,$$

it follows that also

$$\lim_{\theta \to +\infty} S(\theta)z = \lim_{\theta \to +\infty} S(\theta)S(t)y = \lim_{t \to +\infty} S(\theta + t)y = x.$$

This means that $z \in \mathcal{M}^s(x)$, and therefore that $S(t)\mathcal{M}^s(x) \subseteq \mathcal{M}^s(x)$.

2. To show the negative invariance of $\mathcal{M}^u(x)$, fix $t > 0$ and $z \in \mathcal{M}^u(x)$. Since $S(t)$ is invertible, $y := S(t)^{-1}z$ is well defined in $\mathcal{X}$. We claim that $y \in \mathcal{M}^u(x)$. Indeed, recalling proposition 2.5,

$$\lim_{\theta \to +\infty} S(\theta)^{-1}y = \lim_{\theta \to +\infty} S(\theta)^{-1}S(t)^{-1}z = \lim_{\theta \to +\infty} S(\theta + t)^{-1}z = x.$$

Consequently, $z = S(t)y \in S(t)\mathcal{M}^u(x)$. In accord with definition 2.8, this means that $\mathcal{M}^u(x)$ is negatively invariant, as claimed. $\qquad\square$

Other examples of invariant sets are the complete orbits of a semiflow:

PROPOSITION 2.26
Let S be a semiflow on $\mathcal{X}$, and $x \in \mathcal{X}$. Then any complete orbit γ through x is invariant. If in addition S is a flow, then for all x, $y \in \mathcal{X}$,

$$\gamma(x) \cap \gamma(y) \neq \emptyset \iff \gamma(x) = \gamma(y). \tag{2.28}$$

PROOF   1. Let γ be a complete orbit, and $z \in \gamma$. Then γ is the image of a function $u \colon \mathcal{T}_* \to \mathcal{X}$ satisfying (2.21), and there exists $s \in \mathcal{T}_*$ such that $z = u(s)$. Hence, by (2.21), for all $t \geq 0$

$$z = S(t)u(s - t) \in S(t)\gamma.$$

This implies that $\gamma \subseteq S(t)\gamma$ for $t \geq 0$. Conversely, if $z \in S(t)\gamma$, then $z = S(t)u(s)$ for some $s \in T_*$. Then, again by (2.21), $z = u(t+s) \in \gamma$, and $S(t)\gamma \subseteq \gamma$.

2. To prove (2.28), fix x and $y \in \mathcal{X}$. By the characterization of complete orbits given by proposition 2.13, and recalling that $S(t)^{-1} = S(-t)$,

$$\gamma(x) = \bigcup_{t \in T} \{S(t)x\}, \quad \gamma(y) = \bigcup_{t \in T} \{S(t)y\}.$$

Thus, if $\gamma(x) \cap \gamma(y)$ is nonempty, there are $z \in \gamma(x) \cap \gamma(y)$, and $t_x, t_y \in T$ such that

$$z = S(t_x)x = S(t_y)y.$$

Hence,

$$x = S(t_y - t_x)y \in \gamma(y),$$

and, in the same way,

$$y = S(t_x - t_y)x \in \gamma(x).$$

Let now $a \in \gamma(x)$. Then, $a = S(t_a)x$ for some $t_a \in T$, and

$$a = S(t_a + t_y - t_x)y \in \gamma(y).$$

This means that $\gamma(x) \subseteq \gamma(y)$. We prove in the same way that $\gamma(y) \subseteq \gamma(x)$; hence, $\gamma(x) = \gamma(y)$. The converse of the statement is obvious. □

We remark that (2.28) generalizes the familiar fact that two distinct orbits cannot intersect, unless they coincide. For flows generated by ODEs, this is a consequence of general uniqueness theorems.

Other sets that are invariant, at least under certain conditions on S, are the ω-limit sets. In fact, it is precisely this invariance property, together with an additional "regularity" assumption on S, that will allow us to construct an attractor for the semiflow.

We start by showing the positive invariance of ω-limit sets.

PROPOSITION 2.27

Let S be a semiflow on $\mathcal{X}$, and $\mathcal{B} \subseteq \mathcal{X}$. Then for all $t \geq 0$ in T, $\omega(S(t)\mathcal{B}) = \omega(\mathcal{B})$, and $S(t)\omega(\mathcal{B}) \subseteq \omega(\mathcal{B})$.

PROOF 1. Fix $t \geq 0$ in T, and let $z \in \omega(S(t)\mathcal{B})$. By proposition 2.15, there are sequences $(t_n)_{n \in \mathbb{N}} \subseteq T$ and $(z_n)_{n \in \mathbb{N}} \subseteq S(t)\mathcal{B}$, such that $t_n \to +\infty$, and $S(t_n)z_n \to z$. For each $n \in \mathbb{N}$, let $x_n \in \mathcal{B}$ be such that $S(t)x_n = z_n$. Then, $S(t_n + t)x_n \to z$, so $z \in \omega(\mathcal{B})$. This shows that $\omega(S(t)\mathcal{B}) \subseteq \omega(\mathcal{B})$.

2. Conversely, if $z \in \omega(\mathcal{B})$ there are sequences $(t_n)_{n \in \mathbb{N}} \subseteq T$ and $(x_n)_{n \in \mathbb{N}} \subseteq \mathcal{B}$, such that $t_n \to +\infty$ and $S(t_n)x_n \to z$. Since $\theta_n := t_n - t \to +\infty$, $\theta_n \geq 0$ for all large enough n. Since $S(t)x_n \in S(t)\mathcal{B}$ and

$$S(\theta_n)S(t)x_n = S(t_n)x_n \to z,$$

by proposition 2.15 again we conclude that $z \in \omega(S(t)\mathcal{B})$. This shows that $\omega(\mathcal{B}) \subseteq \omega(S(t)\mathcal{B})$.

3. The proof of the positive invariance of $\omega(\mathcal{B})$ is similar. Given $z \in \omega(\mathcal{B})$, let $(t_n)_{n \in \mathbb{N}}$ and $(x_n)_{n \in \mathbb{N}}$ be as above. Since $\tau_n := t_n + t \to +\infty$,

$$S(\tau_n)x_n = S(t)S(t_n)x_n \to S(t)z\,;$$

therefore, $S(t)z \in \omega(x)$.   □

The second statement of proposition 2.27 shows that $\omega(\mathcal{B})$ is positively invariant. We will soon present various sufficient conditions for $\omega(\mathcal{B})$ to be invariant; these conditions all require some degree of compactness of different subsets of $\omega(\mathcal{B})$. Before proceeding, though, we need to explain what we mean when we say that two sets attract each other, and to recall the definition of relatively compact and precompact sets.

DEFINITION 2.28   *Let S be a semiflow on $\mathcal{X}$. Given two subsets $\mathcal{A}$, $\mathcal{B} \subset \mathcal{X}$, we say that $\mathcal{A}$ ATTRACTS $\mathcal{B}$ if*

$$\lim_{t \to +\infty} \partial(S(t)\mathcal{B}, \mathcal{A}) = 0, \tag{2.29}$$

where ∂ is the semidistance in $\mathcal{X}$, introduced in (2.2).

DEFINITION 2.29   *A set $\mathcal{C} \subset \mathcal{X}$ is said to be RELATIVELY COMPACT if its closure $\overline{\mathcal{C}}$ is compact. Also, $\mathcal{C}$ is said to be PRECOMPACT if it can be completed into a compact set.*

Thus, in a Banach space precompact sets are automatically compact.

We can then present the most restrictive condition for the invariance of ω-limit sets:

THEOREM 2.30
Let S be a semiflow on $\mathcal{X}$, and $\mathcal{B} \subseteq \mathcal{X}$ be such that $\omega(\mathcal{B})$ is compact, and attracts $\mathcal{B}$. Then $\omega(\mathcal{B})$ is invariant under S.

PROOF   By proposition 2.27, we only need to show that $\omega(\mathcal{B}) \subseteq S(t)\omega(\mathcal{B})$ for all $t > 0$. Let $x \in \omega(\mathcal{B})$, and $(t_n)_{n \in \mathbb{N}} \subseteq \mathcal{T}$, $(x_n)_{n \in \mathbb{N}} \subseteq \mathcal{B}$, such that $t_n \to +\infty$ and $S(t_n)x_n \to x$. Since the sequence $(x_n)_{n \in \mathbb{N}}$ is attracted by $\omega(\mathcal{B})$, and $t_n - t \to +\infty$,

$$\lim_{m \to \infty} d(S(t_m - t)x_m, \omega(\mathcal{B})) = 0.$$

Recalling (2.1), this means that, given any $\varepsilon > 0$, there is $m_0 \in \mathbb{N}$ such that for all $m \geq m_0$, there is $z_m \in \omega(\mathcal{B})$ with the property that

$$d(S(t_m - t)x_m, z_m) < \varepsilon. \tag{2.30}$$

Since $\omega(\mathcal{B})$ is compact, the sequence $(z_m)_{m\in\mathbb{N}}$ contains a subsequence $(z_{m_k})_{k\in\mathbb{N}}$, converging to some element $z \in \omega(\mathcal{B})$. Then, (2.30) implies that $S(t_{m_k} - t)x_{m_k} \to z$ as $k \to \infty$. Since $S(t)$ is continuous,

$$S(t_{m_k})x_{m_k} = S(t)S(t_{m_k} - t)x_{m_k} \to S(t)z\,;$$

but since $S(t_{m_k})x_{m_k} \to x$, we conclude that $x = S(t)z \in S(t)\omega(\mathcal{B})$. This shows that $\omega(\mathcal{B}) \subseteq S(t)\omega(\mathcal{B})$, as claimed. ∎

Theorem 2.30 can somewhat be relaxed, by assuming that $\omega(\mathcal{B})$ be only compact "asymptotically":

THEOREM 2.31

Let S be a semiflow on $\mathcal{X}$. Assume there are a nonempty subset $\mathcal{B} \subseteq \mathcal{X}$ and a number $\tau \geq 0$ such that the set $\mathcal{B}_\tau$ (recall definition (2.20)) is relatively compact in $\mathcal{X}$. Then the set $\omega(\mathcal{B})$ is nonempty, compact and invariant. If in addition $\mathcal{B}$ is connected by arcs and $\mathcal{T}$ is either $\mathbb{R}$ or $\mathbb{R}_{\geq 0}$, $\omega(\mathcal{B})$ is connected by arcs. Moreover, for all bounded set $\mathcal{G} \subseteq \mathcal{B}$,

$$\lim_{t\to+\infty} \partial(S(t)\mathcal{G}, \omega(\mathcal{B})) = 0. \tag{2.31}$$

PROOF　1. Since $\mathcal{B}$ is nonempty, it contains an element x. The sequence $(y_n)_{n\in\mathbb{N}}$ defined by $y_n = S(n + \tau)x$ takes values in $\mathcal{B}_\tau$, thus in the compact set $\overline{\mathcal{B}_\tau}$. Hence, there is a subsequence $(y_{n_k})_{k\in\mathbb{N}}$ converging to a point $y \in \overline{\mathcal{B}_\tau}$. By proposition 2.15, $y \in \omega(\mathcal{B})$, so $\omega(\mathcal{B})$ is not empty.

2. We know from proposition 2.27 that $\omega(\mathcal{B})$ is positively invariant; thus, to prove its invariance it is sufficient to show that $\omega(\mathcal{B}) \subseteq S(t)\omega(\mathcal{B})$ for all $t > 0$. Fix $t > 0$ and $z \in \omega(\mathcal{B})$, and let $(z_m)_{m\in\mathbb{N}} \subset \mathcal{B}$, $(t_m)_{m\in\mathbb{N}} \subseteq \mathcal{T}$ be such that $S(t_m)z_m \to z$, as per proposition 2.15. Let $\theta_m := t_m - t$. Then $\theta_m \geq \tau$ for all sufficiently large m, and $S(\theta_m)z_m \in \mathcal{B}_\tau$. Thus, there is a subsequence $(S(\theta_{m_k})z_{m_k})_{k\in\mathbb{N}}$ converging to an element $y \in \overline{\mathcal{B}_\tau}$. By proposition 2.15, $y \in \omega(\mathcal{B})$; since $S(t)$ is continuous,

$$S(t)y = \lim_{k\to+\infty} S(t_{m_k})z_{m_k} = z.$$

Thus, $z \in S(t)\omega(\mathcal{B})$. This concludes the proof of the invariance of $\omega(\mathcal{B})$.

3. The compactness of $\omega(\mathcal{B})$ follows from the inclusions

$$\omega(\mathcal{B}) = \bigcap_{s\geq 0} \overline{\mathcal{B}_s} \subseteq \bigcap_{s\geq \tau} \overline{\mathcal{B}_s} \subseteq \overline{\mathcal{B}_\tau},$$

the last set being compact.

4. If $\mathcal{B}$ is connected by arcs and if $\mathcal{T}$ is either $\mathbb{R}$ or $\mathbb{R}_{\geq 0}$, then for all $s \geq 0$ the set $\mathcal{B}_s$ is connected by arcs. Indeed, if $u, v \in \mathcal{B}_s$, there exist $t_u, t_v \geq s$ and $x, y \in \mathcal{B}$ such that $u = S(t_u)x$ and $v = S(t_v)x$. Without loss of generality let $t_u \leq t_v$. By assumption there is a continuous mapping $p: [0, 1] \to \mathcal{B}$ with $p(0) = x$ and $p(1) = y$. The mapping

$\tau \mapsto S(t_u + \tau(t_v - t_u))p(\tau)$, $\tau \in [0,1]$, is continuous with values in $\mathcal{B}_s$. Thus u and v are connected in $\mathcal{B}_s$ by the continuous arc

$$\{z \in \mathcal{X} : z = S(t_u + \tau(t_v - t_u))p(\tau), \tau \in [0,1]\}.$$

It follows that $\overline{\mathcal{B}_s}$, and therefore $\omega(\mathcal{B})$, is also connected.

5. Finally, to prove (2.31) we argue by contradiction. Thus, we assume there are: a bounded set $\mathcal{G} \subseteq \mathcal{B}$, a number $\varepsilon_0 > 0$, and sequences $(t_n)_{n \in \mathbb{N}} \subseteq T$, $(g_n)_{n \in \mathbb{N}} \subseteq \mathcal{G}$, such that $t_n \to +\infty$, and for each $n \in \mathbb{N}$,

$$d(S(t_n)g_n, \omega(\mathcal{B})) \geq \varepsilon_0. \tag{2.32}$$

Since $t_n \to +\infty$, there exists a number N such that $t_n \geq \tau$ for all $n \geq N$. Thus,

$$S(t_n)g_n \in S(t_n)\mathcal{G} \subseteq S(t_n)\mathcal{B} \subseteq \mathcal{B}_\tau.$$

Since $\mathcal{B}_\tau$ is relatively compact, there is a subsequence $(S(t_{n_k})g_{n_k})_{k \in \mathbb{N}}$, converging to a limit $z \in \overline{\mathcal{B}_\tau}$. By proposition 2.15, $z \in \omega(\mathcal{B})$. From (2.32) we obtain then the contradiction

$$0 < \varepsilon_0 \leq d(S(t_{n_k})g_{n_k}, \omega(\mathcal{B})) = \inf_{y \in \omega(\mathcal{B})} d(S(t_{n_k})g_{n_k}, y) \leq d(S(t_{n_k})g_{n_k}, z) \to 0.$$

This concludes the proof of theorem 2.31. $\qquad\qquad\Box$

The importance of theorem 2.31 resides in the fact that it describes a class of sets, namely the sets $\omega(\mathcal{B})$, $\mathcal{B} \subseteq \mathcal{X}$, which share some of the properties we have stated to be characteristic of an attractor. Each of these sets is in fact nonempty, compact, invariant, and attracts some orbits (at least, the orbits starting from bounded subsets of $\mathcal{B}$). Before proceeding, we mention that a similar result holds for α-limit sets. In fact, using the characterization of these sets given by proposition 2.17, we can prove

THEOREM 2.32

Let S be a flow on $\mathcal{X}$, and assume there is a nonempty subset $\mathcal{Y} \subseteq \mathcal{X}$ such that $S(-t)\mathcal{Y} \neq \emptyset$ for all $t \geq 0$. Assume further that there is a number $\tau > 0$ such that the set

$$\bigcup_{t \geq \tau} S(-t)\mathcal{Y}$$

is relatively compact in $\mathcal{X}$. Then the set $\alpha(\mathcal{Y})$ is nonempty, compact and invariant, and for all bounded subset $\mathcal{G} \subseteq \mathcal{B}$,

$$\lim_{t \to -\infty} \partial(S(t)\mathcal{G}, \alpha(\mathcal{Y})) = 0.$$

We omit the proof of this result.

2.4   Attractors

In this section we present the definition and the main properties of the global at-tractor of a semiflow S on $\mathcal{X}$. We only consider attractors that are compact, although it is possible to consider noncompact attractors.

2.4.1   Attracting Sets

We start by recalling, from definition 2.28, that a subset $\mathcal{A} \subseteq \mathcal{X}$ attracts another subset $\mathcal{B} \subset \mathcal{X}$ if (2.29) holds, i.e. if

$$\lim_{t \to +\infty} \partial(S(t)\mathcal{B}, \mathcal{A}) = 0. \tag{2.33}$$

For example, (2.31) shows that if the semiflow S is uniformly compact for large t, the ω-limit set of a bounded set $\mathcal{B}$ attracts all orbits originating in $\mathcal{B}$. Another, fundamental example of attracting set is given by

THEOREM 2.33
Let S be a semiflow on $\mathcal{X}$. Assume there is a nonempty subset $\mathcal{B} \subseteq \mathcal{X}$, which is compact and positively invariant. Then, the ω-limit set $\mathcal{A} := \omega(\mathcal{B})$ is a compact, invariant set, which attracts all bounded sets $\mathcal{G} \subseteq \mathcal{B}$ (including $\mathcal{B}$ itself).

PROOF   Since $\mathcal{B}$ is positively invariant and compact, for any $\tau \geq 0$

$$\overline{\mathcal{B}_\tau} = \overline{\bigcup_{t \geq \tau} S(t)\mathcal{B}} \subseteq \overline{\bigcup_{t \geq \tau} \mathcal{B}} \subseteq \overline{\mathcal{B}} = \mathcal{B}.$$

Hence, $\mathcal{B}_\tau$ is relatively compact in $\mathcal{X}$, and we conclude by applying theorem 2.31. ☐

We proceed now to define the attractor of a semiflow:

DEFINITION 2.34   *A subset $\mathcal{A} \subseteq \mathcal{X}$ is an* ATTRACTOR *for the semiflow S on $\mathcal{X}$ if $\mathcal{A}$ is compact and invariant, and there is a neighborhood $\mathcal{U}$ of $\mathcal{A}$ in $\mathcal{X}$ such that (2.33) holds for any bounded subset $\mathcal{B} \subset \mathcal{U}$. The largest neighborhood $\mathcal{U}$ of $\mathcal{A}$ such that (2.33) holds is called the* BASIN OF ATTRACTION *of $\mathcal{A}$ (fig. 2.4).*

In other words, $\mathcal{A}$ attracts all orbits starting sufficiently close to it. We remark that (2.33) carries no information on the rate of convergence of the orbits to the attractor. However, this rate is uniform for all orbits starting in the same set $\mathcal{B}$; i.e., the rate of convergence does not depend on the particular initial value u_0, as long as $u_0 \in \mathcal{B}$. In chapter 4 we shall instead introduce another class of attracting sets, namely the EXPONENTIAL ATTRACTORS. As their name implies, the rate of

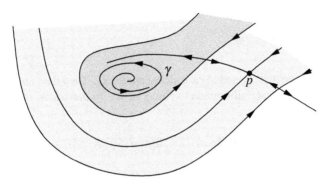

Figure 2.4: Basins of Attraction.

convergence of orbits to these sets is exponential; the same will be true for the other class of attracting sets we introduce in chapter 5, that is, the INERTIAL MANIFOLDS.

When $\mathcal{X}$ is infinite dimensional, we may have to work in subspaces $\mathcal{X}_1 \hookrightarrow \mathcal{X}$ with a stronger topology (one possible reason being, as we have remarked above, to show precompactness of one of the sets $\mathcal{B}_\tau$ in $\mathcal{X}$). In this case, we give

DEFINITION 2.35 *Let $\mathcal{X}_1$ be a subspace of $\mathcal{X}$, endowed with a stronger topology generated by a distance d_1 in $\mathcal{X}_1$. Define a corresponding semidistance ∂_1 as in (2.2). A set $\mathcal{A} \subseteq \mathcal{X}_1$ is an* ATTRACTOR IN $\mathcal{X}_1$ *for the semiflow S on $\mathcal{X}$ if $\mathcal{A}$ is an attractor of S in $\mathcal{X}$, if it is invariant in $\mathcal{X}_1$, and if there is a neighborhood $\mathcal{U}$ of $\mathcal{A}$ in $\mathcal{X}_1$ such that for any subset $\mathcal{B} \subset \mathcal{U}$, bounded in $\mathcal{X}_1$,*

$$\lim_{t \to +\infty} \partial_1(S(t)\mathcal{B}, \mathcal{A}) = 0.$$

When a semiflow is defined by an evolution equation in $\mathcal{X}$, verification of the invariance of $\mathcal{A}$ in $\mathcal{X}_1$ requires that orbits originating in $\mathcal{X}_1$ remain in $\mathcal{X}_1$ for all $t > 0$. For systems defined by PDEEs, this translates into a regularity property of the differential equation (see the examples of chapter 3).

2.4.2 Global Attractors

In many situations, it is possible to show that a semiflow admits an attractor, to which *all* orbits converge. This motivates the following

DEFINITION 2.36 *Let S be a semiflow on $\mathcal{X}$, and $\mathcal{A} \subset \mathcal{X}$ an attractor for S. $\mathcal{A}$ is called a* MAXIMAL *(or* GLOBAL, *or* UNIVERSAL*)* ATTRACTOR *if its basin of attraction is all of $\mathcal{X}$.*

We now give some preliminary results that characterize global attractors, and give some sort of justification to this terminology.

PROPOSITION 2.37
Global attractors are unique.

PROOF   Assume that $\mathcal{A}_1$ and $\mathcal{A}_2$ are global attractors. Then they are both bounded and invariant, and attract all bounded subsets of $\mathcal{X}$. In particular, they attract each other; thus, from (2.33),

$$\lim_{t\to+\infty} \partial(S(t)\mathcal{A}_1,\mathcal{A}_2) = \lim_{t\to+\infty} \partial(S(t)\mathcal{A}_2,\mathcal{A}_1) = 0.$$

The invariance of $\mathcal{A}_1$ and $\mathcal{A}_2$ implies then that $\partial(\mathcal{A}_1,\mathcal{A}_2) = \partial(\mathcal{A}_2,\mathcal{A}_1) = 0$. By (2.3) of definition 2.1, then, $\mathrm{dist}(\mathcal{A}_1,\mathcal{A}_2) = 0$. Since both $\mathcal{A}_1$ and $\mathcal{A}_2$ are compact, by corollary 2.3 it follows that $\mathcal{A}_1 = \mathcal{A}_2$.   □

Next, we show that global attractors are indeed maximal, with respect to set inclusion, among compact invariant sets of $\mathcal{X}$ (thus, in particular, among attractors).

PROPOSITION 2.38
Let S be a semiflow on $\mathcal{X}$. Let $\mathcal{A}_2$ be a compact invariant set, and assume that $\mathcal{A}_1 \subseteq \mathcal{A}_2$ is a global attractor. Then $\mathcal{A}_1 = \mathcal{A}_2$.

PROOF   Since $\mathcal{A}_1 \subseteq \mathcal{A}_2$, $\partial(\mathcal{A}_1,\mathcal{A}_2) = 0$. Since $\mathcal{A}_2$ is invariant,

$$0 \leq \partial(\mathcal{A}_2,\mathcal{A}_1) = \partial(S(t)\mathcal{A}_2,\mathcal{A}_1).$$

Since $\mathcal{A}_2$ is bounded, it is attracted by $\mathcal{A}_1$, so

$$\lim_{t\to+\infty} \partial(S(t)\mathcal{A}_2,\mathcal{A}_1) = 0.$$

This implies that $\partial(\mathcal{A}_2,\mathcal{A}_1) = 0$. Thus, $\mathrm{dist}(\mathcal{A}_1,\mathcal{A}_2) = 0$; by corollary 2.3, it follows that $\mathcal{A}_1 = \mathcal{A}_2$.   □

Finally, we give a characterization of the structure of a global attractor, which will be of fundamental importance in chapter 3.

PROPOSITION 2.39
Let S be a semiflow S on $\mathcal{X}$, and assume that S admits a global attractor A. Let $x \in \mathcal{X}$. Then $x \in \mathcal{A}$ if and only if there exists a complete orbit $\gamma(x)$ through x, contained in A.

PROOF   Assume first that $x \in \mathcal{A}$. Since $\mathcal{A}$ is invariant, the forward orbit $\gamma_+ := \{S(t)x : t \geq 0\}$ is contained in $\mathcal{A}$. To extend this orbit to $t \leq 0$, we first remark that, since $S(1)\mathcal{A} = \mathcal{A}$, there exists $y_1 \in \mathcal{A}$ such that $S(1)y_1 = x$. Likewise, there exists

$y_2 \in \mathcal{A}$ such that $S(1)y_2 = y_1$, and we can proceed inductively to construct a sequence $(y_n)_{n \in \mathbb{N}} \subset \mathcal{A}$, such that

$$y_0 = x, \quad S(1)y_{n+1} = y_n, \quad n \geq 0.$$

It is immediate to check that $S(n)y_n = x$ for all $n \geq 0$. We now define a function $u: \mathcal{T}_* \to \mathcal{X}$ by $u(t) := S(t+n)y_n$, where n is any integer such that $n \geq -t$. This function is well defined, because if also $p \geq -t$, and for instance $p > n$, then

$$y_n = S(1)y_{n+1} = S(2)y_{n+2} = \cdots = S(p-n)y_p,$$

and therefore

$$S(t+n)y_n = S(t+n)S(p-n)y_p = S(t+p)y_p.$$

The function u is obviously continuous, and $u(0) = x$. By the invariance of $\mathcal{A}$, the image $\gamma := \mathrm{im}(u)$ of u is included in $\mathcal{A}$. If $t \geq 0$, taking $n = 0$ we deduce that $u(t) = S(t)y_0 = S(t)x$, which means that γ contains the forward orbit $\gamma_+(x)$. To conclude that γ is in fact a complete orbit, we must verify the semigroup property (2.21), i.e. that $u(t+s) = S(t)u(s)$ for all $t \geq 0$ and $s \in \mathcal{T}_*$, $s < -t$. Given then $t \geq 0$ and $s < -t$, choose an integer $m \geq -s$. Then $m \geq -s - t$ as well, and

$$u(s+t) = S(s+t+m)y_m = S(t)S(s+m)y_m = S(t)u(s).$$

Thus, γ is a complete orbit through x, contained in $\mathcal{A}$. The converse statement of the theorem is obvious. $\qquad\Box$

2.4.3 Compactness

As theorem 2.33 shows, in order to use theorem 2.31 for the construction of an attractor it is desirable to start from a compact subset of $\mathcal{X}$, which is positively invariant. Our next goal is to show that we can choose, as one such set, the closure of one of the sets $\mathcal{B}_\tau$ introduced in theorem 2.31, for some $\tau \geq 0$. Thus, it is essential to verify that these sets are relatively compact, as required in theorem 2.31.

To this end, if the system is finite dimensional it is sufficient to show that one such set $\mathcal{B}_\tau$ is bounded in $\mathbb{R}^n$. If instead the dimension of $\mathcal{X}$ is infinite, a sufficient condition is to show that there exists one such set $\mathcal{B}_\tau$ which is bounded in a subspace $\mathcal{X}_1$, compactly imbedded in $\mathcal{X}$. This strategy is relatively common, and is the one we shall follow in chapter 3. With this in mind, we can easily determine a result, which can be immediately applied to show the required relative compactness of some set $\mathcal{B}_\tau$, in the case the semiflow is generated either by a finite dimensional system, or by an infinite dimensional system corresponding to evolution equations of "parabolic" type.

DEFINITION 2.40 *Let S be a semiflow on $\mathcal{X}$. S is* UNIFORMLY COMPACT FOR LARGE t *if for any bounded set $\mathcal{G} \subseteq \mathcal{X}$ there exists $T > 0$, depending on $\mathcal{G}$, such that the set $\mathcal{G}_T$, defined as in (2.20), is relatively compact in $\mathcal{X}$.*

The following result is then almost a restatement of theorem 2.31:

COROLLARY 2.41

Assume the semiflow S on X is uniformly compact for large t. For any nonempty bounded set $\mathcal{B} \subseteq \mathcal{X}$, the set $\omega(\mathcal{B})$ is a nonempty, compact, invariant subset of $\mathcal{X}$, such that for all bounded subset $\mathcal{G} \subseteq \mathcal{B}$,

$$\lim_{t \to +\infty} \partial(S(t)\mathcal{G}, \omega(\mathcal{B})) = 0.$$

2.5 Dissipativity

Corollary 2.41 states that if the semiflow S is uniformly compact for large t, then ω-limit sets of bounded sets are attracting. Another key ingredient for the existence of a compact global attractor is the existence of an ABSORBING SET for the semiflow S. We shall in fact see that this set provides the natural "starting point" for the actual construction of the attractor of a semiflow.

DEFINITION 2.42 *Let S be a semiflow on X. A subset $\mathcal{B} \subset \mathcal{X}$ is said to be* ABSORBING, *relative to a neighborhood $\mathcal{U}$ of $\mathcal{B}$ in $\mathcal{X}$, if for all bounded sets $\mathcal{G} \subset \mathcal{U}$ there exists $T > 0$, depending on $\mathcal{G}$, such that $S(t)\mathcal{G} \subseteq \mathcal{B}$ for all $t \geq T$.*

In other words, all orbits originating in $\mathcal{G}$ enter $\mathcal{B}$ and, after possibly leaving it for a finite number of times, eventually remain in $\mathcal{B}$ forever.

DEFINITION 2.43 *Let S be a semiflow on X. S is called* DISSIPATIVE, *if it admits a nonempty, bounded absorbing set, relative to all of $\mathcal{X}$.*

This definition of dissipativity coincides with the definition of bounded dissipativity (see e.g. Hale, [Hal88], where the notion of dissipativity is specialized for the attractors of points, bounded sets or compact sets). On the other hand, it may not agree with some other notions of dissipativity more common in physics.

We now show that dissipativity is a natural property of semiflows that have a regularizing effect in time.

PROPOSITION 2.44

Let S be a semiflow on X, uniformly compact for large t. Assume S has a nonempty, bounded absorbing set $\mathcal{B} \subseteq \mathcal{X}$, relative to all $\mathcal{X}$. Then, S admits a compact, positively invariant absorbing set.

PROOF Since $\mathcal{B}$ is bounded, by definition 2.40 there is $T > 0$ such that the set $\mathcal{B}_T$, as defined in (2.20), is relatively compact. We proceed then to show that the compact set $\overline{\mathcal{B}_T}$ is positively invariant and absorbing, relative to $\mathcal{X}$. Let $\mathcal{G} \subset \mathcal{X}$ be bounded. Since $\mathcal{B}$ is absorbing, there is $T_2 > 0$ such that for all $t \geq T_2$, $S(t)\mathcal{G} \subseteq \mathcal{B}$. Let $T_1 := T + T_2$. For each $t \geq T_1$, decompose $t = \theta + T_2$. Then $t - T_2 = \theta \geq T$, and therefore

$$S(t)\mathcal{G} = S(t - T_2)S(T_2)\mathcal{G} \subseteq S(t - T_2)\mathcal{B} \subseteq \mathcal{B}_T \subseteq \overline{\mathcal{B}_T} \,.$$

This shows that $\overline{\mathcal{B}_T}$ is absorbing. To show that $\overline{\mathcal{B}_T}$ is positively invariant, fix $t \geq 0$ and $x \in \overline{\mathcal{B}_T}$. There exists then a sequence $(x_k)_{k \in \mathbb{N}} \subseteq \mathcal{B}_T$ such $x_k \to x$. Since $x_k \in \mathcal{B}_T$, for each k there is $t_k \geq T$ such that $x_k \in S(t_k)\mathcal{B}$; since $t + t_k \geq t_k \geq T$,

$$S(t)x_k \in S(t)S(t_k)\mathcal{B} = S(t + t_k)\mathcal{B} \subseteq \mathcal{B}_T \,.$$

Then, by the continuity of the operator $S(t)$, $S(t)x = \lim S(t)x_k \in \overline{\mathcal{B}_T}$. Thus, $\overline{\mathcal{B}_T}$ is positively invariant. □

As we have stated, absorbing sets are a natural "starting point" for the construction of attractors. To justify this assertion, we show that the existence of an absorbing set is a necessary condition for the existence of a global attractor (recall that, in accord with definition 2.34, the attractors we consider are compact).

PROPOSITION 2.45
Let S be a semiflow on $\mathcal{X}$, admitting a global attractor $\mathcal{A}$. Then S also admits a bounded, positively invariant absorbing set $\mathcal{B}$, such that $\mathcal{A} \subseteq \mathcal{B}$. In particular, S is dissipative.

PROOF Since $\mathcal{A}$ is compact, the set

$$\mathcal{B}_1 := \bigcup_{x \in \mathcal{A}} B(x, 1) \tag{2.34}$$

is bounded, and therefore attracted by $\mathcal{A}$. Thus, there is $T_1 > 0$ such that for all $t \geq T_1$,

$$\partial(S(t)\mathcal{B}_1, \mathcal{A}) = \sup_{x \in S(t)\mathcal{B}_1} d(x, \mathcal{A}) \leq \tfrac{1}{2} \,.$$

Since $\mathcal{A}$ is compact, for each $t \geq T_1$ and $x \in S(t)\mathcal{B}_1$ there is $y \in \mathcal{A}$ such that

$$d(x, \mathcal{A}) = d(x, y) \leq \tfrac{1}{2} \,.$$

Thus, $x \in B(y, 1) \subset \mathcal{B}_1$, and we conclude that $S(t)\mathcal{B}_1 \subseteq \mathcal{B}_1$ if $t \geq T_1$. Define then

$$\mathcal{B}_2 := \bigcup_{0 \leq t \leq T_1} S(t)\mathcal{B}_1 \,, \quad \mathcal{B} := \mathcal{B}_1 \cup \mathcal{B}_2 \,. \tag{2.35}$$

We claim that $\mathcal{B}$ is bounded, positively invariant, and absorbs all bounded sets of $\mathcal{X}$. The boundedness of $\mathcal{B}$ follows from that of $\mathcal{B}_1$ and $\mathcal{B}_2$, the latter being a consequence of the continuity of the map $t \mapsto S(t)$ on the compact interval $[0, T_1]$. To show that $\mathcal{B}$ is positively invariant, fix $t > 0$ and $z \in S(t)\mathcal{B}$. Then $z = S(t)x$ for some $x \in \mathcal{B}$. If $x \in \mathcal{B}_2$, there are $t_2 \in [0, T_1]$ and $b \in \mathcal{B}_1$ such that $x = S(t_2)b$; if $x \in \mathcal{B}_1$, we set $t_2 = 0$, $b = x$. In either case,

$$z = S(t)x = S(t + t_2)b \in S(t + t_2)\mathcal{B}_1 \subseteq \mathcal{B}_i \subseteq \mathcal{B},$$

with $i = 1$ or $i = 2$ according to whether $t + t_2 \geq T_1$ or not. Hence, $S(t)\mathcal{B} \subseteq \mathcal{B}$. To show that $\mathcal{B}$ is absorbing, consider any bounded set $\mathcal{G} \subseteq \mathcal{X}$. Repetition of the argument at the beginning of this proof shows that there is $T > 0$ such that for all $t \geq T$, $S(t)\mathcal{G} \subseteq \mathcal{B}_1 \subseteq \mathcal{B}$. Finally, from (2.34) and the second of (2.35) we conclude that $\mathcal{A} \subseteq \mathcal{B}_1 \subseteq \mathcal{B}$.   □

We conclude this section by pointing out that in definition 2.42, as well as in definitions 2.40 and 2.47 (below), the reference to bounded sets $\mathcal{G}$ can be interpreted in the spirit of our trying to control the effect of possible errors in the determination of the initial data, as mentioned in chapter 1. Indeed, even if the "true" initial value u_0 of an orbit may only be known within a certain approximation, all these approximations will (hopefully!) lie in an explicitly identifiable bounded set of $\mathcal{X}$. For example, when we approximate the number π up to three exact decimal digits, we are in fact considering numbers in the ball $B(\pi, 10^{-3})$ of $\mathbb{R}$ (i.e., in the interval $]\pi - 10^{-3}, \pi + 10^{-3}[$).

2.6   Absorbing Sets and Attractors

We now come to the most important part of this chapter, where we show that if the semiflow S satisfies some compactness assumptions, the existence of an absorbing set is also sufficient for the existence of an attractor.

2.6.1   Attractors of Compact Semiflows

Our first result concerns semiflows that have a regularizing effect in time. In the same spirit of corollary 2.41, we claim

THEOREM 2.46
Assume that the semiflow S on $\mathcal{X}$ is uniformly compact for large t, and that it admits a nonempty, bounded absorbing set $\mathcal{B}$, relative to a neighborhood $\mathcal{U}$ of $\mathcal{B}$ in $\mathcal{X}$. Then the ω-limit set $\mathcal{A} := \omega(\mathcal{B})$ is an attractor for S. More precisely, $\mathcal{A}$ is the global attractor for S in $\mathcal{U}$.

PROOF   By the uniform compactness of S for large t, there is $T_0 > 0$ such that the set $\mathcal{B}_{T_0}$, defined as in (2.20), is relatively compact in $\mathcal{X}$. By corollary 2.41 we know that $\mathcal{A} = \omega(\mathcal{B})$ is nonempty, compact and invariant. We can now proceed along the lines of the proof of theorem 2.31 (we cannot apply this theorem directly, because we want to show that $\mathcal{A}$ attracts all the bounded subsets of $\mathcal{U}$, and not just the subsets of $\mathcal{B}$). Arguing by contradiction, assume that $\mathcal{A}$ does not attract all bounded sets of $\mathcal{U}$. There is then a bounded set $\mathcal{G} \subseteq \mathcal{U}$ for which $\partial(S(t)\mathcal{G}, \mathcal{A})$ does not vanish as $t \to +\infty$. Thus, there also are $\delta > 0$, and a sequence $(t_n)_{n\in\mathbb{N}}$, such that $t_n \to +\infty$ and for all n,

$$\partial(S(t_n)\mathcal{G}, \mathcal{A}) = \sup_{x\in S(t_n)\mathcal{G}} d(x, \mathcal{A}) \geq \delta > 0. \tag{2.36}$$

Since $\mathcal{B}$ is absorbing, there is $T_1 > 0$ such that for all $t \geq T_1$, $S(t)\mathcal{G} \subseteq \mathcal{B}$. Since $t_n \to +\infty$, there also is n_0 such that for all $n \geq n_0$, $t_n \geq T_0 + T_1$. By (2.36), for all n there is $x_n \in S(t_n)\mathcal{G}$ such that $d(x_n, \mathcal{A}) \geq \frac{\delta}{2}$. Let $b_n \in \mathcal{G}$ be such that $x_n = S(t_n)b_n$, and set $z_n := S(T_1)b_n$. Then $z_n \in S(T_1)\mathcal{G} \subseteq \mathcal{B}$. If $n \geq n_0$, then $t_n \geq T_1$ and $t_n - T_1 \geq T_0$; therefore

$$x_n = S(t_n)b_n = S(t_n - T_1)S(T_1)b_n = S(t_n - T_1)z_n$$
$$\in S(t_n - T_1)\mathcal{B} \subseteq \bigcup_{t\geq T_0} S(t)\mathcal{B} = \mathcal{B}_{T_0}.$$

Thus, the sequence $(x_n)_{n\in\mathbb{N}}$ is precompact, and therefore admits a subsequence $(x_{n_k})_{k\in\mathbb{N}}$, with $x_{n_k} = S(t_{n_k})b_{n_k}$, converging to an element $\bar{x} \in \overline{\mathcal{B}_{T_0}}$. Letting $\tau_k := t_{n_k} - T_1$, we deduce as before that, as $k \to +\infty$,

$$S(\tau_k)z_{n_k} = S(t_{n_k} - T_1)S(T_1)b_{n_k} = S(t_{n_k})b_{n_k} = x_{n_k} \to \bar{x}.$$

Consequently, since $z_{n_k} \in \mathcal{B}$, $\bar{x} \in \omega(\mathcal{B}) = \mathcal{A}$ by proposition 2.15. On the other hand,

$$d(\bar{x}, \mathcal{A}) = \lim_{k\to+\infty} d(x_{n_k}, \mathcal{A}) \geq \frac{\delta}{2},$$

and we reach a contradiction. It follows that $\mathcal{A}$ attracts all bounded subsets of $\mathcal{U}$, as claimed. Finally, to show that $\mathcal{A}$ is the maximal attractor in $\mathcal{U}$, assume that $\mathcal{A}_1 \supseteq \mathcal{A}$ is also a compact attractor of the bounded sets of $\mathcal{U}$, as claimed. Since $\mathcal{A}_1$ is invariant and $\mathcal{B}$ is absorbing, as above we can determine $T_1 > 0$, depending on $\mathcal{A}_1$, such that for all $t \geq T_1$, $\mathcal{A}_1 = S(t)\mathcal{A}_1 \subseteq \mathcal{B}$. Thus, by corollary 2.19,

$$\omega(\mathcal{A}_1) \subseteq \omega(\mathcal{B}) = \mathcal{A}.$$

Since $\mathcal{A}_1$ is closed, its invariance also implies that

$$\omega(\mathcal{A}_1) = \bigcap_{s\geq 0} \overline{\bigcup_{t\geq s} S(t)\mathcal{A}_1} = \bigcap_{s\geq 0} \overline{\bigcup_{t\geq s} \mathcal{A}_1} = \mathcal{A}_1.$$

We conclude then that $\mathcal{A}_1 = \omega(\mathcal{A}_1) \subseteq \mathcal{A}$ and, therefore, $\mathcal{A}_1 = \mathcal{A}$. □

In the next chapter we shall see how theorem 2.46 can be applied to establish the existence of attractors for the semiflow generated by a semilinear heat equation. In this case, this procedure is quite natural, since the requirement that the semiflow be uniformly compact for large t is a consequence of the smoothing effect of parabolic operators.

2.6.2 A Generalization

In contrast to the "parabolic" situation, the semiflow generated by semilinear dissipative wave equations does not have a smoothing effect for large t; therefore, theorem 2.46 cannot be directly applied to these equations. In this case, we can resort to a different type of result, in which the requirement that the semiflow be uniformly compact for large t can be relaxed, requiring that it be so only up to a uniformly decaying perturbation. More precisely, we give

DEFINITION 2.47 *Let $S = (S(t))_{t \in T}$ be a family of continuous operators on $\mathcal{X}$. S is* UNIFORMLY DECAYING TO 0 *if for all bounded sets $\mathcal{G} \subseteq \mathcal{X}$,*

$$\lim_{t \to +\infty} \sup_{x \in \mathcal{G}} d(S(t)x, 0) = 0. \tag{2.37}$$

We consider then a semiflow $S = (S(t))_{t \in T}$ which admits a decomposition

$$S = S_1 + S_2, \tag{2.38}$$

where S_1 and S_2 are families of continuous operators on $\mathcal{X}$ (i.e., they satisfy (2.7)), but not necessarily semiflows, and S_1, S_2 are, respectively, uniformly compact for large t and uniformly decaying to 0. We first have

PROPOSITION 2.48
Assume the semiflow S admits a decomposition as in (2.38), where the families of continuous operators S_1 and S_2 are, respectively, uniformly compact for large t and uniformly decaying to 0. Then for any sequences $(x_n)_{n \in \mathbb{N}}$ bounded in $\mathcal{X}$, and $(t_n)_{n \in \mathbb{N}} \subseteq T$, with $t_n \to +\infty$,

$$S(t_n)x_n \text{ converges} \quad \Longleftrightarrow \quad S_1(t_n)x_n \text{ converges and } S_2(t_n)x_n \to 0.$$

In either case, $\lim S(t_n)x_n = \lim S_1(t_n)x_n$.

PROOF Let $(x_n)_{n \in \mathbb{N}}$ and $(t_n)_{n \in \mathbb{N}}$ be as stated. Since $(x_n)_{n \in \mathbb{N}}$ is bounded, and S_2 decays to 0, in accord to (2.37)

$$\lim_{t \to +\infty} \sup_{n \in \mathbb{N}} d(S_2(t)x_n, 0) = 0.$$

Thus, given $\varepsilon > 0$ there exists $T > 0$ such that for all $t \geq T$ and all $n \in \mathbb{N}$, $d(S_2(t)x_n, 0) \leq \varepsilon$. Since $t_n \to +\infty$, there is n_0 such that $t_n \geq T$ for all $n \geq n_0$. Thus, for these n, $d(S_2(t_n)x_n, 0) \leq \varepsilon$. This shows that $S_2(t_n)x_n \to 0$. Then, $S_1(t_n)x_n = S(t_n)x_n - S_2(t_n)x_n$ converges. The rest of the proof follows immediately. $\Box$

The next result generalizes part of corollary 2.41 to the case when S is not itself uniformly compact for large t, but is a uniformly vanishing perturbation of a uniformly compact family of operators.

PROPOSITION 2.49
Assume the semiflow S on X admits a decomposition as in proposition 2.48. Then for any nonempty bounded set B ⊆ X, the set ω(B) is nonempty, compact and invariant, and attracts B.

PROOF   We start by showing that

$$\omega_1(\mathcal{B}) := \bigcap_{s \geq 0} \overline{\bigcup_{t \geq s} S_1(t)\mathcal{B}} = \omega(\mathcal{B}).$$

Indeed, if $x \in \omega(\mathcal{B})$, by proposition 2.15 there are sequences $(x_n)_{n \in \mathbb{N}} \subset \mathcal{B}$ and $(t_n)_{n \in \mathbb{N}} \subseteq \mathcal{T}$, such that $t_n \to +\infty$ and $S(t_n)x_n \to x$. By proposition 2.48, $S_1(t_n)x_n \to x$ as well, so $x \in \omega_1(\mathcal{B})$. The converse is proven similarly, recalling that $\mathcal{B}$ is bounded. Now, since S_1 is uniformly compact for large t, proceeding as in the proof of theorem 2.31 we deduce that $\omega(\mathcal{B}) = \omega_1(\mathcal{B})$ is nonempty and compact, and attracts $\mathcal{B}$. The invariance of $\omega(\mathcal{B})$ is then a consequence of theorem 2.30.   □

The importance of the next result is analogous to that of theorem 2.46, of which it can be seen as an extension. As we shall see in the next chapter, this theorem will in fact be applicable to certain types of dissipative evolution equations of hyperbolic type, to which theorem 2.46 cannot be applied, due to the lack of a regularizing effect of the corresponding solution operator.

THEOREM 2.50
Assume that the semiflow S on X admits a decomposition (2.38) as in proposition 2.48, and that there is a nonempty, bounded absorbing set B, relative to a neighborhood U of B in X. Then the ω-limit set A := ω(B) is a compact attractor for S. More precisely, A is the global attractor for S in U.

PROOF   Since $\mathcal{B}$ is bounded, by proposition 2.49 its ω-limit set $\omega(\mathcal{B}) = \mathcal{A}$ is nonempty, compact and invariant. The rest of the proof proceeds exactly as in that of theorem 2.46, with the only difference that now only the sequence $(S_1(t_n)b_n)_{n \in \mathbb{N}}$ can be said to be precompact. However, there still is a subsequence $(S_1(t_{n_k})b_{n_k})_{k \in \mathbb{N}}$, converging to some $\bar{x} \in \mathcal{X}$, and proposition 2.48 guarantees that, in fact, $S(t_{n_k})b_{n_k} \to \bar{x}$ as well.   □

2.7   Attractors via α-Contractions

In this section we present an alternative procedure to show that the set $\omega(\mathcal{B})$ is a compact attractor, which does not require a decomposition like (2.38). Our presentation follows [EM93], where this method was used to show the existence of an attractor for the semiflow generated by a PDEE modelling the evolution of an extensible beam.

2.7.1   Measuring Noncompactness

In the sequel, we denote by $\mathcal{E}$ a complete metric space, with distance d. If $\mathcal{M} \subseteq \mathcal{E}$, we denote by diam($\mathcal{M}$) the diameter of $\mathcal{M}$, that is,

$$\text{diam}(\mathcal{M}) := \sup\{d(x,y)\colon x,y \in \mathcal{M}\}. \tag{2.39}$$

DEFINITION 2.51   *Let $\mathcal{A} \subseteq \mathcal{E}$. $\mathcal{A}$ is* TOTALLY BOUNDED *if, given any $\varepsilon > 0$, it is possible to find a finite number of points $\{x_1, \ldots, x_N\}$ in $\mathcal{E}$ such that*

$$\mathcal{A} \subseteq \bigcup_{i=1}^{N} B(x_i, \varepsilon).$$

This union is called a FINITE BALL-COVERING *of $\mathcal{A}$.*

Compactness can be expressed in terms of total boundedness, since a subset of $\mathcal{E}$ is compact if and only if it is complete and totally bounded. Thus, a closed subset can fail to be compact only if it is not totally bounded. Since we want to consider semiflows S which are not necessarily compact for large t, but do possess a bounded absorbing set, we need to somehow control the failure of compactness due to the lack of total boundedness. This can be done by means of the measures of compactness introduced by Kuratowski in [Kur66, ch. 3].

Given a subset $\mathcal{M} \subseteq \mathcal{E}$, we denote by $\mathcal{I}(\mathcal{M})$ the subset of $\mathbb{R}_{>0}$ consisting of all the positive numbers β such that $\mathcal{M}$ has a finite covering of sets, each having diameter not exceeding β. That is, $\beta \in \mathcal{I}(\mathcal{M})$ if and only if $\beta > 0$ and there are sets $\mathcal{M}_1, \ldots, \mathcal{M}_m \subset \mathcal{X}$, such that diam($\mathcal{M}_j$) $\leq \beta$ for $j = 1, \ldots, m$, and

$$\mathcal{M} \subseteq \bigcup_{j=1}^{m} \mathcal{M}_j.$$

We can then define the notion of a measure of compactness:

DEFINITION 2.52   *Let $2^{\mathcal{E}}$ denote power set of $\mathcal{E}$ (that is, the set of its subsets). A* MEASURE OF COMPACTNESS *on $\mathcal{E}$ is the map $\alpha\colon 2^{\mathcal{E}} \to [0, +\infty]$ defined by*

$$\mathcal{E} \supseteq \mathcal{A} \mapsto \alpha(\mathcal{A}) := \begin{cases} +\infty & \text{if } \mathcal{A} \text{ has no finite covering}, \\ \inf \mathcal{I}(\mathcal{A}) & \text{otherwise}. \end{cases}$$

The main properties of measures of compactness are summarized in

PROPOSITION 2.53
Let α be a measure of compactness on $\mathcal{E}$. Then

  1. *If $\mathcal{A} \subset \mathcal{E}$ is bounded, $\alpha(\mathcal{A}) < +\infty$;*

  2. *If $\mathcal{A} \subseteq \mathcal{B}$, $\alpha(\mathcal{A}) \leq \alpha(\mathcal{B})$ (monotonicity);*

  3. *If $\alpha(\mathcal{A}) = 0$, then $\mathcal{A}$ is totally bounded;*

  4. *If $\mathcal{A}_1 \supseteq \mathcal{A}_2 \supseteq \cdots \supseteq \mathcal{A}_n \supseteq \cdots$ is a decreasing sequence of nonempty closed sets such that $\alpha(\mathcal{A}_n) \to 0$ as $n \to +\infty$, then the set*

$$\mathcal{A} := \bigcap_{n \geq 1} \mathcal{A}_n$$

is compact.

PROOF   1) and 2) are immediate consequences of definition 2.52; note that $\mathcal{I}(\mathcal{A}) \supseteq \mathcal{I}(\mathcal{B})$ if $\mathcal{A} \subseteq \mathcal{B}$. To show 3), let $\varepsilon > 0$, and consider $\beta \in \mathcal{I}(\mathcal{A}) \cap]0, \varepsilon[$. There is then a finite covering $\{\mathcal{C}_1, \ldots, \mathcal{C}_n\}$ of $\mathcal{A}$, with $\mathrm{diam}(\mathcal{C}_i) \leq \beta$. Choosing points $x_i \in \mathcal{C}_i$, we see that the union of the balls $B(x_i, \varepsilon)$ covers $\mathcal{A}$. Thus, $\mathcal{A}$ is totally bounded. Finally, to prove 4), note that, from 2),

$$0 \leq \alpha(\mathcal{A}) \leq \alpha(\mathcal{A}_n) \to 0.$$

Thus, $\alpha(\mathcal{A}) = 0$, so that, by 3), $\mathcal{A}$ is totally bounded. Since $\mathcal{A}$ is also complete, because it is closed, $\mathcal{A}$ is compact.   □

   We now introduce the notion of α-contraction, and the main result, which guarantees that $\omega(\mathcal{B})$ is indeed a compact attractor for the semiflow S.

DEFINITION 2.54   *Let $\mathcal{B} \subseteq \mathcal{E}$. A continuous map $T : \mathcal{B} \to \mathcal{B}$ is an α-CONTRACTION on $\mathcal{B}$, if there exists a number $q \in]0, 1[$ such that for every subset $\mathcal{A} \subseteq \mathcal{B}$,*

$$\alpha(T(\mathcal{A})) \leq q\,\alpha(\mathcal{A}). \tag{2.40}$$

   We have then the following fundamental result:

THEOREM 2.55
Assume that $\mathcal{B} \subseteq \mathcal{E}$ is closed and bounded, and that $T : \mathcal{B} \to \mathcal{B}$ is an α-contraction on $\mathcal{B}$. Consider the semiflow generated by the iterations of T, i.e. $S = (T^n)_{n \in \mathbb{N}}$. Then $\omega(\mathcal{B})$, if nonempty, is a compact, invariant set, which attracts $\mathcal{B}$.

PROOF   1. For $n \in \mathbb{N}$, set $\mathcal{A}_n := \overline{T^n(\mathcal{B})}$. Clearly, $\mathcal{A}_n \supseteq \mathcal{A}_{n+1}$ for each n. We show that, as a consequence,

$$\omega(\mathcal{B}) = \bigcap_{n \geq 0} \mathcal{A}_n =: \mathcal{A}.$$

To see this, note first that, since for all $n \in \mathbb{N}$ we obviously have

$$T^n(\mathcal{B}) \subseteq \bigcup_{m \geq n} T^m(\mathcal{B}),$$

we immediately deduce that

$$\mathcal{A} \subseteq \omega(\mathcal{B}) = \bigcap_{n \geq 0} \overline{\bigcup_{m \geq n} T^m(\mathcal{B})}.$$

Conversely, let $z \in \omega(\mathcal{B})$. By proposition 2.16 (see (2.24)), there are sequences $(n_j)_{j \in \mathbb{N}}$ and $(z_j)_{j \in \mathbb{N}} \subseteq \mathcal{B}$, such that $n_j \to \infty$ and $T^{n_j} z_j \to z$ as $j \to \infty$. Now, for each $n \in \mathbb{N}$ there is $j_n \in \mathbb{N}$ such that $n_j \geq n$ for all $j \geq j_n$. Hence, for $j \geq j_n$,

$$T^{n_j} z_j \in T^{n_j}(\mathcal{B}) \subseteq \mathcal{A}_{n_j} \subseteq \mathcal{A}_n.$$

Letting $j \to \infty$, it follows that $z \in \mathcal{A}_n$ for all $n \in \mathbb{N}$. Consequently, $z \in \mathcal{A}$, and $\omega(\mathcal{B}) = \mathcal{A}$.

2. Since $\mathcal{B}$ is bounded, by (1) of proposition 2.53 there is $M > 0$ such that $\alpha(\mathcal{B}) \leq M$. A repeated application of (2.40) yields then

$$\alpha(\mathcal{A}_n) = \alpha(T^n(\mathcal{B})) \leq q^n \alpha(\mathcal{B}) \leq q^n M;$$

thus, $\alpha(\mathcal{A}_n) \to 0$. Since each $\mathcal{A}_n$ is closed, part (4) of proposition 2.53 implies that $\omega(\mathcal{B}) = \mathcal{A}$ is compact.

3. To see that $\omega(\mathcal{B})$ attracts $\mathcal{B}$, we show that

$$\lim_{n \to \infty} d(T^n x, \omega(\mathcal{B})) = 0,$$

uniformly in $x \in \mathcal{B}$; that is, we claim that for all $\varepsilon > 0$ there exists N such that for all integer $n \geq N$ and all $x \in \mathcal{B}$,

$$d(T^n x, \omega(\mathcal{B})) < \varepsilon.$$

Proceeding by contradiction, assume there is $\varepsilon_0 > 0$ such that for all integers j it is possible to find another integer $n_j \geq j$, and a point $x_j \in \mathcal{B}$, such that

$$d(T^{n_j}(x_j), \omega(\mathcal{B})) \geq \varepsilon_0. \tag{2.41}$$

This process defines a bounded sequence $\zeta_* := (T^{n_j} x_j)_{j \in \mathbb{N}} \subset \mathcal{B}$. If we can show that ζ_* contains a convergent subsequence, we reach the desired contradiction, because by (2.41) the limit z of this subsequence would on the one hand be in $\omega(\mathcal{B})$ (by proposition 2.15), and on the other would satisfy $d(z, \omega(\mathcal{B})) \geq \varepsilon_0$. To show that ζ_* does contain a convergent subsequence, let Σ be the subset of $\mathcal{B}$ consisting of all the sequences of the form $\zeta = (T^{m_j} x_j)_{j \in \mathbb{N}}$, with $x_j \in \mathcal{B}$, $m_j \in \mathbb{N}$ and $m_j \to \infty$ as $j \to \infty$. Since $\alpha(\zeta) \leq \alpha(\mathcal{B})$ for all $\zeta \in \Sigma$,

$$0 \leq \alpha_0 := \sup_{\zeta \in \Sigma} \alpha(\zeta) < +\infty.$$

We claim that $\alpha_0 = 0$. Otherwise, we could first choose a number θ such that $0 < \theta < (1-q)\alpha_0$, and then a sequence $\zeta_0 \in \Sigma$ such that $\alpha_0 - \theta < \alpha(\zeta_0)$. Let $\zeta_0 = (T^{m_j}x_j)_{j\in\mathbb{N}}$. Since $m_j \to \infty$, there is $j_0 \in \mathbb{N}$ such that $m_j \geq 1$ for all $j \geq j_0$. Consider then the sequence $\zeta_1 := (T^{m_j-1}x_j)_{j\geq j_0}$. Since ζ_1 can be written as $\zeta_1 = (T^{n_k}y_k)_{k\in\mathbb{N}}$, with $n_k = m_{j_0+k} - 1 \to \infty$ as $k \to \infty$, and $y_k = x_{j_0+k} \in \mathcal{B}$, it follows that $\zeta_1 \in \Sigma$; therefore, $\alpha(\zeta_1) \leq \alpha_0$. Next, setting

$$\tilde{\zeta}_0 := T\zeta_1 = (T^{m_j}x_j)_{j\geq j_0},$$

we see that the sequence $T\zeta_1$ coincides with the sequence ζ_0, deprived of its first j_0 terms. We now check that dropping this finite number of terms does not affect the measure of α-compactness of ζ_0. Indeed, from part (2) of proposition 2.53 we first have

$$\alpha(\tilde{\zeta}_0) \leq \alpha(\zeta_0).$$

To show the opposite inequality it is sufficient to show that

$$\mathcal{I}(\tilde{\zeta}_0) \subseteq \mathcal{I}(\zeta_0).$$

Now, if $\beta \in \mathcal{I}(\tilde{\zeta}_0)$ and $\mathcal{C}_1,\ldots,\mathcal{C}_r$ is a finite covering of $\tilde{\zeta}_0$, such that $\text{diam}(\mathcal{C}_i) \leq \beta$, the addition to this covering of the j_0 balls $B(Tx_i, \frac{1}{2}\beta)$, $0 \leq i \leq j_0$, produces a finite covering of ζ_0 with sets whose diameter does not exceed β. Thus, $\beta \in \mathcal{I}(\zeta_0)$, as claimed.

In conclusion, we have the chain of inequalities

$$\alpha_0 - \theta < \alpha(\zeta_0) = \alpha(\tilde{\zeta}_0) = \alpha(T\zeta_1) \leq q\,\alpha(\zeta_1) \leq q\alpha_0 < \alpha_0 - \theta,$$

which yields a contradiction. This means that $\alpha_0 = 0$ and, therefore, $\alpha(\zeta) = 0$ for all $\zeta \in \Sigma$. In particular, $\alpha(\zeta_*) = 0$, which implies, by part (3) of proposition 2.53, that ζ_* is totally bounded. Hence, ζ_* is compact, and contains a convergent subsequence, as claimed. Finally, the invariance of $\omega(\mathcal{B})$ follows from theorem 2.30 (in its discrete version). This concludes the proof of theorem 2.55.   □

We now proceed to extend theorem 2.55 from the discrete case to the continuous one. The corresponding result will then provide the desired alternative to theorems 2.46 and 2.50.

THEOREM 2.56
Assume that S is a continuous semiflow on $\mathcal{X}$, admitting a bounded, positively invariant absorbing set $\mathcal{B}$, and that there exists $t_* > 0$ such that the operator $S_* := S(t_*)$ is an α-contraction on $\mathcal{B}$. Let

$$\mathcal{A}_* := \bigcap_{n\geq 0} \overline{\bigcup_{m\geq n} S_*^m(\mathcal{B})} = \omega_*(\mathcal{B})$$

be the ω-limit set of $\mathcal{B}$ under the map S_*, and set

$$\mathcal{A} := \bigcup_{0\leq t\leq t_*} S(t)\mathcal{A}_*.$$

Assume further that for all $t \in [0, t_*]$, $S(t)$ is Lipschitz continuous from $\mathcal{B}$ into $\mathcal{B}$, with Lipschitz constant $L(t)$, $L : [0, t_*] \to]0, +\infty[$ being a bounded function. Then $\mathcal{A} = \omega(\mathcal{B})$, and this set is the global attractor of S in $\mathcal{B}$.

PROOF 1. To show that $\mathcal{A}$ is compact, we note that the function $F : [0, +\infty[\times \mathcal{B} \to \mathcal{B}$ defined by $F(t, x) := S(t)x$ is continuous on $[0, t_*] \times \mathcal{A}_*$. To see this, we set

$$L_* := \sup_{0 \le t \le t_*} L(t), \tag{2.42}$$

and fix $(t_0, x_0) \in [0, t_*] \times \mathcal{A}_*$. Since the map $t \mapsto S(t)x$ is continuous for each $x \in \mathcal{X}$, given $\eta > 0$ there is $\delta_1 > 0$ such that if $|t - t_0| \le \delta_1$,

$$d(S(t)x_0, S(t_0)x_0) \le \tfrac{1}{2}\eta ; \tag{2.43}$$

note that δ_1 depends on η and, possibly, on (t_0, x_0). Let

$$\delta := \min\{\delta_1, \tfrac{1}{2}L_*^{-1}\eta\},$$

then, if (t, x) is such that

$$(d(x, x_0))^2 + |t - t_0|^2 \le \delta^2,$$

by (2.43) we have that

$$\begin{aligned} d(S(t)x, S(t_0)x_0) &\le d(S(t)x, S(t)x_0) + d(S(t)x_0, S(t_0)x_0) \\ &\le L(t)\, d(x, x_0) + d(S(t)x_0, S(t_0)x_0) \\ &\le L_*\delta + \tfrac{1}{2}\eta \le \eta . \end{aligned}$$

This shows the continuity of F. It is then immediate to verify that

$$\mathcal{A} = \mathcal{A}_1 := F([0, t_*] \times \mathcal{A}_*) ;$$

thus, $\mathcal{A}$ is compact, because F is continuous and $[0, t_*] \times \mathcal{A}_*$ is compact.

2. We show that $\mathcal{A}$ attracts all bounded subsets of $\mathcal{B}$. Let $\mathcal{G} \subseteq \mathcal{B}$ be bounded, and fix $t \ge t_*$. Given any $x \in S(t)\mathcal{G}$ and $a_* \in \mathcal{A}_*$, let $g \in \mathcal{G}$ be such that $x = S(t)g$, and decompose $t = nt_* + \theta_t$, for suitable $n \in \mathbb{N}$ and $\theta_t \in [0, t_*]$. Let $\bar{a} := S(\theta_t)a_*$. Then, $\bar{a} \in \mathcal{A}$, and recalling (2.42) we can estimate

$$\begin{aligned} d(x, \bar{a}) = d(S(\theta_t + t - \theta_t)g, S(\theta_t)a_*) &\le L_* d(S(t - \theta_t)g, a_*) \\ &\le L_* d(S(nt_*)g, a_*) = L_* d(S_*^n g, a_*) . \end{aligned}$$

From this, it follows that

$$\inf_{a \in \mathcal{A}} d(x, a) \le d(x, \bar{a}) \le L_* d(S_*^n g, a_*)$$

and, since a_* is arbitrary in $\mathcal{A}_*$,

$$\inf_{a \in \mathcal{A}} d(x, a) \le L_* \inf_{a_* \in \mathcal{A}_*} d(S_*^n g, a_*) . \tag{2.44}$$

Since $g \in \mathcal{G} \subseteq \mathcal{B}$, and $\mathcal{B}$ is positively invariant, $S_*^n g \in \mathcal{B}$. Thus, recalling the definition (2.2) of semidistance, we can proceed from (2.44) with

$$\inf_{a \in \mathcal{A}} d(x,a) \le L_* \sup_{b \in S_*^n \mathcal{B}} \inf_{a_* \in \mathcal{A}_*} d(b,a_*) = L_* \partial(S_*^n \mathcal{B}, \mathcal{A}_*). \tag{2.45}$$

Since (2.45) is true for arbitrary $x \in \mathcal{S}(t)\mathcal{G}$, it follows that

$$\sup_{x \in S(t)\mathcal{G}} \inf_{a \in \mathcal{A}} d(x,a) = \partial(S(t)\mathcal{G}, \mathcal{A}) \le L_* \partial(S_*^n \mathcal{B}, \mathcal{A}_*). \tag{2.46}$$

Since $\mathcal{A}_*$ attracts $\mathcal{B}$ under S_*, (2.46) implies that $\mathcal{A}$ attracts $\mathcal{G}$ under S, as claimed.

3. We now show that $\mathcal{A} = \omega(\mathcal{B})$. Let $a \in \mathcal{A}$. There are then $\theta \in [0, t_*]$ and $a_* \in \mathcal{A}_*$, such that $a = S(\theta)a_*$. Since $\mathcal{A}_* = \omega_*(\mathcal{B})$, by proposition 2.16 there are sequences $(m_j)_{j \in \mathbb{N}} \subseteq \mathbb{N}$ and $(z_j)_{j \in \mathbb{N}} \subseteq \mathcal{B}$, such that $m_j \to \infty$ and $S_*^{m_j} z_j \to a_*$ as $j \to \infty$. Let $t_j := \theta + m_j t_*$. Then, $t_j \to \infty$, and

$$a = S(\theta)a_* = \lim_{j \to \infty} S(\theta + m_j t_*)z_j = \lim_{j \to \infty} S(t_j)z_j.$$

Thus, again by proposition 2.15, $a \in \omega(\mathcal{B})$. This proves that $\mathcal{A} \subseteq \omega(\mathcal{B})$. Conversely, let $z \in \omega(\mathcal{B})$. Then, there are sequences $(t_j)_{j \in \mathbb{N}} \subseteq \mathcal{T}$ and $(z_j)_{j \in \mathbb{N}} \subseteq \mathcal{B}$, such that $t_j \to \infty$ and $S(t_j)z_j \to z$ as $j \to \infty$. For each $j \in \mathbb{N}$, we can write $t_j = m_j t_* + \theta_j$, with $m_j \in \mathbb{N}$, $\theta_j \in [0, t_*]$, and $m_j \to \infty$ as $j \to \infty$. Since $\mathcal{B}$ is positively invariant, $S(\theta_j)z_j =: \tilde{z}_j \in \mathcal{B}$ for all j. Hence,

$$z = \lim_{j \to \infty} S(t_j)z_j = \lim_{j \to \infty} S(m_j t_*)S(\theta_j)z_j = \lim_{j \to \infty} S_*^{m_j} \tilde{z}_j.$$

This means that $z \in \omega_*(\mathcal{B}) = \mathcal{A}_*$. Since $\mathcal{A}_* \subseteq \mathcal{A}$, it follows that $\omega(\mathcal{B}) \subseteq \mathcal{A}$. Thus, $\mathcal{A} = \omega(\mathcal{B})$.

4. Since $\mathcal{A}$ is compact and attracts $\mathcal{B}$, and $\omega(\mathcal{B}) = \mathcal{A}$, theorem 2.30 implies that $\mathcal{A}$ is invariant. This ends the proof of theorem 2.56. $\square$

In conclusion, theorem 2.56 provides an alternative way to establish the existence of an attractor for a continuous semiflow, if we can choose t_* so that the operator $S(t_*)$ is an α-contraction, and $S(t)$ is a Lipschitz continuous map on $\mathcal{X}$, for all $t \in [0, t_*]$. In applications, this is often achieved by means of the intermediate results described next, which we will use for the dissipative evolution equations we consider in chapter 3.

2.7.2 A Route to α-Contractions

In the light of our previous remark, it is clearly of interest to be able to give some sufficient conditions for a continuous map $T: \mathcal{B} \to \mathcal{B}$, defined on a closed, bounded set $\mathcal{B} \subset \mathcal{X}$, to be an α-contraction. To this end, note first that this is obviously the case if T is a strict contraction, and also if T is compact, because in this case $\alpha(T(\mathcal{B})) = 0$, and we can take arbitrary $q \in {]}0, 1{[}$ in (2.40). This is exactly the situation we had in theorems 2.46 and 2.50. In fact, in both these cases we can choose $T = S(t_*)$, with $t_* > 0$ arbitrary in theorem 2.46 (where T is compact), and, in theorem 2.50, t_* so large that $S_2(t_*)$ is a strict contraction.

We now show that a combination of these two effects still produces an α-contraction.

DEFINITION 2.57   *A pseudometric δ in $\mathcal{X}$ is* PRECOMPACT *with respect to the topology induced by the metric d of $\mathcal{X}$ if every sequence which is bounded relatively to the distance d has a Cauchy subsequence relative to δ.*

PROPOSITION 2.58
Let δ be a precompact pseudometric, and let $\mathcal{B} \subset \mathcal{X}$ be bounded. Then $\mathcal{B}$ is totally bounded with respect to δ. That is, for all $\varepsilon > 0$ there is a finite covering of $\mathcal{B}$, consisting of δ-balls

$$B_\delta(x_i, \varepsilon) := \{y \in \mathcal{X} : \delta(x_i, y) < \varepsilon\}, \quad i = 1, \ldots, m.$$

PROOF   We proceed by contradiction. If there were $\varepsilon_0 > 0$ such that there is no finite covering of $\mathcal{B}$ by means of δ-balls $\mathcal{B}_1 = B_\delta(x_1, \varepsilon_0), \ldots, \mathcal{B}_m = B_\delta(x_m, \varepsilon_0)$, arguing inductively we could construct a sequence $(x_i)_{i \in \mathbb{N}}$, contained in $\mathcal{B}$, and therefore bounded, such that $x_{m+1} \notin \mathcal{B}_m$ for each m. In particular, this would imply that $\delta(x_m, x_{m+1}) \geq \varepsilon_0$. Consequently, $(x_m)_{m \in \mathbb{N}}$ could not be a Cauchy sequence relative to δ. But this sequence must be a Cauchy sequence, since it is bounded and δ is precompact.   ☐

We now show that if T fails to be contractive only because of a precompact pseudometric, it is still an α-contraction.

PROPOSITION 2.59
Let $\mathcal{B} \subset \mathcal{X}$ be bounded, δ a precompact pseudometric in $\mathcal{X}$, and $T : \mathcal{B} \to \mathcal{B}$ be a continuous map. Suppose T satisfies the estimate

$$d(Tx, Ty) \leq q d(x, y) + \delta(x, y) \tag{2.47}$$

for all $x, y \in \mathcal{B}$ and some $q \in \,]0, 1[$ independent of x and y. Then T is an α-contraction.

PROOF   We will show that (2.40) holds, with q from (2.47). Let $\mathcal{A} \subseteq \mathcal{B}$, and $\alpha_0 = \alpha(\mathcal{A})$ (note that $\alpha(\mathcal{A}) \leq \alpha(\mathcal{B}) < +\infty$, because $\mathcal{B}$ is bounded). By proposition 2.58, given $\varepsilon > 0$ we can cover $\mathcal{B}$ with a finite number of δ-balls $\mathcal{B}_1, \ldots, \mathcal{B}_n$ of radius ε. By the definition of α_0, we can also cover $\mathcal{A}$ with a finite number of sets $\mathcal{C}_j$, $j = 1, \ldots, m$, with $\text{diam}(\mathcal{C}_j) \leq \alpha_0 + \varepsilon$ for all j. Clearly,

$$T(\mathcal{A}) \subseteq \bigcup_{\substack{i=1,\ldots,n \\ j=1,\ldots,m}} T(\mathcal{B}_i \cap \mathcal{C}_j).$$

Recalling (2.47), we can estimate the diameter of each set of this covering of $T(\mathcal{A})$ as follows. For $x, y \in \mathcal{B}_i \cap \mathcal{C}_j$ we compute that

$$d(Tx, Ty) \leq qd(x,y) + \delta(x,y) \leq q(\alpha_0 + \varepsilon) + \varepsilon \leq q\alpha_0 + 2\varepsilon.$$

This means that

$$\operatorname{diam}(T(\mathcal{B}_i \cap \mathcal{C}_j)) \leq q\alpha_0 + 2\varepsilon = q\alpha(\mathcal{A}) + 2\varepsilon.$$

Since ε is arbitrary, we conclude that

$$\alpha(T(\mathcal{A})) \leq \sup_{i,j} \operatorname{diam}(T(\mathcal{B}_i \cap \mathcal{C}_j)) \leq q\alpha(\mathcal{A}),$$

from which (2.40) follows. $\square$

2.8   Fractal Dimension

As we have mentioned in chapter 1, in many situations it is possible to show that the attractor of a semiflow has finite dimension. Since in general attractors have a highly irregular structure (they may well be fractal sets), it is often necessary to consider their FRACTAL DIMENSION, whose definition we briefly recall in this section.

Following [Fal85, Fal90, EFNT94], we introduce the following

DEFINITION 2.60   *Let $\mathcal{X}$ be a separable Hilbert space, and $\mathcal{K} \subset \mathcal{X}$ be a compact subset. For $\delta > 0$, denote by $N_\delta(\mathcal{K})$ the smallest number of sets of diameter at most equal to δ which can cover $\mathcal{K}$. The FRACTAL DIMENSION of $\mathcal{K}$ is the number*

$$\dim_{\mathrm{F}}(\mathcal{K}) := \limsup_{\delta \to 0} \frac{\ln N_\delta(\mathcal{K})}{-\ln \delta}. \tag{2.48}$$

(We include the possibility that $\dim_{\mathrm{F}}(\mathcal{K}) = +\infty$ for some sets $\mathcal{K}$.)

We recall that $\dim_{\mathrm{F}}(\mathcal{K})$, as defined in (2.48), is also known as the UPPER BOX-COUNTING dimension of $\mathcal{K}$. There are corresponding definitions of lower box-counting dimension and of box-counting dimensions of $\mathcal{K}$, obtained by replacing, in (2.48), lim sup respectively by lim inf and lim.

The fractal dimension of a set is always larger than or equal its Hausdorff dimension. Further properties of the fractal dimension are given in the following proposition.

PROPOSITION 2.61
Let $\dim_{\mathrm{F}}$ be defined as in (2.48).

1. If $\mathcal{K}_1$, $\mathcal{K}_2$ are two compact sets of $\mathcal{X}$, then:

$$\mathcal{K}_1 \subseteq \mathcal{K}_2 \quad \Longrightarrow \quad \dim_F(\mathcal{K}_1) \leq \dim_F(\mathcal{K}_2) \qquad (2.49)$$

(i.e., the fractal dimension is monotonic). Moreover,

$$\dim_F(\mathcal{K}_1 \cup \mathcal{K}_2) \leq \max(\dim_F(\mathcal{K}_1), \dim_F(\mathcal{K}_2)), \qquad (2.50)$$
$$\dim_F(\mathcal{K}_1 \times \mathcal{K}_2) \leq \dim_F(\mathcal{K}_1) + \dim_F(\mathcal{K}_2). \qquad (2.51)$$

2. If $f \colon \mathcal{X} \to \mathcal{X}$ is Lipschitz continuous, then for any compact set $\mathcal{K} \subset \mathcal{X}$,

$$\dim_F(f(\mathcal{K})) \leq \dim_F(\mathcal{K}). \qquad (2.52)$$

3. If $\mathcal{M} \subseteq \mathbb{R}^N$ is a smooth, N-dimensional compact manifold of $\mathbb{R}^N$, then

$$\dim_F(\mathcal{M}) = N.$$

PROOF   1. The monotonicity of $\dim_F$ is immediate, since if $\mathcal{K}_1 \subseteq \mathcal{K}_2$, then any covering of $\mathcal{K}_2$ is automatically a covering of $\mathcal{K}_1$. In particular, $N_\delta(\mathcal{K}_1) \leq N_\delta(\mathcal{K}_2)$, and (2.49) follows after division of this inequality by $-\ln \delta$, which is positive if $0 < \delta < 1$.

2. To prove (2.50), fix $\delta \in \,]0, 1[$ and cover $\mathcal{K}_1$ and $\mathcal{K}_2$ with exactly $N_\delta(\mathcal{K}_1)$ and $N_\delta(\mathcal{K}_2)$ balls of diameter δ. Then, the union of these coverings is a covering of $\mathcal{K}_1 \cup \mathcal{K}_2$, consisting of $N_\delta(\mathcal{K}_1) + N_\delta(\mathcal{K}_1)$ balls of diameter δ. Since $N_\delta(\mathcal{K}_1 \cup \mathcal{K}_2)$ is the minimum number of balls of diameter at most δ that are necessary to cover $\mathcal{K}_1 \cup \mathcal{K}_2$, it follows that, setting $\tilde{N}_\delta := \max(N_\delta(\mathcal{K}_1), N_\delta(\mathcal{K}_2))$,

$$N_\delta(\mathcal{K}_1 \cup \mathcal{K}_2) \leq N_\delta(\mathcal{K}_1) + N_\delta(\mathcal{K}_2) \leq 2\tilde{N}_\delta.$$

Thus,

$$\frac{\ln N_\delta(\mathcal{K}_1 \cup \mathcal{K}_2)}{-\ln \delta} \leq \frac{\ln 2 + \ln \tilde{N}_\delta}{-\ln \delta},$$

from which (2.50) follows, letting $\delta \to 0$.

3. To prove (2.51), fix $\delta \in \,]0, \frac{1}{2}[$, and cover $\mathcal{K}_1$ and $\mathcal{K}_2$ with exactly $N_\delta(\mathcal{K}_1)$ and $N_\delta(\mathcal{K}_2)$ balls of diameter δ. Then, the product of these coverings is a covering of $\mathcal{K}_1 \times \mathcal{K}_2$, consisting of $N_\delta(\mathcal{K}_1) \cdot N_\delta(\mathcal{K}_1)$ balls of diameter at most 2δ. It follows that

$$N_{2\delta}(\mathcal{K}_1 \times \mathcal{K}_2) \leq N_\delta(\mathcal{K}_1) \cdot N_\delta(\mathcal{K}_2),$$

and, therefore,

$$\frac{\ln N_{2\delta}(\mathcal{K}_1 \times \mathcal{K}_2)}{-\ln(2\delta)} \leq \frac{\ln N_\delta(\mathcal{K}_1) + \ln N_\delta(\mathcal{K}_2)}{-\ln(2\delta)}.$$

Multiplying and dividing the right side of this inequality by $-\ln \delta$, and letting $\delta \to 0$, we obtain (2.51).

4. To prove (2.52), fix again $\delta \in \,]0,1[$ and cover $\mathcal{K}$ with exactly $N_\delta(\mathcal{K})$ balls of diameter δ. Then $f(\mathcal{K})$ can be covered by the images of these balls, which have diameter at most $L\delta$. Hence, $N_{L\delta}(f(\mathcal{K})) \leq N_\delta(\mathcal{K})$ and, therefore,

$$\dim_F(f(\mathcal{K})) = \limsup_{L\delta \to 0} \frac{\ln N_{L\delta}(f(\mathcal{K}))}{-\ln(L\delta)} \leq \limsup_{\delta \to 0} \frac{\ln \delta}{\ln L + \ln \delta} \frac{\ln N_\delta(\mathcal{K})}{-\ln \delta}$$

$$= \limsup_{\delta \to 0} \frac{\ln N_\delta(\mathcal{K})}{-\ln \delta} = \dim_F(\mathcal{K}).$$

Thus, (2.52) holds. □

2.9 A Priori Estimates

The discussion of Lorenz' equations in section 2.2.2 highlighted the importance of obtaining A PRIORI ESTIMATES on the solution of a system of differential equations. These estimates provide uniform bounds on a local solution, which allow us to extend this local solution to a global one. For this purpose, it is in general sufficient to establish an estimate of the form

$$\|u(t)\|_X \leq \varphi(t), \qquad t \in [0, \tau[, \tag{2.53}$$

where u is a local solution of the system defined on a (maximal) interval $[0, \tau[$, and $\varphi \colon \mathbb{R} \to \mathbb{R}_{\geq 0}$ is continuous and globally defined. For example, estimate (2.18) of proposition 2.7 for the local solutions of Lorenz' equations is of the form (2.53), with $\varphi(t) = \|u_0\|_X e^{\Lambda t}$. However, estimates like (2.53) are in general not sufficient to ensure the existence of an attractor. Indeed, if an attractor exists, proposition 2.45 implies that the system is dissipative, that is, the semiflow admits a bounded absorbing set. In particular, the dissipativity of a system forces all solutions to be bounded eventually (once the orbits enter the absorbing set). It follows that the boundedness of solutions is a necessary condition for dissipativity. Therefore, as a first step towards the existence of an attractor, it is desirable to improve estimate (2.53), with a function φ which is bounded; that is, to obtain an estimate of the form

$$\|u(t)\|_X \leq M, \tag{2.54}$$

where $M > 0$ is independent of $t \in \mathbb{R}_{\geq 0}$. For example, in proposition 2.65 below, we will see that if σ and b are positive, we can improve (2.18) into a uniform estimate of the form (2.54).

In this section we present two results that are commonly used in order to obtain a priori estimates, respectively of the form (2.53) and (2.54). These results are known, respectively, as the GRONWALL'S INEQUALITY and the EXPONENTIAL INEQUALITY.

2.9.1   Integral and Differential Inequalities

In this section we prove one version of the GRONWALL'S INEQUALITY, and a COMPARISON THEOREM for linear differential inequalities.

PROPOSITION 2.62 (*Gronwall's inequality*)
Let u, v and w be continuous functions defined on an interval $[a,b] \subset \mathbb{R}$, with v nonnegative and w continuously differentiable. If

$$u(t) \leq w(t) + \int_a^t u(s)v(s)\,\mathrm{d}s \tag{2.55}$$

for all $t \in [a,b]$, then also

$$u(t) \leq w(t) + \int_a^t \exp\left(\int_\tau^t v(s)\,\mathrm{d}s\right) v(\tau)w(\tau)\,\mathrm{d}\tau. \tag{2.56}$$

If in addition w is nonnegative and increasing, then

$$u(t) \leq w(t)\exp\left(\int_a^t v(s)\,\mathrm{d}s\right). \tag{2.57}$$

PROOF   The proof of proposition 2.62 is immediate; we report it for completeness. Let $U(t)$ denote the right side of (2.55). Then

$$U'(t) = w'(t) + u(t)v(t) \leq w'(t) + U(t)v(t). \tag{2.58}$$

Let $E(t,\tau) := \exp(\int_\tau^t v(s)\,\mathrm{d}s)$. Since

$$\frac{\partial}{\partial \tau}E(t,\tau) = -E(t,\tau)v(\tau),$$

after changing t into τ throughout we can rewrite (2.58) as

$$\frac{\mathrm{d}}{\mathrm{d}\tau}(E(t,\tau)U(\tau)) \leq w'(\tau)E(t,\tau).$$

We integrate this inequality with respect to $\tau \in [a,t]$. Using integration by parts and the fact that $E(t,t) = 1$, we obtain

$$U(t) \leq E(t,a)U(a) + \int_a^t E(t,\tau)w'(\tau)\,\mathrm{d}\tau$$

$$= E(t,a)w(a) + E(t,t)w(t) - E(t,a)w(a) - \int_a^t \frac{\partial}{\partial \tau}E(t,\tau)w(\tau)\,\mathrm{d}\tau$$

$$= w(t) + \int_a^t E(t,\tau)v(\tau)w(\tau)\,\mathrm{d}\tau. \tag{2.59}$$

Since $u(t) \leq U(t)$, (2.56) follows. If w is nonnegative and increasing, we can proceed from (2.59) with

$$u(t) \leq w(t)\left(1 + \int_a^t E(t,\tau)v(\tau)\,d\tau\right) = w(t)\left(1 - \int_a^t \frac{\partial}{\partial \tau}E(t,\tau)\,d\tau\right)$$
$$= w(t)\left(1 - E(t,t) + E(t,a)\right) = w(t)E(t,a),$$

from which (2.57) follows. ☐

To state our next result, we denote by $AC([a,b];\mathbb{R})$ the space of all absolutely continuous functions defined on a closed interval $[a,b]$. Recall that if $u \in AC([a,b];\mathbb{R})$, then u is differentiable almost everywhere in $[a,b]$ (see e.g. theorem A.4).

PROPOSITION 2.63 (Linear differential inequality)
Let $u \in AC([a,b];\mathbb{R})$, and α, $\beta \in L^1(a,b;\mathbb{R})$. If for almost every $t \in [a,b]$,

$$u'(t) \leq \alpha(t) + \beta(t)u(t),$$

then for all $t \in [a,b]$

$$u(t) \leq u(a)\exp\left(\int_a^t \beta(s)\,ds\right) + \int_a^t \exp\left(\int_\tau^t \beta(s)\,ds\right)\alpha(\tau)\,d\tau. \qquad (2.60)$$

In particular, if α and β are constant, and $\beta \neq 0$, then for $t \in [a,b]$

$$u(t) \leq e^{(t-a)\beta}u(a) + \frac{\alpha}{\beta}\left(e^{(t-a)\beta} - 1\right). \qquad (2.61)$$

PROOF Setting $E(t,\tau) := \exp(\int_\tau^t \beta(s)\,ds)$, we have

$$\frac{d}{dt}(E(a,t)u(t)) = E(a,t)u'(t) - \beta(t)E(a,t)u(t) \leq E(a,t)\alpha(t)$$

for almost every $t \in [a,b]$. Hence,

$$E(a,t)u(t) - u(a) \leq \int_a^t E(a,\tau)\alpha(\tau)\,d\tau,$$

from which, for all $t \in [a,b]$,

$$u(t) \leq E(t,a)u(a) + \int_a^t E(t,\tau)\alpha(\tau)\,d\tau.$$

This is (2.60), from which (2.61) follows immediately. Note that if $\beta = 0$, the corresponding trivial inequality (for constant α)

$$u(t) \leq u(a) + \alpha(t-a)$$

follows from (2.61) by letting $\beta \to 0$ for each fixed t. ☐

2.9.2   Exponential Inequality

The bounds provided by proposition 2.62 and 2.63 clearly depend on the interval $[a,b]$; in particular, on its endpoint b. A more favorable situation is when we can establish a linear EXPONENTIAL INEQUALITY on the solution: in this case, we can obtain a uniform bound independent of t.

PROPOSITION 2.64

Let $\mathcal{X}$ be a Banach space, S a semiflow on $\mathcal{X}$, and u_0, $\kappa \in \mathcal{X}$. Assume that the function $\varphi \colon [0,\infty[\to \mathbb{R}$ defined by

$$\varphi(t) := \|S(t)u_0 - \kappa\|^2$$

is absolutely continuous. Assume further that there exist positive numbers m, M such that for almost all $t \geq 0$,

$$\varphi'(t) + m\,\varphi(t) \leq M. \qquad (2.62)$$

Let $\rho := M/m$. Then the closed ball $\overline{B}(\kappa, \sqrt{\rho})$ is positively invariant, and for all $\eta > 0$, each closed ball $\overline{B}(\kappa, \sqrt{\rho+\eta})$ is positively invariant and absorbing.

PROOF   Proposition 2.63 with $\alpha(t) \equiv M$, $\beta(t) \equiv -m$ and $a = 0$ implies that, for all $t \geq 0$,

$$\varphi(t) \leq e^{-tm}\varphi(0) + \rho(1 - e^{-tm}). \qquad (2.63)$$

In particular, if $\varphi(0) \leq \rho + \eta$, $\eta \geq 0$, we deduce that for $t \geq 0$

$$\varphi(t) \leq e^{-tm}(\rho + \eta) + \rho(1 - e^{-tm}) \leq \rho + \eta;$$

that is, $\overline{B}(\kappa, \sqrt{\rho+\eta})$ is positively invariant. To show that when $\eta > 0$ this ball is also absorbing, let $\mathcal{G} \subset \mathcal{X}$ be bounded. There exists then $R > \sqrt{\rho+\eta}$ such that $\mathcal{G} \subseteq \overline{B}(\kappa, R)$. Let $u_0 \in \mathcal{G}$. Then $\varphi(0) \leq R^2$, and (2.63) implies

$$\varphi(t) \leq e^{-tm}R^2 + \rho(1 - e^{-tm}).$$

Thus, $\varphi(t) \leq \rho + \eta$ for all $t \geq T$, with $T = 0$ if $R^2 - \rho \leq \eta$, and $T > 0$ defined by

$$T = \tfrac{1}{m}\ln\frac{R^2 - \rho}{\eta}$$

otherwise. Thus, $S(t)\mathcal{G} \subseteq \overline{B}(\kappa, \sqrt{\rho+\eta})$ for $t \geq T$, and we conclude that $\overline{B}(\kappa, \sqrt{\rho+\eta})$ is absorbing.   ☐

By way of illustration, we apply this result to prove the existence of an absorbing ball for Lorenz' equations.

PROPOSITION 2.65
For all positive values of σ and b, the semiflow S defined by Lorenz' equations (1.50) admits a family of bounded, positively invariant absorbing balls in $\mathbb{R}^3$. Thus, if σ and b are positive, Lorenz' equations are dissipative.

PROOF We choose $\kappa := (0,0,r+\sigma) \in \mathbb{R}^3$, and set $u_0 := (x_0, y_0, z_0)$. Since $\sigma, b > 0$, we can estimate (dropping the argument t)

$$\frac{d}{dt}|S(t)u_0 - \kappa|^2 = 2(x-0)\dot{x} + 2(y-0)\dot{y} + 2(z-r-\sigma)\dot{z}$$
$$= 2x(-\sigma x + \sigma y) + 2y(rx - y - xz) + 2(z-r-\sigma)(xy - bz)$$
$$= -2\sigma x^2 - 2y^2 - 2bz(z-r-\sigma)$$
$$\leq -2\sigma x^2 - 2y^2 - b(z-r-\sigma)^2 + b(r+\sigma)^2.$$

Thus, setting $m := \min\{2, 2\sigma, b\}$ and $M := b(r+\sigma)^2$, we obtain the differential inequality

$$\frac{d}{dt}|S(t)u_0 - \kappa|^2 \leq -m|S(t)u_0 - \kappa|^2 + M.$$

This means that the function $t \mapsto \varphi(t) := |S(t)u_0 - \kappa|$ satisfies (2.62). Applying proposition 2.64 we conclude that every ball $\overline{B}(\kappa, R)$ in $\mathbb{R}^3$, with $\kappa = (0,0,r+\sigma)$ and $R > \sqrt{M/m}$, is absorbing and positively invariant for S. □

In a similar way, it is possible to show that for all positive values of k the solution operator defined by Duffing's equations (1.53) admits a bounded, positively invariant absorbing set in $\mathbb{R}^2$, even in the nonautonomous case $\lambda \neq 0$. We can establish this by means of an analogous a priori estimate on the function

$$E(x,y) := \tfrac{1}{2}k^2x^2 + kxy + y^2 + \tfrac{1}{2}x^4$$

of the solution of (1.53). Note that E is positive definite, and its first three terms are the square of an equivalent norm in $\mathbb{R}^2$, as we immediately see by Schwarz' inequality.

We start by multiplying the second equation in (1.53) by $2y$ and kx, and adding the resulting identities. Using also the identity

$$kx\dot{y} = \frac{d}{dt}(kxy) - k\dot{x}y = \frac{d}{dt}(kxy) - ky^2,$$

and resorting to weighted Cauchy-Schwarz inequalities of the form

$$Axy \leq \eta y^2 + C_\eta x^2 \leq \eta y^2 + \eta x^4 + C'_\eta,$$

we arrive at an exponential inequality of the type

$$\frac{d}{dt}E(x,y) + \alpha E(x,y) \leq C,$$

where $C > 0$ depends on $|\lambda|$. Thus, by proposition 2.64 we deduce that for each λ there exists $R > 0$ (depending on $|\lambda|$), such that the set

$$\mathcal{C}_R := \left\{ (x,y) \in \mathbb{R}^2 : \tfrac{1}{2}k^2x^2 + kxy + y^2 + \tfrac{1}{2}x^4 \leq R^2 \right\}$$

is positively invariant and absorbing. Note that $\mathcal{C}_R$ is not a ball of $\mathbb{R}^2$.

Chapter 3

Attractors for Semilinear Evolution Equations

In this chapter we apply the results described in chapter 2, in particular theorems 2.46 and 2.50, to show the existence of a global attractor for the semiflows generated by two very simple dissipative evolution equations. These are, respectively, a semilinear version of the heat equation, and of the dissipative wave equation. As such, they are a model of, respectively, a parabolic and of a hyperbolic equation. In chapter 7 we shall apply the same methods to other systems of PDEEs arising from various models in mathematical physics.

3.1 PDEEs as Dynamical Systems

3.1.1 The Model IBV Problems

Let $\Omega \subset \mathbb{R}^n$ be a bounded domain with smooth boundary $\partial\Omega$. We wish to investigate the existence of global attractors, in suitable function spaces, for the semiflows generated by the following two "model" initial-boundary value problems (IBVPs in short) on the cylinder $\Omega \times]0, +\infty[$:

1) The IBVP for the semilinear heat equation

$$u_t - \Delta u + g(u) = f \tag{3.1}$$

in $\Omega \times]0, +\infty[$, together with the initial and boundary conditions

$$u(0, \cdot) = u_0 \quad \text{in} \quad \{0\} \times \Omega, \tag{3.2}$$

$$u\big|_{\partial\Omega} = 0 \quad \text{in} \quad]0, +\infty[\times \partial\Omega; \tag{3.3}$$

2) The IBVP for the semilinear dissipative wave equation

$$\varepsilon u_{tt} + u_t - \Delta u + g(u) = f, \qquad \varepsilon > 0, \tag{3.4}$$

in $\Omega \times]0, +\infty[$, together with the initial and boundary conditions

$$u(0, \cdot) = u_0, \quad u_t(0, \cdot) = u_1 \quad \text{in} \quad \{0\} \times \Omega, \tag{3.5}$$

$$u\big|_{\partial\Omega} = 0 \qquad\qquad\qquad \text{in }]0,+\infty[\times\partial\Omega. \qquad (3.6)$$

CONVENTION 3.1   *We refer to problem (3.1) + (3.2) + (3.3) as*

problem (P) ,

and to problem (3.4) + (3.5) + (3.6) as

problem (H_ε) .

We can regard (3.4) as a perturbation of (3.1); or, alternatively, (3.1) as a "reduced version" of (3.4), corresponding to the "limit case" $\varepsilon = 0$. Indeed, problems (P) and (H_ε) can be studied together, in the framework of the so-called SINGULAR PER-TURBATION theory (see e.g. Lions, [Lio73]). We shall not explore this issue in any detail, however, except for a brief mention, in section 3.6, of a result on the UPPER SEMICONTINUITY of the global attractors of these problems, as $\varepsilon \to 0$. On the other hand, while it is true that the basic well-posedness results for problem (H_ε) do hold for arbitrary $\varepsilon > 0$, we are able to establish most results on the existence of different sorts of attracting sets only when ε is sufficiently small.

For specific choices of the nonlinearity g, equations (3.1) and (3.4) describe various models of interest in mathematical physics. In particular, (3.1) is a general model for so-called REACTION-DIFFUSION equations, or CHAFEE-INFANTE equations; equations of this type were first studied by Chafee and Infante in [CI74]. Equation (3.4) is generally known as a GORDON TYPE equation; these include, in particular, the so-called SINE-GORDON and the KLEIN-GORDON equation of quantum mechanics, corresponding respectively to $g(u) = \sin u$ and $g(u) = u^3 - u$. For an extensive review of results on both problems (P) and (H_ε), we refer e.g. to Sell-You, [SY02, scts. 5.1, 5.2].

Problems (P) and (H_ε) are qualitatively very different. This is reflected mainly in the fact that the "space" operator in equation (3.1), i.e. the Laplacian $-\Delta$, generates an analytic semigroup, while, in contrast, when (3.4) is transformed into a first order system in a suitable product phase space, the semigroup generated by the corresponding "space" operator is only C^0 (see section A.3 for the relevant definitions and properties of semigroups). A major consequence of this difference is that the "parabolic" semigroup is compact for large t, while the "hyperbolic" one is not. Thus, the semiflow generated by (3.1) is asymptotically compact, while the one generated by (3.4) is so only up to a uniformly decaying perturbation. The dissipative effects that force this decay in problem (H_ε) are caused by the term u_t, and it will be clear from our arguments that the results we establish for (3.4) do not carry over to the nondissipative case, where the dissipation term u_t is not present.

In both problems, we assume that $f: \mathbb{R} \times \Omega \to \mathbb{R}$; that is, f is independent of u. Later on, we will restrict ourselves to the autonomous case, i.e. when $f: \Omega \to \mathbb{R}$ depends only on the space variable x. For reasons of simplicity, we shall only consider the simple (but nontrivial) model of the quantum mechanics equations, corresponding to the case when $n \leq 3$ and

$$g(u) := k(u^3 - u), \qquad k > 0 \tag{3.7}$$

(in this and the next chapter, we take $k = 1$). However, most of the results we describe in this and the next chapters can be established, with quite similar methods, for more general nonlinearities g, satisfying suitable growth restrictions. For example, for problem (P) the existence of a global attractor can be established if g satisfies the condition

$$\limsup_{|r| \to +\infty} \frac{-g(r)}{r} \leq 0 \tag{3.8}$$

(see e.g. Sell-You, [SY02, sct. 5.1]), while for problem (H_ε) the existence of a global attractor can be established if, in addition to (3.8), g and its antiderivative

$$G(u) := \int_0^u g(v) \, dv$$

are such that for all $r \in \mathbb{R}$,

$$-C_1(r^2 + 1) \leq G(r) \leq C_2(rg(r) + 1), \tag{3.9}$$

and, if $n \geq 2$,

$$|g'(r)| \leq C_3(1 + |r|^\rho), \tag{3.10}$$

where C_1, C_2, C_3 are positive constants independent of r, $0 < \rho \leq \frac{2}{n-2}$ if $n \geq 3$, and $\rho > 0$ is arbitrary if $n = 2$. For this result, we refer e.g. to Babin and Vishik, [BV92, ch. 1.8], or Temam, [Tem88, ch. IV.3]).

In a sense we explain later, these assumptions on g (which are verified in particular by our choice (3.7); take for example, $C_1 = C_2 = 1$ in (3.9) and $C_3 = 3$, $\rho = 2$ in (3.10), for $n = 3$) are "not too strong", and allow us to solve both problems (P) and (H_ε) in an appropriate weak sense, so that we can define corresponding solution operators in a suitable Hilbert space $\mathcal{X}$. Moreover, theorem 2.46 can be applied to the solution operator defined by (3.1), and we can deduce the existence of a global attractor for the semiflow generated by the parabolic problem. In contrast, for the hyperbolic problem (H_ε) the cubic growth of g is critical, with respect to the space dimension $n = 3$, in the sense that, in this case, the presence of a stronger growth (i.e. with larger exponent) would not allow us to establish a suitable weak solution theory for problem (H_ε). On the other hand, when g is cubic we can proceed to show that, at least if ε is sufficiently small, theorem 2.50 can be applied to the solution operator defined by (3.4), and deduce the existence of a global attractor also for the semiflow generated by the hyperbolic problem (H_ε). Still, the procedure we follow is typical, and in fact almost the same for both problems; indeed, our choice of the nonlinearity (3.7), with $n \leq 3$, is motivated only by the goal of providing an outline of the general arguments, in one of the simplest nontrivial settings.

Examples of "parabolic" equations that can be treated in the same way include reaction-diffusion equations, the Cahn-Hilliard equations, the 2-dimensional Navier-Stokes equations, the Chafee-Infante equations, etc. Furthermore, the same procedure could be followed for quasilinear parabolic evolution equations of monotone type

$$u_t - \operatorname{div} \zeta(\nabla u) = f, \tag{3.11}$$

with $\zeta : \mathbb{R}^n \to \mathbb{R}^n$ strongly monotone, including the so called p-Laplacian operator

$$\operatorname{div} \zeta(\nabla u) = \sum_{i=1}^{n} \partial_i(|\partial_i u|^{p-2} \partial_i u), \qquad p \geq 2.$$

(This operator is a generalization of the usual Laplace operator in $\mathbb{R}^n$, to which it reduces when $p = 2$.) In section 6.4 of chapter 6, we will prove the existence of a global attractor for a quasilinear evolution equation of the type (3.11), which models the quasi-stationary Maxwell's equations in a ferromagnetic medium. Unfortunately, we do not know if this result carries over to the hyperbolic perturbation of (3.11), that is, to an hyperbolic dissipative quasilinear equation of the type

$$\varepsilon u_{tt} + u_t - \operatorname{div} \zeta(\nabla u) = f.$$

On the other hand, we can treat far fewer models of "hyperbolic" equations. Among these, the sine-Gordon equation, types of Klein-Gordon equations similar to (3.4), and several types of Kirchoff, von Kármán and Cahn-Hilliard equations, including various models of beam and thin plate equations.

We shall present a number of these examples in chapter 6, with the goal of showing that, for all these systems of PDEEs, we can follow a similar procedure, which allows us to apply at least one of theorems 2.46, 2.50, or 2.56, and, consequently, deduce the existence of a global attractor for the semiflows corresponding to each of these systems.

3.1.2   Construction of the Attractors

The main steps of the procedure we want to describe, leading to the existence of a global attractor for the semiflows generated by evolution equations like (3.1) or (3.4), can be summarized by the following sequence of incremental results.

1. Solution of the problem.   In this step, we transform the problem into an abstract evolution equation in an Hilbert space $\mathcal{X}$, consisting of functions of the space variable. We give a precise definition of what we mean by a solution of the problem in $\mathcal{X}$; in turn, this allows us to define a corresponding semiflow S on $\mathcal{X}$, generated by the differential equation.

Formally, in the parabolic case (3.1), S will be defined by

$$[0, \infty[\times \Omega \ni (t, x) \mapsto (S(t)u_0)(x) := u(t, x, u_0), \tag{3.12}$$

where the right side of (3.12) is the value at the point (t,x) of the solution u of problem (P). In the hyperbolic case (3.4), $\mathcal{X}$ is the product of two spaces, each again a space of functions of the space variable, and the semiflow S on $\mathcal{X}$ will be formally defined by

$$[0,\infty[\, \times \Omega \ni (t,x) \mapsto (S(t)(u_0,u_1))(x) := (u(t,x,u_0,u_1),u_t(t,x,u_0,u_1)), \quad (3.13)$$

where $u(\cdot,\cdot,u_0,u_1)$ is the solution of problem (H_ε), and $u_t(\cdot,\cdot,u_0,u_1)$ is its time derivative. In general, the solution operator S will be a semiflow only when the system is autonomous, i.e. when the source term f in the equations is independent of t. As we have indicated, the fact that S is actually a semiflow is a consequence of the well-posedness in the large of the Cauchy problems (P) and (H_ε) in the corresponding spaces $\mathcal{X}$. We carry out this part, first by obtaining local solutions of the problem, and then by establishing a priori estimates which allow us to extend the local solutions to global ones.

2. Absorbing sets.   In this step we show the existence of a bounded, positively invariant absorbing set $\mathcal{B}$ for S in $\mathcal{X}$. As we have seen in the finite dimensional examples of chapter 2, this part can be carried out by refining the a priori estimates established in Step 1. Following the ideas of proposition 2.65 on the absorbing set of Lorenz' equations, the most effective way to obtain an absorbing set is to establish a linear differential inequality on the square of a norm of u, as in (2.62). In a certain sense, the possibility of doing so is characteristic of dissipative systems; for example, differential inequalities like (2.62) are generally not available for nondissipative wave equations of the form

$$u_{tt} - \Delta u = F(t,x,u)\,.$$

3. Compactness of the semiflow.   In this step we establish suitable regularizing properties of the semiflow S. In the parabolic case, this step usually exploits the smoothing effect of parabolic operators, which allows us to deduce directly that S is uniformly compact in $\mathcal{X}$ for large t. In the hyperbolic case, we show instead that, at least if ε is sufficiently small, S admits a decomposition $S = S_1 + S_2$ as in proposition 2.48, with S_1 uniformly compact for large t and S_2 uniformly decaying to 0. This will indeed be possible for equation (3.4); however, there are other types of equations, such as the beam equation, the von Kármán equation, or the perturbed nonviscous Cahn-Hilliard equation, for which we do not know how to implement this step. In this case, we would resort to the method of α-contractions, presented in section 2.7.

4. Conclusion.   We apply theorems 2.46 or 2.50 (or theorem 2.56 for α-contractions), to deduce the existence of a global attractor $\mathcal{A}$ for S in $\mathcal{X}$. This attractor gives a description of the long-time behavior of the solutions of the equations, independently of their initial values.

   Of course, these steps can be supplemented by others, concerning further properties of these attractors, such as their regularity, finite dimensionality, geometrical, topological or differential structure, etc. For instance, regularity is often established by proving that $\mathcal{A}$ is contained in a subspace $\mathcal{X}_1 \hookrightarrow \mathcal{X}$. If the injection $\mathcal{X}_1 \hookrightarrow \mathcal{X}$ is compact, compactness of $\mathcal{A}$ in $\mathcal{X}$ can also be obtained by showing that $\mathcal{A}$ is bounded in $\mathcal{X}_1$. As an illustration, in section 3.5 we shall show a regularity result of this type for problem (H$_\varepsilon$).

REMARK 3.2   In sections 3.3 and 3.4 we shall present a direct implementation of step 1, for both problems (P) and (H$_\varepsilon$). That is, we shall give a constructive procedure, based on an approximation technique, which yields the solutions to both problems, and allows us to generate the corresponding semiflows. An alternative approach, which is more in analogy to the classical methods of ODEs, is provided by the theory of SEMIGROUPS (see section A.3). In this setting, we transform the problem into an abstract evolution equation of the form

$$U_t + AU = F(U), \tag{3.14}$$

in an Hilbert space $\mathcal{X}$, consisting of functions of the space variable. In fact, problem (P) is already in the form (3.14), while for problem (H$_\varepsilon$), $\mathcal{X}$ is a product space, and A is a formal matrix, acting on the vector $U := (u, u_t)^\top$, in accord with (3.13). When both problems are transformed into the form (3.14), the linear operator A is densely defined, with domain

$$\mathrm{dom}(A) := \{u \in \mathcal{X} : Au \in \mathcal{X}\},$$

and generates at least a C^0-SEMIGROUP $(e^{-tA})_{t \geq 0}$ on $\mathcal{X}$. This allows us to transform (3.14) into an integral equation, and, in analogy to the classical procedure in ODEs, consider the corresponding solutions of the initial value problem. In particular, we have the following

DEFINITION 3.3   *Let $U_0 \in \mathcal{X}$, and $T > 0$. A function $U \in C([0,T]; \mathcal{X})$ is a* MILD SOLUTION *of equation (3.14), with initial value $U(0) = U_0$, if U satisfies the integral equation*

$$U(t) = e^{-tA}U_0 + \int_0^t e^{-(t-\theta)A} F(U(\theta)) \, d\theta, \quad 0 \leq t \leq T. \tag{3.15}$$

Note that (3.15) is a straightforward generalization of the familiar Duhamel's formula for the solution of the system of ODEs

$$\dot{u} + Au = F(u),$$

in which A is a constant matrix in $\mathbb{R}^N$.

The following result (see theorem A.52, whose statement we repeat here for convenience) shows that the initial value problem for (3.14) is well posed in the class of mild solutions.

THEOREM 3.4
Assume that $F: \mathcal{X} \to \mathcal{X}$ is globally Lipschitz continuous. Then for all $U_0 \in \mathcal{X}$, the initial value problem for (3.14), with initial value $U(0) = U_0$, has a unique mild

solution u. Moreover, for all $T > 0$ the map

$$\mathcal{X} \ni U_0 \mapsto U \in C([0,T];\mathcal{X})$$

is Lipschitz continuous.

In particular, theorem 3.4 implies that the evolution equation (3.14) generates a continuous semiflow S on $\mathcal{X}$. We shall refer to this approach again in section 5.6.1 of chapter 5, when we construct inertial manifolds for general evolution equations of the form (3.14).

For a comprehensive introduction to semigroup theory, we refer e.g. Pazy, [Paz83], or Engel and Nagel, [EN00]; we report the basic results we need in section A.3. □

3.2 Functional Framework

In this section we introduce the functional space framework in which we study problems (P) and (H_ε). We also recall some well known facts on the Laplace operator Δ, associated to homogeneous Dirichlet boundary conditions. Throughout the rest of this chapter, we assume that Ω is a bounded domain of $\mathbb{R}^n$, with a smooth (i.e. at least Lipschitz) boundary $\partial\Omega$.

3.2.1 Function Spaces

We consider the following spaces and adopt the following notations:

1. $\mathcal{H} := L^2(\Omega)$, with norm $\|\cdot\|$ and scalar product $\langle\cdot,\cdot\rangle$.

2. $\mathcal{V} := H_0^1(\Omega)$, with norm $\|u\|_{\mathcal{V}} = \|\nabla u\|$; this is justified by Poincaré's inequality

$$\exists \lambda_1 > 0 \, \forall u \in \mathcal{V}: \quad \sqrt{\lambda_1}\|u\| \le \|\nabla u\| \tag{3.16}$$

   (see (A.73)). The corresponding scalar product in $\mathcal{V}$ is then

$$\langle u,v\rangle_{\mathcal{V}} := \langle \nabla u, \nabla v\rangle.$$

3. $\mathcal{V}' := H^{-1}(\Omega)$, with norm $\|\cdot\|_{\mathcal{V}'}$. $\mathcal{V}'$ is the topological dual of $\mathcal{V}$. We denote the duality pairing between $\mathcal{V}'$ and $\mathcal{V}$ by $(\cdot,\cdot)_{\mathcal{V}'\times\mathcal{V}}$.

4. $\mathcal{D} := H^2(\Omega) \cap H_0^1(\Omega)$. Since the Poincaré type inequality

$$\sqrt{\lambda_1}\|\nabla u\| \le \|\Delta u\| \tag{3.17}$$

   holds for all $u \in \mathcal{D}$, we can, and will, choose in $\mathcal{D}$ the norm $\|u\|_{\mathcal{D}} := \|\Delta u\|$. To prove (3.17), note that if $u \in \mathcal{D}$, then

$$\|\nabla u\|^2 = \langle -\Delta u, u\rangle \le \|\Delta u\| \, \|u\|,$$

   so (3.17) follows by (3.16).

5. $L: \mathcal{H} \to \mathcal{H}$ is the unbounded operator formally defined by $Lu = -\Delta u$, with domain
$$\text{dom}(L) = \mathcal{D}.$$

Δ is the usual Laplace operator, formally defined by

$$\Delta := \sum_{j=1}^{n} \frac{\partial^2}{\partial x_j^2}.$$

Note that, in choosing $\mathcal{D}$ as the domain of L, we are automatically imposing homogeneous Dirichlet boundary conditions on $\partial\Omega$, in the weak sense of $H_0^1(\Omega)$.

6. For $1 \leq p \leq +\infty$, $|\cdot|_p$ denotes the norm in $L^p(\Omega)$, and $q \in [1, +\infty]$ denotes the conjugate index of p, i.e. $q := 1$ if $p = +\infty$, $q := +\infty$ if $p = 1$, and $q := \frac{p}{p-1}$ otherwise.

7. If $\mathcal{X}$ is a function space on Ω and $f: [0,T] \to \mathcal{X}$, we implicitly define a function on $[0,T] \times \Omega$, with value $f(t)(x)$ for $(t,x) \in [0,T] \times \Omega$. We denote this function again by f; that is, we set

$$f(t,x) := f(t)(x), \qquad (t,x) \in [0,T] \times \Omega.$$

Similarly, given a function $g: [0,T] \times \Omega \to \mathbb{R}$, we define a function from $[0,T]$ into $\mathcal{X}$, with value $g(t,\cdot)$ at $t \in [0,T]$. We denote this function again by g; that is, we set

$$g(t)(x) := g(t,x), \qquad (t,x) \in [0,T] \times \Omega.$$

Likewise, we denote by $\Delta f(t)$ (respectively, $\nabla f(t)$) the function with value $\Delta f(t,x)$ (respectively, $\nabla f(t,x)$) at $x \in \Omega$.

8. Finally, for $T > 0$ we set

$$\mathcal{W}(T) := \{ f \in L^2(0,T;\mathcal{V}): f_t \in L^2(0,T;\mathcal{V}') \},$$

and recall that

$$\mathcal{W}(T) \hookrightarrow C([0,T];\mathcal{H}). \tag{3.18}$$

Imbedding (3.18) is part of the following result, whose proof can be found e.g. in Tanabe, [Tan79] (Lemma 5.5.1):

PROPOSITION 3.5
Let $\mathcal{V}$ and $\mathcal{H}$ be separable Hilbert spaces, with $\mathcal{V} \hookrightarrow \mathcal{H}$ continuously and densely. If $u \in \mathcal{W}(T)$, u can be modified on a set of measure 0, in such a way that $u \in C([0,T];\mathcal{H})$. If also $v \in \mathcal{W}(T)$, the function $t \mapsto \langle u(t), v(t) \rangle_{\mathcal{H}}$ is absolutely continuous on $[0,T]$, and satisfies the identity

$$\frac{d}{dt} \langle u(t), v(t) \rangle_{\mathcal{H}} = (u_t(t), v(t))_{\mathcal{V}' \times \mathcal{V}} + (v_t(t), u(t))_{\mathcal{V}' \times \mathcal{V}}$$

in $D'(]0,T[)$ *and, in fact, for almost every* $t \in [0,T]$. *In particular, taking* $v = u$ *we deduce that for almost all* $t \in [0,T]$,

$$\frac{d}{dt}\|u(t)\|_{\mathcal{H}}^2 = 2\left(u_t(t),u(t)\right)_{\mathcal{V}' \times \mathcal{V}}.$$

3.2.2   Orthogonal Bases

As we recall in theorem A.76, the eigenvalue problem for L

$$\begin{cases} -\Delta w_j = \lambda_j w_j \\ w_j|_{\partial\Omega} = 0 \end{cases} \tag{3.19}$$

admits an unbounded sequence of positive eigenvalues $(\lambda_j)_{j \in \mathbb{N}}$. This sequence can be ordered so that

$$0 < \lambda_1 \le \lambda_2 \le \cdots \le \lambda_j \le \cdots, \qquad \lambda_j \to +\infty. \tag{3.20}$$

For each $j \in \mathbb{N}$, the corresponding eigenvector w_j is in $C^\infty(\Omega) \cap C(\overline{\Omega})$, and the sequence $(w_j)_{j \in \mathbb{N}}$ is a complete orthogonal system in $L^2(\Omega)$. This means that each $u \in L^2(\Omega)$ has a uniquely determined Fourier series expansion with respect to the w_j's; more precisely,

$$u = \sum_{j=1}^\infty \langle u, w_j \rangle w_j, \tag{3.21}$$

with the series (3.21) converging in $L^2(\Omega)$. Since rescaled eigenvectors are still eigenvectors, we can assume that the $(w_j)_{j \in \mathbb{N}}$ are in fact an orthonormal system, i.e. that $\|w_j\| = 1$ for all $j \in \mathbb{N}$. Then, Parseval's formula holds: for all $u \in \mathcal{H}$,

$$\|u\|^2 = \sum_{j=1}^\infty \langle u, w_j \rangle^2 \|w_j\|^2 = \sum_{j=1}^\infty \langle u, w_j \rangle^2. \tag{3.22}$$

The particular choice (3.19) of eigenvectors assures the orthogonality (but, of course, not the orthonormality) of the sequence $(w_j)_{j \in \mathbb{N}}$ also in $\mathcal{V}$. In fact, for arbitrary j and k we compute that

$$\langle \nabla w_j, \nabla w_k \rangle = \langle -\Delta w_j, w_k \rangle = \lambda_j \langle w_j, w_k \rangle; \tag{3.23}$$

thus, if $j \ne k$,

$$\langle \nabla w_j, \nabla w_k \rangle = 0,$$

while if $j = k$

$$\langle \nabla w_j, \nabla w_j \rangle = \|\nabla w_j\|^2 = \lambda_j \|w_j\|^2 = \lambda_j. \tag{3.24}$$

As a consequence, in analogy to (3.22) we have that for all $u \in \mathcal{V}$,

$$\|\nabla u\|^2 = \sum_{j=1}^\infty \langle u, w_j \rangle^2 \|\nabla w_j\|^2 = \sum_{j=1}^\infty \lambda_j \langle u, w_j \rangle^2. \tag{3.25}$$

3.2.3 Finite Dimensional Subspaces

We will construct solutions to problems (P) and (H_ε) as limits of sequences of approximating solutions, in the framework of a suitable Galerkin scheme. In the implementation of this type of scheme, the starting step is the projection of the problem into a sequence of finite dimensional subspaces of $\mathcal{V}$, on each of which the PDEE is reduced to a finite system of ODEs. We will recall this procedure when we consider problem (P). This step requires the choice of a so-called TOTAL BASIS of $\mathcal{V}$. As we recall in definition A.10, this is a countable set of linearly independent vectors w_j, such that

$$\overline{\bigcup_{m=1}^{\infty} \mathcal{V}_m} = \mathcal{V}$$

(closure in $\mathcal{V}$), where for $m \in \mathbb{N}$,

$$\mathcal{V}_m := \operatorname{span}\{w_1, \ldots, w_m\}. \tag{3.26}$$

Total bases of $\mathcal{V}$ exist, because $\mathcal{V}$ is separable. Among these, a particularly convenient one is the sequence of eigenvectors $(w_j)_{j\in\mathbb{N}}$ of L, defined in (3.19). Referring then to the series expansion (3.21), for each $N \in \mathbb{N}$ we define a projection $P_N : \mathcal{V} \to \mathcal{V}_N$ by setting $P_0 := 0$ and, if $N \geq 1$, by

$$P_N(u) := \sum_{j=1}^{N} \langle u, w_j \rangle w_j \qquad \text{if} \quad u = \sum_{j=1}^{\infty} \langle u, w_j \rangle w_j. \tag{3.27}$$

We also set

$$Q_N := I - P_N;$$

that is, $Q_N(u)$ is the tail of the series (3.21). Both P_N and Q_N are clearly orthogonal projections in $\mathcal{H}$; because of (3.23), they are also orthogonal in $\mathcal{V}$. It is also worth recalling that the family $(P_N)_{n\geq 1}$ is monotone; that is, if $n \leq m$, then $P_n \leq P_m$. This inequality is to be understood in the operator sense, that is, that

$$\langle P_n x, x \rangle \leq \langle P_m x, x \rangle$$

for all $x \in \mathcal{X}$. To see this, it is sufficient to note that, because of the orthogonality of the sequence $(w_j)_{j\in\mathbb{N}}$, from (3.21) and (3.27) we have

$$\langle P_n x, x \rangle = \sum_{j=1}^{n} \langle x, w_j \rangle^2 \leq \sum_{j=1}^{m} \langle x, w_j \rangle^2 = \langle P_m x, x \rangle.$$

The following result, which generalizes Poincaré's inequality (3.16), will be used in chapters 4 and 5:

PROPOSITION 3.6
For all $N \in \mathbb{N}$ and $u \in \mathcal{V}$,

$$\|\nabla(Q_N(u))\|^2 \geq \lambda_{N+1} \|Q_N(u)\|^2, \tag{3.28}$$

$$\|\nabla(P_N(u))\|^2 \le \lambda_N \|P_N(u)\|^2. \tag{3.29}$$

PROOF   Let $u = \sum\limits_{j=0}^{\infty} \langle u, w_j \rangle w_j$, as in (3.21). Then (3.28) follows from (3.22), (3.25) and (3.20), since

$$\|\nabla(Q_N(u))\|^2 = \sum_{j=N+1}^{\infty} \lambda_j \langle u, w_j \rangle^2 \ge \lambda_{N+1} \sum_{j=N+1}^{\infty} \langle u, w_j \rangle^2 = \lambda_{N+1} \|Q_N(u)\|^2.$$

This proves (3.28); the proof of (3.29) is similar. Note that (3.16) corresponds to (3.28) with $N = 0$. $\square$

3.3   The Parabolic Problem

In this section we illustrate how to implement the steps we have listed in section 3.1.2 for the model parabolic IBVP (3.1). We choose $\mathcal{X} := \mathcal{H} = L^2(\Omega)$, and set

$$\mathcal{W}_0(T) := \{f \in \mathcal{W}(T) \colon f(T, \cdot) = 0\};$$

note that $\mathcal{W}_0(T)$ is well defined, because by proposition 3.5 the trace $f(T, \cdot)$ makes sense at least in $\mathcal{X}$. We need the following generalization of proposition 3.5:

PROPOSITION 3.7
Let $\mathcal{V}$ and $\mathcal{H}$ be as in proposition 3.5, and assume that

$$\begin{aligned} u &\in L^2(0, T; \mathcal{V}) \cap L^p(]0, T[\times \Omega) =: \mathcal{Y}, \\ u_t &\in L^2(0, T; \mathcal{V}') + L^q(]0, T[\times \Omega) =: \mathcal{Z}, \end{aligned} \tag{3.30}$$

for some $p \in]1, +\infty[$, with $q = \frac{p}{p-1}$. Then u can be modified on a set of measure 0 so that $u \in C([0, T]; \mathcal{H})$. Moreover, the function $t \mapsto \|u(t)\|_{\mathcal{H}}^2$ is absolutely continuous on $[0, T]$, and if $u_t = g_1 + g_2$, with $g_1 \in L^2(0, T; \mathcal{V}')$ and $g_2 \in L^q(]0, T[\times \Omega)$ (which is the meaning of the second of (3.30)), then for almost all $t \in [0, T]$,

$$\frac{\mathrm{d}}{\mathrm{d}t} \|u(t)\|_{\mathcal{H}}^2 = 2(g_1(t), u(t))_{\mathcal{V}' \times \mathcal{V}} + 2\langle g_2(t), u(t) \rangle_{\mathcal{H}}. \tag{3.31}$$

PROOF   We briefly sketch the proof of proposition 3.7 for the sake of completeness. We recall that $\mathcal{Z} = \mathcal{Y}'$, via the duality

$$(g_1 + g_2, u)_{\mathcal{Z} \times \mathcal{Y}} := \int_0^T (g_1(t), u(t))_{\mathcal{V}' \times \mathcal{V}} \, \mathrm{d}t + \int_0^T \langle g_2(t), u(t) \rangle_{\mathcal{H}} \, \mathrm{d}t;$$

note that the product g_2u belongs to $L^1(]0,T[\times\Omega)$. If $u \in C^1([0,T];\mathcal{H})$, from the identity

$$\frac{d}{dt}\|u(t)\|_{\mathcal{H}}^2 = 2\langle u_t(t),u(t)\rangle_{\mathcal{H}} = 2\left(g_1(t)+g_2(t),u(t)\right)_{\mathcal{Y}'\times\mathcal{Y}}$$
$$= 2\left(g_1(t),u(t)\right)_{\mathcal{V}'\times\mathcal{V}}+2\langle g_2(t),u(t)\rangle_{\mathcal{H}}, \tag{3.32}$$

we deduce that for $0 \le s < t \le T$

$$\|u(t)\|_{\mathcal{H}}^2 = \|u(s)\|_{\mathcal{H}}^2 + 2\int_s^t \left(g_1(\theta),u(\theta)\right)_{\mathcal{V}'\times\mathcal{V}}d\theta + 2\int_s^t \langle g_2(\theta),u(\theta)\rangle_{\mathcal{H}}d\theta. \tag{3.33}$$

Choosing s so that

$$\|u(s)\|_{\mathcal{H}}^2 = \frac{1}{T}\int_0^T \|u(\theta)\|_{\mathcal{H}}^2 d\theta,$$

we deduce from (3.33) that

$$\|u(t)\|_{\mathcal{H}}^2 \le \frac{1}{T}\int_0^T \|u(\theta)\|_{\mathcal{H}}^2 d\theta + 2\int_0^T \|g_1(\theta)\|_{\mathcal{V}'}\|u(\theta)\|_{\mathcal{V}}d\theta$$
$$+ 2\int_0^T |g_2(\theta)|_q|u(t)|_p d\theta.$$

From this, we obtain that

$$\|u\|_{C([0,T];\mathcal{H})} \le \frac{1}{T}\|u\|_{L^2(0,T;\mathcal{H})}^2 + 2\|u_t\|_{\mathcal{Z}}\|u\|_{\mathcal{Y}}.$$

We can then easily conclude that

$$\{u \in \mathcal{Y} : u_t \in \mathcal{Y}'\} \hookrightarrow C([0,T];\mathcal{H})$$

by a density argument. Likewise, (3.31) follows from (3.32), noting that its right side is in $L^1(0,T)$. □

3.3.1 Step 1: The Solution Operator

We now prove a global existence and well-posedness result for problem (P), which allows us to define the associated solution operator. If the equation is autonomous, i.e. if f only depends on x, this solution operator is in fact a continuous semiflow.

DEFINITION 3.8 *Let $T > 0$, $u_0 \in \mathcal{H}$, $f \in L^2(0,T;\mathcal{V}')$. A function u is a* WEAK SOLUTION *of problem (P), with $g(u) = u^3 - u$, on the interval $[0,T]$ if*

$$u \in C([0,T];\mathcal{H})\cap L^2(0,T;\mathcal{V})\cap L^4(]0,T[\times\Omega),$$

and for all $\varphi \in \mathcal{W}_0(T)\cap L^4(]0,T[\times\Omega)$,

$$\int_0^T \left(-\langle u,\varphi_t\rangle + \langle\nabla u,\nabla\varphi\rangle + \langle u^3 - u,\varphi\rangle\right)dt = \int_0^T (f,\varphi)_{\mathcal{V}'\times\mathcal{V}}dt + \langle u_0,\varphi(0)\rangle. \tag{3.34}$$

Note that the left side of (3.34) makes sense: in fact, by theoremA.58, the product $u^3 \varphi$ is integrable if $u \in L^4(]0,T[\times\Omega)$ and $\varphi \in L^4(]0,T[\times\Omega)$.

Weak solvability of problem (P) is then assured by

THEOREM 3.9

For all $T > 0$, $u_0 \in \mathcal{H}$ and $f \in L^2(0,T;\mathcal{V}')$, there exists a unique weak solution u of problem (P). This solution depends continuously on the initial value u_0, and the function $t \mapsto \|u(t)\|_{\mathcal{H}}$ is absolutely continuous on $[0,T]$. If in addition $u_0 \in \mathcal{V}$ and $f \in L^2(0,T;\mathcal{H})$, then

$$u \in C([0,T];\mathcal{V}) \cap L^2(0,T;\mathrm{H}^2(\Omega)), \quad u_t \in L^2(0,T;\mathcal{H}), \tag{3.35}$$

and the function $t \mapsto \|u(t)\|_{\mathcal{V}}$ is absolutely continuous on $[0,T]$.

PROOF   1. We first remark that if problem (P) does admit a weak solution u as claimed, then (3.34) implies that the identity

$$u_t = f + u - u^3 + \Delta u \tag{3.36}$$

holds at least in $D'(0,T;\mathcal{V}')$. Thus, reading (3.36) as

$$u_t = (f + u + \Delta u) - u^3 =: g_1 - g_2 \in L^2(0,T;\mathcal{V}') + L^{4/3}(]0,T[\times\Omega),$$

by proposition 3.7 we deduce that $u \in C([0,T];\mathcal{H})$, and that the function $t \mapsto \|u(t)\|_{\mathcal{H}}$ is absolutely continuous. Finally, if u enjoys the additional regularity described in (3.35), then proposition 3.7, applied to each partial derivative $\partial_k u$, $1 \le k \le n$, with $p = q = 2$, implies that the function $t \mapsto \|u(t)\|_{\mathcal{V}}$ is also absolutely continuous.

2. A proof of the existence and regularity part of theorem 3.9 can be given by means of a standard Galerkin approximation method, as e.g. in Lions, [Lio69], or Temam, [Tem88, sct. III.1.1]. We sketch the details of the proof of the existence claim for completeness.

We refer to the finite dimensional subspaces $\mathcal{V}_m$ of $\mathcal{V}$ introduced in section 3.2.3 and, for each $m \in \mathbb{N}$, project problem (P) on $\mathcal{V}_m$. That is, we look for a function $(t,x) \mapsto u^m(t,x)$ (here, u^m does not mean u to the power m), with $u^m(t,\cdot) \in \mathcal{V}_m$ for $t \in [0,T]$, determined as the solution of the system of m ODEs

$$\begin{cases} \langle u_t^m(t) - \Delta u^m(t) + (u^m(t))^3 - u^m(t) - f(t), w_j \rangle = 0, \\ j = 1, \dots, m, \end{cases} \tag{3.37}$$

supplemented by the initial conditions

$$u^m(0) = u_0^m := P_m(u_0) \in \mathcal{V}_m. \tag{3.38}$$

Thus, u^m should have the form

$$u^m(t,x) = \sum_{j=1}^m \gamma_j(t) w_j(x), \tag{3.39}$$

and (3.37) is a system of ODEs for the scalar components $\gamma_1, \ldots, \gamma_m$ of u^m ((3.39) means that we solve the projected PDE by separation of variables).

The Cauchy problem (3.37)+(3.38) can be solved by means of Carathéodory's theorem A.5, which provides a local solution

$$u^m \in AC([0, t_m]; \mathcal{V}_m)$$

of (3.37), defined on some interval $[0, t_m] \subseteq [0, T]$ (that is, each $\gamma_j \in AC([0, t_m])$). We then extend each u^m to all of $[0, T]$, by means of the a priori estimates that follow. Before presenting these estimates, we introduce two conventions, that we follow not only here, but almost always in the remainder of these notes.

CONVENTION 3.10 *In writing estimates, we often omit, for convenience, one of the variables t and x, or both. This means that, for example, if $u: [0, T] \to L^2(\Omega)$, we write $\|u\|$ instead of $\|u(t, \cdot)\|$.*

CONVENTION 3.11 *Unless otherwise specified, we denote by C a generic constant, which may be different from line to line of the same estimate, or even within the same line. This means that, for example, we identify quantities like C, C^2, e^C, etc. On the other hand, when a constant depends on a parameter in a crucial way (e.g., if the constant is unbounded as the parameter vanishes), we shall always indicate this dependence explicitly.*

We proceed then to multiply each equation (3.37) by $\gamma_j(t)$. Summing all the resulting identities for $j = 1, \ldots, m$, we obtain

$$\frac{d}{dt} \|u^m\| + 2\|\nabla u^m\|^2 + 2|u^m|_4^4 = 2\langle f + u^m, u^m \rangle \leq \|f\|_{\mathcal{V}'}^2 + \|\nabla u^m\|^2 + 2\|u^m\|^2.$$

We integrate this inequality on $[0, t]$, $0 < t \leq t_m$:

$$\|u^m(t)\|^2 + \int_0^t \|\nabla u^m(\theta)\|^2 \, d\theta + 2 \int_0^t |u^m(\theta)|_4^4 \, d\theta$$

$$\leq \|u_0^m\|^2 + \int_0^T \|f(\theta)\|_{\mathcal{V}'}^2 \, d\theta + 2 \int_0^t \|u^m(\theta)\|^2 \, d\theta. \qquad (3.40)$$

Recalling that the sequence $(u_0^m)_{m \in \mathbb{N}}$ is bounded in $\mathcal{H}$ (because it converges to u_0), by Gronwall's inequality (proposition 2.62) we derive from (3.40) the estimate

$$\|u^m\|_{L^\infty(0, t_m; \mathcal{H})} + \|u^m\|_{L^2(0, t_m; \mathcal{V})} + \|u^m\|_{L^4(]0, t_m[\times \Omega)}^2 \leq M_1 e^{t_m}, \qquad (3.41)$$

where

$$M_1^2 := \sup_{m \geq 1} \|u_0^m\|^2 + \int_0^T \|f(\theta)\|_{\mathcal{V}'}^2 \, d\theta$$

is finite. Since

$$M_1 e^{t_m} \leq M_1 e^T =: M,$$

we obtain from (3.41) an estimate independent of both t_m and m (however, M depends on T, as is typical of a priori estimates obtained by means of Gronwall's inequality). The fact that M is independent of t_m allows us to extend u^m to all of $[0,T]$, by a usual continuation argument (see theorem A.2). The fact that M is independent of m means that the sequence $(u^m)_{m\in\mathbb{N}}$ is bounded in each of the spaces

$$L^\infty(0,T;\mathcal{H}), \quad L^2(0,T;\mathcal{V}), \quad L^4(]0,T[\times\Omega). \tag{3.42}$$

Hence, by theorem A.16, there is a subsequence of $(u^m)_{m\in\mathbb{N}}$, which we still denote by $(u^m)_{m\in\mathbb{N}}$, converging weakly to a limit u in each of the spaces of (3.42). Moreover, since the sequence $((u^m)^3)_{m\in\mathbb{N}}$ is bounded in $L^{4/3}(]0,T[\times\Omega)$, there is a further subsequence, which we again keep denoting by $(u^m)_{m\in\mathbb{N}}$, such that $(u^m)^3$ converges to a limit χ weakly in $L^{4/3}(]0,T[\times\Omega)$. Fixing j in (3.37) and letting $m\to+\infty$, we deduce that u and χ solve the equation

$$u_t - \Delta u + \chi - u = f.$$

Thus, to complete the proof of the existence part of theorem 3.9 it is sufficient to show that $\chi = u^3$. To this end, we fist show that

$$u^m \to u \quad \text{in } L^2(0,T;\mathcal{H}) \text{ strongly}. \tag{3.43}$$

To obtain this, we recall from theorem A.74 that the Sobolev imbedding $H^1(\Omega)\hookrightarrow L^4(\Omega)$ holds, since $n\leq 3$. In turn, this implies that

$$L^4(0,T;\mathcal{V})\hookrightarrow L^4(]0,T[\times\Omega)$$

and, therefore, that the dual imbedding

$$L^{4/3}(]0,T[\times\Omega)\hookrightarrow L^{4/3}(0,T;\mathcal{V}')$$

holds. From this it follows that the sequence $((u^m)^3)_{m\in\mathbb{N}}$ is also bounded in the space $L^{4/3}(0,T;\mathcal{V}')$. In fact, we have that

$$\|(u^m)^3\|_{L^{4/3}(0,T;\mathcal{V}')} \leq C\|(u^m)^3\|_{L^{4/3}(]0,T[\times\Omega)} = C\|u^m\|^3_{L^4(]0,T[\times\Omega)},$$

and this last sequence is bounded, since $(u^m)_{m\in\mathbb{N}}$ is bounded in $L^4(]0,T[\times\Omega)$.

Now, (3.37) implies that

$$u_t^m = P_m f + u^m - (u^m)^3 + \Delta u^m,$$

and, therefore, that also the sequence $(u_t^m)_{m\in\mathbb{N}}$ is bounded in $L^{4/3}(0,T;\mathcal{V}')$. From theorem A.82, we have that the injection

$$\{u \in L^2(0,T;\mathcal{V}): u_t \in L^{4/3}(0,T;\mathcal{V}')\}\hookrightarrow L^2(0,T;\mathcal{H})$$

is compact; hence, (3.43) follows. As a consequence of (3.43), $(u^m)^3 \to u^3$ in $L^1(]0,T[\times\Omega)$ strongly. This follows from the estimate

$$\|(u^m)^3 - u^3\|_{L^1(]0,T[\times\Omega)} \leq \|(u^m)^2 + u^m u + u^2\|_{L^2(]0,T[\times\Omega)}\|u^m - u\|_{L^2(]0,T[\times\Omega)},$$

and from the fact that the sequence $((u^m)^2)_{m \in \mathbb{N}}$ is bounded in $L^2(]0, T[\times \Omega)$, since

$$\|(u^m)^2\|_{L^2(]0,T[\times \Omega)} = \|u^m\|_{L^4(]0,T[\times \Omega)}^2 .$$

Therefore, for all $\psi \in L^\infty(]0, T[\times \Omega)$,

$$I_m := \int_0^T \int_\Omega (u^m(t,x))^3 \, \psi(t,x) \, dx \, dt \; \longrightarrow \; \int_0^T \int_\Omega (u(t,x))^3 \, \psi(t,x) \, dx \, dt . \tag{3.44}$$

On the other hand, since evidently $\psi \in L^4(]0, T[\times \Omega)$, and $(u^m)^3 \to \chi$ weakly in the dual space $L^{4/3}(]0, T[\times \Omega)$,

$$I_m \to \int_0^T \int_\Omega \chi(t,x) \, \psi(t,x) \, dx \, dt . \tag{3.45}$$

From (3.44) and (3.45) we deduce that for each $\psi \in L^\infty(]0, T[\times \Omega)$,

$$\int_0^T \int_\Omega (u(t,x))^3 \, \psi(t,x) \, dx \, dt = \int_0^T \int_\Omega \chi(t,x) \, \psi(t,x) \, dx \, dt .$$

Since $L^\infty(]0, T[\times \Omega)$ is isomorphic to the dual of $L^1(]0, T[\times \Omega)$, by the Hahn-Banach theorem (see theorem A.13), we conclude that $\chi = u^3$ in $L^1(]0, T[\times \Omega)$. This completes the proof of the existence part of theorem 3.9.

    3. To show the well-posedness of problem (P), we consider the difference $z = u - v$ of two solutions of (3.1), which solves the equation

$$z_t = z + \Delta z - (u^3 - v^3) . \tag{3.46}$$

Since

$$z + \Delta z \in L^2(0, T; \mathcal{V}'), \quad u^3 - v^3 \in L^{4/3}(]0, T[\times \Omega) ,$$

by proposition 3.7 we obtain

$$\frac{d}{dt}\|z\|^2 = 2 \, (z + \Delta z, z) - 2 \langle u^3 - v^3, u - v \rangle$$

almost everywhere in t. Since $2 \langle u^3 - v^3, u - v \rangle \geq 0$ (by monotonicity), we obtain that, for almost all t,

$$\frac{d}{dt}\|z(t)\|^2 \leq 2\|z(t)\|^2 - 2\|\nabla z(t)\|^2 \leq 2\|z(t)\|^2 . \tag{3.47}$$

Applying the comparison theorem (proposition 2.63), we deduce from (3.47) that

$$\|z(t)\|^2 \leq \|z(0)\|^2 e^{2t} ,$$

that is,

$$\|u(t) - v(t)\| \leq \|u(0) - v(0)\| \, e^t . \tag{3.48}$$

Thus, uniqueness and well-posedness in the large follow. ⬜

We can now define the solution operator generated by problem (P). Indeed, since T is arbitrary in theorem 3.9, we have that if $f \in L^2_{loc}(0,+\infty;\mathcal{V}')$, then $u \in C([0,+\infty[;\mathcal{H})$; thus, theorem 3.9 defines a unique global solution of problem (P). If f is independent of t, i.e. if $f(t) \equiv f \in \mathcal{V}'$, theorem 3.9 defines, by means of (3.12), a continuous semiflow S on $\mathcal{X} = \mathcal{H}$. This is the semiflow generated by problem (P). In particular, note that inequality (3.48) shows that each operator $S(t)$, $t \geq 0$, is Lipschitz continuous in $\mathcal{X}$.

Finally, we would like to mention a possible modification of the setting of problem (P), motivated by the fact that in order to proceed with the construction of the attractor for the semiflow S so defined, we shall need to assume that $f(t) \equiv f \in \mathcal{H}$ (as opposed to $\mathcal{V}'$). Thus, in theorem 3.9 we could assume that $f \in L^2(0,T;\mathcal{H})$; as we now show, this would allow us to simplify the choice of the space of test functions $\mathcal{W}_0(T)$ into

$$\tilde{\mathcal{W}}_0(T) := \{f \in L^2(0,T;\mathcal{V}): f_t \in L^2(0,T;\mathcal{H}), \ f(T,\cdot) = 0\}. \tag{3.49}$$

In fact, in this case we have the imbeddings

$$\tilde{\mathcal{W}}_0(T) \hookrightarrow C([0,T];H^{1/2}(\Omega)), \quad H^{1/2}(\Omega) \hookrightarrow L^3(\Omega),$$

the second of which holds because $n \leq 3$ (see theorems A.81 and A.74). Thus, we deduce that $\tilde{\mathcal{W}}_0(T) \hookrightarrow C([0,T];L^3(\Omega))$. By means of the interpolation inequality

$$|\varphi|_4 \leq C|\varphi|_6^{1/2}|\varphi|_3^{1/2},$$

(see theorem A.61), we deduce that $\tilde{\mathcal{W}}_0(T) \hookrightarrow L^4(]0,T[\times\Omega)$ as well. Consequently, if $\varphi \in \tilde{\mathcal{W}}_0(T)$, the product $u^3\varphi$ is again in $L^1(]0,T[\times\Omega)$, and equation (3.34) makes sense.

3.3.2 Step 2: Absorbing Sets

In this step we prove the existence of a bounded absorbing set $\mathcal{B}$ for S in $\mathcal{X}$, by establishing suitable a priori estimates on u in $\mathcal{X}$ (recall that $\mathcal{X} = \mathcal{H} = L^2(\Omega)$).

PROPOSITION 3.12
Assume $f \in C_b([0,+\infty[;\mathcal{X})$. There exist positive constants m_1 and λ_1, such that for all $u_0 \in \mathcal{X}$ and $t \geq 0$,

$$\|u(t)\|^2 \leq \left(\|u_0\|^2 - \tfrac{m_1}{2\lambda_1}\right)e^{-2\lambda_1 t} + \tfrac{m_1}{2\lambda_1}. \tag{3.50}$$

Consequently, there exists $R > 0$ such that the ball $B(0,R)$ in $\mathcal{X}$ is absorbing and positively invariant for S.

PROOF   Formally multiplying equation (3.1) by $2u$ (as in the proof of uniqueness in theorem 3.9, this can be justified by means of proposition 3.5), we obtain, dropping the argument t for convenience:

$$\frac{d}{dt}\|u\|^2 + 2\|\nabla u\|^2 + 2|u|_4^4 = 2\langle f+u,u\rangle \le C_1 + \|f\|^2 + |u|_4^4, \qquad (3.51)$$

where C_1 is a constant depending only on Ω. Recalling Poincaré's inequality (3.16), we deduce from (3.51)

$$\frac{d}{dt}\|u\|^2 + 2\lambda_1\|u\|^2 \le C_1 + \sup_{t\ge 0}\|f(t)\|^2 =: m_1. \qquad (3.52)$$

Inequality (3.52) is a linear differential inequality, like (2.62); since the function $t \mapsto \|u(t)\|^2$ is absolutely continuous, estimate (3.50) follows from (3.52), via proposition 2.64. We deduce then that for any $R > \sqrt{m_1/(2\lambda_1)}$, the ball $B(0,R)$ in $\mathcal{X}$ is absorbing and positively invariant for S. In particular, we find that

$$S(t)u_0 \in B(0,R)$$

for all $t \ge T_0$, with $T_0 = 0$ if $\|u_0\| \le R$ or, if $\|u_0\| > R$,

$$T_0 = \frac{1}{2\lambda_1}\ln\frac{\|u_0\|^2 - \frac{m_1}{2\lambda_1}}{R^2 - \frac{m_1}{2\lambda_1}}.$$

This concludes the proof of proposition 3.12; we mention that in chapter 4, section 4.3.1, we shall show the existence of a positively invariant ball $\mathcal{B}_1$, bounded and absorbing in $\mathcal{V}$. $\qquad\square$

3.3.3 Step 3: Compactness of the Solution Operator

We now establish the uniform compactness of the solution operator S for large t, by means of further a priori estimates on the solution of (3.1). These estimates show that $u(t) \in \mathcal{V}$ if t is sufficiently large. We assume again that $f \in C_b([0,+\infty[;\mathcal{X})$. In the sequel, we denote by C, C_1, C_2, $\ldots$ any generic positive constant, possibly depending on f or Ω, but not on u (recall convention 3.11). We formally multiply (3.1) in $\mathcal{X}$ by $-(e^t-1)\Delta u(t)$ and obtain

$$(e^t-1)\langle\nabla u_t(t),\nabla u(t)\rangle + (e^t-1)\|\Delta u(t)\|^2 - (e^t-1)\langle u^3(t),\Delta u(t)\rangle$$
$$= -(e^t-1)\langle f(t)+u(t),\Delta u(t)\rangle. \qquad (3.53)$$

This procedure is formal, because we do not know that $\Delta u(t) \in L^2(\Omega)$; in fact, this is even more than what we are trying to establish. To proceed rigorously, we should establish the estimates that follow for the Galerkin approximations of u, and then realize that the final estimate we obtain can be carried over to u itself. Proceeding from (3.53), we obtain

$$\frac{d}{dt}\left((e^t-1)\|\nabla u(t)\|^2\right) + 2(e^t-1)\|\Delta u(t)\|^2 + 6(e^t-1)\langle u^2(t)\nabla u(t),\nabla u(t)\rangle$$

$$= e^t \|\nabla u(t)\|^2 - 2(e^t - 1)\langle f(t) + u(t), \Delta u(t)\rangle$$
$$= \|\nabla u(t)\|^2 - (e^t - 1)\langle 2f(t) + 3u(t), \Delta u(t)\rangle$$
$$\leq \|\nabla u(t)\|^2 + C_2(e^t - 1)\left(\|f(t)\|^2 + \|u(t)\|^2\right) + (e^t - 1)\|\Delta u(t)\|^2 \,.$$

From this, estimating $\|u(t)\|$ by (3.50), we obtain

$$\frac{\mathrm{d}}{\mathrm{d}t}\left((e^t - 1)\|\nabla u(t)\|^2\right) + (e^t - 1)\|\Delta u(t)\|^2$$
$$\leq \|\nabla u(t)\|^2 + C_2(e^t - 1)\left(m_1 + \|u_0\|^2 + \tfrac{m_1}{\lambda_1}\right)\,. \tag{3.54}$$

Integration of (3.51) yields, in particular, that

$$2\int_0^t \|\nabla u(s)\|^2\,\mathrm{d}s \leq \|u_0\|^2 + m_1 t\,;$$

thus, from (3.54) we obtain (neglecting a positive term at its left side)

$$(e^t - 1)\|\nabla u(t)\|^2 \leq C_2\,e^t\left(m_1 + \|u_0\|^2 + \tfrac{m_1}{2\lambda_1}\right) =: C_3 e^t\,.$$

From this we deduce that, if e.g. $t > \ln 2$,

$$\|\nabla u(t)\|^2 \leq \frac{C_3\,e^t}{e^t - 1} \leq 2C_3\,.$$

This proves the asserted uniform compactness of the solution operator S for large t. For future reference, we remark that if $u_0 \in \mathcal{V}$, the same estimates (in fact, simpler, because we do not need the factor $e^t - 1$) would yield the regularity result

$$u \in C_b([0, +\infty[; \mathcal{V}) \cap L^2(0, +\infty; H^2(\Omega))$$

for the solution of problem (P).

3.3.4 Step 4: Conclusion

If problem (P) is autonomous, we can now deduce the existence of the global attractor for S, as a consequence of theorem 2.46. In conclusion, we have:

THEOREM 3.13
Let $u_0, f \in L^2(\Omega)$. The initial boundary value problem (P) defines a semiflow S in $\mathcal{X} = L^2(\Omega)$, which admits a global attractor in $\mathcal{X}$. This attractor is the set

$$\mathcal{A} := \omega(\mathcal{B}) = \bigcap_{s \geq 0} \overline{\bigcup_{t \geq s} S(t)\mathcal{B}}$$

(closure in $\mathcal{X}$), where $\mathcal{B}$ is the absorbing ball $B(0, R)$ determined in proposition 3.12.

Much more information is available on the properties of the global attractor $\mathcal{A}$ of the semiflow S generated by problem (P) obtained in theorem 3.13. Among these, we mention the following.

1. Structure of the Attractor.   In addition to all stationary solutions of problem
(P), that is, the solutions of the nonlinear elliptic boundary value problem

$$\begin{cases} -\Delta u + u^3 - u = f, \\ \qquad u|_{\partial\Omega} = 0, \end{cases} \tag{3.55}$$

$\mathcal{A}$ also contains the unstable manifolds (in $\mathcal{X}$) of all these stationary solutions (see
e.g. Babin-Vishik, [BV92, ch. 3]).

2. Attractors via α-contractions.   The existence of $\mathcal{A}$ can also be established by
means of theorems 2.56 and 2.55. Indeed, as we have already remarked, since S is
compact for large t, the operator $T = S(t_*)$ is an α-contraction if t_* is large enough.

3. Regularity of the Attractor.   The regularizing effect of the heat operator would
allow us to show that $\mathcal{A}$ is in fact contained and bounded in $\mathcal{V} \cap H^2(\Omega)$; thus, $\mathcal{A}$ is
compact not only in $\mathcal{X}$, but also in $\mathcal{V}$. For a proof of this result on the regularity of
the attractor, see e.g. Temam, [Tem88, ch. 3], or Babin-Vishik, [BV92, sct. 3.3].

4. Finite dimensionality of the Attractor.   The fractal dimension $\dim_F(\mathcal{A})$ of $\mathcal{A}$
is finite. While this can be proven directly (see e.g. Temam, [Tem88, sct. VI.2]), we
give an alternative proof of this result in (4.87) of section 4.5.6 in the next chapter,
where we prove the existence of an exponential attractor $\mathcal{E}$ for S. The finite dimen-
sionality of $\mathcal{A}$ is in fact a consequence of that of $\mathcal{E}$, and of the fact that $\mathcal{A} \subseteq \mathcal{E}$, which
implies that $\dim_F(\mathcal{A}) \leq \dim_F(\mathcal{E})$.

3.3.5 Backward Uniqueness

The heat equation cannot be solved backward in time directly. Indeed, even in
the autonomous case (i.e., with f independent of t), the usual change of variable
$v(t,x) := u(-t,x), t < 0$, transforms the heat equation (3.1) into the equation

$$v_t + \Delta v = f + g(v),$$

in which $+\Delta$ is a negative operator in $L^2(\Omega)$. Consequently, problem (P) actually
defines a semiflow only. Nevertheless, we can extend S to a flow, by means of proposition 2.5. This requires that each operator $S(t)$, $t > 0$, be invertible; in turn, this requires the semiflow generated by the heat equation to satisfy a BACKWARD UNIQUE-
NESS property, in the sense of the following

DEFINITION 3.14   *The semiflow S satisfies the* BACKWARD UNIQUENESS PROP-
ERTY *if whenever x, y $\in \mathcal{X}$ and t > 0 are such that S(t)x $=$ S(t)y, then x $=$ y.*

To this end, we first need the following property of the nonlinear term of the
equation:

PROPOSITION 3.15
The function $u \mapsto u^3$ is locally Lipschitz continuous from $\mathcal{V}$ into $\mathcal{H}$.

PROOF First we note that $u^3 \in \mathcal{H}$ if $u \in \mathcal{V}$. In fact,

$$\int_{\Omega} |u^3|^2 \, dx = \int_{\Omega} |u|^6 \, dx = |u|_6^6 \leq C \|u\|_{\mathcal{V}}^6,$$

where the constant $C > 0$ is determined by the imbedding $H_0^1(\Omega) \hookrightarrow L^6(\Omega)$. In the same way, we also see that, if u and $v \in \mathcal{V}$, the products u^2, uv and v^2 are in $L^3(\Omega)$. Thus, if u and v are in a ball of $\mathcal{V}$ of radius R, by Minkowski's inequality (see theorem A.57) we obtain

$$\|u^3 - v^3\|^2 = \int_{\Omega} |u - v|^2 |u^2 + uv + v^2|^2 \, dx$$

$$\leq \left(\int_{\Omega} |u - v|^6 \, dx \right)^{1/3} \left(\int_{\Omega} |u^2 + uv + v^2|^3 \, dx \right)^{2/3}$$

$$= |u - v|_6^2 |u^2 + uv + v^2|_3^2 \leq C \|u - v\|_{\mathcal{V}}^2 \left(\|u\|_{\mathcal{V}}^4 + \|v\|_{\mathcal{V}}^4 \right)$$

$$\leq C(R) \|u - v\|_{\mathcal{V}}^2.$$

This shows the asserted local Lipschitz continuity. ▯

We shall use proposition 3.15 also in the next section, for the proof of the uniqueness of solutions to problem (H_{ε}).

We can now prove the asserted backward uniqueness result:

PROPOSITION 3.16
Let $T > 0$, and assume that problem (P) has two solutions u, v, with

$$u, v \in C([0, T]; \mathcal{V}) \cap L^2(0, T; H^2(\Omega)), \quad u_t, v_t \in L^2(0, T; \mathcal{H}),$$

and $u(T, \cdot) = v(T, \cdot)$. Then, $u \equiv v$ on $[0, T]$.

PROOF We follow Temam, [Tem88, ch. 3.6]. Consider the difference $z = u - v$, which solves equation (3.46). Arguing by contradiction, assume that there is $t_0 \in [0, T[$ such that $z(t_0, \cdot) \neq 0$. Then, by continuity, there is a largest interval $[t_0, t_1] \subset [0, T[$ such that

$$z(t, \cdot) \neq 0 \quad \text{if } t_0 \leq t < t_1, \quad z(t_1, \cdot) = 0.$$

Let $M := \max_{t_0 \leq t \leq t_1} \|z(t, \cdot)\|$. On the interval $[t_0, t_1[$, the function

$$t \mapsto \ln \frac{M}{\|z(t, \cdot)\|}$$

is well defined, nonnegative and differentiable. Using equation (3.46), we compute that (omitting the variables; recall convention 3.10)

$$
\frac{\mathrm{d}}{\mathrm{d}t}\ln\frac{M}{\|z\|} = \frac{\mathrm{d}}{\mathrm{d}t}\left(-\tfrac{1}{2}\ln\|z\|^2\right) = -\tfrac{1}{2}\frac{2\langle z, z_t\rangle}{\|z\|^2} = -\frac{1}{\|z\|^2}\langle z, z - (u^3 - v^3) + \Delta z\rangle
$$

$$
= -1 + \frac{\|\nabla z\|^2}{\|z\|^2} + \frac{1}{\|z\|^2}\langle u^3 - v^3, z\rangle. \tag{3.56}
$$

Since $u, v \in C([0,T];\mathcal{V})$, by proposition 3.15 we can estimate

$$
\langle u^3 - v^3, z\rangle \leq C\|\nabla z\|\,\|z\|, \tag{3.57}
$$

with C depending on u and v. Setting

$$
\Lambda(t) := \frac{\|\nabla z(t, \cdot)\|^2}{\|z(t, \cdot)\|^2}, \tag{3.58}
$$

we deduce from (3.56) and (3.57) that

$$
\frac{\mathrm{d}}{\mathrm{d}t}\ln\frac{M}{\|z(t, \cdot)\|} \leq -1 + \Lambda(t) + C(\Lambda(t))^{1/2} \leq \tfrac{1}{4}C^2 + 2\Lambda(t). \tag{3.59}
$$

Integrating (3.59) in $[t_0, t]$, $t_0 < t < t_1$, we obtain

$$
0 \leq \ln\frac{M}{\|z(t, \cdot)\|} \leq \ln\frac{M}{\|z(t_0, \cdot)\|} + \tfrac{1}{4}C(t - t_0) + 2\int_{t_0}^{t}\Lambda(s)\,\mathrm{d}s =: \varphi(t). \tag{3.60}
$$

We shall show that φ remains bounded as $t \to t_1^-$; as a consequence, (3.60) yields the desired contradiction, since $\|z(t, \cdot)\| \to 0$ as $t \to t_1^-$.

To prove that φ is bounded, it is sufficient to show that Λ satisfies the differential inequality

$$
\frac{\mathrm{d}\Lambda}{\mathrm{d}t} \leq C\Lambda \tag{3.61}
$$

for almost all $t \in [t_0, t_1[$, with $C > 0$ independent of t. Indeed, (3.61) implies that

$$
\Lambda(t) \leq \Lambda(t_0)e^{C(t - t_0)},
$$

so that

$$
\int_{t_0}^{t}\Lambda(s)\,\mathrm{d}s \leq \Lambda(t_0)\tfrac{1}{C}\left(e^{C(t_1 - t_0)} - 1\right).
$$

To show (3.61), we first observe that (omitting again the variables, as in convention 3.10)

$$
\langle -\Delta z - \Lambda z, z\rangle = \langle -\Delta z - \Lambda z, \Lambda z\rangle = 0 \tag{3.62}
$$

(recall that Λ is independent of x). Next, recalling (3.46), we compute that

$$\frac{1}{2}\frac{d\Lambda}{dt} = \frac{1}{\|z\|^4}\left(\langle\nabla z,\nabla z_t\rangle\|z\|^2 - \|\nabla z\|^2\langle z,z_t\rangle\right)$$

$$= \frac{1}{\|z\|^2}\left(\langle-\Delta z,z_t\rangle - \Lambda\langle z,z_t\rangle\right)$$

$$= \frac{1}{\|z\|^2}\langle-\Delta z - \Lambda z, z - (u^3 - v^3) + \Delta z\rangle.$$

By (3.62) and (3.57), we can proceed with

$$\frac{1}{2}\frac{d\Lambda}{dt} = \frac{1}{\|z\|^2}\langle-\Delta z - \Lambda z, \Delta z + \Lambda z - (u^3 - v^3)\rangle$$

$$= \frac{1}{\|z\|^2}\left(-\|\Delta z + \Lambda z\|^2 + \langle\Delta z + \Lambda z, u^3 - v^3\rangle\right)$$

$$\le \frac{1}{\|z\|^2}\left(-\tfrac{1}{2}\|\Delta z + \Lambda z\|^2 + \tfrac{1}{2}\|u^3 - v^3\|^2\right)$$

$$\le \frac{1}{2\|z\|^2}C\|\nabla z\|^2,$$

from which (3.61) follows. This concludes the proof of the backward uniqueness for solutions of (3.1). □

3.4 The Hyperbolic Problem

In this section we consider the hyperbolic IBVP (H_ε). As mentioned in (3.13), the underlying space for the semiflow S generated by (3.4) is now the product space $\mathcal{X} := \mathcal{V} \times \mathcal{H}$, and each orbit is the image of a pair $(u(\cdot,u_0,u_1), u_t(\cdot,u_0,u_1))$. We also establish regularity results in the subspace $\mathcal{X}_1 := \mathcal{D} \times \mathcal{V}$.

We will consider in $\mathcal{X}$ various norms, having different weights depending on the parameter ε. All these norms will be equivalent to the graph norm in $\mathcal{X} = \mathcal{V} \times \mathcal{H}$, defined by

$$\|(u,v)\|_{\mathcal{X}}^2 := \|u\|_{\mathcal{V}}^2 + \varepsilon\|v\|_{\mathcal{H}}^2 = \|\nabla u\|^2 + \varepsilon\|v\|^2, \tag{3.63}$$

which is itself weighted with respect to ε. For this norm we have a result analogous to proposition 3.5:

PROPOSITION 3.17
Let $\mathcal{V}$ and $\mathcal{H}$ be as above, and $\varepsilon > 0$. Assume that the function $u \in L^2(0,T;\mathcal{V})$ is such that $u_{tt} \in L^2(0,T;\mathcal{V}')$ and $\varepsilon u_{tt} - \Delta u \in L^2(0,T;\mathcal{H})$. Then the function

$$R(t) := \varepsilon\|u_t(t)\|^2 + \|\nabla u(t)\|^2$$

is absolutely continuous on $[0,T]$, *and the identity*

$$\frac{\mathrm{d}}{\mathrm{d}t}\left(\varepsilon\|u_t(t)\|^2 + \|\nabla u(t)\|^2\right) = 2\langle\varepsilon u_{tt}(t) - \Delta u(t), u_t(t)\rangle \qquad (3.64)$$

holds in $\mathrm{D}'(0,T)$ *and, in fact, for almost all* $t \in [0,T]$.

A proof of this result when $\varepsilon = 1$ can be found in Temam, [Tem88, lem. II.4.1]; the case of variable ε follows in the same way. Note that the absolute continuity of R follows from (3.64), and the fact that both R and the function $t \mapsto 2\langle\varepsilon u_{tt}(t) - \Delta u(t), u_t(t)\rangle$ are in $\mathrm{L}^1(0,T)$.

3.4.1 Step 1: The Solution Operator

In this section we prove a global existence and well posedness result for problem (H_ε), which allows us to define the associated solution operator for any $\varepsilon > 0$. If the equation is autonomous, i.e. if f only depends on x, this solution operator is in fact a continuous semiflow. For simplicity, we consider in detail only the case when $0 < \varepsilon \le 1$.

DEFINITION 3.18 *Let* $T > 0$, $u_0 \in \mathcal{V}$, $u_1 \in \mathcal{H}$, $f \in \mathrm{L}^2(0,T;\mathcal{H})$. *A function* u *is a* WEAK SOLUTION *of problem* (H_ε), *with* $g(u) = u^3 - u$, *on the interval* $[0,T]$ *if* $u \in \mathrm{C}([0,T];\mathcal{V}) \cap \mathrm{C}^1([0,T];\mathcal{H})$, $u(0,\cdot) = u_0$, *and for all* $\varphi \in \tilde{W}_0(T)$ *(the space defined in (3.49))*,

$$\int_0^T \left(-\varepsilon\langle u_t, \varphi_t\rangle + \langle u_t, \varphi\rangle + \langle\nabla u, \nabla\varphi\rangle + \langle u^3 - u, \varphi\rangle\right) \mathrm{d}t$$
$$= \int_0^T \langle f, \varphi\rangle \mathrm{d}t + \varepsilon\langle u_1, \varphi(0)\rangle. \qquad (3.65)$$

Note that the left side of (3.65) makes sense, since $u^3 \in \mathrm{L}^2(0,T;\mathcal{H})$ if u belongs to $\mathrm{C}([0,T];\mathcal{V})$. In fact, we now have

$$\int_0^T \|u^3(t)\|^2 \mathrm{d}t \le \int_0^T |u(t)|_6^6 \mathrm{d}t \le \int_0^T \|u(t)\|_{\mathcal{V}}^6 \mathrm{d}t \le \|u\|_{\mathrm{C}([0,T];\mathcal{V})}^4 \int_0^T \|u(t)\|_{\mathcal{V}}^2 \mathrm{d}t.$$

We consider in $\mathcal{X}$ an ε-weighted norm whose square is defined by

$$E_0(u,v) := \varepsilon\|v\|^2 + \varepsilon\langle u, v\rangle + \tfrac{1}{2}\|u\|^2 + \|\nabla u\|^2. \qquad (3.66)$$

This is indeed an equivalent norm, since by Schwarz' inequality

$$\tfrac{1}{2}\|(u,v)\|_{\mathcal{X}}^2 \le E_0(u,v) \le \alpha\|(u,v)\|_{\mathcal{X}}^2, \qquad (3.67)$$

where

$$\alpha := \max\left\{\tfrac{3}{2}, \tfrac{1}{\lambda_1} + 1\right\}, \qquad (3.68)$$

with λ_1 as in (3.16). In analogy to proposition 3.17, we have

PROPOSITION 3.19
Let $\mathcal{V}$ and $\mathcal{H}$ be as above, and $0 < \varepsilon \leq 1$. Assume that the function $u \in L^2(0,T;\mathcal{V})$ is such that $u_{tt} \in L^2(0,T;\mathcal{V}')$, and $\varepsilon u_{tt} - \Delta u \in L^2(0,T;\mathcal{H})$. Then the function $t \mapsto E_0(u(t), u_t(t))$ is absolutely continuous on $[0,T]$, and satisfies the identity

$$\frac{d}{dt} E_0(u(t), u_t(t)) = \langle \varepsilon u_{tt}(t) - \Delta u(t), 2u_t(t) + u(t) \rangle$$
$$+ \varepsilon \|u_t(t)\|^2 + \langle u_t(t), u(t) \rangle - \|\nabla u(t)\|^2$$

in $D'(0,T)$ and, in fact, almost everywhere in $[0,T]$.

We have then the following result:

THEOREM 3.20
For all $\varepsilon \in]0,1]$, $f \in L^2(0,T;\mathcal{H})$, $u_0 \in \mathcal{V}$ and $u_1 \in \mathcal{H}$, there exists a unique weak solution u of problem (H_ε). u depends continuously on the initial data u_0, u_1, and the function $t \mapsto E_0(u(t), u_t(t))$ is absolutely continuous on $[0,T]$. If in addition $f_t \in L^2(0,T;\mathcal{H})$, $u_0 \in \mathcal{D}$ and $u_1 \in \mathcal{V}$, then

$$u \in C([0,T];\mathcal{D}) \cap C^1([0,T];\mathcal{V}) \cap C^2([0,T];\mathcal{H}),$$

and the function
$$t \mapsto \|\Delta u(t)\|^2 + \|u_t(t)\|^2 + \|\varepsilon u_{tt}(t)\|^2$$

is absolutely continuous on $[0,T]$.

PROOF   A proof of the existence and regularity part can again be given by means of a standard Galerkin approximation method, as in Lions, [Lio69], or Temam, [Tem88, ch. IV]. Since the procedure is similar to the one we followed for the parabolic problem (P), we do not repeat the details. In particular, the absolute continuity of the map $t \mapsto E_0(u(t), u_t(t))$ is a consequence of proposition 3.19.

It is worth noting explicitly that in order to obtain uniform estimates on the norm E_0, the standard practice of multiplying the equation of (3.4) by $2u_t$ in $L^2(\Omega)$ is supplemented by a second multiplication of the equation by λu, where λ is chosen conveniently. In our case, we choose either $\lambda = 1$ or $\lambda = \frac{1}{\varepsilon}$, according to the type of estimate that we need to establish. Moreover, we shall make constant use of the Sobolev imbeddings and interpolation inequalities reported in theorems A.74 and A.61.

To prove the uniqueness of the solutions of (3.4), and the continuity of each operator $S(t)$, we consider two solutions of (3.4), whose difference z satisfies the equation

$$\varepsilon z_{tt} + z_t - \Delta z = g(v) - g(u), \tag{3.69}$$

in the distributional sense of definition (3.18). Since the function $u \mapsto u^3$ is monotone, by propositions 3.19 and 3.15 we obtain (dropping the argument t and using different constants C)

$$
\begin{aligned}
\frac{d}{dt} E_0(z, z_t) &= -(2 - \varepsilon)\|z_t\|^2 - \|\nabla z\|^2 - \langle g(u) - g(v), 2z_t + z \rangle \\
&\leq 2\|g(u) - g(v)\| \|z_t\| + \|z\|^2 \leq C\|\nabla z\|^2 + \|z_t\|^2 + (1 + C)\|z\|^2 \\
&\leq C(\|\nabla z\|^2 + \|z\|^2) \leq C_1 E_0(z, z_t),
\end{aligned}
$$

where the constants C and C_1 depend on u and v. Applying proposition 2.63, we obtain

$$
E_0(z(t), z_t(t)) \leq E_0(z(0), z_t(0)) e^{C_1 t}, \tag{3.70}
$$

which implies the uniqueness and well-posedness in the large of solutions of (3.4). $\square$

Since T is arbitrary in theorem 3.20, we obtain that if $f \in L^2_{loc}(0, +\infty; \mathcal{H})$, then

$$
u \in C([0, +\infty[; \mathcal{V}) \cap C^1([0, +\infty[; \mathcal{H}).
$$

Thus, theorem 3.20 defines a unique global solution of problem (H_ε). If f is independent of t, theorem 3.20 shows that problem (H_ε) generates a continuous semiflow S on the space $\mathcal{X} = \mathcal{V} \times \mathcal{H}$, defined by (3.13). In particular, each operator $S(t)$ is Lipschitz continuous, since from (3.70) we deduce that, for each fixed $t > 0$,

$$
E_0(S(t)(u_0, u_1) - S(t)(v_0, v_1)) \leq E_0((u_0, u_1) - (v_0, v_1)) e^{C_1 t}. \tag{3.71}
$$

We also remark that the change of variable $t \to -t$ transforms the equation of (3.4) into

$$
\varepsilon u_{tt} - u_t - \Delta u + g(u) = \tilde{f},
$$

with $\tilde{f}(t, x) := f(-t, x)$. It is then easy to see that theorem 3.20 still holds, provided that f is defined in $[-T, 0]$. Thus, if $f \in C_b(\mathbb{R}; \mathcal{H})$, the solution of (3.4) is defined for all time. If in particular $f(t) \equiv f \in \mathcal{H}$ is independent of time, S is a flow. This is in sharp contrast to the parabolic problem, where S was only a semiflow. In fact, each operator $S(t)$ is an isomorphism of $\mathcal{X}$ into itself, and also of $\mathcal{X}_1$ into itself. (This last assertion can be proved by means of further a priori estimates on the solution of (3.4); see e.g. Temam, [Tem88, ch. IV], for details.)

3.4.2 Step 2: Absorbing Sets

As in section 3.3.2, we now proceed to establish time-independent a priori estimates on the solution u of (3.4). In order to take care of the nonlinear term u^3, for $(u, v) \in \mathcal{X}$ we introduce the function

$$
N_0(u, v) := E_0(u, v) + \tfrac{1}{2}|u|_4^4 = \varepsilon\|v\|^2 + \varepsilon\langle u, v \rangle + \tfrac{1}{2}\|u\|^2 + \|\nabla u\|^2 + \tfrac{1}{2}|u|_4^4. \tag{3.72}
$$

Because of (3.67), N_0 is positive definite. We claim:

PROPOSITION 3.21
Assume $f \in C_b([0,+\infty[;\mathcal{H})$. There exists $R_0 > 0$ such that the set

$$\mathcal{B}_0 := \{(u,v) \in \mathcal{X}: N_0(u,v) \le R_0^2\}$$

is bounded, absorbing and positively invariant for the solution operator S in $\mathcal{X} = \mathcal{V} \times \mathcal{H}$. (Note that $\mathcal{B}_0$ is not a ball of $\mathcal{X}$.)

PROOF   Multiplying the equation of (3.4) in $\mathcal{H}$ by $2u_t + u$ (this is justified by proposition 3.19), we obtain (dropping the argument t)

$$\frac{\mathrm{d}}{\mathrm{d}t}\left(\varepsilon\|u_t\|^2 + \varepsilon\langle u_t,u\rangle + \tfrac{1}{2}\|u\|^2 + \|\nabla u\|^2 + \tfrac{1}{2}|u|_4^4\right)$$
$$+ (2-\varepsilon)\|u_t\|^2 + \|\nabla u\|^2 + |u|_4^4 = \langle f+u, 2u_t+u\rangle. \qquad (3.73)$$

Setting $F := \sup_{t \ge 0}\|f(t)\|^2$, we estimate

$$\langle f, 2u_t+u\rangle \le 4F + \tfrac{1}{4}\|u_t\|^2 + F + \tfrac{1}{4}\|u\|^2 \le C_1 + \tfrac{1}{4}\|u_t\|^2 + \tfrac{1}{4}|u|_4^4;$$
$$\langle u, 2u_t+u\rangle \le \tfrac{1}{4}\|u\|^2 + \tfrac{1}{4}\|u_t\|^2 + \|u\|^2 \le C_2 + \tfrac{1}{4}|u|_4^4 + \tfrac{1}{4}\|u_t\|^2.$$

Thus, from (3.73) and (3.67) we obtain that

$$\frac{\mathrm{d}}{\mathrm{d}t}N_0(u,u_t) + \tfrac{1}{2}\left(\|u_t\|^2 + \|\nabla u\|^2 + |u|_4^4\right) \le C_3.$$

Recalling then (3.67), and that $\varepsilon \le 1 < \alpha$, we deduce the linear differential inequality

$$\frac{\mathrm{d}}{\mathrm{d}t}N_0(u,u_t) + \tfrac{1}{2\alpha}N_0(u,u_t) \le C.$$

Since the function $t \mapsto N_0(u(t),u_t(t))$ is absolutely continuous, by proposition 2.64 we deduce that for all $t \ge 0$,

$$N_0(u(t),u_t(t)) \le (N_0(u_0,u_1) - 2\alpha C)e^{-t/2\alpha} + 2\alpha C. \qquad (3.74)$$

This implies not only the boundedness of the function $t \mapsto (u(t),u_t(t))$ on $[0,+\infty[$, but also that, if $R_0^2 > 2\alpha C$, each corresponding set $\mathcal{B}_0$ is as claimed. In particular, if $N_0(u_0,u_1) > 2\alpha C$, we have that

$$N_0(u(t),u_t(t)) \in \mathcal{B}_0$$

for all $t \ge T_0$, where $T_0 = 0$ if $N_0(u_0,u_1) \le R_0^2$, and

$$T_0 := 2\alpha \ln\frac{N_0(u_0,u_1) - 2\alpha C}{R_0^2 - 2\alpha C}$$

otherwise. This ends the proof of proposition 3.21.   ☐

To conclude, we mention that in section 4.4.1 of chapter 4 we will show that if ε is sufficiently small, S also admits a positively invariant set $\mathcal{B}_1$, bounded and absorbing in $\mathcal{X}_1$. On the other hand, if $\varepsilon > 1$ we can still establish the existence of an absorbing set for S, with the same proof; however, instead of (3.66), we have to consider a differently weighted norm, whose square is defined by

$$(u,v) \mapsto \varepsilon \|v\|^2 + \langle u,v \rangle + \tfrac{1}{2\varepsilon}\|u\|^2 + \|\nabla u\|^2$$

(otherwise, E_0 is not necessarily positive definite), and the a priori estimates of proposition 3.21 are obtained by multiplying the equation of (3.4) by $2u_t$ and $\tfrac{1}{\varepsilon}u$.

3.4.3   Step 3: Compactness of the Solution Operator

Since in the hyperbolic case the solution operator S does not enjoy any smoothing property for $t > 0$ (on the contrary, S is an isomorphism in $\mathcal{X}$), we proceed instead to establish a decomposition of S like in proposition 2.48, under the condition that ε be sufficiently small. That is, we construct families S_1 and S_2, with S_1 uniformly compact for large t, S_2 uniformly decaying to 0, and $S = S_1 + S_2$. This construction is presented in Babin-Vishik, [BV92, sct. II.6], for the case when $\varepsilon = 1$ and g is a more general nonlinearity, satisfying conditions (3.9) and (3.10). Here, the smallness of ε allows us to considerably simplify their proofs.

The family $S_2 = (S_2(t))_{t \geq 0}$ of operators on $\mathcal{X} = V \times \mathcal{H}$ is defined as follows. Given $(u_0, u_1) \in \mathcal{X}$ and $t \geq 0$, $S_2(t)(u_0, u_1)$ is the pair of functions in $\mathcal{X}$ defined by

$$\Omega \ni x \mapsto (S_2(t)(u_0,u_1))(x) := (v(t,x,u_0,u_1), v_t(t,x,u_0,u_1)),$$

where $v(\cdot,\cdot,u_0,u_1)$ is the solution of the IBVP

$$\begin{cases} \varepsilon v_{tt} + v_t - \Delta v + v^3 = 0 \\ v(0,\cdot) = u_0, \quad v_t(0,\cdot) = u_1 \\ v_{|\partial\Omega} = 0. \end{cases} \tag{3.75}$$

Given this function v, we define a second function $w(\cdot,\cdot,u_0,u_1)$ as the solution of the IBVP

$$\begin{cases} \varepsilon w_{tt} + w_t - \Delta w = f + v^3 - g(v+w) \\ w(0,\cdot) = 0, \quad w_t(0,\cdot) = 0 \\ w_{|\partial\Omega} = 0. \end{cases} \tag{3.76}$$

Since w depends on v, which in turn depends on (u_0,u_1), problem (3.76) defines a family $S_1 = (S_1(t))_{t \geq 0}$ of operators on $\mathcal{X}$, by

$$\Omega \ni x \mapsto (S_1(t)(u_0,u_1))(x) := (w(t,x,u_0,u_1), w_t(t,x,u_0,u_1)),$$

for $(u_0,u_1) \in \mathcal{X}$ and $(t,x) \in [0,+\infty[\times \Omega$. If v and w solve problems (3.75) and (3.76), we deduce that the function $u := w + v$ solves problem (H_ε). Hence, the decomposition $S = S_1 + S_2$ does hold. We proceed then to show that the operators S_1 and S_2 so defined on $\mathcal{X}$ satisfy the requirements of the decomposition (2.38). Note that S_2 is a semiflow, but S_1 is not, because problem (3.76) is not autonomous.

We now prove the uniform decay of S_2 on bounded sets of $\mathcal{X}$.

PROPOSITION 3.22

Let $(u_0,u_1) \in \mathcal{V} \times \mathcal{H}$. Problem (3.75) has, for all $\varepsilon \in]0,1]$, a unique solution

$$v \in C([0,+\infty[;\mathcal{V}) \cap C^1([0,+\infty[;\mathcal{H}).$$

If α is defined as in (3.68), v satisfies the estimate

$$N_0(v(t),v_t(t)) \le N_0(u_0,u_1)e^{-t/2\alpha}. \tag{3.77}$$

PROOF Only (3.77) needs to be proven. The procedure is identical to the one we followed to obtain (3.74). Multiplying the equation in (3.75) by $2v_t$ and v, and adding the resulting identities, we now obtain

$$\frac{d}{dt}N_0(v,v_t) + \frac{1}{2\alpha}N_0(v,v_t) \le 0,$$

from which (3.77) follows. ◻

The uniform compactness of S_1 for large t, at least when ε is sufficiently small, is a consequence of the second part of the following proposition.

PROPOSITION 3.23

Let $(u_0,u_1) \in \mathcal{V} \times \mathcal{H}$, and let v be the function provided by proposition 3.22. Problem (3.76) has, for all $\varepsilon \in]0,1]$, a unique solution $w \in C([0,+\infty[;\mathcal{V}) \cap C^1([0,+\infty[;\mathcal{H})$. If in addition $f_t \in C_b([0,+\infty[;\mathcal{H})$, there exists $\varepsilon_0 \in]0,1]$ such that for all $\varepsilon \in]0,\varepsilon_0]$ and all $t \ge 0$,

$$w(t) \in H^{5/4}(\Omega), \quad w_t(t) \in H^{1/4}(\Omega). \tag{3.78}$$

Moreover, there is $M > 0$, independent of w and ε, such that for all $t \ge 0$,

$$\varepsilon \|w_t(t)\|^2_{H^{1/4}(\Omega)} + \|w(t)\|^2_{H^{5/4}(\Omega)} \le M^2. \tag{3.79}$$

PROOF Only (3.78) and (3.79) need to be proved. In the estimates that follow, in accord with our convention 3.11, we denote by C, C_1, $\ldots$, various generic positive constants which are independent of t and of the right side of (3.79). In particular, we recall that the norms defined by $E_0(v,v_t)$ and $E_0(w,w_t)$ are bounded with respect to

t, because of propositions 3.21 and 3.22. More precisely, consider the absorbing set $\mathcal{B}_0$ for S whose existence is assured by proposition 3.21. Recalling (3.74) and (3.77), and of course that $w = u - v$, by (3.67) we deduce that there exists C_1, depending on the norm (3.63) of (u_0, u_1) in $\mathcal{X}$, such that for all $t \geq 0$

$$E_0(v(t), v_t(t)) + E_0(w(t), w_t(t)) \leq C_1^2. \tag{3.80}$$

Note that, because of (3.67), and $\varepsilon \leq 1$, C_1 can be chosen independent of ε.

We now show that w, solution of (3.76), is in fact more regular than v, namely that (3.78) holds for each $t \geq 0$. This is essentially due to the fact that the initial values in (3.76) are (evidently!) more regular than (u_0, u_1), and that the highest order term v^3 does not appear in the equation of (3.76). To achieve this, we establish an additional estimate on a higher norm for w, which we define by resorting to the fractional powers of the operator L introduced in (5) of section 3.2.1. More precisely, recalling that L is self-adjoint and positive, we can define its fractional powers on $\mathcal{H}$ by means of its spectral basis (see section 3.2, as well as section A.5.5). In particular, we have the imbeddings

$$\mathrm{dom}(L^{i/8}) \hookrightarrow H^{i/4}(\Omega) \hookrightarrow L^{12/(6-i)}(\Omega), \tag{3.81}$$

the first of which follows from theorem A.78, and the second, which is valid for $0 \leq i < 6$ (recall that $n \leq 3$), from theorem A.69. We consider then in $\mathcal{X}_1 = \mathcal{D} \times V$ a norm which has a different weight with respect to ε: more precisely, for $(u, v) \in \mathcal{X}_1$ we define

$$E_\varepsilon(u, v) := \varepsilon \|L^{1/8} v\|^2 + \langle L^{1/8} v, L^{1/8} u \rangle + \tfrac{1}{2\varepsilon} \|L^{1/8} u\|^2 + \|L^{5/8} u\|^2.$$

Note that E_ε is indeed the square of an equivalent norm in $H^{5/4}(\Omega) \times H^{1/4}(\Omega)$. In fact, from (A.30) and (A.71) we immediately obtain that

$$\lambda_1 \|L^{1/8} u\|^2 \leq \|L^{5/8} u\|^2; \tag{3.82}$$

consequently, we easily deduce that, for all $(u, v) \in \mathcal{X}_1$,

$$\tfrac{1}{2} \left(\varepsilon \|L^{1/8} v\|^2 + \|L^{5/8} u\|^2 \right) \leq E_\varepsilon(u, v) \leq \tfrac{\alpha}{\varepsilon} \left(\varepsilon \|L^{1/8} v\|^2 + \|L^{5/8} u\|^2 \right), \tag{3.83}$$

where α is as in (3.68).

Our goal is to show that the function $t \mapsto E_\varepsilon(w(t), w_t(t))$ is bounded, uniformly in t and ε. To this end, we multiply the equation in (3.76) by $2L^{1/4} w_t + \tfrac{1}{\varepsilon} L^{1/4} w$. Setting

$$\Phi := f + v^3 - g(v + w), \tag{3.84}$$

we obtain (omitting the variable t)

$$\frac{\mathrm{d}}{\mathrm{d}t} E_\varepsilon(w, w_t) + \|L^{1/8} w_t\|^2 + \tfrac{1}{\varepsilon} \|L^{5/8} w\|^2 = \langle \Phi, 2L^{1/4} w_t + \tfrac{1}{\varepsilon} L^{1/4} w \rangle. \tag{3.85}$$

We rewrite (3.85) as

$$\frac{d}{dt}\left(E_\varepsilon(w,w_t) - 2\langle\Phi, L^{1/4}w\rangle\right) + \|L^{1/8}w_t\|^2 + \frac{1}{\varepsilon}\|L^{5/8}w\|^2 - \frac{1}{\alpha}\langle\Phi, L^{1/4}w\rangle$$
$$= -2\langle\Phi_t, L^{1/4}w\rangle + \left(\frac{1}{\varepsilon} - \frac{1}{\alpha}\right)\langle\Phi, L^{1/4}w\rangle =: \Lambda. \tag{3.86}$$

To estimate Λ, we show that there exist positive constants C_2 and C_3, depending on f and C_1 of (3.80), but not on ε, such that

$$2\langle\Phi, L^{1/4}w\rangle \leq C_2, \tag{3.87}$$
$$2\langle\Phi_t, L^{1/4}w\rangle \leq \frac{3}{2}C_3 + \frac{1}{2\varepsilon}\|L^{5/8}w\|^2 + \frac{1}{2}\|L^{1/8}w_t\|^2. \tag{3.88}$$

Indeed, since

$$\Phi = f - 3v^2w - 3vw^2 - w^3 + v + w, \tag{3.89}$$

at first we have that

$$2\langle\Phi, L^{1/4}w\rangle \leq C(\|f\| + \|v^2w + vw^2 + w^3\| + \|v + w\|)\|L^{1/4}w\|$$
$$\leq C(\|f\| + |v|_6^2|w|_6 + |v|_6|w|_6^2 + |w|_6^3 + \|v + w\|)\|\nabla w\|. \tag{3.90}$$

From this, (3.87) follows, recalling (3.80). Next, we compute from (3.89) that

$$\Phi_t = f_t - (6vw + 3w^2 - 1)v_t - (3v^2 + 6vw + 3w^2 - 1)w_t. \tag{3.91}$$

Therefore, setting $F_1 := \max_{t\geq 0}\|f_t(t)\|$, we can estimate

$$2\langle\Phi_t, L^{1/4}w\rangle \leq 2F_1\|L^{1/4}w\| + C\left(|v|_6|w|_{12} + |w|_8^2 + 1\right)|v_t|_2|L^{1/4}w|_4$$
$$+ C\left(|v|_6^2 + |v|_6|w|_6 + |w|_6^2 + 1\right)|w_t|_{12/5}|L^{1/4}w|_4. \tag{3.92}$$

We now resort to the interpolation inequality

$$|w|_8 \leq C|w|_6^{1/2}|w|_{12}^{1/2}$$

(see theorem A.61), as well as to the imbedding inequalities (3.81) with $i = 3$, $i = s$ and $i = 1$, which yield

$$|L^{1/4}w|_4 \leq C\|L^{1/4}w\|_{H^{3/4}(\Omega)} \leq C\|L^{1/4}w\|_{\text{dom}(L^{3/8})} \leq C\|L^{5/8}w\|,$$
$$|w|_{12} \leq C\|w\|_{H^{5/4}(\Omega)} \leq C\|L^{5/8}w\|,$$
$$|w_t|_{12/5} \leq C\|w_t\|_{H^{1/4}(\Omega)} \leq C\|L^{1/8}w_t\|.$$

Thus, recalling also (3.82), we can proceed from (3.92) with

$$2\langle\Phi_t, L^{1/4}w\rangle \leq \left(\frac{2F_1}{\lambda_1} + C\|v_t\| + C\|w_t\|\right)\|L^{5/8}w\|$$
$$+ C(\|\nabla v\| + \|\nabla w\|)\|v_t\|\|L^{5/8}w\|^2$$

$$+C\left(\|\nabla v\|+\|\nabla w\|\right)\|\nabla w\|\,\|L^{5/8}w\|\,\|L^{1/8}w_t\| \tag{3.93}$$
$$=: R.$$

Recalling (3.80), we can estimate

$$R \le C\frac{1}{\sqrt{\varepsilon}}\|L^{5/8}w\|+C\frac{1}{\sqrt{\varepsilon}}\|L^{5/8}w\|^2+C\|L^{5/8}w\|\,\|w_t\|$$
$$\le \tfrac{3}{2}C^2+\tfrac{1}{6\varepsilon}\|L^{5/8}w\|^2+C\frac{1}{\sqrt{\varepsilon}}\|L^{5/8}w\|^2+\tfrac{1}{6\varepsilon}\|L^{5/8}w\|^2+\tfrac{3}{2}C^2\varepsilon\|L^{1/8}w_t\|^2. \tag{3.94}$$

At this point, we define

$$\varepsilon_0 := \min\{\tfrac{1}{36C^2},1\}.$$

Then, if $\varepsilon \le \varepsilon_0$ (this is our smallness assumption on ε), we deduce from (3.94) that

$$R \le \tfrac{3}{2}C_3+\tfrac{1}{2\varepsilon}\|L^{5/8}w\|^2+\tfrac{1}{2}\|L^{1/8}w_t\|^2\,;$$

inserting this into (3.93), we obtain (3.88), with $C_3 = \tfrac{3}{2}C^2$ (independent of ε).

Inserting (3.87) and (3.88) into (3.86), we obtain

$$\frac{\mathrm{d}}{\mathrm{d}t}\left(E_\varepsilon(w,w_t)-2\langle\Phi,L^{1/4}w\rangle\right)+\tfrac{1}{2}\|L^{1/8}w_t\|^2+\tfrac{1}{2\varepsilon}\|L^{5/8}w\|^2-\tfrac{1}{\alpha}\langle\Phi,L^{1/4}w\rangle \le C_4\,;$$

recalling (3.83) we deduce the linear differential inequality

$$\frac{\mathrm{d}}{\mathrm{d}t}\left(E_\varepsilon(w,w_t)-2\langle\Phi,L^{1/4}w\rangle\right)+\tfrac{1}{2\alpha}\left(E_\varepsilon(w,w_t)-2\langle\Phi,L^{1/4}w\rangle\right) \le C_4. \tag{3.95}$$

Because of the initial conditions of (3.76), by proposition 2.63 we obtain that for all $t \ge 0$,

$$E_\varepsilon(w(t),w_t(t))-2\langle\Phi(t),L^{1/4}w(t)\rangle \le 2\alpha C_4\,;$$

therefore, by (3.87),

$$E_\varepsilon(w(t),w_t(t)) \le 2\alpha C_4+C_2.$$

Recalling (3.83) and (3.67) we conclude that, for all $t \ge 0$,

$$\varepsilon\|L^{1/8}w_t(t)\|^2+\|L^{5/8}w(t)\|^2 \le 2(2\alpha C_4+C_2) =: C_5,$$

with C_5 independent of t and ε. This means that the orbit

$$\{(w(t),w_t(t)): t \ge 0\} = \{S_1(t)(u_0,u_1): t \ge 0\}$$

lies in a bounded set of

$$\mathrm{dom}(L^{5/8}) \times \mathrm{dom}(L^{1/8}) = \mathrm{H}_0^{5/4}(\Omega) \times \mathrm{H}^{1/4}(\Omega)\,;$$

thus, (3.78) and (3.79) follow, and the proof of proposition 3.23 is complete. □

Since $\mathrm{H}^{5/4}(\Omega) \times \mathrm{H}^{1/4}(\Omega)$ is compactly imbedded in $\mathcal{X}$, we conclude that the family S_1 is uniformly compact for large t.

3.4.4 Step 4: Conclusion

If problem (H_ε) is autonomous, we can now deduce the existence of the global attractor for S, as a consequence of theorem 2.50. In conclusion, we have:

THEOREM 3.24
Let $u_0 \in H_0^1(\Omega)$, and u_1, $f \in L^2(\Omega)$. If ε is sufficiently small, the initial boundary value problem (H_ε) defines a semiflow S in $\mathcal{X} = H_0^1(\Omega) \times L^2(\Omega)$. S admits a global attractor in $\mathcal{X}$, given by the set

$$\mathcal{A} = \mathcal{A}^\varepsilon = \omega(\mathcal{B}_0) = \bigcap_{s \geq 0} \overline{\bigcup_{t \geq s} S(t)\mathcal{B}_0},$$

where $\mathcal{B}_0$ is the absorbing set described in proposition 3.21.

As in the parabolic case, the global attractor $\mathcal{A}^\varepsilon$ so obtained contains, in addition to all stationary solutions of problem (H_ε), that is, the solutions of the elliptic problem (3.55), also the unstable manifolds (in $\mathcal{X}$) of all these stationary solutions (see e.g. Babin-Vishik, [BV92, sct. III.3]). The existence of $\mathcal{A}^\varepsilon$ can also be established by means of α-contractions, and $\mathcal{A}^\varepsilon$ is contained and bounded in the "more regular" space $\mathcal{X}_1 = \mathcal{D} \times \mathcal{V}$. We show these two results in the next sections. Finally, the fractal dimension of $\mathcal{A}^\varepsilon$ is finite.

3.4.5 Attractors via α-Contractions

In this section we show how to establish the existence of the global attractor, again if ε is sufficiently small, by means of theorem 2.56 (note that $\omega(\mathcal{B}_0)$ is not empty, since it contains the stationary solutions of (3.4)).

By proposition 3.21 the semiflow S admits a bounded, positively invariant absorbing set $\mathcal{B}_0$. Recalling proposition 2.59, to apply theorem 2.56 it is sufficient to find an appropriate pseudometric δ on $\mathcal{X}$, and a number $t_* > 0$, such that condition (2.47) holds, with $T = S(t_*)$. This is the goal of our next argument, for which the positive invariance of $\mathcal{B}_0$ is essential.

For fixed $T > 0$ we define in $\mathcal{X} \times \mathcal{X}$ the function

$$\delta_T((u,v),(\bar{u},\bar{v})) := \left(\int_0^T \|P_1(S(t)(u,v)) - P_1(S(t)(\bar{u},\bar{v}))\|^2 \, dt \right)^{1/2}, \qquad (3.96)$$

where P_1 is the projection from $\mathcal{V} \times \mathcal{H}$ onto $\mathcal{V}$. We explicitly remark that, although $P_1(S(t)(u,v))$ and $P_1(S(t)(\bar{u},\bar{v}))$ are in $\mathcal{V}$, in (3.96) we consider the norm of their difference in $\mathcal{H}$. We claim:

PROPOSITION 3.25
For each $T > 0$, δ_T is a pseudometric on $\mathcal{X}$, precompact on $\mathcal{B}_0$ with respect to the norm of $\mathcal{X}$ defined by E_0.

PROOF   The finiteness of δ_T follows from the invariance of $\mathcal{B}_0$, since if $(u,v) \in \mathcal{B}_0$, then so is the whole arc $\{S(t)(u,v) \colon 0 \leq t \leq T\}$. In particular, this means that the set $\{u(t) \colon 0 \leq t \leq T\}$ is bounded in $\mathcal{V}$, and therefore in $\mathcal{H}$. The other conditions of a pseudometric are immediately verified. To see that δ_T is precompact on $\mathcal{B}_0$, we need to show that, given any sequence $((u^n, v^n))_{n \in \mathbb{N}} \subset \mathcal{B}_0$ (thus, bounded), there is a subsequence, still denoted $((u^n, v^n))_{n \in \mathbb{N}}$, which converges relative to δ_T. Because of (3.96), this amounts to show that the subsequence $(u^n)_{n \in \mathbb{N}}$ converges in $L^2(0, T; \mathcal{H})$. Thus, let $((u^n, v^n))_{n \in \mathbb{N}} \subset \mathcal{B}_0$. Since $\mathcal{B}_0$ is positively invariant, the corresponding orbits $\{S(t)(u^n, v^n) \colon t \geq 0\}$ remain in $\mathcal{B}_0$. Since $\mathcal{B}_0$ is bounded, (3.67) implies that, setting

$$w^n(t) := P_1(S(t)(u^n, v^n)),$$

there is $M > 0$, independent of ε, such that for all $t \geq 0$ and $n \in \mathbb{N}$,

$$\|\nabla w^n(t)\|^2 + \varepsilon \|w_t^n(t)\|^2 \leq M^2. \tag{3.97}$$

We now recall from theorem A.82 that, since the injection of $\mathcal{V}$ into $\mathcal{H}$ is compact, the injection of

$$\tilde{\mathcal{W}}(T) := \{u \in L^2(0, T; \mathcal{V}) \colon u_t \in L^2(0, T; \mathcal{H})\}$$

into $L^2(0, T; \mathcal{H})$ is also compact. Hence, since (3.97) implies that $(u^n)_{n \in \mathbb{N}}$ is bounded in $\tilde{\mathcal{W}}(T)$, there is a subsequence of $(u^n)_{n \in \mathbb{N}}$ which converges in $L^2(0, T; \mathcal{H})$, as desired.   □

Our final step is to estimate the difference of two solutions of (3.4) in a sharper way than (3.70), so as to be able to apply proposition 2.59. To this end, letting $U_0 := (u_0, u_1)$ and $\overline{U}_0 := (\bar{u}_0, \bar{u}_1)$, we claim:

PROPOSITION 3.26
There are $\varepsilon_0 \in \,]0, 1]$ and $K > 0$, such that for all $\varepsilon \leq \varepsilon_0$ and $U_0, \overline{U}_0 \in \mathcal{B}_0$, $t \geq 0$,

$$E_\varepsilon(S(t)U_0 - S(t)\overline{U}_0) \leq e^{-t/2\alpha} E_\varepsilon(U_0 - \overline{U}_0) + \tfrac{1}{\varepsilon} K^2 (\delta_t(U_0, \overline{U}_0))^2, \tag{3.98}$$

with δ_t defined in (3.96), and α in (3.68).

PROOF   Let $S(t)U_0 =: (u(t), u_t(t))$ and $S(t)\overline{U}_0 =: (\bar{u}(t), \bar{u}_t(t))$. Since U_0 and $\overline{U}_0$ are in $\mathcal{B}_0$, which is positively invariant, $S(t)U_0$ and $S(t)\overline{U}_0$ remain in $\mathcal{B}_0$ for all $t \geq 0$. Since $\mathcal{B}_0$ is also bounded, there is $M > 0$, independent of ε, such that for all $t \geq 0$,

$$\|\nabla u(t)\| + \|\nabla \bar{u}(t)\| \leq M. \tag{3.99}$$

Let $z(t) := u(t) - \bar{u}(t)$. Multiplying (3.69) by $2z_t + \tfrac{1}{\varepsilon}z$, we obtain (omitting, as usual, the variable t)

$$\frac{\mathrm{d}}{\mathrm{d}t} E_\varepsilon(z(t), z_t(t)) + \|z_t(t)\|^2 + \tfrac{1}{\varepsilon}\|\nabla z(t)\|^2$$

$$= \langle g(\bar{u}(t)) - g(u(t)), 2z_t(t) + \tfrac{1}{\varepsilon}z(t) \rangle. \qquad (3.100)$$

Since g is Lipschitz continuous from $\mathcal{V}$ into $\mathcal{H}$ on bounded sets of $\mathcal{V}$ (this is a consequence of proposition 3.15), we deduce from (3.99) that there is $K > 0$, independent of t and ε, such that

$$2\|g(\bar{u}(t)) - g(u(t))\| \, \|z_t(t)\| \le 2K\|\nabla z(t)\| \, \|z_t(t)\| \le 2K^2\|\nabla z(t)\|^2 + \tfrac{1}{2}\|z_t(t)\|^2,$$

$$\tfrac{1}{\varepsilon}\|g(\bar{u}(t)) - g(u(t))\| \, \|z(t)\| \le \tfrac{1}{\varepsilon}K\|\nabla z(t)\| \, \|z(t)\| \le \tfrac{1}{4\varepsilon}\|\nabla z(t)\|^2 + \tfrac{1}{\varepsilon}K^2\|z(t)\|^2.$$

Consequently, we obtain from (3.100)

$$\frac{\mathrm{d}}{\mathrm{d}t}E_\varepsilon(z(t), z_t(t)) + \|z_t(t)\|^2 + \tfrac{1}{\varepsilon}\|\nabla z(t)\|^2$$

$$\le \left(2K^2 + \tfrac{1}{4\varepsilon}\right)\|\nabla z(t)\|^2 + \tfrac{1}{2}\|z_t(t)\|^2 + \tfrac{1}{\varepsilon}K^2\|z(t)\|^2.$$

Setting then $\varepsilon_0 := \tfrac{1}{8K^2}$, we deduce that if $\varepsilon \le \varepsilon_0$,

$$\frac{\mathrm{d}}{\mathrm{d}t}E_\varepsilon(z(t), z_t(t)) + \tfrac{1}{2}\|z_t(t)\|^2 + \tfrac{1}{2\varepsilon}\|\nabla z(t)\|^2 \le \tfrac{1}{\varepsilon}K^2\|z(t)\|^2. \qquad (3.101)$$

We easily verify that

$$E_\varepsilon(z, z_t) \le \alpha \left(\|z_t\|^2 + \tfrac{1}{\varepsilon}\|\nabla z\|^2\right);$$

hence, we obtain from (3.101) the linear differential inequality

$$\frac{\mathrm{d}}{\mathrm{d}t}E_\varepsilon(z(t), z_t(t)) + \tfrac{1}{2\alpha}E_\varepsilon(z(t), z_t(t)) \le \tfrac{1}{\varepsilon}K^2\|z(t)\|^2.$$

By proposition 2.63, we deduce that

$$E_\varepsilon(z(t), z_t(t)) \le \mathrm{e}^{-t/2\alpha}E_\varepsilon(z(0), z_t(0)) + \tfrac{1}{\varepsilon}K^2\int_0^t \mathrm{e}^{-(t-\theta)/2\alpha}\|z(\theta)\|^2\,\mathrm{d}\theta,$$

from which (3.98) follows. $\qquad\qquad\qquad\qquad\qquad\qquad\qquad\qquad\square$

We can finally conclude the proof of the existence of a global attractor, by means of theorem 2.56 and proposition 2.59. In fact, choosing $t_* > 0$ such that $q := \mathrm{e}^{-t_*/2\alpha} < 1$, from (3.98) we see that the operator $T = S(t_*)$ and the pseudometric δ_{t_*} satisfy condition (2.47) of proposition 2.59. Hence, T is an α-contraction. The fact that $\omega(\mathcal{B}_0)$ is the desired attractor for S follows then from theorem 2.56.

3.5 Regularity

In this section, we prove a result concerning the regularity of the global attractor of the hyperbolic problem (H_ε), assuming again that the parameter ε is sufficiently small.

THEOREM 3.27

Let ε_0 be as in proposition 3.23. For $\varepsilon \in]0, \varepsilon_0]$, let $\mathcal{A}^\varepsilon$ be the global attractor of problem (H_ε), given by theorem 3.24. There is $\varepsilon_1 \in]0, \varepsilon_0]$ such that if $\varepsilon \leq \varepsilon_1$, $\mathcal{A}^\varepsilon$ is contained in a bounded set of $\mathcal{X}_1$. This set is independent of ε.

PROOF We follow an idea presented in Grasselli-Pata, [GP02]; for an alternative proof, see e.g. Temam, [Tem88, sct. IV.6]. We proceed in three steps. At first we show, with methods similar to those of section 3.4.3, that the attractor $\mathcal{A}^\varepsilon$ is contained in the subspace $\mathcal{X}_{1/4} := H^{5/4}(\Omega) \times H^{1/4}(\Omega)$, and bounded with respect to the norm defined by

$$\|(u,v)\|_{\mathcal{X}_{1/4}}^2 := \varepsilon \|L^{1/8} v\|^2 + \|L^{5/8} u\|^2 , \quad (u,v) \in \mathcal{X}_{1/4} .$$

Then, we improve the estimates of solutions in $\mathcal{X}_{1/4}$, and show that if ε is sufficiently small, a technique analogous to that of section 3.4.3 allows us to conclude that, in fact, $\mathcal{A}^\varepsilon \subseteq \mathcal{X}_1$, and is bounded with respect to the norm whose square is defined by

$$E_1(u,v) := E_0(\nabla u, \nabla v), \quad (u,v) \in \mathcal{X}_1 .$$

Finally, we establish the bound

$$\|\nabla u_1\|^2 + \|\Delta u_0\|^2 \leq M^2 , \tag{3.102}$$

where $M > 0$ is independent of ε and $(u_0, u_1) \in \mathcal{A}^\varepsilon$.

1. Let $(u_0, u_1) \in \mathcal{A}^\varepsilon$. Since $\mathcal{A}^\varepsilon = \omega(\mathcal{B})$, by proposition 2.15 there are sequences $((\varphi_n, \psi_n))_{n \in \mathbb{N}} \subseteq \mathcal{B}$ and $(t_n)_{n \in \mathbb{N}} \subseteq]0, +\infty[$ such that $S(t_n)(\varphi_n, \psi_n) \to (u_0, u_1)$ in $\mathcal{X}$ and $t_n \to +\infty$ as $n \to \infty$. Decompose

$$S(t_n) = S_1(t_n) + S_2(t_n)$$

as in section 3.4.3. Since $\mathcal{B}$ is bounded in $\mathcal{X}$, (3.77) implies that there is $R > 0$, independent of ε, such that for all $n \in \mathbb{N}$,

$$N_0(S_2(t_n)(\varphi_n, \psi_n)) \leq R e^{-t_n/2\alpha} , \tag{3.103}$$

with N_0 defined in (3.72). Since the function f is independent of t, (3.79) holds, so the sequence

$$(S_1(t_n)(\varphi_n, \psi_n))_{n \in \mathbb{N}}$$

is bounded in $\mathcal{X}_{1/4}$. Since $\mathcal{X}_{1/4}$ is compactly imbedded into $\mathcal{X}$, there is a subsequence, which we still denote by $(S_1(t_n)(\varphi_n, \psi_n))_{n \in \mathbb{N}}$, converging to some element $(\bar{u}_0, \bar{u}_1)$ weakly in $\mathcal{X}_{1/4}$, and strongly in $\mathcal{X}$. Because of (3.103), $S(t_n)(\varphi_n, \psi_n) \to (\bar{u}_0, \bar{u}_1)$ in $\mathcal{X}$; hence, $(u_0, u_1) = (\bar{u}_0, \bar{u}_1) \in \mathcal{X}_{1/4}$. Since (u_0, u_1) was arbitrary in $\mathcal{A}^\varepsilon$, we conclude that $\mathcal{A}^\varepsilon \subseteq \mathcal{X}_{1/4}$. Moreover, since we can now say that $S_1(t_n)(\varphi_n, \psi_n) \to (u_0, u_1)$ weakly in $\mathcal{X}_{1/4}$,

$$\|(u_0, u_1)\|_{\mathcal{X}_{1/4}} \leq \liminf_{n \to \infty} \|S_1(t_n)(\varphi_n, \psi_n)\|_{\mathcal{X}_{1/4}} , \tag{3.104}$$

and this shows that $\mathcal{A}^\varepsilon$ is bounded in $\mathcal{X}_{1/4}$.

2. Set now, as in section 3.4.3,

$$(u(t), u_t(t)) := S(t)(u_0, u_1),$$
$$(v(t), v_t(t)) := S_2(t)(u_0, u_1),$$
$$(w(t), w_t(t)) := S_1(t)(u_0, u_1).$$

In the first part of this proof we have shown that $(u, u_t) \in C_b(\mathbb{R}; \mathcal{X}_{1/4})$. Since $(u_0, u_1) \in \mathcal{X}_{1/4}$, we can deduce, as in the proof of proposition 3.23, that $(v, v_t) \in C_b([0, +\infty[; \mathcal{X}_{1/4})$. Indeed, replacing the function Φ of (3.84) by $-v^3$, the same estimates that in that proof led to the exponential inequality (3.95) now lead to the inequality

$$\frac{d}{dt}(E_\varepsilon(v, v_t) - 2\langle v^3, L^{1/4}v\rangle) + \tfrac{1}{2\alpha}(E_\varepsilon(v, v_t) - 2\langle v^3, L^{1/4}v\rangle) \le C_4.$$

From this, we obtain that for all $t \ge 0$,

$$E_\varepsilon(v(t), v_t(t)) \le 2\langle v^3(t), L^{1/4}v(t)\rangle + \max\{E_\varepsilon(u_0, u_1) - 2\langle u_0^3, L^{1/4}u_0\rangle, 2\alpha C_4\}.$$

Recalling that (u_0, u_1) is in a bounded set of $\mathcal{X}_{1/4}$, we easily conclude that $(v, v_t) \in C_b([0, +\infty[; \mathcal{X}_{1/4})$ as claimed. It follows then that also

$$(w, w_t) = (u - v, u_t - v_t) \in C_b([0, +\infty[; \mathcal{X}_{1/4}).$$

3. We can now bootstrap the argument of the first part of this proof. We first show that if ε is sufficiently small, then, as a consequence of the fact that (u_0, u_1) is in a bounded set of $\mathcal{X}_{1/4}$, $(w, w_t) \in C_b([0, +\infty[; \mathcal{X}_1)$. To this end, we (formally) multiply the equation of (3.76) in $\mathcal{H}$ by $-2\Delta w_t - \Delta w$: dropping as usual the argument t, we obtain

$$\frac{d}{dt}(\varepsilon \|\nabla w_t\|^2 + \varepsilon\langle \nabla w, \nabla w_t\rangle + \tfrac{1}{2}\|\nabla w\|^2 + \|\Delta w\|^2 + 2\langle \Phi, \Delta w\rangle)$$
$$+ \|\nabla w_t\|^2 + \|\Delta w\|^2 + \tfrac{2}{\beta_1}\langle \Phi, \Delta w\rangle$$
$$= \langle 2\Phi_t + \left(\tfrac{2}{\beta_1} - 1\right)\Phi, \Delta w\rangle, \tag{3.105}$$

with Φ as in (3.84) and $\beta_1 := 2(1 + \tfrac{1}{\lambda_1})$. To estimate the right side of (3.105), we first note that, as in (3.90), recalling (3.80),

$$\langle \Phi, \Delta w\rangle \le C\|\Delta w\| \le C + \tfrac{1}{6}\|\Delta w\|^2. \tag{3.106}$$

Next, recalling (3.91) (with $f_t \equiv 0$),

$$2\langle \Phi_t, \Delta w\rangle \le C(|v|_{12}|w|_\infty + |w|_{12}|w|_\infty + 1)|v_t|_{12/5}\|\Delta w\|$$
$$+ C(|v|_{12}^2 + |v|_{12}|w|_{12} + |w|_{12}^2 + 1)|w_t|_3\|\Delta w\|. \tag{3.107}$$

Recalling (3.81) and theorem A.61, we have the imbeddings

$$H^{5/4}(\Omega) \hookrightarrow L^{12}(\Omega), \quad H^{1/4}(\Omega) \hookrightarrow L^{12/5}(\Omega)$$

as well as the interpolation inequalities

$$|w|_\infty \le C\|\Delta^2 w\|^{1/2}|w|_6^{1/2} + C|w|_6,$$
$$|w_t|_3 \le C\|\nabla w_t\|^{1/2}\|w_t\|^{1/2} + C\|w_t\|.$$

By ellipticity (see theorem A.77), we can further estimate

$$|w|_\infty \le C\|w\|_{H^2(\Omega)}^{1/2}|w|_6^{1/2} + C|w|_6^{1/2} \le C(\|\Delta w\|^{1/2} + \|w\|^{1/2})|w|_6^{1/2} + C|w|_6^{1/2}.$$

Thus, recalling that both (v, v_t), $(w, w_t) \in C_b([0, +\infty[; \mathcal{X}_{1/4})$, we obtain from (3.107) that, for δ and $\eta > 0$,

$$2\langle \Phi_t, \Delta w \rangle \le C(1 + \delta\|\Delta w\|)\|\Delta w\| + C\|\nabla w_t\|^{1/2}\|\Delta w\|$$
$$\le C + \delta\|\Delta w\|^2 + \eta\|\nabla w_t\|^2 + \eta\|\Delta w\|^2, \tag{3.108}$$

where C denotes various constants, depending on δ, η and, of course, on the uniform bounds on (v, v_t) and (w, w_t) in $\mathcal{X}_{1/4}$, but not on ε. Choosing δ and η sufficiently small, we obtain from (3.106) and (3.108), inserted into (3.105), that

$$\frac{\mathrm{d}}{\mathrm{d}t}\left(\varepsilon\|\nabla w_t\|^2 + \varepsilon\langle\nabla w, \nabla w_t\rangle + \tfrac{1}{2}\|\nabla w\|^2 + \|\Delta w\|^2 + 2\langle\Phi, \Delta w\rangle\right)$$
$$+ \tfrac{1}{2}\|\nabla w_t\|^2 + \tfrac{1}{2}\|\Delta w\|^2 + \tfrac{2}{\beta_1}\langle\Phi, \Delta w\rangle \le C. \tag{3.109}$$

As in (3.83), we easily verify that if $\varepsilon \le \min\{\varepsilon_0, \tfrac{1}{3}\} =: \varepsilon_1$,

$$\Psi := \varepsilon\|\nabla w_t\|^2 + \varepsilon\langle\nabla w, \nabla w_t\rangle + \tfrac{1}{2}\|\nabla w\|^2 + \|\Delta w\|^2 + 2\langle\Phi, \Delta w\rangle$$
$$\le \beta_1\left(\tfrac{1}{2}\|\nabla w_t\|^2 + \|\Delta w\|^2 + \tfrac{2}{\beta_1}\langle\Phi, \Delta w\rangle\right);$$

hence, (3.109) yields the exponential inequality

$$\Psi' + \tfrac{1}{\beta_1}\Psi \le C.$$

From this, we obtain that for all $t \ge 0$,

$$\varepsilon\|\nabla w_t(t)\|^2 + \varepsilon\langle\nabla w(t), \nabla w_t(t)\rangle + \tfrac{1}{2}\|\nabla w(t)\|^2 + \|\Delta w(t)\|^2$$
$$\le 2\langle\Phi(t), \Delta w(t)\rangle + C\beta_1 \le 2\|\Phi(t)\|^2 + \tfrac{1}{2}\|\Delta w(t)\|^2 + C\beta_1.$$

Since the function $t \mapsto \|\Phi(\cdot, t)\|$ is uniformly bounded in t, by Schwarz' inequality we conclude that there is $M_1 > 0$ such that for all $t \ge 0$ and $\varepsilon \le \varepsilon_1$,

$$\varepsilon\|\nabla w_t(\cdot, t)\|^2 + \|\Delta w(\cdot, t)\|^2 \le M_1^2. \tag{3.110}$$

This shows that $(w, w_t) \in C_b([0, +\infty[; \mathcal{X}_1)$ as claimed.

4. We now conclude as in the first part of this proof. Given arbitrary $(u_0, u_1) \in \mathcal{A}^\varepsilon$, we construct sequences $((\varphi_n, \psi_n))_{n \in \mathbb{N}} \subseteq \mathcal{B}$ and $(t_n)_{n \in \mathbb{N}}$ as before. Estimate (3.110) implies that a subsequence of $(S_1(\varphi_n, \psi_n))_{n \in \mathbb{N}}$ converges to a limit $(\bar{u}_0, \bar{u}_1)$ weakly in $\mathcal{X}_1$, and strongly in $\mathcal{X}$. Again, (3.103) allows us to conclude that $(u_0, u_1) = (\bar{u}_0, \bar{u}_1) \in \mathcal{X}_1$. Finally, as in (3.104), we deduce from (3.110) that (u_0, u_1) is in a bounded set of $\mathcal{X}_1$. Since (u_0, u_1) is arbitrary in $\mathcal{A}^\varepsilon$, this implies that $\mathcal{A}^\varepsilon$ is bounded in $\mathcal{X}_1$.

5. We now proceed to show that, in fact, $\mathcal{A}^\varepsilon$ can be bounded in $\mathcal{X}_1$ independently of ε. Let $(u_0, u_1) \in \mathcal{A}^\varepsilon$. By proposition 2.39, (u_0, u_1) lies on a complete orbit $(u(t), u_t(t))_{t \in \mathbb{R}}$, contained in $\mathcal{A}^\varepsilon$; without loss of generality, we can assume that $(u_0, u_1) = (u(0), u_t(0))$. Since all the constants appearing in the proof of the boundedness of $\mathcal{A}^\varepsilon$ in $\mathcal{X}_{1/4}$ and $\mathcal{X}_1$, including in particular the constant R of (3.103), the boundedness implied by (3.104), and M_1 of (3.110), depend only on the norm of u in $C_b(\mathbb{R}; \mathcal{V})$, we deduce that the estimate

$$\varepsilon \|\nabla u_1\|^2 + \|\Delta u_0\|^2 \leq C_2 \tag{3.111}$$

holds uniformly with respect to ε and $(u_0, u_1) \in \mathcal{A}^\varepsilon$. This provides part of (3.102); to remove the dependence of the term with u_1 on ε, we now prove that there exists a positive constant C_3, independent of ε, such that for all $t \in \mathbb{R}$ and $\varepsilon \in]0, \varepsilon_1]$,

$$\varepsilon \|u_{tt}(t)\|^2 + \|\nabla u_t(t)\|^2 \leq C_3. \tag{3.112}$$

To this end, we first note that, since $\mathcal{A}^\varepsilon$ is invariant, from (3.111) we deduce that for all $t \in \mathbb{R}$,

$$\varepsilon \|\nabla u_t(t)\|^2 + \|\Delta u(t)\|^2 \leq C_2. \tag{3.113}$$

As a preliminary step, we establish the estimate

$$\|u_t(t)\| \leq C, \tag{3.114}$$

where, here and in the sequel, we denote by C a generic positive constant, independent of $\varepsilon \in]0, \varepsilon_1]$ and $t \in \mathbb{R}$. Multiplying equation (3.4) by $2u_t$ we obtain, by (3.113),

$$\varepsilon \frac{d}{dt} \|u_t\|^2 + \|u_t\|^2 \leq \|f + \Delta u + u - u^3\|^2$$
$$\leq C \left(\|f\|^2 + \|\Delta u\|^2 + \|u\|^2 + \|\nabla u\|^6 \right) \leq C. \tag{3.115}$$

Integrating (3.115) on an arbitrary interval $[t_0, t] \subset \mathbb{R}$, and recalling (3.113) again, we obtain

$$\|u_t(t)\|^2 \leq C \left(\tfrac{1}{\varepsilon} e^{-(t-t_0)/\varepsilon} + 1 \right),$$

from which we deduce (3.114) by letting $t_0 \to -\infty$.

We now differentiate equation (3.4) with respect to t, and multiply the resulting equation by $2u_{tt} + u_t$, to obtain

$$\frac{\mathrm{d}}{\mathrm{d}t}\left(\varepsilon\|u_{tt}\|^2 + \varepsilon\langle u_{tt}, u_t\rangle + \tfrac{1}{2}\|u_t\|^2 + \|\nabla u_t\|^2\right) + (2-\varepsilon)\|u_{tt}\|^2 + \|\nabla u_t\|^2$$
$$= -\langle(3u^2-1)u_t, 2u_{tt}+u_t\rangle =: R_1. \tag{3.116}$$

Combining the imbedding and elliptic estimates from theorems A.74 and A.77, and recalling (3.80) and (3.114), we can estimate the right side of (3.116) by

$$R_1 \le 2\left(|3u^2-1|_\infty \|u_t\|\right)\left(\|u_{tt}\| + \|u_t\|\right)$$
$$\le C\left(1+\|u\|_2^2\right)\left(1+\|u_{tt}\|\right) \le \tfrac{1}{2}\|u_{tt}\|^2 + C. \tag{3.117}$$

We now denote by $\Phi_1(u_t, u_{tt})$ the term under differentiation in (3.116), and note that, as in (3.67),

$$\Phi_1(u_t, u_{tt}) \ge \tfrac{1}{\alpha}\left(\varepsilon\|u_{tt}\|^2 + \|\nabla u_t\|^2\right). \tag{3.118}$$

Replacing (3.117) into (3.116), and recalling (3.114), we obtain, as usual, the inequality

$$\frac{\mathrm{d}}{\mathrm{d}t}\Phi_1(u_t, u_{tt}) + \tfrac{1}{\alpha}\Phi_1(u_t, u_{tt}) \le C.$$

Integrating this inequality on an arbitrary interval $[t_0, t] \subset \mathbb{R}$, we obtain

$$\Phi_1(u_t(t), u_{tt}(t)) \le \mathrm{e}^{-(t-t_0)/\alpha}\Phi_1(u_t(t_0), u_{tt}(t_0)) + \alpha C. \tag{3.119}$$

From (3.4) for $t = t_0$ we have

$$\varepsilon u_{tt}(t_0) = f(t_0) + u(t_0) - (u(t_0))^3 + \Delta u(t_0) - u_t(t_0);$$

therefore, because of (3.113) and (3.114),

$$\Phi_1(u_t(t_0), u_{tt}(t_0)) \le \tfrac{1}{\varepsilon}C.$$

Replacing this into (3.119) we obtain then that

$$\Phi_1(u_t(t), u_{tt}(t)) \le C\left(\tfrac{1}{\varepsilon}\mathrm{e}^{-(t-t_0)/\alpha} + 1\right). \tag{3.120}$$

Letting $t_0 \to -\infty$, and recalling (3.118), we can finally deduce (3.112) from (3.120).

6. We can now conclude the proof of theorem 3.27: indeed, (3.102) follows from (3.111), and (3.112) with $t = 0$. □

REMARK 3.28   Following the proof of theorem 3.27, we realize that the requirement that ε be small is essential only in order to show that the boundedness of $\mathcal{A}^\varepsilon$ in $\mathcal{X}_1$ is uniform with respect to ε. In fact, it is possible to show that each attractor

$\mathcal{A}^{\varepsilon}$ is bounded in $\mathcal{X}_1$. To this end, it is sufficient to modify the estimates we have established, multiplying the equation e.g. by $2u_t + \frac{\lambda}{\varepsilon}u$, or by $2u_t + \lambda u$, with λ sufficiently small. Indeed, both Temam, [Tem88], and Grasselli-Pata, [GP02], prove the boundedness of $\mathcal{A}^{\varepsilon}$ in $\mathcal{X}_1$ for $\varepsilon = 1$. ☐

On the other hand, if ε is small, we can actually prove much more than mere boundedness in $\mathcal{X}_1$:

THEOREM 3.29
In the same conditions of theorem 3.27, if $\varepsilon \leq \varepsilon_1$ each attractor $\mathcal{A}^{\varepsilon}$ is compact in $\mathcal{X}_1$.

PROOF Let $(\varphi_n, \psi_n)_{n \in \mathbb{N}}$ be a sequence in $\mathcal{A}^{\varepsilon}$. Since $\mathcal{A}^{\varepsilon}$ is bounded in $\mathcal{X}_1$, there is a subsequence, still denoted by $(\varphi_n, \psi_n)_{n \in \mathbb{N}}$, converging to a pair (φ_*, ψ_*), weakly in $\mathcal{X}_1$ and strongly in $\mathcal{X}$. We must show that

$$(\varphi_n, \psi_n) \to (\varphi_*, \psi_*) \quad \text{strongly in } \mathcal{X}_1.$$

1. By proposition 2.39, for each $n \in \mathbb{N}$ there is a complete orbit $(u^n(t), u_t^n(t))_{t \in \mathbb{R}}$ passing through (φ_n, ψ_n), and completely contained in $\mathcal{A}^{\varepsilon}$; without loss of generality, we can assume that

$$(u^n(0), u_t^n(0)) = (\varphi_n, \psi_n). \tag{3.121}$$

From (3.112) and (3.113) we deduce that, for all $T > 0$, the functions u^n, u_t^n and u_{tt}^n are, respectively, in a bounded set of $L^2(-T, T; H^2(\Omega))$, $L^2(-T, T; H^1(\Omega))$, and $L^2(-T, T; L^2(\Omega))$ (this set depends on ε, but here we consider ε as a fixed parameter). Hence, by theorem A.16, for each ε there are a function u and a subsequence, still denoted by $(u^n)_{n \in \mathbb{N}}$, such that

$$u^n \to u \quad \text{in} \quad L^2(-T, T; H^2(\Omega)) \quad \text{weakly}, \tag{3.122}$$

$$u_t^n \to u_t \quad \text{in} \quad L^2(-T, T; H^1(\Omega)) \quad \text{weakly}, \tag{3.123}$$

$$u_{tt}^n \to u_{tt} \quad \text{in} \quad L^2(-T, T; L^2(\Omega)) \quad \text{weakly}. \tag{3.124}$$

We recall from theorem A.82 that for each $j \in \mathbb{N}$, the injections

$$\{u \in L^2(-T, T; H^{j+1}(\Omega)) : u_t \in L^2(-T, T; H^j(\Omega))\} \hookrightarrow L^2(-T, T; H^j(\Omega)) \tag{3.125}$$

are compact. Taking $j = 1$, (3.125), (3.122) and (3.123) imply that

$$u^n \to u \quad \text{in} \quad L^2(-T, T; H^1(\Omega)) \quad \text{strongly}. \tag{3.126}$$

Likewise, taking $j = 0$ in (3.125), (3.123) and (3.124) imply that

$$u_t^n \to u_t \quad \text{in} \quad L^2(-T, T; L^2(\Omega)) \quad \text{strongly}. \tag{3.127}$$

We now show that we can deduce, from (3.126), that

$$(u^n)^3 \to u^3 \quad \text{in} \quad L^2(-T,T;L^2(\Omega)) \quad \text{strongly}. \tag{3.128}$$

In fact, we can estimate

$$\int_{-T}^{T} \|(u^n)^3 - u^3\|^2 \, dt \leq \int_{-T}^{T} |(u^n)^2 + u^n u + u^2|_\infty^2 \|u^n - u\|^2 \, dt ; \tag{3.129}$$

since u^n is bounded in $L^\infty(-T,T;H^2(\Omega))$, and this space is continuously imbedded into $L^\infty(\Omega \times]-T,T[)$, (3.128) follows from (3.129).

2. Recall now the definition (3.18) of weak solution of (3.4): in particular, u^n solves (3.65). Then, by (3.126), (3.127) and (3.128), letting $n \to +\infty$ we deduce that u is also a solution of (3.65). We now show that

$$(u(t), u_t(t)) = S_\varepsilon(t)(\varphi_*, \psi_*), \quad t \in \mathbb{R}. \tag{3.130}$$

To this end, we recall from theorem A.81 that the space

$$\{u \in L^2(-T,T;H^2(\Omega)) : u_t \in L^2(-T,T;L^2(\Omega))\}$$

is continuously injected in $C([-T,T];H^1(\Omega))$, with

$$\max_{-T \leq t \leq T} \|u^n(t) - u(t)\|_1$$

$$\leq C \|u^n - u\|_{L^2(-T,T;H^2(\Omega))}^{1/2} \|u_t^n - u_t\|_{L^2(-T,T;L^2(\Omega))}^{1/2} + \|u^n - u\|_{L^2(-T,T;L^2(\Omega))} \, .$$

Hence, (3.122) and (3.127) imply that

$$u^n \to u \quad \text{in} \quad C([-T,T];H^1(\Omega)) \quad \text{strongly}. \tag{3.131}$$

From theorem A.81, we also have the trace estimate

$$\max_{-T \leq t \leq T} \|u_t^n(t) - u_t(t)\| \tag{3.132}$$

$$\leq C \|u^n - u\|_{L^2(-T,T;H^1(\Omega))}^{1/4} \|u_{tt}^n - u_{tt}\|_{L^2(-T,T;L^2(\Omega))}^{3/4} + \|u^n - u\|_{L^2(-T,T;L^2(\Omega))} \, .$$

Hence, by (3.131), we conclude that also

$$u_t^n \to u_t \quad \text{in} \quad C([-T,T];L^2(\Omega)) \quad \text{strongly}. \tag{3.133}$$

From (3.131) and (3.133) we deduce that $u(0) = \varphi_*$ and $u_t(0) = \psi_*$; hence, (3.130) follows, by the uniqueness of solutions to (H_ε) with the same initial values.

3. Consider now the difference $z^n := u^n - u$, which solves the equation

$$\varepsilon z_{tt}^n + z_t^n - \Delta z^n = z^n - ((u^n)^3 - u^3). \tag{3.134}$$

Multiplying (3.134) in $L^2(\Omega)$ by $-\Delta(2z_t^n + z^n)$, we obtain

$$\frac{d}{dt} E_0(\nabla z^n, \nabla z_t^n) + (2 - \varepsilon)\|\nabla z_t^n\|^2 + \|\Delta z^n\|^2 = ((u^n)^3 - u^3 - z^n, -2\Delta(z_t^n + z^n))$$

$$= 2\left(\nabla((u^n)^3 - u^3 - z^n), \nabla z_t^n\right) - \left((u^n)^3 - u^3 - z^n, \Delta z_t^n\right) =: R_2\,, \tag{3.135}$$

where E_0 is as in (3.66). We can estimate the right side of (3.135) by

$$R_2 \leq (2+\lambda_1)\|\nabla((u^n)^3 - u^3 - z^n)\|^2 + \tfrac{1}{2}\|\nabla z_t^n\|^2 + \tfrac{1}{2}\|\Delta z^n\|^2\,;$$

thus, recalling (3.67), we obtain from (3.135) that

$$\frac{\mathrm{d}}{\mathrm{d}t}E_0(\nabla z^n, \nabla z_t^n) + \tfrac{1}{\alpha}E_0(\nabla z^n, \nabla z_t^n) \leq (2+\lambda_1)\|\nabla((u^n)^3 - u^3 - z^n)\|^2\,. \tag{3.136}$$

We now write

$$\nabla((u^n)^3 - u^3 - z^n) = \nabla\left(((u^n)^2 + u^n u + u^2 - 1)z^n\right)\,,$$

and recall that, since $n \leq 3$, $\mathrm{H}^2(\Omega)$ is an algebra (see theorem A.69). Hence, $(u^n)^2 + u^n u + u^2$ is in a bounded set of $\mathrm{H}^2(\Omega)$, and we easily see that

$$\|\nabla((u^n)^3 - u^3 - z^n)\| \leq C\|\nabla z^n\|\,,$$

for suitable constant C depending only on $\mathcal{A}^\varepsilon$. Thus, we further obtain from (3.136) that

$$\frac{\mathrm{d}}{\mathrm{d}t}E_0(\nabla z^n, \nabla z_t^n) + \tfrac{1}{\alpha}E_0(\nabla z^n, \nabla z_t^n) \leq C(2+\lambda_1)\|\nabla z^n\|^2\,. \tag{3.137}$$

From (3.131) we deduce that for all $T > 0$ and all $\eta > 0$ there is $n_0 \in \mathbb{N}$ such that for all $n \geq n_0$ and $t \in [-T,T]$,

$$\alpha C(2+\lambda_1)\|\nabla z^n(\cdot,t)\|^2 \leq \eta\,.$$

Thus, integrating (3.137) in $[-T,T]$, for $n \geq n_0$, we obtain that

$$E_0(\nabla z^n(t), \nabla z_t^n(t)) \leq \mathrm{e}^{-T/\alpha}E_0(\nabla z^n(-T), \nabla z_t^n(-T)) + \eta$$
$$\leq C_1\mathrm{e}^{-T/\alpha} + \eta\,, \tag{3.138}$$

with C_1 again depending only on $\mathcal{A}^\varepsilon$, because (z^n, z_t^n) is in a bounded set of $\mathcal{X}_1$. Given then $\eta > 0$, we can therefore predetermine $T > 0$ so large that $C_1\mathrm{e}^{-T/\alpha} \leq \eta$; with this choice of T, we determine n_0 as above, and (3.138) implies that for all $n \geq n_0$ and $t \in [-T,T]$,

$$E_0(\nabla z^n(t), \nabla z_t^n(t)) \leq 2\eta\,. \tag{3.139}$$

Recalling (3.67), and choosing $t = 0$ in (3.139), we deduce that

$$u^n(0) \to u(0) \quad \text{in } \mathrm{H}^2(\Omega)\,, \quad u_t^n(0) \to u_t(0) \quad \text{in } \mathrm{H}^1(\Omega)\,.$$

Since $(u^n(0), u_t^n(0)) = (\varphi_n, \psi_n)$ and $(u(0), u_t(0)) = (\varphi_*, \psi_*)$, we conclude that (φ_n, ψ_n) converges strongly in $\mathcal{X}_1$ to (φ_*, ψ_*), as desired. This concludes the proof of theorem 3.29. □

3.6   Upper Semicontinuity of the Global Attractors

1.   The hyperbolic problem (H_ε) can be seen as a perturbation, for small $\varepsilon > 0$, of the reduced parabolic problem (P). In this context, a natural question is that of the convergence of the solutions of (H_ε), which we now rename u^ε to emphasize their dependence on ε, to the solution u of (P). Another, related question is whether the corresponding global attractors $\mathcal{A}^\varepsilon$ and $\mathcal{A}$, obtained in theorems 3.24 and 3.13, can be compared, and in what sense.

In this section, we briefly consider some aspects of the latter question, proving the so-called UPPER SEMICONTINUITY of the global attractors as $\varepsilon \to 0$. As a by-product, we also obtain some results on the convergence $u^\varepsilon \to u$, limited to motions contained in the attractors $\mathcal{A}^\varepsilon$ and $\mathcal{A}$.

We note that, in general, the convergence $u^\varepsilon \to u$ can be expected to be singular in time, because of the loss of the initial condition on u_t; we refer to Lions, [Lio73], for an extensive presentation of results on singular perturbation problems of this type. In contrast, we shall see that if we restrict our attention to motions in the attractors, it is possible to transform the singular convergence problem into a regular convergence one. On the other hand, the question of the convergence of the global attractors $\mathcal{A}^\varepsilon$ to $\mathcal{A}$ as $\varepsilon \to 0$ is also complicated by the obvious fact that $\mathcal{A}^\varepsilon$ is in the product space $H_0^1(\Omega) \times L^2(\Omega)$, while $\mathcal{A}$ is contained in $H_0^1(\Omega)$ only. One way to solve this difficulty is to consider the projections $P_1 \mathcal{A}^\varepsilon$ of $\mathcal{A}^\varepsilon$ into $H^1(\Omega)$, and to show that

$$\lim_{\varepsilon \to 0} \partial(P_1 \mathcal{A}^\varepsilon, \mathcal{A}) = 0,$$

where ∂ is the semidistance in $H^1(\Omega)$ defined in (2.2) (see e.g. the survey article [EM95]). Another possibility, which is the one we present here, is to construct a "natural" imbedding of $\mathcal{A}$ into a set $\mathcal{A}^0 \subseteq \mathcal{X} = H_0^1(\Omega) \times L^2(\Omega)$, and then show that

$$\lim_{\varepsilon \to 0} \partial(\mathcal{A}^\varepsilon, \mathcal{A}^0) = 0,$$

where ∂ is now the semidistance in $\mathcal{X}$.

In this section, we loosely follow the presentation of Hale, [Hal88, sct. 4.10]; for more results on this type of question, and related ones, such as for instance the LOWER SEMICONTINUITY of the attractors, we refer to Hale-Raugel, [HR88, HR90], or to Grasselli-Pata, [GP02].

2.   We recall the definition of upper and lower semicontinuity of a family of sets.

DEFINITION 3.30   *Let $\mathcal{X}$ be a complete metric space, $\Lambda \subseteq \mathbb{R}$, and $(\mathcal{C}^\lambda)_{\lambda \in \Lambda}$ a family of subsets of $\mathcal{X}$. Let $\lambda_0 \in \Lambda$. Then:*

1. $(\mathcal{C}^\lambda)_{\lambda \in \Lambda}$ is UPPER SEMICONTINUOUS *at λ_0 if*

$$\lim_{\lambda \to \lambda_0} \partial(\mathcal{C}^\lambda, \mathcal{C}^{\lambda_0}) = 0, \qquad (3.140)$$

2. $(\mathcal{C}^\lambda)_{\lambda \in \Lambda}$ *is* LOWER SEMICONTINUOUS *at* λ_0 *if*

$$\lim_{\lambda \to \lambda_0} \partial(\mathcal{C}^{\lambda_0}, \mathcal{C}^\lambda) = 0.$$

3. $(\mathcal{C}^\lambda)_{\lambda \in \Lambda}$ *is* CONTINUOUS *at* λ_0 *if it is upper and lower semicontinuous at* λ_0.

3. We proceed then to prove the upper semicontinuity of the attractors $\mathcal{A}^\varepsilon$ as $\varepsilon \to 0$. Let $\mathcal{A}$ be the attractor of the semiflow S generated by the parabolic problem (P). As we remarked after theorem 3.13, $\mathcal{A}$ is bounded in $H^2(\Omega)$; hence, we can introduce the set

$$\mathcal{A}^0 := \{(u,v) \in \mathcal{X} : u \in \mathcal{A}, \ v = f + u - u^3 + \Delta u\}, \tag{3.141}$$

which we consider as a "natural" imbedding of $\mathcal{A}$ in $\mathcal{X}$.

We have then the following result:

THEOREM 3.31
Let ε_1 be as in theorem 3.27 and, for $0 < \varepsilon \le \varepsilon_1$, let $\mathcal{A}^\varepsilon$ be the global attractor of the semiflow S_ε generated by the hyperbolic problem (H_ε). Let $\mathcal{A}^0$ be as in (3.141). The family $(\mathcal{A}^\varepsilon)_{0 \le \varepsilon \le \varepsilon_1}$ is upper semicontinuous at $\varepsilon = 0$, with respect to the topology of $\mathcal{X}$.

PROOF   Recalling (3.140), we must show that

$$\sup_{(u,v) \in \mathcal{A}^\varepsilon} \inf_{(\bar{u},\bar{v}) \in \mathcal{A}^0} \left(\|\nabla(u - \bar{u})\|^2 + \|v - \bar{v}\|^2 \right)^{1/2} \to 0 \tag{3.142}$$

as $\varepsilon \to 0$. We reason by contradiction. Assuming (3.142) did not hold, we could find $\eta_0 > 0$, and sequences $(\varepsilon_n)_{n \in \mathbb{N}} \subset \,]0, \varepsilon_1]$, $((\varphi_n, \psi_n))_{n \in \mathbb{N}} \subseteq \mathcal{A}^{\varepsilon_n}$, such that $\varepsilon_n \to 0$, and for all $n \in \mathbb{N}$,

$$\inf_{(\bar{u},\bar{v}) \in \mathcal{A}^0} \left(\|\nabla(\varphi_n - \bar{u})\|^2 + \|\psi_n - \bar{v}\|^2 \right) \ge \eta_0^2. \tag{3.143}$$

By (3.102), we have the uniform estimate

$$\|\psi_n\|_1^2 + \|\varphi_n\|_2^2 \le M^2,$$

with M independent of n; thus, there is a subsequence, still denoted by $((\varphi_n, \psi_n))_{n \in \mathbb{N}}$, converging to a limit (φ_*, ψ_*) weakly in $\mathcal{X}_1$ and, by compactness, strongly in $\mathcal{X}$. We now claim that $(\varphi_*, \psi_*) \in \mathcal{A}^0$: if true, this would contradict (3.143).

We now proceed in analogy to the proof of theorem 3.29. By proposition 2.39, for each $n \in \mathbb{N}$ there is a complete orbit

$$(u^n(t), u_t^n(t))_{t \in \mathbb{R}} = (S_{\varepsilon_n}(t)(\varphi_n, \psi_n))_{t \in \mathbb{R}} \tag{3.144}$$

contained in $\mathcal{A}^{\varepsilon_n}$ and passing through (φ_n, ψ_n). In particular, we can assume that (3.121) holds, for u^n defined as in (3.144). From (3.112) and (3.113) we have the uniform estimates

$$\varepsilon_n \|u_{tt}^n(t)\|^2 + \|u_t^n(t)\|_1^2 + \|u^n(t)\|_2^2 \leq M_2^2,\tag{3.145}$$

with M_2 independent of t and ε_n. From this it follows that, for all $T > 0$, the functions u^{ε_n}, $u_t^{\varepsilon_n}$ and $\sqrt{\varepsilon_n}u_{tt}^{\varepsilon_n}$ are, respectively, in a bounded set of $L^\infty(-T,T;H^2(\Omega))$, $L^\infty(-T,T;H^1(\Omega))$ and $L^\infty(-T,T;L^2(\Omega))$. Consequently, there is a function u, and a subsequence, still denoted $(\varepsilon_n)_{n\in\mathbb{N}}$, such that

$$u^{\varepsilon_n} \to u \quad \text{in } L^2(-T,T;H^2(\Omega)) \text{ weakly}^*,$$
$$u_t^{\varepsilon_n} \to u_t \quad \text{in } L^2(-T,T;H^1(\Omega)) \text{ weakly}^*,$$
$$\varepsilon_n u_{tt}^{\varepsilon_n} \to 0 \quad \text{in } L^2(-T,T;L^2(\Omega)) \text{ weakly}^*.$$

Proceeding exactly as in theorem 3.29, we show that u is a weak solution of the parabolic problem (P) on $\mathbb{R}$. By the analogous of (3.131), $\varphi_n = u^n(0) \to u(0)$ in $H^1(\Omega)$; hence, $u(0) = \varphi_*$, and, therefore, $u(0) \in H^2(\Omega)$. Moreover, since u is a complete orbit of S through φ_*, by proposition 2.39 $\varphi_* \in \mathcal{A}$. By (3.145),

$$\|\varepsilon_n u_{tt}^n(0)\| = \sqrt{\varepsilon_n}\,\|\sqrt{\varepsilon_n}u_{tt}^n(0)\| \leq \sqrt{\varepsilon_n}\,M_2;$$

hence, $\varepsilon_n u_{tt}^n(0) \to 0$ in $L^2(\Omega)$. Consequently,

$$\begin{aligned}
u_t^n(0) &= f + u^n(0) - (u^n(0))^3 + \Delta u^n(0) - \varepsilon_n u_{tt}^n(0)\\
&= f + \varphi_n - \varphi_n^3 + \Delta\varphi_n - \varepsilon_n u_{tt}^n(0)\\
&\to f + \varphi_* - \varphi_*^3 + \Delta\varphi_*
\end{aligned}\tag{3.146}$$

in $L^2(\Omega)$ weakly. Since $u_t^n(0) = \psi_n$, from (3.146) we deduce that

$$\psi_* = f + \varphi_* - \varphi_*^3 + \Delta\varphi_*.\tag{3.147}$$

Since $\varphi_* \in \mathcal{A}$, (3.147) implies that $(\varphi_*, \psi_*) \in \mathcal{A}^0$, as claimed. Having thus reached a contradiction with (3.143), the proof of theorem 3.31 is complete. Note that, comparing (3.146) with (3.147), we have that

$$u_t^{\varepsilon_n}(0) \to u_t(0).$$

This means that, as mentioned above, the convergence $u^{\varepsilon_n} \to u$ is not singular at $t = 0$.

$\square$

Chapter 4

Exponential Attractors

In this chapter we give the definition of EXPONENTIAL ATTRACTOR for a semiflow S defined on a Hilbert space $\mathcal{X}$, and describe an explicit construction of the exponential attractor when the semiflow satisfies a geometric property, called the DISCRETE SQUEEZING PROPERTY. As an application, we show that the continuous semiflows defined by the initial-boundary value problems (P) and (H$_\varepsilon$) (the latter at least for small ε) satisfy the discrete squeezing property. Therefore, these systems admit an exponential attractor.

4.1 Introduction

1. In the following, $\mathcal{X}$ is, as usual, a Banach space on $\mathbb{R}$, with norm $\|\cdot\|$.

DEFINITION 4.1 *Let $\mathcal{G}$ be an arbitrary subset of $\mathcal{X}$, and $\mathcal{T}$ be one of the sets $\mathbb{R}$, $\mathbb{R}_{\geq 0}$, $\mathbb{N}$ or $\mathbb{Z}$. A subset $\mathcal{E} \subset \mathcal{G}$ is an EXPONENTIAL ATTRACTOR for the semiflow $S = (S(t))_{t \in \mathcal{T}}$ in $\mathcal{G}$, if $\mathcal{E}$ is a compact, positively invariant set which has finite fractal dimension and attracts each bounded set $\mathcal{B} \subseteq \mathcal{G}$ exponentially. The last requirement means that there are positive numbers c and k, depending on $\mathcal{B}$, such that for all $t \in \mathcal{T}_{\geq 0}$,*

$$\partial(S(t)\mathcal{B}, \mathcal{E}) \leq ce^{-kt}, \tag{4.1}$$

where ∂ is the semidistance in $\mathcal{X}$ introduced in definition 2.1.

2. A large class of dynamical systems is known to admit an exponential attractor; for example, the semiflows generated by the 2-dimensional Navier-Stokes equations with periodic boundary conditions, the Kuramoto-Sivashinski equations, the Chafee-Infante and the Cahn-Hilliard equations (in any space dimension), and the "original" Burger's equation all admit an exponential attractor; we will see some examples in chapter 6. In contrast, much fewer systems are known to have an inertial manifold; for example, even for the 2-dimensional Navier-Stokes equations the existence of an inertial manifold is open.

3. Exponential attractors share with attractors the property of being attracting sets, but they do not have to be invariant. However, in contrast to (2.33), which carried no information on the rate of convergence of the orbits to the attractor, (4.1) gives an explicit information on the rate of convergence of the orbits to the exponential attractor. The fact that this rate is exponential makes exponential attractors much more robust than global attractors for numerical analysis. Consequently, exponential attractors are generally more suitable for the qualitative study of the long-time behavior of a semiflow. For example, it is proven in Eden, Foias, Nicolaenko and Temam, [EFNT94, ch. 4], that, in the framework of Galerkin approximations schemes, approximate and exact exponential attractors are continuous with respect to the Hausdorff distance of definition 2.1.

4. When the global attractor $\mathcal{A}$ and an exponential attractor $\mathcal{E}$ exist, then $\mathcal{A}$ is contained in $\mathcal{E}$. Indeed, since $\mathcal{A}$ is bounded, it is attracted by $\mathcal{E}$. Recalling (2.33), this means that

$$\lim_{t \to +\infty} \partial(S(t)\mathcal{A}, \mathcal{E}) = 0. \tag{4.2}$$

Since $\mathcal{A}$ is invariant, (4.2) implies that $\partial(\mathcal{A}, \mathcal{E}) = 0$. Thus, since $\mathcal{E}$ is compact, proposition 2.2 implies that

$$\mathcal{A} \subseteq \mathcal{E}.$$

5. One of the assumptions under which we will prove the existence of an exponential attractor is that the semiflow S admits a compact, positively invariant absorbing set $\mathcal{B}$. In this case, S also admits the global attractor $\mathcal{A} = \omega(\mathcal{B})$. Then, as we have seen, $\mathcal{A} \subseteq \mathcal{E}$ and, therefore, $\mathcal{A}$ has finite fractal dimension, with $\dim_F(\mathcal{A}) \leq \dim_F(\mathcal{E})$ by (2.49). Thus, $\mathcal{A}$ can be imbedded in $\mathbb{R}^N$ and it can be proven that $N \geq 2\dim_F(\mathcal{A}) + 1$. Moreover, there exists a N-dimensional system of ODEs

$$\dot{x} = F(x) \tag{4.3}$$

(with F not necessarily continuous) such that this imbedding of $\mathcal{A}$ is the global attractor (respectively, this imbedding of $\mathcal{E}$ is the exponential attractor) of the generalized semiflow ("generalized" because F may not be continuous) generated by (4.3) (see [EFNT94, thms. 10.2, 10.3]). We refer to [EFNT94] for more details.

6. Exponential attractors can in fact be seen, in some sense, as intermediate objects between attractors and inertial manifolds, which we study in detail in chapter 5. Essentially, inertial manifolds are positively invariant submanifolds of $\mathcal{X}$, which also attract orbits exponentially. The basic difference between inertial manifolds and exponential attractors is that exponential attractors need not have a smooth structure, and inertial manifolds need not be compact. Because exponential attractors are not required to have a smooth structure, we can establish existence results for exponential attractors under conditions on the semiflow that are much less restrictive than those required for the existence of an inertial manifold. In fact, as shown in [EFNT94], not only there are semiflows, generated by PDEEs, for which the existence of an exponential attractor is known, while that of an inertial manifold is not, but there actually are semiflows that admit an exponential attractor, and not an inertial manifold. Indeed, in chapter 7 we present an example due to Mora and Solà-Morales, [MSM87], of a semiflow generated by an IBVP for a hyperbolic equation like (3.4), which does not admit any C^1 inertial manifold. Of course, there are much simpler examples of two-dimensional ODEs which do not have an inertial manifold; for example, the system

$$\dot{x} = x, \quad \dot{y} = 2y,$$

whose orbits admit two unstable manifolds, but no inertial manifold.

4.2   The Discrete Squeezing Property

In this section we give the definition of the DISCRETE SQUEEZING PROPERTY, which is the main ingredient for the construction of an exponential attractor. The discrete squeezing property for the semiflow S generated by an evolution equation of the form

$$u_t + Au = f(u) \tag{4.4}$$

is usually defined by means of orthogonal projections in $\mathcal{X}$. Roughly speaking, the discrete squeezing property translates a dichotomy principle, whereby either the system is exponentially contracting on a fixed compact set $\mathcal{B} \subseteq \mathcal{X}$, or the evolution of the difference of two solutions originating in $\mathcal{B}$, when expressed as a Fourier series with respect to the eigenvectors of the operator A in (4.4), can be controlled by a finite number of terms of the series. In other words, in this series the tail can be dominated by its complementary finite sum. In more suggestive terms, this is often expressed by saying that if the system is not already exponentially contracting from the outset, then the higher modes of the difference of any two solutions can be dominated by the lower modes.

4.2.1   Orthogonal Projections

As usual, we denote by $\| \cdot \|$, $\langle \cdot, \cdot \rangle$ and I, respectively the norm, the scalar product and the identity in $\mathcal{X}$.

Recall that a PROJECTION in $\mathcal{X}$ is a linear map $P \colon \mathcal{X} \to \mathcal{X}$ such that $P^2 = P$. Suppose there exists a closed subspace $\mathcal{W} \subset \mathcal{X}$ such that $\mathcal{X}$ can be decomposed into the direct sum of $\mathcal{W}$ and its orthogonal complement $\mathcal{W}^\perp$, i.e.

$$\mathcal{X} = \mathcal{W} \oplus \mathcal{W}^\perp .$$

Then, every $x \in \mathcal{X}$ can be expressed in a unique way as a sum

$$x = w + z, \quad w \in \mathcal{W}, \quad z \in \mathcal{W}^\perp, \tag{4.5}$$

with $\langle w, z \rangle = 0$ (by orthogonality). Consequently,

$$\|x\|^2 = \|w\|^2 + \|z\|^2. \tag{4.6}$$

We define then the ORTHOGONAL PROJECTION $P\colon \mathcal{X} \to \mathcal{W}$ by

$$P(x) = w, \tag{4.7}$$

with w as in (4.5). This makes sense because w is unique. P is indeed a projection, since for all $x \in \mathcal{X}$,

$$P^2(x) = P(w) = w = P(x);$$

that is, $P^2 = P$. Note that P is linear, and (4.6) can be written as

$$\|x\|^2 = \|P(x)\|^2 + \|(I - P)(x)\|^2. \tag{4.8}$$

From (4.8) we deduce that orthogonal projections are Lipschitz continuous: indeed,

$$\|P(x) - P(y)\|^2 = \|P(x-y)\|^2 = \|x-y\|^2 - \|(I-P)(x-y)\|^2 \leq \|x-y\|^2,$$

and analogously for $I - P$. In fact, orthogonal projections are contractions, albeit not strict ones.

Finally, we recall that orthogonal projections allow us to define a special cone in $\mathcal{X}$, namely

$$\mathcal{C}_P := \{x \in \mathcal{X} \colon \|(I-P)(x)\| \leq \|P(x)\|\}, \tag{4.9}$$

or, equivalently (this is an immediate consequence of (4.8))

$$\mathcal{C}_P = \{x \in \mathcal{X} \colon \|x\| \leq \sqrt{2}\|P(x)\|\}. \tag{4.10}$$

4.2.2 Squeezing Properties

To define the squeezing properties we want to introduce, we consider projections (4.7) defined into a subspace $\mathcal{W} \subseteq \mathcal{X}$ which is finite dimensional. In this case, the opening of the cone $\mathcal{C}_P$ is controlled by a projection of finite rank.

We start with the definition of squeezing property for a map:

DEFINITION 4.2 *A map* $S\colon \mathcal{X} \to \mathcal{X}$ *satisfies the* SQUEEZING PROPERTY *relative to a subset* $\mathcal{B} \subseteq \mathcal{X}$ *if there is a finite dimensional closed subspace* $\mathcal{X}_N \subset \mathcal{X}$ *and a corresponding orthogonal projection* $P_N\colon \mathcal{X} \to \mathcal{X}_N$, *such that the following condition holds: There exists* $\gamma \in \left]0, \frac{1}{2}\right[$ *such that for any* $u, v \in \mathcal{B}$, *either*

$$\|S(u) - S(v)\| \leq \gamma\|u - v\|, \tag{4.11}$$

or, otherwise, $S(u) - S(v) \in \mathcal{C}_{P_N}$, *i.e.*

$$\|(I - P_N)(S(u) - S(v))\| \leq \|P_N(S(u) - S(v))\|. \tag{4.12}$$

This condition means that either S is contractive on the set $\mathcal{B}$, or that *any* difference $S(u) - S(v)$ is in the cone $\mathcal{C}_{P_N}$. Note that, because of (4.9) [or (4.10)], the condition $S(u) - S(v) \notin \mathcal{C}_{P_N}$ can be expressed by either one of the conditions

$$\|P_N(S(u) - S(v))\| < \|(I - P_N)(S(u) - S(v))\|, \tag{4.13}$$

$$\|S(u) - S(v)\| > \sqrt{2}\|P_N(S(u) - S(v))\|. \tag{4.14}$$

Thus, the discrete squeezing property can also be expressed by the requirement that if $u, v \in \mathcal{B}$ are such that $S(u) - S(v) \notin \mathcal{C}_{P_N}$, i.e. if either (4.13) or (4.14) holds, then S is contractive on $\mathcal{B}$, i.e. (4.11) holds.

We can then extend the definition of squeezing property to a semiflow:

DEFINITION 4.3   *Let $\mathcal{B} \subseteq \mathcal{X}$ be bounded, and $S = (S(t))_{t \in T}$ a semiflow in $\mathcal{X}$. S satisfies the* DISCRETE SQUEEZING PROPERTY *relative to $\mathcal{B}$ if there is $t_* \in T$, with $t_* > 0$, such that the map $S_* := S(t_*)$ satisfies the squeezing property of definition 4.2.*

We will see that, in contrast to the case of inertial manifolds, the discrete squeezing property, i.e. the "cone condition" (4.12), is a dichotomy at the specific time t_* only. That is, this condition neither requires nor implies that the squeezing property of definition 4.2 be satisfied by any of the other maps $S(t)$, either at the intermediate times $0 < t < t_*$, or at all later times $t > t_*$ (that is, the invariance of the cone (4.9) for $t > t_*$). This explains in part the qualification of "discrete" in definition 4.3.

4.2.3 Squeezing Properties and Exponential Attractors

The importance of the discrete squeezing property resides in the fact that it provides the main sufficient condition for the existence of an exponential attractor. The basic result concerns discrete semiflows generated by iterated maps:

THEOREM 4.4
Assume the map $S \colon \mathcal{X} \to \mathcal{X}$ is Lipschitz continuous, and $\mathcal{B} \subseteq \mathcal{X}$ is a nonempty, compact, and positively invariant subset. Assume that S satisfies the squeezing property of definition 4.2, relative to $\mathcal{B}$. Then the discrete semiflow $\tilde{S} := (S^n)_{n \in \mathbb{N}}$ admits an exponential attractor $\mathcal{E}$ in $\mathcal{B}$. Moreover, given any $\theta \in {]}2\gamma, 1{[}$, the fractal dimension of $\mathcal{E}$ (see (2.48)) can be estimated by

$$\dim_F(\mathcal{E}) \le N \max\left\{1, \frac{\ln(\frac{2\sqrt{2}L}{\theta - 2\gamma} + 1)}{-\ln\theta}\right\}, \tag{4.15}$$

where N is the rank of the projection appearing in definition 4.2, and L is the Lipschitz constant of S on $\mathcal{B}$.

We postpone the proof of theorem 4.4 to the end of this chapter. By means of theorem 4.4, we can construct an exponential attractor for a general semiflow:

THEOREM 4.5

Assume the continuous semiflow $S = (S(t))_{t \in T}$ satisfies the discrete squeezing property, relative to a nonempty, compact, positively invariant subset $\mathcal{B} \subseteq \mathcal{X}$. Let t_* be as in definition 4.3, set $S_* := S(t_*)$, and let $\mathcal{E}_*$ be the exponential attractor in $\mathcal{X}$ for the semiflow generated by S_*, as in theorem 4.4. Assume further that S satisfies the following properties:

PS1) For all $t \in [0, t_*]$, $S(t)$ is Lipschitz continuous from $\mathcal{B}$ into $\mathcal{B}$, with Lipschitz constant $L(t)$, $L \colon [0, t_*] \to \,]0, +\infty[$ being a bounded function.

PS2) For all $x \in \mathcal{E}_*$, the map $S(\cdot)x$ is Lipschitz continuous from $[0, t_*]$ into $\mathcal{B}$, with Lipschitz constant $L'(x)$, $L' \colon \mathcal{E}_* \to \,]0, +\infty[$ being also a bounded function.

Then, the set

$$\mathcal{E} := \bigcup_{0 \leq t \leq t_*} S(t)\mathcal{E}_* \tag{4.16}$$

is an exponential attractor for S in $\mathcal{B}$. If in addition $\mathcal{B}$ is an absorbing set which attracts all bounded sets of $\mathcal{X}$, then $\mathcal{E}$ is an exponential attractor in $\mathcal{X}$. Moreover, the fractal dimension of $\mathcal{E}$ (see (2.48)) can be estimated by

$$\dim_F(\mathcal{E}) \leq 1 + N \max\left\{1, \frac{\ln(\frac{2\sqrt{2}L(t_*)}{\theta - 2\gamma} + 1)}{-\ln\theta}\right\}, \tag{4.17}$$

where N is the rank of the projection appearing in definition 4.2, $L(t_*)$ is as in (PS1), and $\theta \in \,]2\gamma, 1[$.

From this theorem, which we prove in the next section, it follows that one of the most direct ways to establish the existence of an exponential attractor for a given semiflow S is to show the existence of a compact, positively invariant set $\mathcal{B}$, relative to which the system satisfies the discrete squeezing property. In most cases, $\mathcal{B}$ is a bounded absorbing set, whose existence is implied by the dissipativity of the system. The compactness of $\mathcal{B}$ is usually proven by showing that $\mathcal{B} \subseteq \mathcal{X}_1$, $\mathcal{X}_1$ a compactly imbedded subspace of $\mathcal{X}$. We will follow this procedure for both problems (P) and (H_ε); however, we mention that, as shown in Eden, Foias and Kalantarov, [EFK98], the assumption of compactness of $\mathcal{B}$ is not actually needed, since in both cases S is an α-contraction, as we saw in chapter 3. This possibility of relaxing the requirement of compactness of $\mathcal{B}$ is of course extremely important for applications to situations when, as for hyperbolic equations, the solution operator is not regularizing. At any rate, once a compact, positively invariant set $\mathcal{B}$ has been determined, we proceed to conveniently choose a time t_*, such that the map $S(t_*)$ satisfies the squeezing property relative to $\mathcal{B}$. The evolution of the system on $[0, t_*]$ is then controlled by continuity, since the system is well posed on compact intervals (this "explains" (4.16)), while for $t > t_*$ the system is essentially finite dimensional. Finally, we remark that condition (PS1) was already assumed in theorem 2.56 (with $\mathcal{E}_*$ replaced by $\mathcal{A}_*$), and is usually a consequence of estimates that show that the Cauchy problem relative to the differential equation that generates the semiflow is well posed. On the other hand, condition (PS2) is a consequence of the fact that, usually, the compactness of $\mathcal{B}$ results from its being bounded in a subspace $\mathcal{X}_1$ compactly imbedded in $\mathcal{X}$. In particular, since the solutions we consider are valued in $\mathcal{B}$, this implies that these solutions are differentiable in t, with derivative still valued in $\mathcal{X}$; therefore, they are locally Lipschitz continuous in t as well.

4.2.4   Proof of Theorem 4.5

Recalling definition 4.1, we need to show that the set $\mathcal{E}$ defined by (4.16) is compact, positively invariant, satisfies (4.1), and has finite fractal dimension.

1.   The compactness of $\mathcal{E}$ is proven exactly as in part 1 of the proof of theorem 2.56.

2.   To show that $\mathcal{E}$ is positively invariant, we first note that for all $t \geq 0$,

$$S(t)\mathcal{E} \subseteq \bigcup_{t \leq \theta \leq t_* + t} S(\theta)\mathcal{E}_* . \tag{4.18}$$

Indeed, let $x \in \mathcal{E}$. There is then $\tau \in [0, t_*]$ such that $x \in S(\tau)\mathcal{E}_*$. Thus, $S(t)x \in S(t + \tau)\mathcal{E}_*$, and if $\theta = t + \tau$, $S(t)x \in S(\theta)\mathcal{E}_*$, with $t \leq \theta \leq t_* + t$. This shows (4.18). Fix now $t > 0$, and consider first the case when $0 \leq t \leq t_*$. From (4.18) we obtain that

$$S(t)\mathcal{E} \subseteq \bigcup_{t \leq \theta \leq t_* + t} S(\theta)\mathcal{E}_* = \underbrace{\left(\bigcup_{t \leq \theta \leq t_*} S(\theta)\mathcal{E}_* \right)}_{:= \mathcal{E}_2} \cup \underbrace{\left(\bigcup_{t_* \leq \theta \leq t_* + t} S(\theta)\mathcal{E}_* \right)}_{:= \mathcal{E}_3} .$$

Obviously, $\mathcal{E}_2 \subseteq \mathcal{E}$. As for $\mathcal{E}_3$, setting $\tau = \theta - t_*$ we can write

$$\mathcal{E}_3 = \bigcup_{0 \leq \tau \leq t} S(\tau)S(t_*)\mathcal{E}_* ;$$

and since $\mathcal{E}_*$ is positively invariant with respect to $S_* = S(t_*)$,

$$\mathcal{E}_3 \subseteq \bigcup_{0 \leq \tau \leq t} S(\tau)\mathcal{E}_* \subseteq \mathcal{E} .$$

Thus, $S(t)\mathcal{E} \subseteq \mathcal{E}$ if $0 \leq t \leq t_*$. If instead $t > t_*$, we can decompose $t = nt_* + \theta$, for some $n \in \mathbb{N}$ and $\theta \in [0, t_*]$. Then, $S(t)\mathcal{E} = S(nt_*)S(\theta)\mathcal{E}$, and by the first part of this argument, $S(\theta)\mathcal{E} \subseteq \mathcal{E}$. Consequently, recalling (4.18),

$$S(t)\mathcal{E} \subseteq S(nt_*)\mathcal{E} \subseteq \bigcup_{nt_* \leq s \leq nt_* + t_*} S(s)\mathcal{E}_* .$$

Decomposing $s = nt_* + \tau$, with $0 \leq \tau \leq t_*$, and recalling that $\mathcal{E}_*$ is positively invariant with respect to $S_* = S(t_*)$, we obtain

$$S(t)\mathcal{E} \subseteq \bigcup_{0 \leq \tau \leq t_*} S(\tau)S(nt_*)\mathcal{E}_* = \bigcup_{0 \leq \tau \leq t_*} S(\tau)S_*^n \mathcal{E}_* \subseteq \bigcup_{0 \leq \tau \leq t_*} S(\tau)\mathcal{E}_* = \mathcal{E}.$$

This completes the proof that $\mathcal{E}$ is positively invariant.

3. We now show that $\mathcal{E}$ attracts all bounded subsets of $\mathcal{B}$ exponentially. Since $\mathcal{E}_*$ is an exponential attractor for S_*^n in $\mathcal{B}$, there are two positive constants c_1 and k_1, depending on $\mathcal{B}$, such that for all $n \in \mathbb{N}$

$$\partial(S_*^n \mathcal{B}, \mathcal{E}_*) \leq c_1 e^{-k_1 n}. \tag{4.19}$$

Let then $\mathcal{G} \subseteq \mathcal{B}$ be bounded, and fix $t \geq t_*$.

Given any $x \in S(t)\mathcal{G}$, and $z_* \in \mathcal{E}_*$, let $g \in \mathcal{G}$ be such that $x = S(t)g$, and decompose $t = nt_* + \theta$, for suitable $n \in \mathbb{N}$, and $\theta \in [0, t_*]$. Let $\bar{z} := S(\theta)z_*$. Then, $\bar{z} \in \mathcal{E}$, and recalling (2.42), we can estimate

$$\|x - \bar{z}\| = \|S(t)g - \bar{z}\| = \|S(\theta)S(nt_*)g - S(\theta)z_*\| \leq L_* \|S_*^n g - z_*\|. \tag{4.20}$$

In fact, recalling that $\mathcal{B}$ is positively invariant with respect to S_*^n, and that $\mathcal{G} \subseteq \mathcal{B}$, we have that $S(nt_*)g = S_*^n g \in S_*^n(\mathcal{B}) \subseteq \mathcal{B}$; moreover, also $z_* \in \mathcal{E}_* \subseteq \mathcal{B}$. From (4.20), it follows that

$$\inf_{z \in \mathcal{E}} \|x - z\| \leq \|x - \bar{z}\| \leq L_* \|S_*^n g - z_*\|.$$

Since z_* is arbitrary in $\mathcal{E}_*$,

$$\inf_{z \in \mathcal{E}} \|x - z\| \leq L_* \inf_{z_* \in \mathcal{E}_*} \|S_*^n g - z_*\|. \tag{4.21}$$

Since $S_*^n g \in \mathcal{B}$, recalling the definition of semidistance, we can proceed from (4.21) with

$$\inf_{z \in \mathcal{E}} \|x - z\| \leq L_* \sup_{b \in S_*^n \mathcal{B}} \inf_{z_* \in \mathcal{E}_*} \|b - z_*\| = L_* \partial(S_*^n \mathcal{B}, \mathcal{E}_*). \tag{4.22}$$

Since (4.22) is true for arbitrary $x \in S(t)\mathcal{G}$, it follows that

$$\sup_{x \in S(t)\mathcal{G}} \inf_{z \in \mathcal{E}} \|x - z\| = \partial(S(t)\mathcal{G}, \mathcal{E}) \leq L_* \partial(S_*^n \mathcal{B}, \mathcal{E}_*). \tag{4.23}$$

Recalling (4.19), we deduce from (4.23) that

$$\partial(S(t)\mathcal{G}, \mathcal{E}) \leq L_* c_1 e^{-k_1 n}. \tag{4.24}$$

Now, from $t = nt_* + \theta \leq nt_* + t_*$, we deduce the inequality $-n \leq 1 - t/t_*$; hence we deduce from (4.24) that

$$\partial(S(t)\mathcal{G}, \mathcal{E}) \leq L_* c_1 e^{2k_1} e^{-k_1 t/t_*}.$$

Thus, (4.1) holds, with $c = L_* c_1 e^{2k_1}$ and $k = k_1/t_*$.

4. Finally, we prove estimate (4.17). To this end, note that $\mathcal{E}$ is the image of $[0,t_*] \times \mathcal{E}_*$ under the map F defined in step 1 of the proof of theorem 2.56. Since F is easily to seen to be Lipschitz continuous on $[0,t_*] \times \mathcal{E}_*$ as a consequence of the assumptions PS1) and PS2) on S), by (2.52) and (2.51) of proposition 2.61 we have

$$\dim_F(\mathcal{E}) \leq \dim_F([0,t_*] \times \mathcal{E}_*) \leq \dim_F(\mathcal{E}_*) + \dim_F([0,t_*]).$$

Recalling (4.15), and that, by part 3 of proposition 2.61, the fractal dimension of an interval is 1, we conclude that (4.17) holds. This ends the proof of theorem 4.5.

4.3 The Parabolic Problem

In this section and the next we proceed to show how theorem 4.5 can be applied to deduce the existence of an exponential attractor for the semiflows defined by the IBV problems (P) and (H_ε). Recall that both these systems have a bounded, positively invariant absorbing ball $\mathcal{B}$, as well as a compact attractor $\mathcal{A}$, as shown in chapter 3.

With the same notations of chapter 3, we consider the parabolic IBVP (P) in $\mathcal{X} = L^2(\Omega) = \mathcal{H}$. In section 3.3 we proved that (P) defines a semiflow S on $\mathcal{X}$, which admits a bounded, positively invariant absorbing set $\mathcal{B}$. We now proceed to show that S admits an exponential attractor $\mathcal{E} \subseteq \mathcal{B}$. As outlined in the previous section, we proceed in two steps, first showing that $\mathcal{B}$ contains a compact, positively invariant set $\mathcal{B}_1$, then showing that S satisfies the discrete squeezing property relative to $\mathcal{B}_1$.

4.3.1 Step 1: Absorbing Sets in $\mathcal{X}_1$

We first show that $\mathcal{B}$ contains an absorbing set bounded in $\mathcal{X}_1 = V = H_0^1(\Omega)$, and therefore compact in $\mathcal{X}$. We claim:

PROPOSITION 4.6
Assume $f \in C_b([0,+\infty[;\mathcal{X})$. The semiflow defined by the IBVP (P) admits a positively invariant absorbing set $\mathcal{B}_1$, bounded in $\mathcal{X}_1$.

PROOF We must show that orbits originating in bounded sets of $\mathcal{X}_1$ enter, and eventually remain, in a set $\mathcal{B}_1$ as stated. To this end, we establish further a priori estimates on the solution of problem (P). Formally multiplying equation (3.1) in $\mathcal{X}$ by $-2\Delta u$ (again, this procedure should be carried out at the level of Galerkin approximations), we obtain (omitting as usual the variable t)

$$\frac{d}{dt}\|\nabla u\|^2 + 2\|\Delta u\|^2 + 6\langle u^2 \nabla u, \nabla u\rangle = 2\langle f + u, -\Delta u\rangle.$$

Neglecting the positive term $6\langle u^2 \nabla u, \nabla u \rangle$ and integrating the last term by parts, we further obtain

$$\frac{\mathrm{d}}{\mathrm{d}t}\|\nabla u\|^2 + 2\|\Delta u\|^2 \leq 2\|f\|\,\|\Delta u\| + 2\|\nabla u\|^2.$$

From this we deduce that

$$\frac{\mathrm{d}}{\mathrm{d}t}\|\nabla u\|^2 + \|\Delta u\|^2 \leq \|f\|^2 + 2\|\nabla u\|^2. \tag{4.25}$$

We now recall estimate (3.51) of chapter 3, which we rewrite as

$$\frac{\mathrm{d}}{\mathrm{d}t}\|u\|^2 + 2\|\nabla u\|^2 + |u|_4^4 \leq \|f\|^2 + 3\|u\|^2 \leq \|f\|^2 + C_\Omega + |u|_4^4.$$

Summing this to (4.25) we obtain

$$\frac{\mathrm{d}}{\mathrm{d}t}(\|u\|^2 + \|\nabla u\|^2) + \|\Delta u\|^2 \leq 2\|f\|^2 + C_\Omega \leq K, \tag{4.26}$$

where $K := 2\sup_{t \geq 0}\|f(\cdot,t)\|^2 + C_\Omega$ is independent of t and u_0. From (4.26) we can deduce a linear differential inequality on the norm of u in $\mathcal{X}_1$. Indeed, from the Poincaré inequalities (3.16) and (3.17) we have

$$\|\Delta u\|^2 \geq \beta(\|u\|^2 + \|\nabla u\|^2) \qquad \text{with } \beta := \frac{\lambda_1^2}{1+\lambda_1}.$$

Inserting this into (4.26) we conclude that

$$\frac{\mathrm{d}}{\mathrm{d}t}(\|u\|^2 + \|\nabla u\|^2) + \beta(\|u\|^2 + \|\nabla u\|^2) \leq K,$$

which is the desired linear differential inequality. We can then conclude the proof of proposition 4.6 as in that of proposition 2.64.   □

4.3.2 Step 2: The Discrete Squeezing Property

We now assume that f is independent of t, i.e. $f(\cdot,t) \equiv f \in \mathcal{H}$, and proceed to show that theorem 4.5 can be applied to the corresponding semiflow S generated by (3.1).

THEOREM 4.7
Let $\mathcal{B}_1$ as in proposition 4.6. The semiflow defined by the IBVP (P) satisfies the discrete squeezing property relative to $\mathcal{B}_1$. Consequently, the semiflow generated by problem (P) admits an exponential attractor in $\mathcal{X}$.

PROOF We consider the spaces $\mathcal{X}_N$ and the corresponding orthogonal projections P_N and Q_N defined in (3.26) and (3.27) of section 3.2.3, and refer to the alternative characterization of the discrete squeezing property "(4.13) $\implies$ (4.11)". In fact, we prove a bit more: namely, that, given any $t_* > 0$ and $\gamma \in]0, \frac{1}{2}[$, there exists an integer N_*, with the property that if u_0, $v_0 \in \mathcal{B}_1$ are such that $S(t_*)u_0 - S(t_*)v_0 \notin \mathcal{C}_{N_*}$, i.e. if

$$\|P_{N_*}(S(t_*)u_0 - S(t_*)v_0)\| < \|Q_{N_*}(S(t_*)u_0 - S(t_*)v_0)\| \tag{4.27}$$

(that is, if (4.13) holds for the operator $S(t_*)$), then (4.11) must hold, i.e.

$$\|S(t_*)u_0 - S(t_*)v_0\| \leq \gamma\|u_0 - v_0\|. \tag{4.28}$$

Therefore, S satisfies the discrete squeezing property, relative to $\mathcal{B}_1$.

To this end, following Eden, Foias, Nicolaenko and Temam, [EFNT94, ch. 3], we study the evolution of the so-called "quotient norm"

$$\Lambda(t) := \frac{\|\nabla z(t)\|^2}{\|z(t)\|^2},$$

introduced in (3.58), where $z(t) := S(t)u_0 - S(t)v_0$ is the difference of the two solutions of (3.1) with initial data u_0 and v_0. Our goal is to establish a series of inequalities for Λ, from which we can deduce (4.28). At first we note that, since the projections P_N and Q_N are orthogonal in both $\mathcal{X}$ and $\mathcal{X}_1$, if an inequality like (4.27) holds for generic $N \in \mathbb{N}$ and $t \geq 0$, then, by (3.28),

$$\Lambda(t) = \frac{\|P_N(z(t))\|_1^2 + \|Q_N(z(t))\|_1^2}{\|P_N(z(t))\|^2 + \|Q_N(z(t))\|^2} \geq \frac{\|Q_N(z(t))\|_1^2}{2\|Q_N(z(t))\|^2} \geq \frac{1}{2}\lambda_{N+1}. \tag{4.29}$$

Next, we recall that Λ satisfies the differential inequality (3.61), i.e.

$$\Lambda' \leq C\Lambda,$$

for almost all $t \geq 0$. Integrating this inequality for $0 < s < t$ yields

$$\Lambda(s) \geq \Lambda(t)e^{-C(t-s)};$$

integrating a second time (with respect to s in the interval $[0,t]$) we obtain

$$\int_0^t \Lambda(s)\,\mathrm{d}s \geq \Lambda(t)\int_0^t e^{C(s-t)}\,\mathrm{d}s = \frac{1}{C}\Lambda(t)(1 - e^{-Ct}). \tag{4.30}$$

Our last step is an estimate of z in terms of the integral at the left side of (4.30). Recalling (3.47), z satisfies, for almost all $t > 0$, the inequality

$$\frac{\mathrm{d}}{\mathrm{d}t}\|z\|^2 + 2\|\nabla z\|^2 = -2\langle g(u) - g(v), z\rangle \leq 2\|z\|^2. \tag{4.31}$$

Thus, recalling the definition of Λ, we obtain from (4.31)

$$\frac{\mathrm{d}}{\mathrm{d}t}\|z(t)\|^2 + 2\Lambda(t)\|z(t)\|^2 \leq 2\|z(t)\|^2;$$

applying proposition 2.63 we obtain, for $t \geq 0$,

$$\|z(t)\|^2 \leq \|z(0)\|^2 \exp\left(2t - 2\int_0^t \Lambda(s)\,ds\right). \tag{4.32}$$

We are now ready to conclude. Replacing (4.30) into (4.32), we obtain that, if $t > 0$ is such that (4.29) holds, then

$$\|z(t)\|^2 \leq \|z(0)\|^2 \exp\left(2t - \tfrac{1}{C}\lambda_{N+1}(1 - e^{-Ct})\right). \tag{4.33}$$

Recalling that $\lambda_N \to +\infty$ as $n \to +\infty$, given $t_* > 0$ and $\gamma \in]0, \tfrac{1}{2}[$ we choose N_* so large that

$$-\tfrac{1}{C}\lambda_{N_*+1}(1 - e^{-Ct_*}) \leq 2\ln\gamma - 2t_* \tag{4.34}$$

(note that the right side of (4.34) is negative). With this choice of N_*, (4.33) implies that

$$\|z(\cdot, t_*)\| \leq \gamma \|z(\cdot, 0)\|$$

that is, (4.28) holds. As we have already remarked, conditions (PS1) and (PS2) of theorem 4.5 hold, because problem (P) is well posed, and the corresponding semiflow is differentiable. Thus, the proof of theorem 4.7 is complete. $\quad\square$

REMARK 4.8 In estimate (4.31), we have made use of the fact that the nonlinearity $g(u) = u^3 - u$ satisfies

$$\langle g(u) - g(v), u - v \rangle \geq -\|u - v\|^2.$$

More generally, the right side of (4.31) is estimated by $2L\|z\|^2$, where L is the Lipschitz constant of g, and (4.33) becomes

$$\|z(t)\|^2 \leq \|z(0)\|^2 \exp\left(2Lt - \tfrac{1}{2C}\lambda_{N+1}(1 - e^{-Ct})\right);$$

note that C also depends on L. N_* is then chosen so that

$$-\tfrac{1}{2}\lambda_{N_*+1}\left(1 - e^{-Ct_*}\right) \leq 2\ln\gamma - 2Lt_*.$$

In any case, the choice of t_* is independent of $\mathcal{B}_1$. Actually, in the proof of theorem 4.7 we see that the set $\mathcal{B}_1$ can be replaced by the set $\mathcal{B}$. In fact, the compact set $\mathcal{B}_1$ is needed only because it is required in theorem 4.5. In conclusion, the only relevant property we have actually needed is the Lipschitz continuity of the nonlinearity g, locally in $\mathcal{V}$. It follows that the local Lipschitz continuity of g in $\mathcal{V}$ is a sufficient condition for the existence of an exponential attractor for the semiflow generated by a semilinear parabolic PDE of the form (3.1). $\quad\square$

4.4 The Hyperbolic Problem

We now turn to the hyperbolic IBV problem (H_ε) in $\mathcal{X} := H_0^1(\Omega) \times L^2(\Omega)$. In section 3.4 we proved that (3.4) defines a semiflow S on $\mathcal{X}$, which admits a bounded, positively invariant absorbing set $\mathcal{B}_0$. We now proceed to show that S admits an exponential attractor $\mathcal{E} \subseteq \mathcal{B}_0$. Exactly as in the previous section, we proceed in two steps, first showing that $\mathcal{B}_0$ contains a compact, positively invariant set $\mathcal{B}_1$, and then that S satisfies the discrete squeezing property relative to $\mathcal{B}_1$. As in chapter 3, we set $\mathcal{H} := L^2(\Omega)$, $\mathcal{V} := H_0^1(\Omega)$, and choose in $\mathcal{X}$ the norm

$$\|(u,v)\|_{\mathcal{X}}^2 := \|\nabla u\|^2 + \varepsilon \|v\|^2.$$

We also consider the subspaces $\mathcal{D} := H^2(\Omega) \cap H_0^1(\Omega)$ and $\mathcal{X}_1 := \mathcal{D} \times \mathcal{V}$. Since we can choose on $\mathcal{D}$ the norm $\|u\|_{\mathcal{D}} = \|\Delta u\|$, we consider in $\mathcal{X}_1$ the norm

$$\|(u,v)\|_{\mathcal{X}_1}^2 := \|\Delta u\|^2 + \varepsilon \|\nabla v\|^2.$$

This norm is equivalent to the norm defined by

$$E_1(u,v) := \varepsilon \|\nabla v\|^2 + \langle \nabla v, \nabla u \rangle + \tfrac{1}{2\varepsilon} \|\nabla u\|^2 + \|\Delta u\|^2.$$

(We recall that to distinguish between pairs of functions (u,v) and their scalar product in $L^2(\Omega)$, the latter is denoted by $\langle u, v \rangle$.)

4.4.1 Step 1: Absorbing Sets in $\mathcal{X}_1$

We first show that, at least if ε is sufficiently small, $\mathcal{B}_0$ contains an absorbing set bounded in $\mathcal{X}_1$, and therefore compact in $\mathcal{X}$. We claim:

PROPOSITION 4.9
Assume f is independent of t, i.e. $f(t) \equiv f \in \mathcal{H}$. Let λ_1 be as in (3.16), and $\varepsilon_1 := \min\{1, \frac{1}{\lambda_1}\}$. For all $\varepsilon \in]0, \varepsilon_1]$, the semiflow defined by the IBVP (3.4) admits a positively invariant absorbing set $\mathcal{B}_1$, bounded in $\mathcal{X}_1$.

PROOF We must show that orbits originating in bounded sets of $\mathcal{X}_1$ enter, and eventually remain, in a set $\mathcal{B}_1$ as stated. To this end, as in the parabolic case, we establish further a priori estimates on the solution of (3.4). Multiplying (formally) the equation of (3.4) in $\mathcal{H}$ by $-2\Delta u_t - \frac{1}{\varepsilon}\Delta u$, and adding the term $\frac{1}{2}\lambda_1\langle f, \Delta u\rangle$ to both sides, we obtain

$$\frac{\mathrm{d}}{\mathrm{d}t}\left(E_1(u,u_t) + 2\langle f, \Delta u\rangle\right) + \|\nabla u_t\|^2 + \tfrac{1}{\varepsilon}\|\Delta u\|^2 + \tfrac{1}{2}\lambda_1\langle f, \Delta u\rangle$$
$$= 2\langle u^3 - u, \Delta u_t\rangle + \langle \rho f + u^3 - u, \tfrac{1}{\varepsilon}\Delta u\rangle, \tag{4.35}$$

with $\rho := \frac{1}{2}\varepsilon\lambda_1 - 1$. Recalling proposition 3.21, we have a bound on $\|\nabla u(\cdot,t)\|$ independent of t and ε. Consequently, we can estimate the right side of (4.35) as follows. At first, using the elliptic estimate

$$\|u(\cdot,t)\|_{H^2(\Omega)} \le C(\|\Delta u(\cdot,t)\| + \|u(\cdot,t)\|)$$

(see theorem A.77), and denoting by C various different positive constants, depending on the uniform bound on $\|\nabla u(\cdot,t)\|$, but not on t nor on ε, we have

$$\begin{aligned}
2\langle u^3 - u, \Delta u_t \rangle = -2\langle \nabla u - 3u^2 \nabla u, \nabla u_t \rangle &\le C(\|\nabla u\| + |u|_6^2 |\nabla u|_6)\|\nabla u_t\| \\
&\le C(1 + \|\nabla u\|^2 \|u\|_2)\|\nabla u_t\| \le C(1 + \|\Delta u\|)\|\nabla u_t\| \\
&\le \tfrac{1}{2}\|\nabla u_t\|^2 + \tfrac{1}{4\varepsilon}\|\Delta u\|^2 + \tfrac{1}{\varepsilon}C
\end{aligned} \tag{4.36}$$

(recall that $\varepsilon \le 1$). Next, analogously:

$$\begin{aligned}
\tfrac{1}{\varepsilon}\langle \rho f + u^3 - u, \Delta u \rangle &\le \tfrac{1}{\varepsilon}(\|\rho f\| + |u|_6^3 + \|u\|)\|\Delta u\| \le \tfrac{1}{\varepsilon}(\|\rho f\| + \|\nabla u\|^3 + \|u\|)\|\Delta u\| \\
&\le C\tfrac{1}{\varepsilon}\|\Delta u\| \le C\tfrac{1}{\varepsilon} + \tfrac{1}{4\varepsilon}\|\Delta u\|^2 .
\end{aligned}$$

Replacing this and (4.36) into (4.35) we obtain

$$\frac{d}{dt}(E_1(u,u_t) + 2\langle f, \Delta u \rangle) + \tfrac{1}{2}(\|\nabla u_t\|^2 + \tfrac{1}{\varepsilon}\|\Delta u\|^2) + \tfrac{1}{2}\lambda_1\langle f, \Delta u \rangle \le \tfrac{1}{\varepsilon}C. \tag{4.37}$$

Again, we check that, since $\varepsilon \le \frac{1}{\lambda_1}$,

$$E_1(u,v) \le \tfrac{2}{\lambda_1}(\|\nabla v\|^2 + \tfrac{1}{\varepsilon}\|\Delta u\|^2)$$

for $(u,v) \in \mathcal{X}_1$. Consequently, from (4.37) we derive the exponential inequality

$$\frac{d}{dt}(E_1(u,u_t) + 2\langle f, \Delta u \rangle) + \tfrac{1}{4}\lambda_1(E_1(u,u_t) + 2\langle f, \Delta u \rangle) \le \tfrac{1}{\varepsilon}C.$$

From this we deduce, by proposition 2.63, that for all $t \ge 0$

$$E_1(u(t),u_t(t)) + 2\langle f, \Delta u(t) \rangle \le \left(E_1(u_0,u_1) + 2\langle f, \Delta u_0 \rangle - \tfrac{4C}{\lambda_1\varepsilon}\right)e^{-\lambda_1 t/4} + \tfrac{4C}{\lambda_1\varepsilon}.$$

As usual, this implies that for all $R > \frac{4C}{\lambda_1\varepsilon}$, the set

$$\mathcal{B}_1 := \{(u,v) \in \mathcal{X}_1 : E_1(u,v) + 2\langle f, \Delta u \rangle \le R\} \cap \mathcal{B}$$

is bounded, absorbing and positively invariant for S. This ends the proof of proposition 4.9. $\square$

4.4.2   Step 2: The Discrete Squeezing Property

We now show that, again if ε is sufficiently small, theorem 4.5 can be applied to the semiflow S generated by (3.4).

THEOREM 4.10
Let $\mathcal{B}_1$ and ε_1 be as in proposition 4.9. There exists $\varepsilon_2 \in]0, \varepsilon_1]$ such that if $\varepsilon \in]0, \varepsilon_2]$, the semiflow defined by the IBVP (3.4) satisfies the discrete squeezing property, relative to $\mathcal{B}_0$. Consequently, the semiflow generated by problem (H_ε) admits an exponential attractor in $\mathcal{X}$.

PROOF   We follow Eden, Milani and Nicolaenko, [EMN92]. Here too, we refer to the alternative characterization of the discrete squeezing property "(4.13) $\Longrightarrow$ (4.11)". We consider the same spaces $\mathcal{X}_N$ and projections P_N, Q_N introduced in (3.26) and (3.27), and define corresponding product projections on $\mathcal{X} = \mathcal{V} \times \mathcal{H}$ by

$$\mathcal{P}_N: \mathcal{X} \to (P_N(\mathcal{V}) \times P_N(\mathcal{H})), \quad \mathcal{Q}_N = I - \mathcal{P}_N$$

in the canonical way, i.e.

$$\mathcal{P}_N(u, v) := (P_N(u), P_N(v)), \quad \mathcal{Q}_N(u, v) = (Q_N(u), Q_N(v)). \tag{4.38}$$

We must find t_* and $N_* \in \mathbb{N}$ such that if (u_0, u_1), $(v_0, v_1) \in \mathcal{B}_1$ are such that

$$S(t_*)(u_0, u_1) - S(t_*)(v_0, v_1) \notin \mathcal{C}_{N_*},$$

i.e. if

$$\|\mathcal{P}_{N_*}(S(t_*)(u_0, u_1) - S(t_*)(v_0, v_1))\|_{\mathcal{X}} < \|\mathcal{Q}_{N_*}(S(t_*)(u_0, u_1) - S(t_*)(v_0, v_1))\|_{\mathcal{X}}, \tag{4.39}$$

then (4.11) must hold for some $\gamma \in]0, \frac{1}{2}[$, i.e.

$$\|S(t_*)(u_0, u_1) - S(t_*)(v_0, v_1)\|_{\mathcal{X}} \le \gamma \|(u_0, u_1) - (v_0, v_1)\|_{\mathcal{X}}. \tag{4.40}$$

In fact, we shall prove slightly more: namely, that for all $t_* > 0$ and $\gamma \in]0, \frac{1}{2}[$ there exist $\varepsilon_2 \in]0, \varepsilon_1]$ such that for all $\varepsilon \in]0, \varepsilon_2]$, there exists $N_* \in \mathbb{N}$ such that (4.40) holds. To this end, we define on $\mathcal{X}$ the function

$$M(u, v) := \langle u, v \rangle_{\mathcal{H}} + \|(u, v)\|_{\mathcal{X}}^2 = \varepsilon \|v\|^2 + \|\nabla u\|^2 + \langle u, v \rangle_{\mathcal{H}},$$

and show that M is the square of an equivalent norm in $\mathcal{Q}_N(\mathcal{X})$.

PROPOSITION 4.11
Assume $\varepsilon \le 1$, and let $N \in \mathbb{N}$ be such that $\varepsilon \lambda_{N+1} \ge 1$ (this is possible since $\lambda_N \to +\infty$). Then for all $(u, v) \in \mathcal{Q}_N(\mathcal{X})$,

$$\|(u, v)\|_{\mathcal{X}}^2 \le 2M(u, v) \le 3\|(u, v)\|_{\mathcal{X}}^2. \tag{4.41}$$

PROOF   This result is a consequence of proposition 3.6 and Schwarz' inequality. Indeed, if $(u,v) \in \mathcal{Q}_N(\mathcal{X})$, by (3.28)

$$\langle u,v \rangle_{\mathcal{H}} \le \tfrac{1}{2\varepsilon}\|u\|^2 + \tfrac{1}{2}\varepsilon\|v\|^2 \le \tfrac{1}{2\varepsilon\lambda_{N+1}}\|\nabla u\|^2 + \tfrac{1}{2}\varepsilon\|v\|^2 .$$

$\square$

We now estimate the difference of two solutions of (3.4), whose orbits are in $\mathcal{B}_0$. If u and v are two such solutions, corresponding to initial values $U_0 := (u_0,u_1)$, $V_0 := (v_0,v_1) \in \mathcal{B}_1$, we set $z(t) := u(\cdot,t) - v(\cdot,t)$, and $Z(t) := (z(t),z_t(t))$. At first, we recall estimate (3.71) of section 3.4.1, which provides a control of the growth of Z on bounded time intervals. More precisely,

$$\|Z(t)\|_{\mathcal{X}} \le \|Z(0)\|_{\mathcal{X}} e^{ct} , \tag{4.42}$$

for suitable $c > 0$ independent of the solutions u and v. Next, we establish a linear differential inequality on Z in $\mathcal{Q}_N(\mathcal{X})$.

PROPOSITION 4.12

Let $N \in \mathbb{N}$ be as in proposition 4.11, and $\varphi_N := \mathcal{Q}_N(z)$, $\Phi_N := (\varphi_N,(\varphi_N)_t) = \mathcal{Q}_N(Z)$. Then $M(\Phi_N)$ satisfies, for almost all $t \ge 0$, the linear differential inequality

$$\frac{\mathrm{d}}{\mathrm{d}t}M(\Phi_N(t)) + \tfrac{1}{2\varepsilon}M(\Phi_N(t)) \le K\|\nabla z(t)\|^2 , \tag{4.43}$$

where K is independent of t and ε.

PROOF   The procedure is similar to the proof of proposition 3.21. Applying $\mathcal{Q}_N$ to the equation satisfied by z, i.e

$$\varepsilon z_{tt} + z_t - \Delta z = g(v) - g(u) ,$$

and noting that $\mathcal{Q}_N$ commutes with $-\Delta$, we see that, abbreviating $\varphi_N = \varphi$, φ satisfies the equation

$$\varepsilon \varphi_{tt} + \varphi_t - \Delta \varphi = \mathcal{Q}_N(g(v) - g(u)) =: G . \tag{4.44}$$

Multiplying (4.44) in $\mathcal{H}$ by $2\varphi_t + \tfrac{1}{\varepsilon}\varphi$ and splitting one of the terms in two, we obtain

$$\frac{\mathrm{d}}{\mathrm{d}t}M(\Phi) + \|\varphi_t\|^2 + \tfrac{1}{2\varepsilon}\langle \varphi_t,\varphi \rangle + \tfrac{1}{\varepsilon}\|\nabla\varphi\|^2 = \underbrace{-\tfrac{1}{2\varepsilon}\langle \varphi_t,\varphi \rangle}_{=:H_1} + \underbrace{\langle G, 2\varphi_t + \tfrac{1}{\varepsilon}\varphi \rangle}_{=:H_2} . \tag{4.45}$$

By Schwarz' inequality and proposition 3.6, we estimate

$$H_1 \le \tfrac{1}{4}\|\varphi_t\|^2 + \tfrac{1}{4\varepsilon^2}\|\varphi\|^2 \le \tfrac{1}{4}\|\varphi_t\|^2 + \tfrac{1}{4\varepsilon^2}\lambda_{N+1}^{-1}\|\nabla\varphi\|^2$$

$$\leq \tfrac{1}{4}\|\varphi_t\|^2 + \tfrac{1}{4\varepsilon}\|\nabla\varphi\|^2, \tag{4.46}$$

where we have used $\varepsilon\lambda_{N+1} \geq 1$. To estimate H_2, we recall that, by proposition 3.21, $u(\cdot,t)$ and $v(\cdot,t)$ remain in a bounded set of $\mathcal{V}$ for all $t \geq 0$, with bounds independent of ε if ε is small. Thus, recalling proposition 3.15, we can estimate

$$H_2 \leq 2\|G\|\left(\|\varphi_t\| + \frac{1}{\varepsilon}\|\varphi\|\right) \leq 2C\|\nabla z\|\left(\|\varphi_t\| + \frac{1}{\varepsilon}\|\varphi\|\right)$$

$$\leq C_1\|\nabla z\|^2 + \tfrac{1}{4}\|\varphi_t\|^2 + \tfrac{1}{4\varepsilon^2}\lambda_{N+1}^{-1}\|\nabla\varphi\|^2, \tag{4.47}$$

where C and C_1 depend on the uniform bounds on $\|\nabla u(\cdot,t)\|$ and $\|\nabla v(\cdot,t)\|$, but not on t nor on ε. Inserting (4.46) and (4.47) into (4.45), we easily deduce (4.43). ☐

We are now ready to conclude the proof of theorem 4.10. From (4.42) and (4.43) we obtain

$$\frac{\mathrm{d}}{\mathrm{d}t}M(\Phi_N) + \tfrac{1}{2\varepsilon}M(\Phi_N) \leq K E_0(z(t)) \leq K E_0(z(0))\mathrm{e}^{ct}.$$

Integrating, we deduce that

$$M(\Phi_N(t)) \leq M(\Phi_N(0))\mathrm{e}^{-t/2\varepsilon} + 2\varepsilon K E_0(z(0))\mathrm{e}^{ct}. \tag{4.48}$$

Thus, recalling (4.41),

$$\|\Phi_N(t)\|_{\mathcal{X}}^2 \leq 3\|\Phi_N(0)\|_{\mathcal{X}}^2\,\mathrm{e}^{-t/2\varepsilon} + 6\varepsilon K E_0(z(0))\mathrm{e}^{ct}$$

$$\leq 3E_0(z(0))(\mathrm{e}^{-t/2\varepsilon} + 2\varepsilon K\mathrm{e}^{ct}). \tag{4.49}$$

Given then $t_* > 0$ and $\gamma \in\,]0,\tfrac{1}{2}[$, we first choose $\varepsilon_2 \in\,]0,\varepsilon_1]$ so small that

$$3(\mathrm{e}^{-t_*/\varepsilon_2} + 2\varepsilon_2 K\mathrm{e}^{ct_*}) \leq \tfrac{1}{2}\gamma^2;$$

then, given $\varepsilon \in\,]0,\varepsilon_2]$, we choose N_* so large that

$$\lambda_{N_*+1} \geq \tfrac{1}{\varepsilon}. \tag{4.50}$$

With these choices, (4.49) implies that

$$\|\Phi_{N_*}(t_*)\|_{\mathcal{X}}^2 \leq \tfrac{1}{2}\gamma^2\,\|Z(0)\|_{\mathcal{X}}^2. \tag{4.51}$$

Recalling that $\Phi_N = \mathcal{Q}_N(Z)$, from (4.39) and (4.51) we deduce that

$$\|Z(t_*)\|_{\mathcal{X}}^2 \leq 2\|\Phi_{N_*}(t_*)\|_{\mathcal{X}}^2 \leq \gamma^2\|Z(0)\|_{\mathcal{X}}^2,$$

that is, (4.40) holds. As a consequence of proposition 4.12, we can then conclude the proof of theorem 4.10, by means of theorem 4.5. Indeed, as we have remarked for the parabolic problem (P), assumptions (PS1) and (PS2) are satisfied, because problem (H_ε) is well posed in $\mathcal{X}$, and the corresponding semiflow S is differentiable. Finally, note that, by (4.50), in general $N_* \to +\infty$ if $\varepsilon \to 0$. ☐

REMARK 4.13   1. As in the parabolic case (see remark 4.8), the proof of theorem 4.10 does not require the existence of the compact, positively invariant absorbing set $\mathcal{B}_1$. This is only required by theorem 4.5, and is the only reason we have introduced proposition 4.9. However, this condition is not a necessary one, as shown in Eden, Foias and Kalantarov, [EFK98].

2. As was the case for the global attractors, a natural question is to know whether any result on the upper or lower semicontinuity would hold for the exponential attractors of problems (H_ε) and (P), whose existence we have established in theorems 4.7 and 4.10. In the case of the global attractors, we were able to answer this question in section 3.6. However, the proof of theorem 3.31 required in an essential way the invariance of the attractors, at least for $t \geq 0$. Since we only know that the exponential attractors are positively invariant for $t \geq 0$, this procedure cannot be applied. On the other hand, the upper semicontinuity of the exponential attractors may follow from some recent results of Fabrie, Galinski, Miranville and Zelik, [FGMZ03], and of Gatti, Grasselli, Miranville and Pata, [GGMP03].   □

4.5 Proof of Theorem 4.4

In this section we prove theorem 4.4, following in large part the procedure presented in chapter 2 of Eden, Foias, Nicolaenko and Temam, [EFNT94].

4.5.1 Outline

We refer to the second alternative characterization of the discrete squeezing property "(4.14) $\Longrightarrow$ (4.11)". More precisely, we assume that the map $S\colon \mathcal{X} \to \mathcal{X}$ is Lipschitz continuous, that $\mathcal{B}$ is a nonempty, compact and positively invariant subset of $\mathcal{X}$, that $\mathcal{X}$ admits an N-dimensional subspace $\mathcal{X}_N$, with corresponding orthogonal projection $P_N\colon \mathcal{X} \mapsto \mathcal{X}_N$, and, finally, that there is some $\gamma \in]0, \frac{1}{2}[$, with the property that, if $u, v \in \mathcal{B}$ are such that

$$\|S(u) - S(v)\| \geq \sqrt{2}\|P_N(S(u)) - P_N(S(v))\|, \tag{4.52}$$

then

$$\|S(u) - S(v)\| \leq \gamma\|u - v\|. \tag{4.53}$$

In the sequel, we omit the dependence on N for convenience, i.e. we write $P_N = P$; also, we write Px instead of $P(x)$, etc.

We know from theorem 2.31 that $\mathcal{A} = \omega(\mathcal{B})$ is the global attractor of S. We construct $\mathcal{E}$ as the union of $\mathcal{A}$ and a set generated by an iterative procedure, at each stage of which we add to $\mathcal{A}$ a set of points which are "bad", in the sense that the orbits generating from these points do not converge exponentially to $\mathcal{A}$. Since exponential convergence is a consequence of the strict contractivity of S, these points must lie in a set where (4.53) doesn't hold. Then, the squeezing property asserts that (4.52) must hold for points in this set. Thus, at each stage, the fundamental step of each iteration consists in adding these "noncontractive" points to the set $S^n(\mathcal{B})$, by covering it with a minimal number of balls of decreasing size. The number of these balls is seen to be not increasing with n, as a consequence of the squeezing property. This provides a control, at each step, of the size of the enlarged set obtained by the covering of $S^n(\mathcal{B})$. The inductive limit of this process produces a set, which is shown to be the desired exponential attractor $\mathcal{E}$.

In this construction, we make essential use of the facts that the global attractor $\mathcal{A}$ is invariant, i.e. $\mathcal{A} = S(\mathcal{A})$, and the subset $\mathcal{B}$ is positively invariant, i.e. $S(\mathcal{B}) \subseteq \mathcal{B}$. We shall also see that the squeezing property implies the exponential convergence to $\mathcal{E}$ of the orbits which start at all other points of $\mathcal{X}$. In addition, the rate of convergence of the orbits in this construction of the exponential attractor depends only on γ; more precisely, the constant k in (4.1) is only required to satisfy the bounds $0 < k < -\ln(2\gamma)$ (recall that $\gamma < \frac{1}{2}$). Finally, we remark that this construction of the exponential attractor is not the only possible one; for alternative constructions, see e.g. chapter 7 of [EFNT94].

4.5.2 The Cone Property

Recalling definition (4.10) of the cone $\mathcal{C}_P$ defined by the orthogonal projection P, we give

DEFINITION 4.14 *Let $\mathcal{C} \subseteq \mathcal{X}$. $\mathcal{C}$ satisfies the* CONE PROPERTY *if for all $x, y \in \mathcal{C}$,*

$$\|x - y\| \leq \sqrt{2}\|Px - Py\|. \tag{4.54}$$

For instance, any singleton $\mathcal{C} = \{x\}$ trivially satisfies the cone property. The following is a more interesting example, which will be the basis of the construction of inertial manifolds in chapter 5.

Example 4.15
The graph of a Lipschitz continuous function $m \colon P(\mathcal{X}) \to (I-P)(\mathcal{X})$, with Lipschitz constant $L \leq 1$, satisfies the cone property. That is, any set of the form

$$\mathcal{C}(m) := \{x \in \mathcal{X} : x = \xi + m(\xi), \ \xi \in P(\mathcal{X})\}, \tag{4.55}$$

with m satisfying the global Lipschitz condition

$$\|m(\xi_1) - m(\xi_2)\| \leq L\|\xi_1 - \xi_2\|, \quad 0 < L \leq 1,$$

satisfies the cone property. Indeed, if $x = \xi + m(\xi)$ and $y = \eta + m(\eta) \in \mathcal{C}(m)$, then, recalling that the projections are orthogonal,

$$\|x - y\|^2 = \|\xi - \eta\|^2 + \|m(\xi) - m(\eta)\|^2 \leq (1 + L^2)\|\xi - \eta\|^2$$

$$\leq 2\|\xi - \eta\|^2 = 2\|Px - Py\|^2\,;$$

i.e., (4.54) holds.   □

Note also that the sets $C(m)$ defined in (4.55) are MAXIMAL with respect to the cone property (4.54), in the sense that if C' is another set satisfying the cone property, and $C(m) \subseteq C'$, then $C(m) = C'$. To show this, let $x \in C'$, and set $y := Px + m(Px)$. Then $y \in C(m) \subseteq C'$, and $Py = Px$. Since C' satisfies the cone property, (4.54) implies

$$\|x - y\| \leq \sqrt{2}\|Px - Py\| = 0. \tag{4.56}$$

Hence, $x = y \in C(m)$, so that $C' \subseteq C(m)$. This motivates the following

DEFINITION 4.16   *Let* $C \subseteq \mathcal{Z} \subseteq \mathcal{X}$. *$C$ is MAXIMAL with respect to the cone property (4.54) in $\mathcal{Z}$ if C satisfies the cone property and, whenever $C \subseteq C' \subseteq \mathcal{Z}$ and C' satisfies the cone property, $C' = C$.*

For example, given $\mathcal{Z} \subseteq \mathcal{X}$, each set $C(m) \cap \mathcal{Z}$, with $C(m)$ defined in (4.55), is maximal with respect to the cone property in $\mathcal{Z}$.

We next show that we can always construct subsets of any given nonempty set, which are maximal with respect to the cone property in this set.

PROPOSITION 4.17
Let $\mathcal{Z} \subseteq \mathcal{X}$, *with* $\mathcal{Z} \neq \emptyset$. *Choose* $z \in \mathcal{Z}$. *Then the set*

$$C := \bigcup \{ \mathcal{D} \subseteq \mathcal{Z} : z \in \mathcal{D}, \, \mathcal{D} \text{ has the cone property} \} \tag{4.57}$$

is maximal with respect to the cone property in $\mathcal{Z}$.

PROOF   Obviously, the singleton $\{z\}$ is in $C \subseteq \mathcal{Z}$, since it trivially satisfies the cone property. Let $C' \supseteq C$ be another subset of $\mathcal{Z}$ which satisfies the cone property. Then $z \in C'$ and, therefore, C' is one of the sets in the union at the right side of (4.57). Hence, $C' \subseteq C$; therefore, $C' = C$.   □

Thus, without loss of generality, we can consider sets which are maximal with respect to the cone property (4.54) in any given nonempty set $\mathcal{Z}$.

PROPOSITION 4.18
Let $C \subseteq \mathcal{Z} \subseteq \mathcal{X}$ *be maximal with respect to the cone property in* $\mathcal{Z}$, *and let* $x \in \mathcal{X}$. *Then* $x \in C$ *if and only if there is* $y \in C$ *such that* $y - x \in C_P$.

PROOF   Assume first that $x \in C$. Taking $y = x$, the cone property (4.54) is trivially satisfied. Conversely, assume there is $y \in C$ such that x and y satisfy (4.54). Then x

and y belong to a larger set $\tilde{C} \subseteq \mathcal{Z}$ satisfying the cone property. Since C is maximal, $\tilde{C} = C$; hence, $x \in C$. ☐

The following two results do not require C to be maximal, but only to satisfy the cone property.

PROPOSITION 4.19
Let $C \subseteq \mathcal{X}$ satisfy the cone property. The orthogonal projection P is injective on C.

PROOF   This is an immediate consequence of (4.56). ☐

PROPOSITION 4.20
Let $C \subseteq \mathcal{X}$ satisfy the cone property. Then, C has no interior.

PROOF   Arguing by contradiction, assume the interior of C contains a point x. There is then $r > 0$ such that $B(x,r) \subset C$. Let $y \in B(x,r)$ be such that $y \neq x$ and $y \neq Py$, and set

$$z := x + \frac{r}{2} \frac{y - Py}{\|y - Py\|}$$

(if $x \neq Px$, we can take $y = x$). Then $z \in B(x,r)$, since $\|z - x\| = \frac{r}{2}$. Thus, both x and z are in C, so by the cone property (4.54),

$$\|z - x\| \leq \sqrt{2}\|Pz - Px\| = \sqrt{2}\left\|Px + \frac{r}{2}\frac{P(y - Py)}{\|y - Py\|} - Px\right\| = \sqrt{2}\frac{r}{2}\left\|\frac{Py - P^2y}{\|y - Py\|}\right\|.$$

Recalling that $P^2 = P$, this implies a contradiction. ☐

4.5.3   The Basic Covering Step

In this section we describe the basic step in the construction of the covering of the global attractor $\mathcal{A}$; this step, repeated inductively, will provide the basis for the construction of an exponential attractor $\mathcal{E}$, containing $\mathcal{A}$.

We adopt the following notations: For $a \in \mathcal{X}$ and $r > 0$, $\overline{B}(a,r)$ is the *closed* ball with center a and radius r. As in (2.39) of section 2.7.1, $\mathrm{diam}(\mathcal{C})$ denotes the diameter of a subset $\mathcal{C} \subset \mathcal{X}$.

In the remaining part of this section, we fix a compact set $\mathcal{K} \subset \mathcal{X}$ and, given $a \in \mathcal{K}$ and $r > 0$ we define the set

$$\mathcal{Z} := S(\mathcal{K} \cap \overline{B}(a,r)). \tag{4.58}$$

Since S is continuous and $\mathcal{K}$ is compact, $\mathcal{Z}$ is also compact.

PROPOSITION 4.21
Let $C \subseteq \mathcal{Z}$ be a set which is maximal with respect to the cone property (4.54). Then C is compact in $\mathcal{X}$.

PROOF   Since C is a subset of the compact set $\mathcal{Z}$, it is sufficient to show that C is closed. Thus, let $(x_m)_{m \in \mathbb{N}} \subset C$ be such that $x_m \to x$. Since $C \subseteq \mathcal{Z}$ and $\mathcal{Z}$ is compact, $x \in \mathcal{Z}$. By proposition 4.18, for each $m \in \mathbb{N}$ there is $y_m \in C$ such that $x_m - y_m \in C_P$. Since $(y_m)_{m \in \mathbb{N}} \subset \mathcal{Z}$, there is a subsequence (y_{m_k}) converging to a limit $y \in \mathcal{Z}$ as $k \to +\infty$. Since P is continuous, by (4.54)

$$\|x - y\| = \lim_{k \to +\infty} \|x_{m_k} - y_{m_k}\| \leq \sqrt{2} \lim_{k \to +\infty} \|Px_{m_k} - Py_{m_k}\| = \sqrt{2} \|Px - Py\|.$$

This means that $x - y \in C_P$. Thus, x and y belong to a larger set $\tilde{C} \subseteq \mathcal{Z}$ satisfying the cone property. Since C is maximal with respect to the cone property, $\tilde{C} = C$; hence, $x \in C$, and C is closed.   ☐

Since C is compact and P is continuous, the set $P(C)$ is also compact. Thus, given any $\rho > 0$, $P(C)$ can be covered by a finite number $K_0(\rho)$ of balls $B(Py_j, \rho)$, $j = 1, \ldots, K_0(\rho)$, $y_j \in C$. Since $Py_j \neq Py_k$ (by proposition 4.19), we can obviously choose these balls so that

$$\|Py_j - Py_k\| > \rho \qquad \text{if } j \neq k. \tag{4.59}$$

We now show that this covering of $P(C)$ can be lifted to a covering of C, and to one of $\mathcal{Z}$. We note explicitly that since C and $\mathcal{Z}$ are compact, they can obviously be covered by a finite number of balls; however, we want to cover C and $\mathcal{Z}$ with a specific covering, for which we have an explicit estimate on the radii of the covering balls (see proposition 4.24 below).

PROPOSITION 4.22
Let C be as in proposition 4.21. Fix $\rho > 0$, and let $y_1, \ldots, y_{K_0(\rho)}$ be as above. Then

$$C \subset \bigcup_{j=1}^{K_0(\rho)} B(y_j, \sqrt{2}\rho).$$

PROOF   Let $y \in C$. Since $Py \in P(C)$, there is $j \in \{1, \ldots, K_0(\rho)\}$ such that $Py \in B(Py_j, \rho)$. Since both y and $y_j \in C$, by (4.54)

$$\|y - y_j\| \leq \sqrt{2} \|Py - Py_j\| < \sqrt{2}\rho ;$$

thus, $y \in B(y_j, \sqrt{2}\rho)$.   ☐

Next, we see that this same covering of C generates a covering of $\mathcal{Z}$.

PROPOSITION 4.23
Given γ as in (4.53), r as in (4.58), and $\rho > 0$, set

$$\rho_1 := 2\gamma r + \sqrt{2}\rho. \tag{4.60}$$

Then

$$\mathcal{Z} \subset \bigcup_{j=1}^{K_0(\rho)} B(y_j, \rho_1).$$

PROOF   Let $z \in \mathcal{Z}$. If $z \in \mathcal{C}$, proposition 4.22 applies, since $\sqrt{2}\rho < \rho_1$. If $z \notin \mathcal{C}$, choose $y \in \mathcal{C}$. Then, by proposition 4.18, $z - y \notin \mathcal{C}_P$, i.e.

$$\|z - y\| \geq \sqrt{2}\|Pz - Py\|. \tag{4.61}$$

Since $z, y \in \mathcal{Z} = S(\mathcal{K} \cap \overline{B}(a, r))$, there are u and $v \in \overline{B}(a, r)$ such that $z = S(u)$ and $y = S(v)$. Then (4.61) implies (4.52) and, therefore, (4.53) holds. Then,

$$\|z - y\| \leq \gamma \|u - v\| \leq \gamma \operatorname{diam}(\overline{B}(a, r)) \leq 2\gamma r. \tag{4.62}$$

Since $y \in \mathcal{C}$, by proposition 4.22 there is $j \in \{1, \ldots, K_0(\rho)\}$ such that $y \in B(y_j, \sqrt{2}\rho)$. Hence, (4.62) implies that

$$\|z - y_j\| \leq \|z - y\| + \|y - y_j\| \leq 2\gamma r + \sqrt{2}\rho = \rho_1,$$

i.e. $z \in B(y_j, \rho_1)$.   ☐

We now proceed to give a first estimate on the number $K_0(\rho)$ of the balls of radius ρ, which cover $\mathcal{Z}$.

PROPOSITION 4.24
Let γ, r and ρ be as in proposition 4.23, and N be the rank of $P = P_N$. Then,

$$K_0(\rho) \leq \left(\frac{2rL}{\rho} + 1\right)^N,$$

where L denotes the Lipschitz constant of S.

PROOF   Note first that if we cover $P(\mathcal{C})$ with other balls, whose centers are exactly at a distance ρ apart, and H is the number of these balls, then $K_0(\rho) \leq H$. In fact, by (4.59) the distance between the centers of these other balls is less than that between the centers Py_j of the "original" balls, so there must be more of the "new" balls. Consider then any such covering, i.e.

$$\bigcup_{h=1}^{H} B(Px_h, \rho) \supset P(\mathcal{C}),$$

with centers $Px_h \in P(\mathcal{C})$ such that $\|Px_k - Px_j\| = \rho$. Then for all $j \neq k$

$$B(Px_j, \tfrac{\rho}{2}) \cap B(Px_k, \tfrac{\rho}{2}) = \emptyset,$$

for otherwise if there were a z in one such intersection, we would deduce the contradiction

$$\rho = \|Px_j - Px_k\| \leq \|Px_j - z\| + \|z - Px_k\| < \rho.$$

This disjoint family is also contained in the $\frac{1}{2}\rho$-neighborhood of $P(\mathcal{C})$, i.e. the set

$$\mathcal{V} := \bigcup_{h=1}^{H} B(Px_h, \tfrac{\rho}{2}) \subseteq \{y \in P(\mathcal{X}) : d(y, P(\mathcal{C})) < \tfrac{\rho}{2}\} =: \mathcal{U}. \qquad (4.63)$$

In fact, if $x \in \mathcal{V}$, there is $j \in \{1, \dots, H\}$ such that $x \in B(Px_j, \tfrac{\rho}{2})$; hence,

$$d(x, P(\mathcal{C})) \leq \|x - Px_j\| < \tfrac{\rho}{2}.$$

Recalling that $\mathcal{C} \subset \mathcal{Z}$, from (4.63) we have the estimate

$$\operatorname{diam}(\mathcal{U}) \leq \operatorname{diam}(P(\mathcal{C})) + \rho \leq \operatorname{diam}(P(\mathcal{Z})) + \rho \leq \operatorname{diam}(\mathcal{Z}) + \rho \qquad (4.64)$$

(the last step because P, being orthogonal, is contractive). To estimate $\operatorname{diam}(\mathcal{Z})$, let z_1 and $z_2 \in \mathcal{Z}$. Since $\mathcal{Z} = S(\mathcal{K} \cap \overline{B}(a, r))$, there are b_1 and $b_2 \in \mathcal{K} \cap \overline{B}(a, r)$ such that $z_i = S(b_i)$, $i = 1, 2$. Recalling that S is Lipschitz continuous with Lipschitz constant L, we have

$$\|z_1 - z_2\| \leq L\|b_1 - b_2\| \leq 2rL.$$

Consequently, we deduce that

$$\operatorname{diam}(\mathcal{Z}) \leq L \operatorname{diam}(\overline{B}(a, r)) = 2rL.$$

Inserting this into (4.64) we deduce that $\operatorname{diam}(\mathcal{U}) \leq 2rL + \rho$, which means that $\mathcal{U}$ is contained in a ball of radius $rL + \tfrac{\rho}{2}$. Since $P(\mathcal{X})$ is N-dimensional, we conclude that

$$\operatorname{vol}(\mathcal{U}) \leq \omega_N \left(rL + \tfrac{\rho}{2}\right)^N, \qquad (4.65)$$

where ω_N denotes the volume of the unit ball of $\mathbb{R}^N$. Now, the family $\mathcal{V}$ is composed of H balls, each of which has volume $\omega_N \left(\tfrac{\rho}{2}\right)^N$. Hence, (4.63) and (4.65) imply that

$$\operatorname{vol}(\mathcal{V}) = H\omega_N \left(\tfrac{\rho}{2}\right)^N \leq \operatorname{vol}(\mathcal{U}) \leq \omega_N \left(rL + \tfrac{\rho}{2}\right)^N.$$

From this we conclude that

$$H \leq \left(\tfrac{2rL}{\rho} + 1\right)^N. \qquad (4.66)$$

Since $K_0(\rho) \leq H$, proposition 4.24 follows. $\qquad\qquad\qquad\qquad \square$

We now show that we can choose a particular value of ρ, which allows us to estimate the number of the balls of the corresponding covering of C and Z independently of r. Indeed, given $\theta \in]2\gamma, 1[$, recalling definition (4.60) we can choose ρ such that $\rho_1 = \theta r$. More precisely, we choose

$$\rho = \frac{(\theta - 2\gamma)r}{\sqrt{2}}. \tag{4.67}$$

With this value of ρ, estimate (4.66) implies that

$$K_0(\rho) \leq \left(\frac{2\sqrt{2}L}{\theta - 2\gamma} + 1 \right)^N.$$

In conclusion, we can summarize the results of this section in the following

THEOREM 4.25
Let $K \subset X$ be a compact set. Given any $a \in K$ and $r > 0$, set

$$Z := S(\overline{B}(a,r) \cap K).$$

There exists a compact subset $C \subseteq Z \subseteq S(K)$ such that C is maximal with respect to the cone property (4.54). Moreover, for any $\theta \in]2\gamma, 1[$ there is a covering of both C and Z, consisting of balls of radius θr and center in C. This covering has the form

$$C \subseteq Z \subset \bigcup_{j=1}^{M} \overline{B}(y_j, \theta r), \quad y_j \in C, \tag{4.68}$$

and consists of M balls, where

$$M \leq H_1(\theta) := \left(\frac{2\sqrt{2}L}{\theta - 2\gamma} + 1 \right)^N. \tag{4.69}$$

PROOF   The existence and compactness of the set C follow from propositions 4.17 and 4.59. Given then $\theta \in]2\gamma, 1[$, choosing ρ as in (4.67) we have from (4.60) that $\rho_1 = \theta r$. With this value of ρ, estimate (4.66) implies (4.69). Noting that the right side of (4.69) is independent of both r and ρ, we conclude that there exists a covering as claimed in (4.68). ⛶

4.5.4   The First and Second Iterates

Let $B \subset X$ be a compact, positively invariant set for the semiflow S, containing the global attractor A. By repeatedly iterating the construction described in the previous section, we define successive coverings of the sets $S^k(B)$, for integer $k \geq 0$, in a way that allows us to keep a suitable control on the radii of the balls covering $S^k(B)$.

More precisely, we want the radius of each ball of the covering of $S^k(\mathcal{B})$ to be equal to $\theta^k R$, where $R > 0$ and $\theta \in \,]2\gamma, 1[$ are fixed. (A specific choice of the radius R will be determined later on, when we prove the estimate on the fractal dimension of $\mathcal{E}$.)

The starting point is the covering of $\mathcal{B}$ itself. Since $\mathcal{B}$ is compact, we can cover it by a finite number M_0 of balls of radius 1; that is, there are points $a_1, \ldots, a_{M_0}$ in $\mathcal{B}$ such that

$$\mathcal{B} \subseteq \bigcup_{j=1}^{M_0} \overline{B}(a_j, 1).$$

Obviously, then, for any $R \geq 1$,

$$\mathcal{B} \subseteq \bigcup_{j=1}^{M_0} (\overline{B}(a_j, R) \cap \mathcal{B}). \tag{4.70}$$

This is the covering step corresponding to $k = 0$; to emphasize this, we rewrite (4.70) as

$$\mathcal{B} \subseteq \bigcup_{j_0=1}^{M_0} (\overline{B}(a_{j_0}, \theta^0 R) \cap S^0(\mathcal{B})). \tag{4.71}$$

We remark that, since $\mathcal{B}$ is compact, we could cover it by a single ball of a sufficiently large radius R. We use the covering (4.70) instead, because in the estimate of the fractal dimension of $\mathcal{E}$ we will need to be able to take R as close to 1 as we want (see step (8) of the proof of theorem 4.26 below).

For the next step, i.e. the covering of $S(\mathcal{B})$, we first deduce from (4.71) that

$$S(\mathcal{B}) \subseteq \bigcup_{j_0=1}^{M_0} S(\overline{B}(a_{j_0}, \theta^0 R) \cap S^0(\mathcal{B})). \tag{4.72}$$

Then, we use theorem 4.25, with $r = R$ and $\mathcal{K} = S^0(\mathcal{B}) = \mathcal{B}$ (which is compact), to cover each set at the right side of (4.72). More precisely, setting

$$\mathcal{B}_{j_0}^{(1)} := S(\overline{B}(a_{j_0}, \theta^0 R) \cap S^0(\mathcal{B})),$$

we can cover each set $\mathcal{B}_{j_0}^{(1)}$, $1 \leq j_0 \leq M_0$, with M_{1,j_0} balls, $M_{1,j_0} \leq H_1(\theta)$ (defined in (4.69)), having radii θR and centers in a subset $\mathcal{E}_{j_0}^{(1)}$ of $\mathcal{B}_{j_0}^{(1)}$, maximal with respect to the cone property (4.54) (that is, $\mathcal{B}_{j_0}^{(1)}$ and $\mathcal{E}_{j_0}^{(1)}$ are, respectively, the sets $\mathcal{Z}$ and $\mathcal{C}$ of theorem 4.25). Explicitly, this means that there are points $a_{j_0 j_1} \in \mathcal{E}_{j_0}^{(1)}$, $1 \leq j_1 \leq M_{1,j_0}$, such that

$$\mathcal{B}_{j_0}^{(1)} \subset \bigcup_{j_1=1}^{M_{1,j_0}} \overline{B}(a_{j_0 j_1}, \theta R).$$

From this and (4.71), it follows that

$$
S^1(\mathcal{B}) \subset \bigcup_{j_0=1}^{M_0} \mathcal{B}_{j_0}^{(1)} \subset \bigcup_{j_0=1}^{M_0} \bigcup_{j_1=1}^{M_{1,j_0}} \overline{B}(a_{j_0 j_1}, \theta R)
$$

$$
\subset \bigcup_{j_0=1}^{M_0} \bigcup_{j_1=1}^{M_{1,j_0}} (\overline{B}(a_{j_0 j_1}, \theta R) \cap S^1(\mathcal{B})). \tag{4.73}
$$

This describes the second step of our construction, corresponding to $k = 1$. In the terminology of [EFNT94], the sets $\mathcal{E}_{j_0}^{(1)}$, $1 \leq j_0 \leq M_0$, make up the "first generation" of the points to be added to the attractor $\mathcal{A}$.

4.5.5   The General Iterate

We now proceed inductively. In analogy to the previous construction, we set

$$
\mathcal{E}_{j_0, j_{-1}}^{(0)} := \mathcal{B}
$$

and, for $k \geq 0$, define the compact set

$$
\mathcal{B}_{j_0 \cdots j_k}^{(k+1)} := S(\overline{B}(a_{j_0 \cdots j_k}, \theta^k R) \cap S^k(\mathcal{B})), \tag{4.74}
$$

where the balls $\overline{B}(a_{j_0 \cdots j_k}, \theta^k R)$ have centers

$$
a_{j_0 \cdots j_k} \in \mathcal{E}_{j_0 \cdots j_{k-1}}^{(k)}, \tag{4.75}
$$

and $\mathcal{E}_{j_0 \cdots j_k}^{(k+1)}$ is the compact set $\mathcal{C}$, maximal with respect to the cone property (4.54) in the set $\mathcal{Z} = \mathcal{B}_{j_0 \cdots j_k}^{(k+1)}$, corresponding, in theorem 4.25, to the choices $r = \theta^k R$ and $\mathcal{K} = S^k(\mathcal{B})$ (which, being the continuous image of a compact set, is compact). Thus, for all $k \geq 0$,

$$
\mathcal{E}_{j_0 \cdots j_k}^{(k+1)} \subseteq \mathcal{B}_{j_0 \cdots j_k}^{(k+1)}, \tag{4.76}
$$

and both these sets are compact. To proceed with the iteration, we apply again theorem 4.25 to cover the set $\mathcal{B}_{j_0 \cdots j_k}^{(k+1)}$ defined in (4.74) by $M_{k+1, j_0 \cdots j_k}$ balls with centers $a_{j_0 \cdots j_{k+1}} \in \mathcal{E}_{j_0 \cdots j_k}^{(k+1)}$ and radii $\theta^{k+1} R$, with

$$
M_{k+1, j_0 \cdots j_k} \leq H_1(\theta) \tag{4.77}
$$

(defined in (4.69)); that is,

$$
\mathcal{B}_{j_0 \cdots j_k}^{(k+1)} \subset \bigcup_{j_{k+1}=1}^{M_{k+1, j_0 \cdots j_k}} \overline{B}(a_{j_0 \cdots j_{k+1}}, \theta^{k+1} R). \tag{4.78}
$$

We now show that this covering satisfies the inclusions

$$\mathcal{E}^{(k+1)} := \bigcup_{j_0=1}^{M_0} \cdots \bigcup_{j_k=1}^{M_{k,j_0\cdots j_{k-1}}} \mathcal{E}^{(k+1)}_{j_0\cdots j_k} \subset S^{k+1}(\mathcal{B}), \tag{4.79}$$

$$S^{k+1}(\mathcal{B}) \subset \bigcup_{j_0=1}^{M_0} \cdots \bigcup_{j_k=1}^{M_{k,j_0\cdots j_{k-1}}} \mathcal{B}^{(k+1)}_{j_0\cdots j_k}$$

$$\subset \bigcup_{j_0=1}^{M_0} \cdots \bigcup_{j_k=1}^{M_{k,j_0\cdots j_{k-1}}} \bigcup_{j_{k+1}=1}^{M_{k+1,j_0\cdots j_k}} \left(\overline{B}(a_{j_0\cdots j_{k+1}}, \theta^{k+1}R) \cap S^{k+1}(\mathcal{B}) \right). \tag{4.80}$$

To show (4.79), let $x \in \mathcal{E}^{(k+1)}$. There are then indices $j_0 \in \{1,\ldots,M_0\}$, …, $j_k \in \{1,\ldots,M_{k,j_0\cdots j_{k-1}}\}$ such that $x \in \mathcal{E}^{(k+1)}_{j_0\cdots j_k}$. Thus, by (4.76) and (4.74),

$$x \in S(\overline{B}(a_{j_0\cdots j_k}, \theta^k R) \cap S^k(\mathcal{B})).$$

Consequently, there is $y \in \overline{B}(a_{j_0\cdots j_k}, \theta^k R) \cap S^k(\mathcal{B})$ such that $x = S(y)$; since $y \in S^k(\mathcal{B})$, it follows that $x \in S^{k+1}(\mathcal{B})$.

To show (4.80), we proceed by induction on k. The step $k = 0$ is (4.73). Then, if (4.80) holds with k replaced by $k-1$, $k \geq 1$, by (4.74) we have

$$S^{k+1}(\mathcal{B}) = S(S^k(\mathcal{B})) \subseteq S\left(\bigcup_{j_0=1}^{M_0} \cdots \bigcup_{j_k=1}^{M_{k,j_0\cdots j_{k-1}}} (\overline{B}(a_{j_0\cdots j_k}, \theta^k R) \cap S^k(\mathcal{B})) \right)$$

$$\subset \bigcup_{j_0=1}^{M_0} \cdots \bigcup_{j_k=1}^{M_{k,j_0\cdots j_{k-1}}} S\left(\overline{B}(a_{j_0\cdots j_k}, \theta^k R) \cap S^k(\mathcal{B}) \right)$$

$$= \bigcup_{j_0=1}^{M_0} \cdots \bigcup_{j_k=1}^{M_{k,j_0\cdots j_{k-1}}} \mathcal{B}^{(k+1)}_{j_0\cdots j_k}.$$

Thus, (4.80) follows by (4.78).

Finally, note that each set $\mathcal{E}^{(k+1)}$, being a finite union of compact sets, is itself compact.

4.5.6 Conclusion

We are almost at the end of the construction of the exponential attractor in $\mathcal{B}$. We recall that we do have a global attractor $\mathcal{A}$ in $\mathcal{B}$, namely the set $\mathcal{A} = \omega(\mathcal{B})$.

The sets $\mathcal{E}^{(k)}$ constructed in (4.79) are defined in terms of sets $\mathcal{C}$ where the cone property holds: this means that the semiflow S needs not be contracting on these sets. Hence, we add all these points to the attractor $\mathcal{A}$. More precisely, setting

$$\mathcal{E}^{(\infty)} := \overline{\bigcup_{k=1}^{\infty} \mathcal{E}^{(k)}} \tag{4.81}$$

(closure in $\mathcal{X}$), and

$$\mathcal{E}_1 := \mathcal{A} \cup \mathcal{E}^{(\infty)},$$

we would like to recognize $\mathcal{E}_1$ as a "good candidate" for the exponential attractor. Indeed, note that $\mathcal{E}_1$ is closed, because so are $\mathcal{A}$ and $\mathcal{E}^{(\infty)}$, and that $\mathcal{E}_1 \subseteq \mathcal{B}$, because $\mathcal{A} \subseteq \mathcal{B}$ and also $\mathcal{E}^{(k)} \subseteq \mathcal{B}$ for all $k \in \mathbb{N}$. The latter claim is a consequence of (4.79) and of the positive invariance of $\mathcal{B}$, which implies that

$$\mathcal{E}^{(k)} \subset S^k(\mathcal{B}) \subset \mathcal{B}. \tag{4.82}$$

Since $\mathcal{B}$ is compact, it follows that $\mathcal{E}^{(\infty)}$ and $\mathcal{E}_1$ are also compact. However, $\mathcal{E}_1$ needs not be positively invariant. To overcome this last problem, we enlarge $\mathcal{E}_1$ by taking all its images under the iterates of S. More precisely, setting

$$\mathcal{G}^{(\infty)} := \bigcup_{j=0}^{\infty} S^j(\mathcal{E}^{(\infty)}), \quad \mathcal{E} := \mathcal{A} \cup \mathcal{G}^{(\infty)}, \tag{4.83}$$

we finally claim

THEOREM 4.26
The set $\mathcal{E}$ defined in (4.83) is an exponential attractor for the semiflow $S = (S^n)_{n \in \mathbb{N}}$, relative to $\mathcal{B}$.

PROOF   Recalling definition 4.1, we must show that $\mathcal{E} \subseteq \mathcal{B}$; that $\mathcal{E}$ is positively invariant and compact; that $\mathcal{E}$ attracts bounded subsets of $\mathcal{B}$ at an exponential rate; and that $\mathcal{E}$ has finite fractal dimension. Note that, by construction, $\mathcal{A} \subseteq \mathcal{E}$.
   1. We show that $\mathcal{E} \subseteq \mathcal{B}$. Indeed, $\mathcal{A} \subseteq \mathcal{B}$ and, since

$$\mathcal{E}^{(\infty)} \subseteq \mathcal{B} \tag{4.84}$$

(which follows from (4.82)), for all $j \in \mathbb{N}$

$$S^j(\mathcal{E}^{(\infty)}) \subseteq S^j(\mathcal{B}) \subseteq \mathcal{B}.$$

Therefore, $\mathcal{G}^{(\infty)} \subseteq \mathcal{B}$ as well.
   2. In preparation for the next steps, for $j \geq 0$ we set

$$\mathcal{L}^{(j)} := \bigcup_{k=1}^{\infty} S^j(\mathcal{E}^{(k)}),$$

and define

$$\mathcal{L}^{(\infty)} := \bigcup_{j=0}^{\infty} \mathcal{L}^{(j)}, \quad \mathcal{E}_2 := \mathcal{A} \cup \mathcal{L}^{(\infty)}.$$

We prove then that $\mathcal{E} = \mathcal{E}_2$, and proceed to show that $\mathcal{E}_2$ satisfies the requirements of the exponential attractor.

3. We show that $\mathcal{E} = \mathcal{E}_2$.

3a. To show that $\mathcal{E} \subseteq \mathcal{E}_2$, let $x \in \mathcal{E}$. If $x \in \mathcal{A}$, there is nothing more to prove. If $x \in \mathcal{G}^{(\infty)}$, there is $j \in \mathbb{N}$ such that $x \in S^j(\mathcal{E}^{(\infty)})$; that is, $x = S^j(y)$ for some $y \in \mathcal{E}^{(\infty)}$. Let

$$(y_m)_{m \in \mathbb{N}} \subset \bigcup_{k=1}^{\infty} \mathcal{E}^{(k)}$$

be a sequence converging to y. Then for all m there is k_m such that $y_m \in \mathcal{E}^{(k_m)}$. By (4.79), $y_m \in S^{k_m}(\mathcal{B})$, so $y_m = S^{k_m}(b_m)$ for some $b_m \in \mathcal{B}$. If the sequence $(k_m)_{m \in \mathbb{N}}$ is unbounded, recalling the characterization of ω-limit sets given in proposition 2.15, we conclude that

$$y = \lim y_m \in \omega(\mathcal{B}) = \mathcal{A}.$$

Consequently, since $\mathcal{A}$ is invariant,

$$x = S^j(y) \in S^j(\mathcal{A}) = \mathcal{A} \subseteq \mathcal{E}_2.$$

If instead $(k_m)_{m \in \mathbb{N}}$ is definitely constant, i.e. if there are m_0 and $k_* \in \mathbb{N}$ such that $k_m \equiv k_*$ for all $m \geq m_0$, then the sequence $(y_m)_{m \geq m_0}$ is all contained in the compact set $\mathcal{E}^{(k_*)}$. Thus, there is a subsequence $(y_{m_r})_{r \in \mathbb{N}}$, with $m_r \geq m_0$, converging to some $y_* \in \mathcal{E}^{(k_*)}$. This implies that $y = y_* \in \mathcal{E}^{(k_*)}$, and, therefore,

$$x = S^j(y) \in S^j(\mathcal{E}^{(k_*)}) \subseteq \mathcal{L}^{(j)} \subseteq \mathcal{L}^{\infty} \subseteq \mathcal{E}_2.$$

It follows that $\mathcal{E} \subseteq \mathcal{E}_2$.

3b. To show that $\mathcal{E}_2 \subseteq \mathcal{E}$, let $x \in \mathcal{E}_2$. If $x \in \mathcal{A}$, there is nothing more to prove. If $x \in \mathrm{L}^{(\infty)}$, there are integers j and k such that $x \in S^j(\mathcal{E}^{(k)})$, so $x = S^j(y)$ for some $y \in \mathcal{E}^{(k)}$. But $\mathcal{E}^{(k)} \subseteq \mathcal{E}^{(\infty)}$, so

$$x \in S^j(\mathcal{E}^{(\infty)}) \subseteq \mathcal{G}^{(\infty)} \subseteq \mathcal{E}.$$

4. We show that $\mathcal{E}_2$ is positively invariant. Let $x \in \mathcal{E}_2$. If $x \in \mathcal{A}$, then

$$S(x) \in S(\mathcal{A}) = \mathcal{A} \subseteq \mathcal{E}_2.$$

If $x \in \mathcal{L}^{(\infty)}$, as before there are integers j and k such that $x \in S^j(\mathcal{E}^{(k)})$. Then

$$S(x) \in S^{j+1}(\mathcal{E}^{(k)}) \subseteq \mathcal{L}^{(j+1)} \subseteq \mathcal{L}^{(\infty)} \subseteq \mathcal{E}_2,$$

and, therefore, $S(\mathcal{E}_2) \subseteq \mathcal{E}_2$. Note that it is precisely to ensure the positive invariance of $\mathcal{E}$ that we had to enlarge $\mathcal{E}_1$ into $\mathcal{E}_2$.

5. We show that $\mathcal{E}_2$ is closed. Let $(a_m)_{m \in \mathbb{N}} \subset \mathcal{E}_2$ be a sequence such that $a_m \to a$, $a \in \mathcal{X}$. If there is a proper subsequence $(a_{m_k})_{k \in \mathbb{N}} \subset \mathcal{A}$, then

$$a = \lim_{k \to +\infty} a_{m_k} \in \mathcal{A} \subset \mathcal{E}_2.$$

Otherwise, at most a finite number of elements of $(a_m)_{m \in \mathbb{N}}$ are in $\mathcal{A}$. We reorder the other elements into a sequence $(y_m)_{m \in \mathbb{N}} \subset \mathcal{L}^{(\infty)}$, with $y_m \to a$. Since each $y_m \in \mathcal{L}^{(\infty)}$, for all m there exist n_m and k_m such that

$$y_m = S^{n_m}(x_m), \quad x_m \in \mathcal{E}^{(k_m)} \subseteq \mathcal{B}. \tag{4.85}$$

Suppose first that the sequence $(n_m)_{m \in \mathbb{N}}$ is unbounded. Then by the already recalled characterization of $\mathcal{A} = \omega(\mathcal{B})$ given in proposition 2.15, (4.85) implies that

$$a = \lim_{m \to +\infty} y_m \in \omega(\mathcal{B}) = \mathcal{A} \subset \mathcal{E}_2.$$

If instead the sequence $(n_m)_{m \in \mathbb{N}}$ is definitely constant, i.e. if there are m_0 and $n_* \in \mathbb{N}$ such that for all $m \geq m_0$, $n_m = n_*$, then for $m \geq m_0$

$$y_m = S^{n_*}(x_m).$$

Since $(x_m)_{m \in \mathbb{N}} \subset \mathcal{B}$ and $\mathcal{B}$ is compact, there is at least a subsequence $(x_{m_j})_{j \in \mathbb{N}}$ converging to some $x \in \mathcal{B}$. Since S^{n_*} is continuous, we have that for all j such that $m_j \geq m_0$

$$y_{m_j} = S^{n_*}(x_{m_j}) \to S^{n_*}(x),$$

and this implies that $a = S^{n_*}(x)$. Consider now the subsequence $(k_{m_j})_{j \in \mathbb{N}}$ of the sequence $(k_m)_{m \in \mathbb{N}}$ appearing in (4.85), and suppose first that this sequence is definitely constant, i.e., that there are j_0 and $k_* \in \mathbb{N}$ such that for all $j \geq j_0$, $k_{m_j} = k_*$. Then for all $j \geq j_0$,

$$x_{m_j} \in \mathcal{E}^{(k_{m_j})} = \mathcal{E}^{(k_*)},$$

and since this set is compact,

$$x = \lim_{j \to +\infty} x_{m_j} \in \mathcal{E}^{(k_*)}.$$

It follows that

$$a = S^{n_*}(x) \in S^{n_*}(\mathcal{E}^{(k_*)}) \subseteq \mathcal{L}^{(n_*)} \subseteq \mathcal{L}^{(\infty)} \subseteq \mathcal{E}_2.$$

The last possibility is that $k_{m_j} \to +\infty$ as $j \to +\infty$. Then, as before,

$$x_{m_j} \in \mathcal{E}^{(k_{m_j})} \subseteq S^{k_{m_j}}(\mathcal{B}) \subseteq \mathcal{B},$$

and therefore, by proposition 2.15 again, $x \in \omega(\mathcal{B}) = \mathcal{A}$, and

$$a = S^{n_*}(x) \in S^{n_*}(\mathcal{A}) = \mathcal{A} \subset \mathcal{E}_2.$$

This concludes the proof that $\mathcal{E}_2$ is closed.

6. We show that $\mathcal{E}_2$ is compact. Since $\mathcal{B}$ is compact and $\mathcal{E}_2$ is closed, it is sufficient to show that $\mathcal{E}_2 \subseteq \mathcal{B}$. But this follows from steps 1 and 3, since we know that $\mathcal{E} \subseteq \mathcal{B}$ and $\mathcal{E}_2 = \mathcal{E}$.

7. We show the exponential convergence of bounded subsets of $\mathcal{B}$ to $\mathcal{E}_2$. More precisely, setting

$$\kappa := -\ln \theta \in \,]0, -\ln(2\gamma)[$$

(recall that $2\gamma < 1$), we show that for any bounded subset $\mathcal{G} \subseteq \mathcal{B}$, the estimate

$$\partial(S^n \mathcal{G}, \mathcal{E}_2) \leq R e^{-\kappa n} \tag{4.86}$$

holds for all $n \in \mathbb{N}$ (compare to (4.1)). To this end, we recall (4.80), which implies that for all $x \in \mathcal{G}$ and $n \in \mathbb{N}$,

$$S^n(x) \in S^n \mathcal{G} \subseteq S^n(\mathcal{B}) \subset \bigcup_{j_0=1}^{M_0} \cdots \bigcup_{j_n=1}^{M_{n,j_1\cdots j_{n-1}}} \left(\overline{B}(a_{j_0\cdots j_n}, \theta^n R) \cap S^n(\mathcal{B}) \right).$$

Thus, there are indices $j_0, \ldots, j_n$ such that

$$S^n(x) \in \overline{B}(a_{j_0\cdots j_n}, \theta^n R).$$

Now, each center $a_{j_0\cdots j_n}$ is in $\mathcal{E}_2$, because by (4.75) and (4.79),

$$a_{j_0\cdots j_n} \in \mathcal{E}^{(n)}_{j_0\cdots j_{n-1}} \subseteq \mathcal{E}^{(n)} = S^0(\mathcal{E}^{(n)}) \subseteq \mathcal{L}^{(0)} \subseteq \mathcal{L}^{(\infty)} \subseteq \mathcal{E}_2.$$

Consequently,

$$d(S^n(x), \mathcal{E}_2) \leq \|S^n(x) - a_{j_0\cdots j_n}\| \leq \theta^n R,$$

and (4.86) follows. This concludes the proof of the exponential rate of convergence to $\mathcal{E}$ of all orbits starting in $\mathcal{G}$.

8. We show that $\mathcal{E}$ has finite fractal dimension, satisfying estimate (4.15). To this end, we use the first of part (3) of proposition 2.61, applied to the compact sets $\mathcal{A}$ and $\mathcal{G}^\infty$ (the compactness of $\mathcal{G}^\infty$ follows from the fact that both sets $\mathcal{A}$ and $\mathcal{A} \cup \mathcal{G}^\infty = \mathcal{E} = \mathcal{E}_2$ are compact).

8a. We first estimate the fractal dimension of the global attractor $\mathcal{A}$. We claim that

$$\dim_F(\mathcal{A}) \leq \frac{\ln H_1(\theta)}{-\ln \theta}, \tag{4.87}$$

where $\theta \in \,]2\gamma, 1[$, and $H_1(\theta)$ is as in (4.69). The proof of (4.87) is based on the observation that the same construction of the coverings of the sets $S^k(\mathcal{B})$ described in sections 4.5.4 and 4.5.5 can be carried out, in the same way, when the set $\mathcal{B}$ is replaced by the attractor $\mathcal{A}$ itself, since $\mathcal{A}$ is also compact and positively invariant. But since $\mathcal{A}$ is in fact invariant, (4.80) shows that for each positive integer k, the set $\mathcal{A} = S^k(\mathcal{A})$ can be covered by $\tilde{M}_k$ balls of radius $\theta^k R$, where

$$\tilde{M}_k := M_0 M_{1,j_0} \cdots M_{k,j_1,\ldots,j_{k-1}}. \tag{4.88}$$

Since $N_\delta(\mathcal{A})$ is the minimum number of balls of diameter at most equal to δ that can cover $\mathcal{A}$, recalling (4.77) it follows from (4.88) that

$$N_{2\theta^k R}(\mathcal{A}) \leq \tilde{M}_k \leq M_0 (H_1(\theta))^k. \tag{4.89}$$

Given then $\delta > 0$, let $k \in \mathbb{N}$ be such that

$$2\theta^{k+1} R < \delta \leq 2\theta^k R. \tag{4.90}$$

Then, since the function $\delta \mapsto N_\delta(\mathcal{A})$ is decreasing, we obtain from (4.89) and (4.90) that

$$\frac{\ln N_\delta(\mathcal{A})}{-\ln \delta} \leq \frac{\ln N_{2\theta^{k+1}R}(\mathcal{A})}{-\ln(2\theta^k R)} \leq \frac{(k+1)\ln H_1(\theta) + \ln M_0}{-\ln(2R) - k\ln\theta}.$$

Hence, recalling (4.90),

$$\dim_F(\mathcal{A}) = \limsup_{\delta \to 0} \frac{\ln N_\delta(\mathcal{A})}{-\ln \delta} \leq \lim_{k \to +\infty} \frac{(k+1)\ln H_1(\theta) + \ln M_0}{-\ln(2R) - k\ln\theta},$$

from which (4.87) follows.

8b. As an intermediate step, we estimate the fractal dimension of $\mathcal{E}^{(\infty)}$. We claim that

$$\dim_F(\mathcal{E}^{(\infty)}) \leq \max\left\{N, \frac{\ln H_1(\theta)}{-\ln\theta}\right\} =: \eta_0, \tag{4.91}$$

where N is the rank of the projection P appearing in definition 4.2, $\theta \in \,]2\gamma, 1[$ and $H_1(\theta)$ is as in (4.69). To prove (4.91), we consider the coverings of the iterates $S^k(\mathcal{B})$ constructed in sections 4.5.5, with R specified as follows. Given $\delta \in \,]0, 1[$, we first choose $n \in \mathbb{N}$ such that

$$2\theta^{n+1} < \delta \leq 2\theta^n, \tag{4.92}$$

and then $R > 0$ such that

$$2\theta^{n+1}R = \delta. \tag{4.93}$$

Note that both n and R depend on δ, but (4.92) implies that, for all $\delta \in \,]0, 1[$,

$$1 < R < \frac{1}{\theta}. \tag{4.94}$$

We proceed then to construct the sets $\mathcal{E}^{(k)}$ as in section 4.5.5, with R as in (4.93). Recalling the definition of $\mathcal{E}^{(\infty)}$ in (4.81), we split

$$\mathcal{E}^{(\infty)} = \underbrace{\left(\bigcup_{k=1}^{n} \mathcal{E}^{(k)}\right)}_{=:\,\mathcal{E}_3} \cup \underbrace{\left(\bigcup_{k=n+1}^{\infty} \mathcal{E}^{(k)}\right)}_{=:\,\mathcal{E}_4} \tag{4.95}$$

(the first union $\mathcal{E}_3$ is automatically closed, being a finite union of compact sets); therefore,

$$N_\delta(\mathcal{E}^{(\infty)}) \leq N_\delta(\mathcal{E}_3) + N_\delta(\mathcal{E}_4). \tag{4.96}$$

Recalling (4.79), the positive invariance of $\mathcal{B}$ implies that, if $k \geq n+1$,

$$\mathcal{E}^{(k)} \subseteq S^k(\mathcal{B}) \subseteq S^{n+1}(\mathcal{B}).$$

Since $S^{n+1}(\mathcal{B})$ is compact, it follows that $\mathcal{E}_4 \subseteq S^{n+1}(\mathcal{B})$, so that, by part 2 of proposition 2.61,

$$N_\delta(\mathcal{E}_4) \le N_\delta(S^{n+1}(\mathcal{B})). \tag{4.97}$$

To estimate $N_\delta(S^{n+1}(\mathcal{B}))$, it is sufficient to recall (4.80), which implies that $S^{n+1}(\mathcal{B})$ can be covered by κ_1 balls of radii $\theta^{n+1}R$, with

$$\kappa_1 := M_0 M_{1,j_0} \cdots M_{n+1,j_0 \cdots j_n}.$$

Thus, as in (4.89), and recalling (4.93),

$$N_\delta(S^{n+1}(\mathcal{B})) = N_{2\theta^{n+1}R}(S^{n+1}(\mathcal{B})) \le M_0 (H_1(\theta))^{n+1}. \tag{4.98}$$

We now turn to the estimate of $N_\delta(\mathcal{E}_3)$. Recalling (4.79) and (4.80), we have

$$\mathcal{E}^{(k+1)} \subset \bigcup_{j_0=1}^{M_0} \cdots \bigcup_{j_{k+1}=1}^{M_{k+1,j_0 \cdots j_k}} \left(\overline{B}(a_{j_0 \cdots j_{k+1}}, \theta^{k+1}R) \cap S^{k+1}(\mathcal{B}) \right), \tag{4.99}$$

with $a_{j_0 \cdots j_{k+1}} \in \mathcal{E}^{(k+1)}_{j_0 \cdots j_k}$ as per (4.75). We now project this covering of $\mathcal{E}^{(k+1)}$ into the N-dimensional space $P(\mathcal{X})$. Denoting by $\overline{B}_N(a,r)$ the closed balls of $P(\mathcal{X})$ with center a and radius r, we immediately verify that for all $a \in \mathcal{X}, r > 0$ and $m \in \mathbb{N}$,

$$P(\overline{B}(a,r) \cap S^m(\mathcal{B})) \subseteq \overline{B}_N(Pa,r) \cap P(S^m(\mathcal{B})); \tag{4.100}$$

hence, we obtain from (4.99) that

$$P(\mathcal{E}^{(k+1)}) \subset \bigcup_{j_0=1}^{M_0} \cdots \bigcup_{j_{k+1}=1}^{M_{k+1,j_0 \cdots j_k}} \left(\overline{B}_N(Pa_{j_0 \cdots j_{k+1}}, \theta^{k+1}R) \cap P(S^{k+1}(\mathcal{B})) \right). \tag{4.101}$$

We further cover each intersection in the sets at the right side of (4.101) by $v_{j_0 \cdots j_{k+1}}$ smaller balls of radius $\frac{\delta}{2\sqrt{2}}$ (these balls are truly smaller, because of (4.93), and $k \le n$). More precisely, there are points

$$\xi_{\ell_{j_0 \cdots j_{k+1}}} \in \overline{B}_N(Pa_{j_0 \cdots j_{k+1}}, \theta^{k+1}R) \cap P(S^{k+1}(\mathcal{B})),$$

with

$$\xi_{\ell_{j_0 \cdots j_{k+1}}} = Py_{\ell_{j_0 \cdots j_{k+1}}}, \quad y_{\ell_{j_0 \cdots j_{k+1}}} \in \mathcal{E}^{(k+1)}_{j_0 \cdots j_k}, \tag{4.102}$$

such that

$$\overline{B}_N(Pa_{j_0 \cdots j_{k+1}}, \theta^{k+1}R) \cap P(S^{k+1}(\mathcal{B}))$$

$$\subseteq \underbrace{\bigcup_{\ell_{j_0 \cdots j_{k+1}}=1}^{v_{j_0 \cdots j_{k+1}}} \left(\overline{B}_N \left(\xi_{\ell_{j_0 \cdots j_{k+1}}}, \frac{1}{2\sqrt{2}} \delta \right) \cap P(S^{k+1}(\mathcal{B})) \right)}_{=: \mathcal{D}^*_{\ell_{j_0 \cdots j_{k+1}}}}. \tag{4.103}$$

The second of (4.102) holds, because the balls $B(a_{j_0\cdots j_{k+1}}, \theta^{k+1}R)$, which cover $\mathcal{B}^{(k+1)}_{j_0\cdots j_k}$ (by (4.78)), have centers in $\mathcal{E}^{(k+1)}_{j_0\cdots j_k}$ (as per (4.75)). In fact, these balls are a lifting of a covering of $\mathcal{E}^{(k+1)}_{j_0\cdots j_k}$, as seen in proposition 4.23 (recall that, at each stage k, we identify $\mathcal{C} = \mathcal{E}^{(k+1)}_{j_0\cdots j_k}$ and $\mathcal{Z} = \mathcal{B}^{(k+1)}_{j_0\cdots j_k}$). Therefore, from (4.101),

$$P(\mathcal{E}^{(k+1)}) \subset \bigcup_{j_0=1}^{M_0} \cdots \bigcup_{j_{k+1}=1}^{M_{k+1,j_0\cdots j_k}} \bigcup_{\ell_{j_0\cdots j_k}=1}^{v_{j_0\cdots j_{k+1}}} \mathcal{D}^{*}_{\ell_{j_0\cdots j_{k+1}}}. \tag{4.104}$$

Since $\dim(P(\mathcal{X})) = N$, we can estimate the number $v_{j_0\cdots j_{k+1}}$ in terms of N and the radii $\frac{\delta}{2\sqrt{2}}$, $\theta^{n+1}R$, as follows. Recall that, in general, if a ball $\overline{B}_N(a,\rho)$ is covered by v balls with centers $b_1,\ldots,b_v \in \overline{B}_N(a,\rho)$ and smaller radius η, from the inclusions

$$\overline{B}_N(a,\rho) \subseteq \bigcup_{j=1}^{v} \overline{B}_N(b_j,\eta) =: \mathcal{U} \subseteq \overline{B}_N(a,\rho+\eta)$$

we deduce that, in particular,

$$\mathrm{vol}(\overline{B}_N(a,\rho)) = \omega_N \rho^N \leq \mathrm{vol}(\mathcal{U}) = v\omega_N\eta^N \leq \mathrm{vol}(\overline{B}_N(a,\rho+\eta)) = \omega_N(\rho+\eta)^N.$$

Thus, we have the estimate

$$v \leq \left(1+\frac{\rho}{\eta}\right)^N = \sum_{j=0}^{N} \binom{N}{j} \left(\frac{\rho}{\eta}\right)^j \leq \left(\frac{\rho}{\eta}\right)^N \sum_{j=0}^{N} \binom{N}{j} = 2^N \left(\frac{\rho}{\eta}\right)^N. \tag{4.105}$$

When applied to

$$\eta = \frac{1}{2\sqrt{2}}\delta < \rho = \theta^{k+1}R,$$

(4.105) yields the estimate

$$v_{j_0\cdots j_{k+1}} \leq C_N \theta^{(k-n)N},$$

with $C_N := 2^{N/2}$. Hence, recalling (4.104), $P(\mathcal{E}^{(k+1)})$ can be covered by $\kappa_{2,k}$ balls of radius $\frac{\delta}{2\sqrt{2}}$, with, as in (4.89),

$$\kappa_{2,k} := M_0 M_{1,j_0} \cdots M_{k+1,j_0\cdots j_k} v_{j_0\cdots j_{k+1}} \leq C_N M_0 (H_1(\theta))^{k+1} \theta^{(k-n)N}. \tag{4.106}$$

We now show that the covering (4.104) of $P(\mathcal{E}^{(k+1)})$ can be lifted to a covering of $\mathcal{E}^{(k+1)}$ itself, by means of the cone property (4.54). More precisely, we show that, if the points $y_{\ell_{j_0\cdots j_{k+1}}}$ are defined as in (4.102), then

$$\mathcal{E}^{(k+1)} \subset \bigcup_{j_0=1}^{M_0} \cdots \bigcup_{j_{k+1}=1}^{M_{k+1,j_0\cdots j_k}} \bigcup_{\ell_{j_0\cdots j_{k+1}}=1}^{v_{j_0\cdots j_{k+1}}} \overline{B}(y_{\ell_{j_0\cdots j_{k+1}}}, \theta^{n+1}R). \tag{4.107}$$

In fact, let $y \in \mathcal{E}^{(k+1)}$. By (4.79), there are indices $j_0 \cdots, j_k$ such that $y \in \mathcal{E}^{(k+1)}_{j_0,\dots,j_k}$. By (4.76) and (4.78), there is another index j_{k+1} such that $y \in \bar{B}(a_{j_0,\dots,j_{k+1}}, \theta^{k+1}R)$. Indeed, by (4.99),

$$y \in \bar{B}(a_{j_0\cdots j_{k+1}}, \theta^{k+1}R) \cap S^{k+1}(\mathcal{B}).$$

By (4.100) and (4.103), there is one index $\ell_{j_0\cdots j_{k+1}} \in \{1,\dots,v_{j_0\cdots j_{k+1}}\}$ such that

$$Py \in \bar{B}_N\left(\xi_{\ell_{j_0\cdots j_{k+1}}}, \tfrac{1}{2\sqrt{2}}\delta\right) \cap P(S^{k+1}(\mathcal{B})).$$

Recalling (4.102), and that the set $\mathcal{E}^{(k+1)}_{j_0\cdots j_k}$ is maximal with respect to the cone property (4.54), we have

$$\|y - y_{\ell_{j_0\cdots j_{k+1}}}\| \leq \sqrt{2}\|Py - Py_{\ell_{j_0\cdots j_{k+1}}}\| = \sqrt{2}\|Py - \xi_{\ell_{j_0\cdots j_{k+1}}}\|$$
$$\leq \sqrt{2}\tfrac{1}{2\sqrt{2}}\delta = \theta^{n+1}R.$$

Thus, (4.107) follows. Recalling (4.106), we conclude that $\mathcal{E}^{(k+1)}$ can be covered by $\kappa_{2,k}$ balls of diameter $2\theta^{n+1}R = \delta$. Consequently, (4.106) implies that

$$N_\delta(\mathcal{E}^{(k+1)}) \leq C_N \theta^{(k-n)N}(H_1(\theta))^{k+1}.$$

This estimate holds for $0 \leq k \leq n-1$; hence, recalling (4.95),

$$N_\delta(\mathcal{E}_3) \leq \sum_{k=1}^{n} N_\delta(\mathcal{E}^{(k)}) \leq C_N \sum_{k=1}^{n} \theta^{(k-1-n)N}(H_1(\theta))^k$$
$$\leq C_N \theta^{-N(1+n)} \sum_{k=1}^{n} \left(\theta^N H_1(\theta)\right)^k. \tag{4.108}$$

Putting (4.96) and (4.108) together with (4.97) and (4.97) and (4.98), we finally obtain that

$$N_\delta(\mathcal{E}^{(\infty)}) \leq C_N \theta^{-N(1+n)} \sum_{k=1}^{n} \left(\theta^N H_1(\theta)\right)^k + M_0(H_1(\theta))^{n+1}. \tag{4.109}$$

To estimate the right side of (4.109), suppose first that $\theta^N H_1(\theta) \leq 1$. In this case, we can proceed from (4.109) with

$$N_\delta(\mathcal{E}^{(\infty)}) \leq C_N \theta^{-N(1+n)} n + M_0(H_1(\theta))^{n+1} \leq \theta^{-N(1+n)}(C_N(n+1) + M_0).$$

By (4.93),

$$\theta^{n+1} = \frac{\delta}{2R}; \tag{4.110}$$

thus,

$$n \leq n+1 = \frac{\ln(\delta/2R)}{\ln\theta},$$

and

$$N_\delta(\mathcal{E}^{(\infty)}) \le \left(\frac{2R}{\delta}\right)^N \left(C_N \frac{\ln(\delta/2R)}{\ln\theta} + M_0\right).$$

Therefore, since $\delta < 1$,

$$\frac{\ln(N_\delta(\mathcal{E}^{(\infty)}))}{-\ln\delta} \le \frac{N(\ln(2R) - \ln\delta)}{-\ln\delta} + \frac{1}{-\ln\delta}\ln\left(C_N \frac{\ln(\delta/2R)}{\ln\theta} + M_0\right)$$

$$=: h_1(\delta). \tag{4.111}$$

Recalling that, by (4.94), R is a bounded function of δ, we conclude from (4.111) that, if $\theta^N H_1(\theta) \le 1$,

$$\dim_F(\mathcal{E}^{(\infty)}) = \limsup_{\delta\to 0} \frac{\ln(N_\delta(\mathcal{E}^{(\infty)}))}{-\ln\delta} \le \lim_{\delta\to 0} h_1(\delta) = N. \tag{4.112}$$

If instead $\theta^N H_1(\theta) > 1$, we proceed from (4.109) with

$$N_\delta(\mathcal{E}^{(\infty)}) \le C_N \theta^{-N(1+n)} n(\theta^N H_1(\theta))^n + M_0(H_1(\theta))^{n+1}$$
$$\le C_N \theta^{-N} n(H_1(\theta))^n + M_0(H_1(\theta))^{n+1}$$
$$\le C_N n(H_1(\theta))^{1+n} + M_0(H_1(\theta))^{n+1}$$
$$= (H_1(\theta))^{n+1}(C_N n + M_0).$$

From this and (4.110) we obtain that

$$\frac{\ln(N_\delta(\mathcal{E}^{(\infty)}))}{-\ln\delta} \le \frac{(\ln\delta - \ln(2R))\ln(H_1(\theta))}{(-\ln\delta)(\ln\theta)} + \frac{1}{-\ln\delta}\ln\left(C_N \frac{\ln(\delta/2R)}{\ln\theta}\ln\theta + M_0\right)$$

$$=: h_2(\delta).$$

We can then conclude, as in (4.112), that

$$\dim_F(\mathcal{E}^{(\infty)}) \le \lim_{\delta\to 0} h_2(\delta) = \frac{\ln(H_1(\theta))}{-\ln\theta}. \tag{4.113}$$

Thus, (4.91) follows from (4.112) and (4.113).

8c. Our last step is the estimate of the fractal dimension of $\mathcal{G}^{(\infty)}$. We claim that also

$$\dim_F(\mathcal{G}^{(\infty)}) \le \eta_0, \tag{4.114}$$

where η_0 is as in (4.91). We prove (4.114) by showing that for all $\eta > \eta_0$,

$$\dim_F(\mathcal{G}^{(\infty)}) \le \eta. \tag{4.115}$$

As in part (8b), given $\delta \in {]0,1[}$ and $\theta \in {]2\gamma, 1[}$, we determine $n \in \mathbb{N}$ and $R \in {]1, \frac{1}{\theta}[}$ such that (4.92) and (4.93) hold. Then, as in (4.95), recalling (4.83) we split

$$\mathcal{G}^{(\infty)} = \underbrace{\left(\bigcup_{j=0}^{n} S^j(\mathcal{E}^{(\infty)})\right)}_{=:\,\mathcal{E}_5} \cup \underbrace{\left(\bigcup_{j=n+1}^{\infty} S^j(\mathcal{E}^{(\infty)})\right)}_{=:\,\mathcal{E}_6},$$

so that

$$N_\delta(\mathcal{G}^{(\infty)}) \leq N_\delta(\mathcal{E}_5) + N_\delta(\mathcal{E}_6). \qquad (4.116)$$

From (4.84) and the positive invariance of $\mathcal{B}$ we have that, if $j \geq n+1$,

$$S^j(\mathcal{E}^{(\infty)}) \subseteq S^j(\mathcal{B}) \subseteq S^{n+1}(\mathcal{B});$$

hence, $\mathcal{E}_6 \subseteq S^{n+1}(\mathcal{B})$ and, therefore,

$$N_\delta(\mathcal{E}_6) \leq N_\delta(S^{n+1}(\mathcal{B})). \qquad (4.117)$$

We can estimate $N_\delta(S^{n+1}(\mathcal{B}))$ exactly as in (4.98), with the number M_0 replaced by

$$\tilde{M}_0 := N_2(\mathcal{E}^{(\infty)}),$$

i.e. by the minimum number of balls of radius 1 that are needed to cover the compact set $\mathcal{E}^{(\infty)}$. Thus, from (4.117),

$$N_\delta(\mathcal{E}_6) \leq N_\delta(S^{n+1}(\mathcal{B})) \leq \tilde{M}_0(H_1(\theta))^{n+1}. \qquad (4.118)$$

We now turn to the estimate of $N_\delta(\mathcal{E}_5)$. To this end, given an index $j \in \{0,\dots,n\}$, we first cover $\mathcal{E}^{(\infty)}$ by exactly $N_{\delta\theta^{-j}}(\mathcal{E}^{(\infty)})$ balls of radius

$$\tfrac{1}{2}\delta\theta^{-j} = \theta^{n+1-j}R.$$

Then, by repeated applications of theorem 4.25, with the choice, at each stage k, $0 \leq k \leq j$, of $r = \theta^{n+1-j+k}R$ and $\mathcal{K} = S^k(\mathcal{E}^{(\infty)})$, we obtain that the set $S^j(\mathcal{E}^{(\infty)})$ can be covered by β_j balls of radius $\theta^{n+1}R = \tfrac{1}{2}\delta$. For example, the step $k = 0$ is as follows. Letting $r_0 := \theta^{n+1-j}R$ and $v_0 := N_{\delta\theta^{-j}}(\mathcal{E}^{(\infty)})$, the chosen covering of $\mathcal{E}^{(\infty)}$ yields the inclusion

$$\mathcal{E}^{(\infty)} \subseteq \bigcup_{i_0=1}^{\beta_0} (\mathcal{E}^{(\infty)} \cap \overline{B}(a_{i_0}, r_0)),$$

with suitable centers $a_1,\dots,a_{\beta_0} \in \mathcal{E}^{(\infty)}$. Thus,

$$S(\mathcal{E}^{(\infty)}) \subseteq \bigcup_{i_0=1}^{\beta_0} S(\mathcal{E}^{(\infty)} \cap \overline{B}(a_{i_0}, r_0)), \qquad (4.119)$$

and theorem 4.25 implies that each set at the right side of (4.119) can in turn be covered by v_1 balls of radius $\theta r_0 = \theta^{n+1-j+1}R$. This yields the step for $k = 1$. Proceeding in this way, we finally cover $S^j(\mathcal{E}^{(\infty)})$ by $v_0 v_1 \cdots v_j =: \beta_j$ balls of radius $\theta^j r_0 = \theta^{n+1}R$. As in (4.89), β_j can be estimated by

$$\beta_j \leq N_{\delta\theta^{-j}}(\mathcal{E}^{(\infty)})(H_1(\theta))^j;$$

since $N_\delta(S^j(\mathcal{E}^{(\infty)})) \leq \beta_j$, we eventually obtain that

$$N_\delta(\mathcal{E}_5) \leq \sum_{j=0}^n N_\delta(S^j(\mathcal{E}^{(\infty)})) \leq \sum_{j=0}^n N_{\delta\theta^{-j}}(\mathcal{E}^{(\infty)})(H_1(\theta))^j. \tag{4.120}$$

Let now η_0 be as in (4.91). We claim that for all $\eta > \eta_0$, there exists $c_\eta \geq 1$ such that for all $\varepsilon \leq 1$,

$$N_\varepsilon(\mathcal{E}^{(\infty)}) \leq c_\eta \left(\tfrac{1}{\varepsilon}\right)^\eta. \tag{4.121}$$

Indeed, by the very definition of the limes superior, i.e., the lowest upper bound, (4.91) implies that for all $\eta > \eta_0$ there is an $\varepsilon_0 > 0$ such that for all $\varepsilon \leq \varepsilon_0$,

$$\frac{\ln(N_\varepsilon(\mathcal{E}^{(\infty)}))}{-\ln\varepsilon} \leq \eta; \tag{4.122}$$

we can clearly restrict ε_0 to be in $]0,1[$. Then: if $\varepsilon \leq \varepsilon_0$, (4.122) implies that

$$N_\varepsilon(\mathcal{E}^{(\infty)}) \leq \left(\tfrac{1}{\varepsilon}\right)^\eta;$$

if instead $\varepsilon_0 < \varepsilon \leq 1$,

$$N_\varepsilon(\mathcal{E}^{(\infty)}) \leq N_{\varepsilon_0}(\mathcal{E}^{(\infty)}) \leq \left(\tfrac{1}{\varepsilon_0}\right)^\eta \leq \left(\tfrac{1}{\varepsilon_0}\right)^\eta \left(\tfrac{1}{\varepsilon}\right)^\eta =: c_\eta \left(\tfrac{1}{\varepsilon}\right)^\eta, \tag{4.123}$$

with $c_\eta := \varepsilon_0^{-\eta}$ depending (only) on η via ε_0, as per (4.122). Since $c_\eta \geq 1$, (4.122) and (4.123) show that (4.121) holds.

Given then δ and j as above, we specify ε as

$$\varepsilon = \tfrac{1}{2}\theta^{-j}\delta =: \varepsilon_j,$$

and note that $\varepsilon_j \leq 1$ for all $j \in \{0, \ldots, n\}$, because of (4.93) and (4.94). Hence, (4.121) implies that, for $0 \leq j \leq n$,

$$N_{\delta\theta^{-j}}(\mathcal{E}^{(\infty)}) \leq N_{\varepsilon_j}(\mathcal{E}^{(\infty)}) \leq c_\eta \left(\tfrac{2\theta^j}{\delta}\right)^\eta.$$

Thus, we obtain from (4.120) that

$$N_\delta(\mathcal{E}_5) \leq c_\eta \sum_{j=0}^n \left(\tfrac{2\theta^j}{\delta}\right)^\eta (H_1(\theta))^j = c_\eta \left(\tfrac{2}{\delta}\right)^\eta \sum_{j=0}^n (\theta^\eta H_1(\theta))^j. \tag{4.124}$$

As in part (8b), we now distinguish two cases. Assume at first that $\theta^\eta H_1(\theta) \leq 1$. Then from (4.116), (4.118) and (4.124) we have

$$N_\delta(\mathcal{G}^{(\infty)}) \leq c_\eta \left(\tfrac{2}{\delta}\right)^\eta (n+1) + \tilde{M}_0(H_1(\theta))^{n+1}. \tag{4.125}$$

We now recall that, from (4.93),

$$n+1 = \frac{\ln(\delta/2R)}{\ln\theta} = \frac{\ln(2R/\delta)}{-\ln\theta};$$

thus, using the identity $x^{\ln y} = y^{\ln x}$ $(x, y > 0)$, we obtain from (4.125) that

$$N_\delta(\mathcal{G}^{(\infty)}) \leq c_\eta \left(\frac{2}{\delta}\right)^\eta \frac{\ln(2R/\delta)}{-\ln\theta} + \tilde{M}_0 \left((H_1(\theta))^{\ln(2R/\delta)}\right)^{1/(-\ln\theta)}$$

$$= \frac{c_\eta}{R^\eta} \left(\frac{2R}{\delta}\right)^\eta \frac{\ln(2R/\delta)}{-\ln\theta} + \tilde{M}_0 \left(\frac{2R}{\delta}\right)^\Theta, \tag{4.126}$$

where

$$\Theta := \frac{\ln(H_1(\theta))}{-\ln\theta}.$$

Since $\Theta \leq \eta_0 \leq \eta$, (4.126) implies that

$$N_\delta(\mathcal{G}^{(\infty)}) \leq \left(\frac{2R}{\delta}\right)^\eta \left(\frac{c_\eta}{R^\eta} \frac{\ln(2R) - \ln\delta}{-\ln\theta} + \tilde{M}_0\right). \tag{4.127}$$

From this, we obtain that

$$\frac{\ln N_\delta(\mathcal{G}^{(\infty)})}{-\ln\delta} \leq \frac{\eta(\ln(2R) - \ln\delta)}{-\ln\delta} + \frac{1}{-\ln\delta}\ln\left(\frac{c_\eta}{R^\eta}\frac{\ln(2R) - \ln\delta}{-\ln\theta} + \tilde{M}_0\right)$$

$$=: h_3(\delta).$$

We conclude then that if $\theta^\eta H_1(\theta) \leq 1$,

$$\dim_F(\mathcal{G}^{(\infty)}) \leq \lim_{\delta \to 0} h_3(\delta) = \eta. \tag{4.128}$$

If instead $\theta^\eta H_1(\theta) \geq 1$, from (4.116), (4.118) and (4.124) we have

$$N_\delta(\mathcal{G}^{(\infty)}) \leq \frac{c_\eta}{R^\eta}\left(\frac{2R}{\delta}\right)^\eta (\theta^\eta H_1(\theta))^{n+1}(n+1) + \tilde{M}_0(H_1(\theta))^{n+1}$$

$$\leq \frac{c_\eta}{R^\eta}\left(\frac{2R}{\delta}\right)^\eta (\theta^\eta H_1(\theta))^{\frac{\ln(2R/\delta)}{-\ln\theta}} \frac{\ln(2R/\delta)}{-\ln\theta} + \tilde{M}_0 (H_1(\theta))^{\frac{\ln(2R/\delta)}{-\ln\theta}}$$

$$= \frac{c_\eta}{R^\eta}\left(\frac{2R}{\delta}\right)^\eta \left(\frac{2R}{\delta}\right)^{\frac{\ln(H_1(\theta))}{-\ln\theta}-\eta} \frac{\ln(2R/\delta)}{-\ln\theta} + \tilde{M}_0\left(\frac{2R}{\delta}\right)^{\frac{\ln(H_1(\theta))}{-\ln\theta}}$$

$$\leq \left(\frac{2R}{\delta}\right)^\Theta \left(\frac{c_\eta}{R^\eta}\frac{\ln(2R) - \ln\delta}{-\ln\theta} + \tilde{M}_0\right).$$

Since $\Theta \leq \eta$, this yields the same estimate as (4.127); hence, we conclude that (4.128) holds also if $\theta^\eta H_1(\theta) \geq 1$. Thus, (4.115) holds, and this concludes the proof of (4.114).

8d. We can now conclude the proof of the estimate of the fractal dimension of the exponential attractor $\mathcal{E}$. In fact, recalling (4.83), (4.15) follows from (4.87) and (4.114), using the first of part (3) of proposition 2.61.

With this, we have finally completed the proof of theorem 4.26: the set $\mathcal{E}$ defined in (4.83) is the desired exponential attractor of the semiflow S, relative to $\mathcal{B}$. □

Finally, since theorem 4.4 follows from theorem 4.26, theorem 4.4 is now also completely proven.

4.6   Concluding Remarks

We conclude this chapter by remarking that the construction of the exponential attractor that we have described is not the only possible one. In fact, in Eden, Foias, Nicolaenko and Temam, [EFNT94, ch. 7], an alternative construction of the exponential attractor is given, which is not based on the cone property. We refer to [EFNT94] for all details; here, we limit ourselves to illustrate one of these alternative ways to obtain an exponential attractor, by considering the following example, which is adapted from [EFNT94, sct. 7.1].

Example 4.27

Let $\mathcal{K} := \{x \in \mathbb{R}^N : \|x\| \leq 1\}$ be the closed unit ball of $\mathbb{R}^N$, and define a map $S \colon \mathcal{K} \to \mathcal{K}$ by

$$S(x) := \frac{x}{1 + \|x\|} \, .$$

Then, it is easily seen by induction that

$$S^n(x) = \frac{x}{1 + n\|x\|} \, ,$$

for all $n \in \mathbb{N}$ and $x \in \mathcal{K}$. Thus, S has the attractor $\mathcal{A} = \{0\}$, but the convergence of $S^n(x)$ to $\mathcal{A}$ is only polynomial. In fact, note that for all $x \in \mathcal{K}$ and $n \geq 1$,

$$\|S^n(x)\| \leq \frac{1}{n} \, ,$$

while for all $x \in \mathcal{K} \setminus \{0\}$ and $n \geq \left\lfloor \frac{1}{\|x\|} \right\rfloor + 1$,

$$\|S^n(x)\| \geq \frac{1}{2n} \, .$$

To construct an exponential attractor $\mathcal{E}$, we proceed in the following way. We start by choosing, as a compact absorbing set $\mathcal{B}$, any closed ball $\overline{B}(0, R)$, $0 < R \leq 1$. Thus, we look for points in $\mathcal{K}$ which attract other points exponentially, without necessarily converging exponentially to 0. Given $\theta \in \,]0, 1[$, we consider the set

$$\mathcal{G} := \{z \in \mathcal{K} : z = k\theta^m x, \ x \in \mathcal{K}, \ k, m \in \mathbb{N}, \ k\theta^m \leq 1\} \, .$$

The set $\overline{\mathcal{G}}$ is evidently compact. We claim that $\overline{\mathcal{G}}$ attracts all subsets $\mathcal{D} \subseteq \mathcal{K}$ exponentially. Indeed, given any $\mathcal{D} \subseteq \mathcal{K}$ and $n \in \mathbb{N}$, for each $x \in \mathcal{D}$ we define $h \in \,]0, +\infty[$ and $k \in \mathbb{N}$ by

$$h := \frac{1}{(1 + n\|x\|)\theta^n} \, , \qquad k := \lfloor h \rfloor \, ,$$

(both h and k depend on x and n). Since $k \leq h$,

$$k\theta^n \leq h\theta^n = \frac{1}{1 + n\|x\|} \leq 1 \, ,$$

which implies that $z := k\theta^n x \in \overline{\mathcal{G}}$. Since also $h < k+1$, and $\|x\| \leq 1$,

$$\|S^n(x) - z\| \leq \|S^n(x) - h\theta^n x\| + \|h\theta^n x - k\theta^n x\| = 0 + (h-k)\theta^n \|x\| \leq \theta^n.$$

Thus, for each $x \in \mathcal{D}$,

$$d(S^n(x), \overline{\mathcal{G}}) = \inf_{z \in \overline{\mathcal{G}}} d(S^n(x), z) \leq \|S^n(x) - z\| \leq \theta^n.$$

Since this estimate is independent of $x \in \mathcal{D}$, and in fact of $\mathcal{D}$, we conclude that for all $\mathcal{D} \subseteq \mathcal{K}$,

$$\partial(S^n(\mathcal{D}), \overline{\mathcal{G}}) \leq \theta^n = e^{-\kappa n},$$

with $\kappa := -\ln \theta$. However, the set $\overline{\mathcal{G}}$ is not yet an exponential attractor, since it needs not be positively invariant. Thus we enlarge it into the set

$$\mathcal{E} = \bigcup_{n=0}^{\infty} S^n(\overline{\mathcal{G}}).$$

This set is now positively invariant, again compact, and attracts all subsets of $\mathcal{K}$ exponentially. The fractal dimension of $\mathcal{E}$ is obviously finite (being at most equal to N); hence, $\mathcal{E}$ is the desired exponential attractor. $\qquad\square$

Chapter 5

Inertial Manifolds

5.1 Introduction

1. Roughly speaking, INERTIAL MANIFOLDS are positively invariant, finite dimensional, exponentially attracting Lipschitz manifolds. In this chapter we give the precise definition of this notion for a continuous semiflow S defined on a Banach space $\mathcal{X}$, and show how an inertial manifold can be constructed, when S satisfies some natural conditions on its geometrical structure. In this chapter, we study in detail inertial manifolds which are graphs of maps $m: \mathcal{X}_1 \to \mathcal{X}$, with $\mathcal{X}_1$ a finite dimensional subspace of $\mathcal{X}$. These manifolds have a degree of smoothness, inherited by the smoothness of the map m (we shall in particular consider Lipschitz continuous maps), and the geometrical structure of the semiflow S can be described quite naturally for manifolds of this type. In particular, we shall consider the CONE INVARIANCE PROPERTY, and one of several versions of the STRONG SQUEEZING PROPERTY. We proceed then to explore the applicability of these results to abstract evolution equations of the form

$$u_t + A u = F(u), \tag{5.1}$$

where $A: \mathcal{X} \to \mathcal{X}$ is a linear, unbounded operator which generates at least a C^0-SEMIGROUP in $\mathcal{X}$ (see section 5.6.1 below). We find that the squeezing property holds if the eigenvalues in the point spectrum of A satisfy some restrictions on their growth, relative to the nonlinearity F; in particular, if they satisfy one type of SPECTRAL GAP CONDITION. For problems of the type considered in chapter 3, we will be able to show that, when the space dimension is $n = 1$, this spectral gap condition always holds for the parabolic problem (P), as well as for the hyperbolic problem (H_ε), if ε is sufficiently small. For higher space dimensions, the situation depends heavily on special features of the problem, such as the geometric properties of the domain Ω. In particular, we will consider in detail a model of the CHAFEE-INFANTE reaction-diffusion equations in one dimension of space, and its (small) hyperbolic perturbation. In chapter 6 we present some other examples of semiflows which admit an inertial manifold; mostly, these semiflows are generated by PDEEs of "parabolic" type.

2. The limitations on the possibility of extending this theory to general dissipative hyperbolic problems are highlighted by a remarkable result of Mora and Solà-Morales ([MSM87]), concerning a one-dimensional version of the dissipative wave equation (3.4). In short, they have shown that if ε is sufficiently large, and the boundary conditions for (3.4) are of Neumann type, the corresponding semiflow does *not* admit any inertial manifold which is of class C^1 and locally invariant in the neighborhood of one of the stationary solutions of the problem. Thus, for general $\varepsilon > 0$ the existence of an inertial manifold in the hyperbolic case is open to question. We present Mora and Solà-Morales result in chapter 7.

3.   When the global attractor $\mathcal{A}$ and a closed inertial manifold $\mathcal{M}$ exist, then $\mathcal{A} \subseteq \mathcal{M}$. This follows in a similar way as the inclusion of the global attractor $\mathcal{A}$ in an exponential attractor $\mathcal{E}$ (see section 4.1). On the other hand, if $\mathcal{G} \subseteq \mathcal{X}$ is a bounded, positively invariant absorbing set, theorems 2.33 and 2.46 imply that the semiflow S admits a global attractor $\mathcal{A} \subseteq \mathcal{G}$, if either $\mathcal{G}$ is compact or if S is uniformly compact for large t. In either case, if S admits both a compact, positively invariant absorbing set $\mathcal{G}$, and a closed inertial manifold $\mathcal{M}$, then the set $\mathcal{E} := \mathcal{M} \cap \mathcal{G}$ is also a compact set, which is positively invariant (being the intersection of two positively invariant sets). If $\mathcal{M}$ possesses a more specific type of attractivity property, called EXPONENTIAL TRACKING PROPERTY, then $\mathcal{E}$ is exponentially attracting (see section 5.2.2). Thus, in this case, $\mathcal{E}$ is an exponential attractor, and $\mathcal{A} \subseteq \mathcal{E} \subseteq \mathcal{M}$.

4.   As we have already mentioned, there are many more systems that admit an exponential attractor than systems that are known to admit an inertial manifold. The main reason for this difference is that all known inertial manifolds are closed, and therefore the existence of a compact absorbing set (which does hold for the systems we have considered so far) also yields directly the existence of an exponential attractor. Moreover, inertial manifolds are much more regular than exponential attractors.

There is also a "practical" reason for this difference; namely, that the available results on the existence of inertial manifolds require two conditions on the geometrical structure of the semiflow, called here respectively the CONE INVARIANCE PROPERTY and the DECAY PROPERTY. These are usually combined together, in the so-called STRONG SQUEEZING PROPERTY. On the other hand, in order to establish the existence of an exponential attractor it is sufficient to assume that the semiflow S satisfies the discrete squeezing property, which is a much weaker condition.

5.   Roughly speaking, the cone invariance and decay properties describe a sort of dichotomy principle, whereby either the difference of two motions can never leave a certain cone (cone invariance property), or, if it does, the distance between the motions decays exponentially (decay property). In contrast, the discrete squeezing property only requires that either the difference of two motions is in a cone at a specific time (as opposed to for all times), or, if not, the distance between the motions decays exponentially (as in the decay property).

For semiflows generated by evolution equations of the general form (5.1), one of the available ways to verify that the semiflow satisfies the cone invariance property is to deduce this property from a condition on the point spectrum of A. This condition is called the SPECTRAL GAP CONDITION; we introduce it in section 5.6.2. Essentially, this condition guarantees that the linear part of the equation (i.e., the term Au) can "dominate", in a sense to be made precise, the nonlinear part (i.e., the term $f(u)$). In most situations of interest, the spectral gap condition is quite difficult if not impossible to verify, while the decay properties are relatively easy to verify. This is not surprising, since the spectral gap condition requires a sufficiently large difference between successive eigenvalues, whereas, for the decay property, one only needs a sufficiently large eigenvalue.

6. As we have stated, we will construct inertial manifolds of the type

$$\mathcal{M} = \{x + m(x) : x \in \mathcal{X}_1\}, \tag{5.2}$$

where $\mathcal{X} = \mathcal{X}_1 \oplus \mathcal{X}_2$ is decomposed into a closed linear subspace $\mathcal{X}_1$ of finite dimension, and its algebraic complement $\mathcal{X}_2$, and where $m \colon \mathcal{X}_1 \to \mathcal{X}_2$ is a Lipschitz continuous map. In contrast to exponential attractors, this type of inertial manifolds allows us to imbed the global attractor $\mathcal{A}$ (and the exponential attractor $\mathcal{E}$) in $\mathbb{R}^N$, with $N = \dim \mathcal{X}_1 = \dim \mathcal{M}$.

Moreover, if A commutes with the continuous projector π_1 from $\mathcal{X}$ onto $\mathcal{X}_1$, the asymptotic behavior of the solution of (4.4) is governed by the N-dimensional INERTIAL FORM SYSTEM

$$\dot{x} = -Ax + \pi_1 f(x + m(x)) \tag{5.3}$$

in $\mathcal{X}_1$. This is the so-called SLAVING PRINCIPLE. In contrast to (4.3), (5.3) is explicitly given and has a Lipschitz continuous right-hand side if f is Lipschitz continuous.

5.2 Definitions and Comparisons

In this section we give the definition of an inertial manifold, and compare various ways that can be used to construct an inertial manifold.

5.2.1 Lipschitz Manifolds and Inertial Manifolds

In the literature on inertial manifolds, the notion of Lipschitz manifold is often not defined precisely. In most applications, there are two types of manifolds $\mathcal{M}$ which are called Lipschitz manifolds: In the first, the manifold $\mathcal{M}$ is the graph of a Lipschitz continuous map m from a closed, linear subspace $\mathcal{X}_1$ of $\mathcal{X}$ into its algebraic complement $\mathcal{X}_2$, as in (5.2). In the second, $\mathcal{M}$ is the graph of a Lipschitz continuous map m from a bounded subset $\mathcal{M}_1$ of $\mathcal{X}_1$ into $\mathcal{X}_2$, where the set $\mathcal{M}_1$ can be closed. In both cases, $\mathcal{X}_1$ is typically finite dimensional; at least for the second case, we need the notion of a Lipschitz manifold with boundary.

We denote by

$$\mathbb{H}^N := \{(x_1, \ldots, x_N) \in \mathbb{R}^N : x_1 \geq 0\}$$

the (first) closed half-space of $\mathbb{R}^N$.

DEFINITION 5.1   *A subset $\mathcal{M} \subseteq \mathcal{X}$ is an N-DIMENSIONAL LIPSCHITZ SUBMAN-IFOLD (WITH BOUNDARY) of $\mathcal{X}$ if it possesses the following two structural characteristics:*

1. *There exists a countable collection of open sets $\mathcal{V}_i \subseteq \mathcal{X}$, $i \in \mathcal{I}$, such that if $\mathcal{U}_i := \mathcal{V}_i \cap \mathcal{M}$, then $\mathcal{M} = \bigcup_{i \in \mathcal{I}} \mathcal{U}_i$.*

2. *There exist open sets $\mathcal{W}_i \subseteq \mathbb{R}^N$ and invertible mappings $\Phi_i : \mathcal{W}_i \to \mathcal{V}_i$, with $\Phi_i(\mathcal{W}_i \cap \mathbb{H}^N) = \mathcal{U}_i$, such that Φ_i and Φ_i^{-1} are Lipschitz continuous.*

The set

$$\partial \mathcal{M} := \bigcup_{i \in \mathcal{I}} \Phi_i(\partial \mathbb{H}^N)$$

is called the BOUNDARY of $\mathcal{M}$.

In particular, we have the so-called trivial Lipschitz submanifolds:

DEFINITION 5.2   *A subset $\mathcal{M} \subseteq \mathcal{X}$ is called an N-DIMENSIONAL TRIVIAL LIP-SCHITZ SUBMANIFOLD (WITH BOUNDARY) of $\mathcal{X}$ if it possesses the following two structural characteristics:*

1. *There exist a closed N-dimensional linear subspace $\mathcal{X}_1$ of $\mathcal{X}$, and a subset $\mathcal{M}_1$ of $\mathcal{X}_1$, such that $\mathcal{M}_1$ is an N-dimensional Lipschitz manifold (with boundary).*

2. *There is an invertible mapping $\varphi : \mathcal{M}_1 \to \mathcal{M}$ such that φ and φ^{-1} are Lipschitz continuous.*

REMARK 5.3   1. A trivial N-dimensional Lipschitz submanifold (with boundary) of $\mathcal{X}$ is indeed an N-dimensional Lipschitz submanifold of $\mathcal{X}$. To show this, in accord with definition 5.2, let $\mathcal{V}_i \subseteq \mathcal{X}$, $\mathcal{W}_i \subseteq \mathbb{R}^N$ be open sets and let $\Phi_i : \mathcal{W}_i \cap \mathbb{H}^N \to \mathcal{U}_i$, $\mathcal{U}_i := \mathcal{V}_i \cap \mathcal{M}_1$, be invertible mappings such that $\mathcal{M}_1 = \bigcup_{i \in \mathcal{I}} \mathcal{U}_i$ and such that Φ_i and Φ_i^{-1} are Lipschitz continuous. Let $\tilde{\mathcal{V}}_i := \varphi(\mathcal{V}_i)$. Then the sets $\tilde{\mathcal{V}}_i$ are open sets, as pre-images of the open sets $\mathcal{V}_i$, and $\mathcal{M} = \bigcup_{i \in \mathcal{I}} \tilde{\mathcal{U}}_i$ with $\tilde{\mathcal{U}}_i := \tilde{\mathcal{V}}_i \cap \mathcal{M}$. Finally, $\tilde{\Phi}_i := \varphi \circ \Phi_i$ maps $\mathcal{W}_i \cap \mathbb{H}^N$ onto $\tilde{\mathcal{U}}_i$ and $\tilde{\Phi}_i$ and $\tilde{\Phi}_i^{-1}$ are Lipschitz continuous.

2. If, in addition, $\mathcal{M}_1$ is closed, then $\mathcal{M}$ is also a closed set. This follows from the fact that $\mathcal{M}$ is the pre-image of the closed set $\mathcal{M}_1$ under the continuous mapping φ^{-1}.

3. We will consider only two particular cases: The first is the global one, in which $\mathcal{M}_1 = \mathcal{X}_1$. The second is when $\mathcal{M}_1$ is a bounded, closed subset of $\mathcal{X}_1$, with Lipschitz boundary.   □

Example 5.4

1. The set

$$\mathcal{M} := \{(x,y,z) \in \mathbb{R}^3 : \max\{|x|,|y|\} \leq 1, z = xy\}$$

is a two dimensional, trivial Lipschitz submanifold of $\mathbb{R}^3$, with boundary. The boundary $\partial \mathcal{M}$ of $\mathcal{M}$ is the set

$$\{(x,y,z) \in \mathbb{R}^3 : z = xy, |x| = 1 \text{ or } |y| = 1\}.$$

2. The set

$$\mathcal{M} := \{(x,y,z) \in \mathbb{R}^3 : \max\{|x|,|y|,|z|\} = 1, z \geq 0\}$$

is a two dimensional, nontrivial Lipschitz submanifold of $\mathbb{R}^3$, with boundary given by the set

$$\{(x,y,z) \in \mathbb{R}^3 : z = 0, |x| = |y| = 1\}.$$

$\square$

We can now introduce the definition of inertial manifold.

DEFINITION 5.5 *Let $\mathcal{T}$ be one of the sets $\mathbb{R}$ or $\mathbb{R}_{\geq 0}$, and let $S = (S(t))_{t \in \mathcal{T}}$ be a continuous semiflow on $\mathcal{X}$. A subset $\mathcal{M} \subset \mathcal{X}$ is an* INERTIAL MANIFOLD *for S if $\mathcal{M}$ is a finite dimensional Lipschitz submanifold (with boundary) of $\mathcal{X}$, which is positively invariant and exponentially attracting. The latter condition means that there is $\eta > 0$ with the property that for all $x \in \mathcal{X}$ there is $K > 0$ such that for all $t \geq 0$,*

$$d(S(t)x, \mathcal{M}) \leq K e^{-\eta t}. \tag{5.4}$$

REMARK 5.6 1. In the naive understanding of Lipschitz manifolds, the difference between manifolds with or without boundary is not usually emphasized.

2. Currently, we are only able to construct trivial Lipschitz manifolds as graphs of a Lipschitz mapping m from a (closed) subset $\mathcal{M}_1$ of a (closed) linear subspace $\mathcal{X}_1$ of $\mathcal{X}$ into the algebraic complement $\mathcal{X}_2$ of $\mathcal{X}_1$, as in (5.2).

3. Inertial manifolds which are trivial Lipschitz manifolds, in the sense of definition 5.2, can be assumed to be closed (at least after extending of m to the closure of its domain.) $\square$

The notion of inertial manifolds for evolution equations goes back to at least Henry, [Hen81], and Mora, [Mor87], and was also studied by Constantin, Foias, Nicoalenko, Sell and Temam, [FST85, FNST85, CFNT86]; see also Temam, [Tem88]. In all these works, the inertial manifolds were constructed as the intersection of a graph of Lipschitz function over a closed, linear, N-dimensional subspace $\mathcal{X}_1$ of $\mathcal{X}$, with a closed ball in $\mathcal{X}$. Thus, these manifolds are compact, and possess a finite fractal dimension (which is N). Together with the exponential convergence of the orbits described in (5.4), this is a property shared with exponential attractors; as we discussed in chapter 4, these properties imply that, when an inertial manifold exists, this set

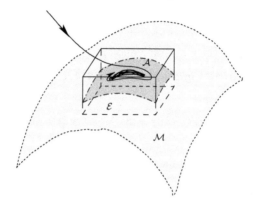

Figure 5.1: Inertial Manifold, Absorbing Set, and Global Attractor

provides an extremely effective way of describing the long-time behavior of the dynamical system, since the latter is asymptotically equivalent to a finite dimensional system (that is, to a system of ODEs). One of the advantages of inertial manifolds over exponential attractors resides in the fact that, when an inertial manifold exists, this system of ODEs has smooth coefficients, inherited from the smoothness of $\mathcal{M}$.

In many situations, it is convenient to strengthen the requirement that motions converge exponentially to $\mathcal{M}$, as in (5.4), and assume the following property:

DEFINITION 5.7   *Let* $\mathcal{M} \subseteq \mathcal{X}$ *be a finite-dimensional Lipschitz manifold, positively invariant with respect to S.* $\mathcal{M}$ *is said to have the* EXPONENTIAL TRACKING PROPERTY, *if there is* $\eta > 0$ *such that for every* $x \in \mathcal{X}$, *there are* $x' \in \mathcal{M}$ *and* $c \geq 0$ *such that for all* $t \geq 0$,

$$\|S(t)x - S(t)x'\|_{\mathcal{X}} \leq ce^{-\eta t} \tag{5.5}$$

(note that $S(t)x' \in \mathcal{M}$ *for all* $t \geq 0$, *because* $\mathcal{M}$ *is positively invariant.) The motion* $t \to S(t)x'$ *in (5.5) is called an* ASYMPTOTIC PHASE *of the motion* $t \to S(t)x$.

The exponential tracking property, which is also called EXISTENCE OF ASYMPTOTIC PHASES, or ASYMPTOTIC COMPLETENESS PROPERTY, was introduced, for example, by Foias, Sell and Titi in [FST89], and by Robinson in [Rob96]. It obviously implies the exponential convergence of the motions to the manifold, because if $x \in \mathcal{X}$ and $x' \in \mathcal{M}$ are as in (5.5), then, since $S(t)x' \in \mathcal{M}$ for all $t \geq 0$,

$$d(S(t)x, \mathcal{M}) = \inf_{y \in \mathcal{M}} \|S(t)x - y\|_{\mathcal{X}} \leq \|S(t)x - S(t)x'\|_{\mathcal{X}} \leq ce^{-\eta t}, \tag{5.6}$$

i.e., (5.4) holds. Of course, the exponential tracking property is much stronger than the exponential convergence of the motions, since (5.5) means that any motion should converge exponentially to a motion which is completely in the manifold $\mathcal{M}$.

In (5.5), the constant c depends on x, both explicitly and via x', which also depends on x. In practice, it is often possible to show that even stronger versions of the exponential tracking property hold. For example, we can require exponential tracking properties with inequalities of the form

$$\|S(t)x - S(t)x'\|_{\mathcal{X}} \leq K_1 \|x - x'\|_{\mathcal{X}} e^{-\eta t},$$
$$\|S(t)x - S(t)x'\|_{\mathcal{X}} \leq K_2 d(x, \mathcal{M}) e^{-\eta t}, \tag{5.7}$$

with K_1 and K_2 independent of $x \in \mathcal{X}$ and $x' \in \mathcal{M}$. For example, (5.7) means that the constant c in (5.5) does not depend on the point $x' \in \mathcal{M}$, but only on the distance of x to the manifold $\mathcal{M}$. In other words, the motion $t \mapsto S(t)x$ admits the motion $t \mapsto S(t)x'$ as an asymptotic phase, and the exponential decay of the difference between these motions is estimated by the initial distance between x and the manifold $\mathcal{M}$, with constants K_2 and η independent of x. Finally, there is no loss in generality in considering $x \in \mathcal{X} \setminus \mathcal{M}$ only, because if $x \in \mathcal{M}$ we can take $x' = x$, and (5.7) is trivially satisfied.

5.2.2   Inertial Manifolds and Exponential Attractors

In general, the existence of an inertial manifold does not imply, and is neither implied by, the existence of an exponential attractor. Moreover, even when both these sets exist, neither is in general contained in the other (although they both contain the global attractor). However, we can construct an exponential attractor from an inertial manifold, by intersecting the latter with a compact, positively invariant absorbing set. More precisely:

PROPOSITION 5.8
Let S be a semiflow on $\mathcal{X}$. Assume that S admits a compact, positively invariant absorbing set $\mathcal{G}$, and a closed, inertial manifold $\mathcal{M}$. Assume further that the stronger version (5.7) of the exponential tracking property holds. Then the set $\mathcal{E} := \mathcal{M} \cap \mathcal{G}$ is an exponential attractor for S in $\mathcal{X}$.

PROOF   Since $\mathcal{M}$ is closed and $\mathcal{G}$ is compact, $\mathcal{E}$ is compact. Since both $\mathcal{M}$ and $\mathcal{G}$ are positively invariant, $\mathcal{E}$ is positively invariant too. Since $\mathcal{E} \subseteq \mathcal{M}$, by (2.49) of proposition 2.61 we have $\dim_F(\mathcal{E}) \leq \dim_F(\mathcal{M})$, so $\mathcal{E}$ has finite fractal dimension. Thus, it only remains to prove that $\mathcal{E}$ attracts all bounded sets of $\mathcal{X}$ exponentially, i.e., that for any bounded set $\mathcal{B} \subseteq \mathcal{X}$, there is $C_{\mathcal{B}} > 0$, depending on $\mathcal{B}$, such that for all $t \geq 0$,

$$\partial(S(t)\mathcal{B}, \mathcal{E}) \leq C_{\mathcal{B}} e^{-\eta t}. \tag{5.8}$$

As an intermediate step, we show that (5.8) holds when $\mathcal{B} = \mathcal{G}$. To this end, we turn to the modified exponential tracking property (5.7). Since the function $x \mapsto d(x, \mathcal{M})$ is continuous, its restriction to the compact set $\mathcal{G}$ is bounded; therefore, by (5.7), there is $K_{\mathcal{G}} > 0$, depending on $\mathcal{G}$, such that for all $g \in \mathcal{G}$, there is $g' \in \mathcal{M}$ with the property that for all $t \geq 0$,

$$\|S(t)g - S(t)g'\|_{\mathcal{X}} \leq K_{\mathcal{G}} e^{-\eta t}. \tag{5.9}$$

Call $\mathcal{G}'$ the subset of $\mathcal{M}$ consisting of all the elements $g' \in \mathcal{M}$ satisfying (5.9) for some $g \in \mathcal{G}$. Then, $\mathcal{G}'$ is bounded. In fact, from (5.9) for $t = 0$ we have

$$\|g'\|_{\mathcal{X}} \leq \|g' - g\|_{\mathcal{X}} + \|g\|_{\mathcal{X}} \leq K_{\mathcal{G}} + \sup_{x \in \mathcal{G}} \|x\|_{\mathcal{X}}.$$

Since $\mathcal{G}$ is absorbing, there is $T_1 \geq 0$, depending on $\mathcal{G}'$, such that for all $t \geq T_1$, $S(t)\mathcal{G}' \subseteq \mathcal{G}$. Since $\mathcal{M}$ is positively invariant, we deduce that if $t \geq T_1$,

$$S(t)\mathcal{G}' \subseteq \mathcal{G} \cap \mathcal{M} = \mathcal{E}.$$

Hence, (5.9) implies that for all $g \in \mathcal{G}$ and $t \geq T_1$,

$$d(S(t)g, \mathcal{E}) \leq \|S(t)g - S(t)g'\|_{\mathcal{X}} \leq K_{\mathcal{G}} e^{-\eta t}. \tag{5.10}$$

Since the right side of (5.10) is independent of g, it follows that, if $t \geq T_1$,

$$\partial(S(t)\mathcal{G}, \mathcal{E}) \leq K_{\mathcal{G}} e^{-\eta t}. \tag{5.11}$$

If instead $0 \leq t \leq T_1$, we estimate

$$\partial(S(t)\mathcal{G}, \mathcal{E}) \leq \max_{0 \leq t \leq T_1} \partial(S(t)\mathcal{G}, \mathcal{E}) =: \delta_1 = \delta_1 e^{\eta t} e^{-\eta t} \leq \delta_1 e^{\eta T_1} e^{-\eta t}. \tag{5.12}$$

Thus, when $\mathcal{B} = \mathcal{G}$, (5.8) follows from (5.12) and (5.11), with

$$C_{\mathcal{G}} := \max\{K_{\mathcal{G}}, \delta_1 e^{\eta T_1}\}.$$

We now show (5.8) for a general bounded set $\mathcal{B} \subseteq \mathcal{X}$. Since $\mathcal{G}$ is absorbing, there is $T_2 \geq 0$, depending on $\mathcal{B}$, such that for all $t \geq T_2$, $S(t)\mathcal{B} \subseteq \mathcal{G}$. Let $t \geq T_2$, and write $t = T_2 + \theta$, $\theta \geq 0$. Then, since

$$S(t)\mathcal{B} = S(\theta)S(T_2)\mathcal{B} \subseteq S(\theta)\mathcal{G},$$

we have that when $t \geq T_2$,

$$\partial(S(t)\mathcal{B}, \mathcal{E}) \leq \partial(S(\theta)\mathcal{G}, \mathcal{E}) \leq C_{\mathcal{G}} e^{-\eta \theta} = C_{\mathcal{G}} e^{-\eta(t - T_2)} = C_{\mathcal{G}} e^{\eta T_2} e^{-\eta t}. \tag{5.13}$$

If instead $0 \leq t \leq T_2$, we proceed as in (5.12), i.e.

$$\partial(S(t)\mathcal{B}, \mathcal{E}) \leq \max_{0 \leq t \leq T_2} \partial(S(t)\mathcal{B}, \mathcal{E}) =: \delta_2 \leq \delta_2 e^{\eta T_2} e^{-\eta t}. \tag{5.14}$$

Thus, (5.8) follows from (5.13) and (5.14), with

$$C_{\mathcal{B}} := e^{\eta T_2} \max\{C_{\mathcal{G}}, \delta_2\}.$$

This concludes the proof of proposition 5.8. □

5.2.3 Methods of Construction of the Inertial Manifold

1. Most available ways to construct an inertial manifold are based on generalizations of methods developed for the construction of unstable, center-unstable or center manifolds for ODEs. For an extensive review, we refer e.g. to Luskin and Sell, [LS89], or to Ninomiya, [Nin93]. Here, we briefly recall the so-called LYAPUNOV-PERRON, the HADAMARD and the integral manifold methods.

The Lyapunov-Perron method was introduced by Lyapunov, [Lya47, Lya92] and Perron, [Per28, Per29, Per30], to prove the existence of stable and unstable manifolds of hyperbolic equilibrium points of systems of ODEs. In this method, the ODEs are transformed into integral equations, and the invariant manifolds are constructed as fixed points of the corresponding integral operator. In [Hen81], Henry gives a generalization of the Lyapunov-Perron method, which leads to a proof of the existence of stable, unstable and center manifolds for semilinear parabolic evolution equations. Further generalizations to the construction of inertial manifolds are presented e.g. in Foias, Sell and Temam, [FST88], Temam, [Tem88], Constantin, Foias, Nicolaenko and Temam, [CFNT86], Foias, Sell and Titi, [FST89], and Demengel and Ghidaglia, [DG91]. Indeed, the Lyapunov-Perron method is arguably one of the most commonly used for the construction of inertial manifolds.

Hadamard's method, introduced in [Had01] and also known as the GRAPH TRANSFORMATION METHOD, has been developed to show the existence of stable and unstable manifolds of fixed points of diffeomorphisms. Its nature is more geometrical than the Lyapunov-Perron method, in that the stable and unstable manifolds (relative to hyperbolic fixed points) are constructed as graphs over the linearized stable and unstable subspaces. An extension of Hadamard's method to infinite dimensional dynamical systems can be found e.g. in Bates and Jones, [BJ89].

The integral manifold method was introduced by Constantin, Foias, Nicoalenko and Temam in [CFNT89]. In this construction, the inertial manifold is defined as the closure of the set $\bigcup_{t \geq 0} S(t)\Gamma$, where Γ is the C^1 boundary of a suitable closed subset of a finite-dimensional, closed, linear subspace of $\mathcal{X}$.

2. In all these three methods, the inertial manifold $\mathcal{M}$ is constructed as the graph of a Lipschitz continuous function m defined on a subset $\mathcal{M}_1$ of a finite dimensional, closed, linear subspace $\mathcal{X}_1$ of $\mathcal{X}$. More precisely, we seek to establish a suitable orthogonal decomposition

$$\mathcal{X} = \mathcal{X}_1 \oplus \mathcal{X}_2, \tag{5.15}$$

where $\mathcal{X}_1$ is a finite dimensional subspace of $\mathcal{X}$ and the orthogonal projections

$$\pi_i \colon \mathcal{X} \to \mathcal{X}_i, \qquad i = 1, 2$$

are continuous, and to construct an inertial manifold $\mathcal{M}$ for the semiflow S, having the form

$$\mathcal{M} = \text{graph}(m) := \{\xi + m(\xi) \colon \xi \in \mathcal{M}_1\}, \tag{5.16}$$

where $m: \mathcal{M}_1 \subseteq \mathcal{X}_1 \to \mathcal{X}_2$ is a Lipschitz continuous function. At least if $\mathcal{M}_1 = \mathcal{X}_1$, $\mathcal{M}$ is a Lipschitz manifold; see figure 5.2. Indeed, let $\mathcal{M}_1 := \mathcal{X}_1$, and define

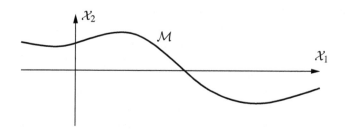

Figure 5.2: A global trivial Lipschitz manifold $\mathcal{M} = \operatorname{graph} m$

$\varphi: \mathcal{M}_1 \to \mathcal{X}$ by $\varphi(x) := x + m(x), x \in \mathcal{X}_1$. Obviously, φ and $\varphi^{-1} = \pi_1$ are Lipschitz continuous.

If $\mathcal{M}_1 \neq \mathcal{X}_1$, in order for $\mathcal{M}$ to be a Lipschitz manifold we need $\mathcal{M}_1$ to be a N-dimensional Lipschitz submanifold of the N-dimensional space $\mathcal{X}_1$. For example, if $\mathcal{M}_1$ is a closed ball in $\mathcal{X}_1$, if $\tilde{m}: \mathcal{X}_1 \to \mathcal{X}_2$ is Lipschitz continuous, and if $m := \tilde{m}\big|_{\mathcal{M}_1}$ is the restriction of $\tilde{m}$ onto $\mathcal{M}_1$, then $\mathcal{M}$ is a Lipschitz manifold with boundary; see figure 5.3. Indeed, the maps $\varphi: \mathcal{M}_1 \to \mathcal{M}$ and $\varphi^{-1}: \mathcal{M} \to \mathcal{M}_1$ defined by

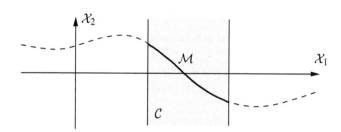

Figure 5.3: A trivial Lipschitz manifold $\mathcal{M} = (\operatorname{graph} m) \cap \mathcal{C}$ as intersection of a global trivial Lipschitz manifold with a cylinder $\mathcal{C}$

$$\varphi(x) := x + m(x), \quad x \in \mathcal{M}_1, \qquad \varphi^{-1} := \pi_1\big|_{\mathcal{C}},$$

where $\mathcal{C} := \{x \in \mathcal{X}: \pi_1 x \in \mathcal{M}_1\}$ is the cylinder in $\mathcal{X}$ with base $\mathcal{M}_1$, are Lipschitz continuous.

3.   Here, we shall follow Hadamard's method; that is, we obtain the function m as a fixed point of a so-called GRAPH TRANSFORMATION MAP, defined on the subspace

$\mathcal{G}_L$ of the Banach space

$$\mathcal{G} := C_b(\mathcal{X}_1; \mathcal{X}_2), \qquad (5.17)$$

consisting of all continuous and bounded functions from $\mathcal{X}_1$ into $\mathcal{X}_2$, which satisfy a global Lipschitz condition of the form

$$\|m(\xi) - m(\eta)\|_{\mathcal{X}} \leq L\|\xi - \eta\|_{\mathcal{X}}, \qquad \xi, \eta \in \mathcal{X}_1, \qquad (5.18)$$

where $\mathcal{X}_1$ is as in (5.15) (we require the boundedness of the functions, in order to make $\mathcal{G}_L$ a proper Banach space of continuous functions). That is, given $L > 0$ we define

$$\mathcal{G}_L := \{\psi \in \mathcal{G}: \|\psi(\xi) - \psi(\eta)\|_{\mathcal{X}} \leq L\|\xi - \eta\|_{\mathcal{X}} \text{ for all } \xi, \eta \in \mathcal{X}_1\}, \qquad (5.19)$$

and we consider the constant L as a parameter for the set $\mathcal{G}_L$, in which we will eventually find the function m, whose graph will be the desired inertial manifold.

4. As noted above, $\mathcal{M}$ is closed, being the pre-image of the closed set $\mathcal{X}_1$ under the Lipschitz map π_1. It follows that if the semiflow S admits a global attractor $\mathcal{A}$ and an inertial manifold $\mathcal{M}$ of the form (5.16), then $\mathcal{A} \subseteq \mathcal{M}$.

On the other hand, inertial manifolds of the form (5.16) need not be compact, since they are not necessarily bounded. To obtain inertial manifolds that are compact, it is sufficient to intersect the inertial manifold (5.16) with a compact, positively invariant absorbing set. More precisely:

PROPOSITION 5.9
Let S be a continuous semiflow on $\mathcal{X}$, and assume that there is a Lipschitz continuous function m as in (5.16), whose graph $\mathcal{M}$ is an inertial manifold for S. Assume that $\mathcal{M}$ satisfies the exponential tracking property (5.7), and that S admits a compact, positively invariant absorbing set $\mathcal{B}$. If $\tilde{\mathcal{M}} := \mathcal{M} \cap \mathcal{B}$ is a nonempty set with Lipschitz boundary, then $\tilde{\mathcal{M}}$ is a compact inertial manifold for S (and also an exponential attractor).

PROOF Since $\mathcal{M}$ is closed, $\tilde{\mathcal{M}} \subseteq \mathcal{B}$, and $\mathcal{B}$ is compact, $\tilde{\mathcal{M}}$ is also compact. Exactly as in the proof of proposition 5.8, we see that $\tilde{\mathcal{M}}$ is positively invariant and exponentially attracting. To show that $\tilde{\mathcal{M}}$ has finite fractal dimension, we note that $\tilde{\mathcal{M}}$ is the image of the compact set $\pi_1(\mathcal{B})$ under the Lipschitz continuous map m. Hence, by proposition 2.61,

$$\dim_F(\tilde{\mathcal{M}}) \leq \dim_F(\pi_1(\mathcal{B})) \leq \dim_F(\mathcal{X}_1).$$

This completes the proof that $\tilde{\mathcal{M}}$ is an inertial manifold for S. Because of proposition 5.8, $\tilde{\mathcal{M}}$ is also an exponential attractor. ⬚

For future reference, we recall here that, since the projections π_i are linear, their continuity is equivalent to the boundedness conditions

$$\|\pi_i\|_{\mathcal{L}(\mathcal{X},\mathcal{X})} := \sup_{\|x\|_{\mathcal{X}} \leq 1} \|\pi_i x\|_{\mathcal{X}} < +\infty, \qquad i = 1, 2. \tag{5.20}$$

Actually, since $\pi_1 + \pi_2 = I_{\mathcal{X}}$ (the identity in $\mathcal{X}$), it is sufficient to assume (5.20) for $i = 1$ or $i = 2$ only.

5.   Finally, we would like to comment on the significance of the exponential attracting property (5.4), in relation to manifolds of the form (5.16). As we have stated above, if S admits an inertial manifold $\mathcal{M}$, then motions on $\mathcal{M}$ are essentially governed by a finite system of ODEs. These motions are such that $S(t)x \in \mathcal{M}$ for all $t \geq 0$; if $\mathcal{M}$ has the form (5.16), this condition translates into the identity

$$\pi_2 S(t)x = m(\pi_1 S(t)x).$$

When instead the motion does not take place on $\mathcal{M}$, i.e. when $S(t)x \notin \mathcal{M}$, we want the difference

$$R(t)x := \pi_2 S(t)x - m(\pi_1 S(t)x),$$

which does no longer vanish, to become negligible as $t \to +\infty$. More specifically, we require this difference to decay exponentially, i.e.

$$\|R(t)x\|_{\mathcal{X}} \leq Ce^{-\eta t}, \tag{5.21}$$

for some positive C and η, the latter independent of x. Now, since

$$\pi_1 S(t)x + m(\pi_1 S(t)x) \in \mathcal{M}, \tag{5.22}$$

we have that, for all $z = \xi + m(\xi) \in \mathcal{M}$,

$$\begin{aligned}
\|R(t)x\|_{\mathcal{X}} &\leq \|\pi_2 S(t)x - m(\xi)\|_{\mathcal{X}} + \|m(\xi) - m(\pi_1 S(t)x)\|_{\mathcal{X}} \\
&\leq \|\pi_2 S(t)x - \pi_2 z\|_{\mathcal{X}} + L\|\xi - \pi_1 S(t)x\|_{\mathcal{X}} \\
&= \|\pi_2(S(t)x - z)\|_{\mathcal{X}} + L\|\pi_1(z - S(t)x)\|_{\mathcal{X}} \\
&\leq (1+L)\|S(t)x - z\|_{\mathcal{X}}
\end{aligned}$$

(recall that L denotes the Lipschitz constant of m). Since z is arbitrary in $\mathcal{M}$, it follows that

$$\|R(t)x\|_{\mathcal{X}} \leq (1+L)d(S(t)x, \mathcal{M}).$$

Thus, (5.21) does hold if $\mathcal{M}$ is exponentially attracting, i.e. if (5.4) holds.

Incidentally, we also note that, conversely, (5.4) follows from (5.21). Indeed, (5.22) also implies that

$$d(S(t)x, \mathcal{M}) \leq \|S(t)x - (\pi_1 S(t)x + m(\pi_1 S(t)x))\|_{\mathcal{X}}$$

$$= \|(S(t)x - \pi_1 S(t)x) - m(\pi_1 S(t)x)\|_{\mathcal{X}}$$
$$= \|\pi_2 S(t)x - m(\pi_1 S(t)x)\|_{\mathcal{X}}$$
$$= \|R(t)x\|_{\mathcal{X}}.$$

In the same way, we see that, for example, the exponential tracking property (5.7) implies the estimate

$$\|\pi_2 S(t)x - m(\pi_1 S(t)x)\|_{\mathcal{X}} \leq Cd(x, \mathcal{M})e^{-\eta t}, \tag{5.23}$$

for all $x \in \mathcal{X}$ and $t \geq 0$.

5.3   Geometric Assumptions on the Semiflow

In this section we introduce two geometric assumptions on the semiflow S, that we first recognize as "natural" conditions that S should satisfy in order to admit an inertial manifold of the form (5.16), and then show to be almost sufficient for the construction of such type of an inertial manifold. These properties are the CONE IN-VARIANCE PROPERTY and the STRONG SQUEEZING PROPERTY. To describe these conditions, it is useful to keep in mind that our goal is the determination of a finite dimensional subspace $\mathcal{X}_1$ of $\mathcal{X}$, and of a function $m \in \mathcal{G}_L$, such that the graph of m is an inertial manifold for S, as in (5.16).

5.3.1   The Cone Invariance Property

We first introduce the CONE INVARIANCE PROPERTY. Given $L > 0$, we denote by $\mathcal{C}_L$ the cone

$$\mathcal{C}_L := \{x \in \mathcal{X}: \|\pi_2 x\|_{\mathcal{X}} \leq L\|\pi_1 x\|_{\mathcal{X}}\}. \tag{5.24}$$

We immediately have

PROPOSITION 5.10
Let $\mathcal{G}$ and $\mathcal{G}_L$ be as in, respectively, (5.17) and (5.19). Let $m \in \mathcal{G}$. Then $m \in \mathcal{G}_L$ if and only if for all $x, y \in \mathrm{graph}(m)$, $x - y \in \mathcal{C}_L$.

PROOF   The proof is immediate. Assume first that $m \in \mathcal{G}_L$. Recalling (5.24), we need to prove that if $x, y \in \mathrm{graph}(m)$,

$$\|\pi_2(y - x)\|_{\mathcal{X}} \leq L\|\pi_1(y - x)\|_{\mathcal{X}}. \tag{5.25}$$

But, by (5.16), $x = \xi + m(\xi)$ and $y = \eta + m(\eta)$ for some ξ and $\eta \in \pi_1 \mathcal{X}$, with $m(\xi)$ and $m(\eta) \in \pi_2 \mathcal{X}$. Consequently, $\pi_2(y - x) = m(\eta) - m(\xi)$ and $\pi_1(y - x) = \eta - \xi$, and (5.18) implies (5.25). Conversely, let $m \in \mathcal{G}$, and assume that for all x, $y \in$ graph(m), $y - x \in \mathcal{C}_L$. We show that m satisfies (5.18). Thus, given ξ and $\eta \in \pi_1 \mathcal{X}$, let $x = \xi + m(\xi)$ and $y = \eta + m(\eta)$. Then x and $y \in$ graph(m), so $y - x \in \mathcal{C}_L$, and we easily conclude as in the first part of the proof. $\qquad\square$

COROLLARY 5.11

Let $m \in \mathcal{G}_L$, and $\mathcal{M} :=$ graph(m). If $\mathcal{M}$ is positively invariant, then for all x_1, $x_2 \in \mathcal{M}$, and all $t \geq 0$, $S(t)x_1 - S(t)x_2 \in \mathcal{C}_L$.

PROOF   Let x_1, $x_2 \in \mathcal{M}$. Since $\mathcal{M}$ is positively invariant, $S(t)x_1$ and $S(t)x_2 \in \mathcal{M}$ for all $t \geq 0$, and $S(t)x_1 - S(t)x_2 \in \mathcal{C}_L$, by proposition 5.10. $\qquad\square$

From proposition 5.10 and corollary 5.11 we deduce that if the semiflow S did have an inertial manifold $\mathcal{M}$ of the form (5.16), that is, if there were $m \in \mathcal{G}_L$ such that $\mathcal{M} =$ graph(m), then for all x_1 and $x_2 \in \mathcal{M}$, both $x_1 - x_2 \in \mathcal{C}_L$ and $S(t)x_1 - S(t)x_2 \in \mathcal{C}_L$ for all $t \geq 0$ (the latter because of the positive invariance of $\mathcal{M}$). We now note that these two conditions can be formulated independently of the actual knowledge of $\mathcal{M}$ (which, of course, is the very manifold we want to construct). It is therefore natural to require this proposition as an *a priori* condition on the semiflow S. Thus, we replace the assumption x_1, $x_2 \in \mathcal{M}$ (which we cannot state if $\mathcal{M}$ is still to be found), with the assumption $x_1 - x_2 \in \mathcal{C}_L$ (which we can state), and require that this new assumption implies that the corresponding difference $S(t)x_1 - S(t)x_2$ be in $\mathcal{C}_L$ for all $t \geq 0$, as was the case in corollary 5.11. This motivates the following

DEFINITION 5.12   Let $L > 0$. The continuous semiflow S satisfies the CONE IN-VARIANCE PROPERTY with parameter L if for all x_1, $x_2 \in \mathcal{X}$, and all $t \geq 0$,

$$x_1 - x_2 \in \mathcal{C}_L \quad \Longrightarrow \quad S(t)x_1 - S(t)x_2 \in \mathcal{C}_L. \tag{5.26}$$

Once again, the cone invariance property means that if the difference $x_1 - x_2$ of two points is in the cone $\mathcal{C}_L$, then the difference $S(t)x_1 - S(t)x_2$ of all successive points on the corresponding orbits starting at x_1 and x_2 remains in the same cone $\mathcal{C}_L$ for all future times. That is, the cone $\mathcal{C}_L$ is invariant with respect to the difference of forward motions (see figure 5.4).

The following result is a direct consequence of the cone invariance property (5.26):

PROPOSITION 5.13

Assume the continuous semiflow S satisfies the cone invariance property with parameter L. Let x_1, $x_2 \in \mathcal{X}$, and assume there exists $T > 0$ such that $S(T)x_1 - S(T)x_2 \notin \mathcal{C}_L$. Then $S(t)x_1 - S(t)x_2 \notin \mathcal{C}_L$ also for all $t \in [0, T]$.

PROOF   We proceed by contradiction. If there were $t' \in [0, T]$ such that $S(t')x_1 - S(t')x_2 \in \mathcal{C}_L$, then, setting $t = T - t' \geq 0$ in (5.26) with x_i replaced by $S(t')x_i$, $i = 1, 2$,

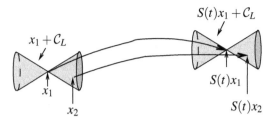

Figure 5.4: The Cone Invariance Property

we would obtain that $S(t)S(t')x_1 - S(t)S(t')x_2 \in \mathcal{C}_L$. But $S(t)S(t') = S(T)$, so this cannot hold.   □

5.3.2   Decay and Squeezing Properties

In this section we introduce the squeezing properties as another set of natural conditions that the continuous semiflow S should satisfy in order to admit an inertial manifold $\mathcal{M}$ of the form (5.16), which also satisfies the modified version (5.7) of the exponential tracking property. To derive these additional conditions, we again start by assuming that S does have an inertial manifold with the desired properties. In addition, we assume that S satisfies the cone invariance property (5.26), and that it admits a positively invariant manifold $\mathcal{M} = \mathrm{graph}(m)$, with $m \in \mathcal{G}_{L'} \subset \mathcal{G}_L$ for some $L' < L$ (that is, the function m satisfies a stronger Lipschitz condition). This additional assumption is not so restrictive, since in most applications it often happens that the parameter L is determined as a solution of an inequality of the form $F(L) > 0$, where $F \colon\,]0, \infty[\to \mathbb{R}$ is a continuous function. Therefore, since we can take $L' < L$ sufficiently close to L and still have $F(L') > 0$, in this case the semiflow does satisfy a second cone invariance property, as stated. (An instance when such a second cone invariance property is explicitly used can be found in Robinson, [Rob93, prop. 3].)

Under these conditions, we claim:

PROPOSITION 5.14
There is a constant $c_2 > 0$ such that for all x, y and $z \in \mathcal{X}$, if $x - z \notin \mathcal{C}_L$ but $y - z \in \mathcal{C}_{L'}$,

$$\|x - z\|_{\mathcal{X}} \le c_2 \|x - y\|_{\mathcal{X}} . \tag{5.27}$$

PROOF   Since $x - z \notin \mathcal{C}_L$, (5.24) implies that

$$L\|\pi_1(x - z)\|_{\mathcal{X}} < \|\pi_2(x - z)\|_{\mathcal{X}} ; \tag{5.28}$$

on the other hand, since $y - z \in \mathcal{C}_{L'}$, again (5.24) implies that

$$\|\pi_2(y - z)\|_{\mathcal{X}} \le L'\|\pi_1(y - z)\|_{\mathcal{X}} .$$

Consequently, we estimate

$$
\begin{aligned}
L\|\pi_1(y-z)\|_{\mathcal{X}} &\leq L\|\pi_1(y-x)\|_{\mathcal{X}} + L\|\pi_1(x-z)\|_{\mathcal{X}} \\
&\leq L\|\pi_1(y-x)\|_{\mathcal{X}} + \|\pi_2(x-z)\|_{\mathcal{X}} \\
&\leq L\|\pi_1(y-x)\|_{\mathcal{X}} + \|\pi_2(x-y)\|_{\mathcal{X}} + \|\pi_2(y-z)\|_{\mathcal{X}} \\
&\leq L\|\pi_1(y-x)\|_{\mathcal{X}} + \|\pi_2(x-y)\|_{\mathcal{X}} + L'\|\pi_1(y-z)\|_{\mathcal{X}}.
\end{aligned}
$$

Since $L > L'$, we obtain

$$
\begin{aligned}
(L-L')\|\pi_1(y-z)\|_{\mathcal{X}} &\leq L\|\pi_1(y-x)\|_{\mathcal{X}} + \|\pi_2(x-y)\|_{\mathcal{X}} \\
&\leq (1+L)\|x-y\|_{\mathcal{X}}.
\end{aligned}
$$

Consequently,

$$
\begin{aligned}
\|x-z\|_{\mathcal{X}} &\leq \|x-y\|_{\mathcal{X}} + \|y-z\|_{\mathcal{X}} \\
&\leq \|x-y\|_{\mathcal{X}} + \|\pi_1(y-z)\|_{\mathcal{X}} + \|\pi_2(y-z)\| \\
&\leq \|x-y\|_{\mathcal{X}} + (1+L')\|\pi_1(y-z)\|_{\mathcal{X}} \\
&\leq \|x-y\|_{\mathcal{X}} + \frac{(1+L')(1+L)}{L-L'}\|x-y\|_{\mathcal{X}},
\end{aligned}
$$

and (5.27) follows. $\qquad\qquad\qquad\qquad\qquad\qquad\qquad\qquad\qquad\qquad$ □

Let now $x_1 \in \mathcal{X} \setminus \mathcal{M}$, and assume there are $x_2 \in \mathcal{M}$ and $t > 0$ such that

$$
S(t)x_1 - S(t)x_2 \notin \mathcal{C}_L. \tag{5.29}
$$

Let $w := \pi_1 S(t)x_1 + m(\pi_1 S(t)x_1)$. Then, $w \in \mathcal{M}$. Since also $S(t)x_2 \in \mathcal{M}$ (because $\mathcal{M}$ is positively invariant), by proposition 5.10 it follows that $S(t)x_2 - w \in \mathcal{C}_{L'}$. Thus, by (5.29), we can choose $x = S(t)x_1$, $y = w$ and $z = S(t)x_2$ in proposition 5.14, and deduce that

$$
\|S(t)x_1 - S(t)x_2\|_{\mathcal{X}} \leq c_2\|S(t)x_1 - w\|_{\mathcal{X}}. \tag{5.30}
$$

Recalling now (5.23), which is a consequence of the exponential tracking property (5.7), and that $\pi_1 S(t)x_1 = \pi_1 w$, we deduce from (5.30) that

$$
\begin{aligned}
\|S(t)x_1 - S(t)x_2\|_{\mathcal{X}} &\leq c_2\|S(t)x_1 - w\|_{\mathcal{X}} = c_2\|\pi_2(S(t)x_1 - w)\|_{\mathcal{X}} \\
&= c_2\|\pi_2 S(t)x_1 - m(\pi_1 S(t)x_1)\|_{\mathcal{X}} \\
&\leq c_2 C d(x_1, \mathcal{M})e^{-\eta t} \leq c_2 C\|x_1 - x_2\|_{\mathcal{X}}e^{-\eta t}, \tag{5.31}
\end{aligned}
$$

where the last step follows because $x_2 \in \mathcal{M}$.

In conclusion, we deduce that if S admitted an inertial manifold $\mathcal{M}$ of the form (5.16), with $m \in \mathcal{G}_{L'}$, $L' < L$, then condition (5.29) would imply that there is $K > 0$, independent of x_1, x_2 and t, such that

$$
\|S(t)x_1 - S(t)x_2\|_{\mathcal{X}} \leq K\|x_1 - x_2\|_{\mathcal{X}}e^{-\eta t}. \tag{5.32}
$$

At this point, we realize that conditions (5.29) and (5.32) can be formulated independently of the requirement that $x_1, x_2 \in \mathcal{M}$. This motivates the following

DEFINITION 5.15   *Let η, $L > 0$, and $\mathcal{C}_L$ as in (5.24). The semiflow S satisfies the* DECAY PROPERTY *with parameters η and L if there is $K > 0$ with the property that whenever $x_1, x_2 \in \mathcal{X}$ and $t > 0$ are such that (5.29) holds, then (5.32) also holds.*

Thus, the decay property means that the difference of motions outside the cone $\mathcal{C}_L$ decays exponentially. For future reference, we remark that, as an immediate consequence of proposition 5.13, if (5.29) and (5.32) hold for a specific $t > 0$, then they both hold for all $t' \in [0,t]$.

Finally, we combine the cone invariance and the decay properties into

DEFINITION 5.16   *In the same conditions of definition 5.15, the semiflow S satisfies the* STRONG SQUEEZING PROPERTY *with parameters η and L if it satisfies the cone invariance property with parameter L, and the decay property with parameters η and L.*

The strong squeezing property was first introduced for the Kuramoto-Sivashinski equations in Foias, Nicolaenko, Sell and Temam, [FNST85, FNST88]. An abstract version of this property was developed in Foias, Sell and Titi, [FST89]; other formulations can be found e.g. in Temam, [Tem88], Constantin, Foias, Nicolaenko and Temam, [CFNT89], Robinson, [Rob93], and Jones and Titi, [JT96]. The term "strong" refers to the fact that, in contrast e.g. to the discrete squeezing property, we require that a cone invariance property hold as well.

5.3.3   Consequences of the Decay Property

In this section we briefly report two immediate consequences of the decay property. We first show that if a semiflow S admits an attractor, and satisfies the decay property, the attractor is smooth, in the sense that it is imbedded into a Lipschitz manifold.

PROPOSITION 5.17
Assume the semiflow S satisfies the decay property with parameters L and η, and admits a global attractor $\mathcal{A}$. There exists then a function $m \in \mathcal{G}_L$ such that $\mathcal{A} \subseteq$ graph(m).

PROOF   Let x_1 and $x_2 \in \mathcal{A}$. Since $\mathcal{A}$ is invariant, for each $t > 0$ there are y_1 and $y_2 \in \mathcal{A}$ such that $x_i = S(t)y_i$, $i = 1, 2$. If $x_1 - x_2 \notin \mathcal{C}_L$, the decay property (5.32) implies that

$$\|x_1 - x_2\|_{\mathcal{X}} = \|S(t)y_1 - S(t)y_2\|_{\mathcal{X}} \leq K\|y_1 - y_2\|_{\mathcal{X}}\, e^{-\eta t}.$$

Since $\mathcal{A}$ is bounded, we deduce that

$$\|x_1 - x_2\|_\mathcal{X} \le R e^{-\eta t},$$

for some R independent of t. Letting $t \to +\infty$, we conclude that $x_1 = x_2$. Thus, we can define a function $m: \pi_1 \mathcal{A} \to \mathcal{X}_2$ by

$$m(\pi_1 x) := \pi_2 x, \qquad x \in \mathcal{A}.$$

Indeed, let $y \in \mathcal{A}$ be such that $\pi_1 y = \pi_1 x$. Then: if $y - x \notin \mathcal{C}_L$, we have shown that $y = x$; if $y - x \in \mathcal{C}_L$, then $\pi_2 y = \pi_2 x$ so that, again, $x = y$. The function m so defined is Lipschitz continuous, as a consequence of proposition 5.10. Indeed, if x, $y \in \text{graph}(m)$, and $x \neq y$, then $x - y \in \mathcal{C}_L$ by the first part of this proof. We can then conclude by extending m to a bounded function, defined on all of $\mathcal{X}_1$, and having the same Lipschitz constant L.   □

Next, we report a technical result which will be used in the next section.

PROPOSITION 5.18

Assume S is a continuous semiflow on $\mathcal{X}$, satisfying the strong squeezing property with parameters L and η. Assume that t_1 and t_2 are such that $0 < t_1 < t_2$ and there are $x_1, x_2 \in \mathcal{X}_1$ such that

$$\pi_1 S(t_1) x_1 = \pi_1 S(t_2) x_2. \tag{5.33}$$

There is then $C_1 > 0$, independent of t_1, t_2, x_1, x_2, such that

$$\|\pi_2 (S(t_1)x_1 - S(t_2)x_2)\|_\mathcal{X} \le C_1 \|\pi_2 (x_1 - S(t_2 - t_1)x_2)\|_\mathcal{X} e^{-\eta t_1}. \tag{5.34}$$

PROOF   If $S(t_1)x_1 - S(t_2)x_2 \in \mathcal{C}_L$, (5.33) implies that $\pi_2 (S(t_1)x_1 - S(t_2)x_2) = 0$; thus, (5.34) holds trivially. Otherwise, we can write that

$$S(t_1)x_1 - S(t_2)x_2 = S(t_1)x_1 - S(t_1)S(t_2 - t_1)x_2 \notin \mathcal{C}_L,$$

and proposition 5.13 implies that also $x_1 - S(t_2 - t_1)x_2 \notin \mathcal{C}_L$. By the decay property (5.32), with $t = t_1$ and x_2 replaced by $S(t_2 - t_1)x_2$, we compute then

$$
\begin{aligned}
\|\pi_2 (S(t_1)x_1 - S(t_2)x_2)\|_\mathcal{X} &= \|S(t_1)x_1 - S(t_2)x_2\|_\mathcal{X} \\
&= \|S(t_1)x_1 - S(t_1)S(t_2 - t_1)x_2\|_\mathcal{X} \le K\|x_1 - S(t_2 - t_1)x_2\|_\mathcal{X} e^{-\eta t_1} \\
&\le K \left(\|\pi_1 (x_1 - S(t_2 - t_1)x_2)\|_\mathcal{X} + \|\pi_2 (x_1 - S(t_2 - t_1)x_2)\|_\mathcal{X} \right) e^{-\eta t_1} \\
&\le K \left(\tfrac{1}{L} + 1 \right) \|\pi_2 (x_1 - S(t_2 - t_1)x_2)\|_\mathcal{X} e^{-\eta t_1},
\end{aligned}
$$

the last step as in (5.28), because $x_1 - S(t_2 - t_1)x_2 \notin \mathcal{C}_L$. Thus, (5.34) follows.   □

5.4   Strong Squeezing Property and Inertial Manifolds

We are now in a position to show that if the semiflow S satisfies the strong squeezing property, it admits an inertial manifold of the form (5.16), which we can construct by Hadamard's graph transformation method. In the sequel, we closely follow Robinson, [Rob01].

5.4.1   Surjectivity and Uniform Boundedness

For our first result, we need

DEFINITION 5.19   *Let S be a continuous semiflow on $\mathcal{X}$; let $\mathcal{G}_L$ be as in (5.19), and $\Phi \subseteq \mathcal{G}_L$. Let π_1 and π_2 be as in (5.32). Then:*

1. *S satisfies the* SURJECTIVITY PROPERTY *with respect to Φ if for all $\varphi \in \Phi$ and $t \geq 0$,*

$$\pi_1 S(t) \operatorname{graph}(\varphi) = \mathcal{X}_1 \,. \tag{5.35}$$

2. *S satisfies the* UNIFORM BOUNDEDNESS PROPERTY *with respect to Φ if the sets $\pi_2 S(t) \operatorname{graph}(\varphi)$ are bounded in $\mathcal{X}$, uniformly with respect to $t \geq 0$ and $\varphi \in \Phi$.*

We have then:

PROPOSITION 5.20
Let S be a semiflow on $\mathcal{X}$. Assume S satisfies the cone invariance property with parameter L, and the surjectivity and uniform boundedness properties with respect to some subset $\Phi \subseteq \mathcal{G}_L$. Then for each $\varphi_0 \in \Phi$ and $t > 0$ there exists a uniquely determined function $\varphi_t \in \mathcal{G}_L$ such that

$$\operatorname{graph}(\varphi_t) = S(t) \operatorname{graph}(\varphi_0) \,. \tag{5.36}$$

PROOF   1. Fix $\varphi_0 \in \Phi$ and $t > 0$. Since S satisfies the surjectivity property, (5.35) implies that

$$\pi_1 S(t) \operatorname{graph}(\varphi_0) = \mathcal{X}_1 \,.$$

Thus, for each $\xi \in \mathcal{X}_1$ there is $x \in S(t) \operatorname{graph}(\varphi_0)$ such that

$$\pi_1 x = \xi \,. \tag{5.37}$$

We claim that x is uniquely determined by ξ (and, of course, by t and φ_0). Indeed, suppose $x' \in S(t) \operatorname{graph}(\varphi_0)$ is also such that $\pi_1 x' = \xi$ as in (5.37). Let z and $z' \in$

graph(φ_0) be such that $x = S(t)z$ and $x' = S(t)z'$. Since $\varphi_0 \in \mathcal{G}_L$, proposition 5.10 implies that $z - z' \in \mathcal{C}_L$. Thus, the cone invariance property (5.26) implies that also $x - x' \in \mathcal{C}_L$, i.e.

$$\|\pi_2(S(t)z - S(t)z')\|_{\mathcal{X}} \leq L\|\pi_1(S(t)z - S(t)z')\|_{\mathcal{X}}. \tag{5.38}$$

But since

$$\pi_1 S(t)z = \pi_1 x = \xi = \pi_1 x' = \pi_1 S(t)z',$$

(5.38) shows that $\pi_2 S(t)z = \pi_2 S(t)z'$. Hence, $S(t)z = S(t)z'$, i.e. $x = x'$ as claimed.

2. From part (1) it follows that we can unambiguously define a function κ from $\mathbb{R}_{\geq 0} \times \mathcal{X}_1 \times \Phi$ into $S(t)\,\mathrm{graph}(\varphi_0)$, by

$$(t, \xi, \varphi_0) \mapsto x := \kappa(t, \xi, \varphi_0). \tag{5.39}$$

We can then define a map $\varphi_t : \mathcal{X}_1 \to \mathcal{X}_2$ by

$$\varphi_t(\xi) := \pi_2 \kappa(t, \xi, \varphi_0), \quad \xi \in \mathcal{X}_1.$$

That is, $\varphi_t(\xi) = \pi_2 x$, where $x \in \mathcal{X}$ is the unique element in $S(t)\,\mathrm{graph}(\varphi_0)$ such that $\pi_1 x = \xi$. In other words,

$$\pi_1 \kappa(t, \xi, \varphi_0) = \xi, \quad \pi_2 \kappa(t, \xi, \varphi_0) = \varphi_t(\xi). \tag{5.40}$$

Note that, by construction,

$$x = \xi + \varphi_t(\xi) \in \mathrm{graph}(\varphi_t).$$

We proceed then to show (5.36). Indeed, let first $z \in \mathrm{graph}(\varphi_t)$. Then, $z = \eta + \varphi_t(\eta)$, for some $\eta \in \mathcal{X}_1$. By (5.40), $z = \kappa(t, \eta, \varphi_0)$; thus, by (5.39), $z \in S(t)\,\mathrm{graph}(\varphi_0)$. Conversely, let $z \in S(t)\,\mathrm{graph}(\varphi_0)$, and set $\xi := \pi_1 z$. Then, ξ uniquely determines $x := \kappa(t, \xi, \varphi_0) \in S(t)\,\mathrm{graph}(\varphi_0) = \mathrm{graph}(\varphi_t)$, with $\pi_1 x = \xi$. Since $z \in S(t)\,\mathrm{graph}(\varphi_0)$, and $\xi = \pi_1 z$, the stated uniqueness implies that $x = z$. Thus, $z = x \in \mathrm{graph}(\varphi_t)$, and (5.36) holds.

3. Finally, we show that $\varphi_t \in \mathcal{G}_L$, i.e. that $\varphi_t : \mathcal{X}_1 \to \mathcal{X}_2$ is bounded and Lipschitz continuous, with the same Lipschitz constant L of φ_0. To show that φ_t is bounded, we must exhibit $R_1 > 0$ such that

$$\|\varphi_t\|_{\mathcal{G}} = \sup_{\xi \in \mathcal{X}_1} \|\varphi_t(\xi)\|_{\mathcal{X}} = \sup_{\xi \in \mathcal{X}_1} \|\pi_2 \kappa(t, \xi, \varphi_0)\|_{\mathcal{X}} \leq R_1. \tag{5.41}$$

Since S satisfies the uniform boundedness property with respect to Φ, there is $R > 0$ such that for all $t > 0$, $\varphi \in \Phi$, and $y \in S(t)\,\mathrm{graph}(\varphi)$,

$$\|\pi_2 y\|_{\mathcal{X}} \leq R. \tag{5.42}$$

But since $\varphi_0 \in \Phi$ and $\kappa(t, \xi, \varphi_0) \in S(t)\,\mathrm{graph}(\varphi_0)$, there is $y \in \mathrm{graph}(\varphi_0)$ such that $\kappa(t, \xi, \varphi_0) = S(t)y$. Hence, (5.41) follows from (5.42), with $R_1 = R$. To show that φ_t is Lipschitz continuous, given ξ_1 and $\xi_2 \in \mathcal{X}_1$, let $x_i = \xi_i + \varphi_t(\xi_i)$, $i = 1, 2$. Then $x_i \in$

graph$(\varphi_t) = S(t)$ graph(φ_0); thus, $x_i = S(t)y_i$ for some $y_i = \eta_i + \varphi_0(\eta_i) \in$ graph(φ_0). Since $\varphi_0 \in \mathcal{G}_L$, $y_1 - y_2 \in \mathcal{C}_L$; hence, by the cone invariance property (5.26), $x_1 - x_2 \in \mathcal{C}_L$ as well. Thus,

$$\|\varphi_t(\xi_1) - \varphi_t(\xi_2)\|_{\mathcal{X}} = \|\pi_2(x_1 - x_2)\|_{\mathcal{X}} \leq L\|\pi_1(x_1 - x_2)\|_{\mathcal{X}} = L\|\xi_1 - \xi_2\|_{\mathcal{X}}.$$

This shows that $\varphi_t \in \mathcal{G}_L$, and concludes the proof of proposition 5.20. □

As a consequence of proposition 5.20, if for $t > 0$ we set

$$\mathcal{M}_t := \text{graph}(\varphi_t), \tag{5.43}$$

then (5.36) implies that

$$\mathcal{M}_t = S(t)\mathcal{M}_0. \tag{5.44}$$

5.4.2 Construction of the Inertial Manifold

We can finally proceed to implement Hadamard's construction of an inertial manifold.

THEOREM 5.21
Let S be a continuous semiflow on $\mathcal{X}$. Assume that S satisfies the strong squeezing property with parameters L and η, as well as the surjectivity and uniform boundedness properties with respect to the set $\Phi := \{0\} \subset \mathcal{G}_L$. Then S admits an inertial manifold $\mathcal{M} \subseteq \mathcal{X}$, of the form (5.16). More precisely, there is $m \in \mathcal{G}_L$ such that $\mathcal{M} := \text{graph}(m)$ is an inertial manifold for S. Moreover, for all $x \in \mathcal{X}$ and $t \geq 0$,

$$d(S(t)x, \mathcal{M}) \leq C_1 \left(d(x, \mathcal{M}) + 2\|m\|_{\mathcal{G}_L}\right) e^{-\eta t}, \tag{5.45}$$

where C_1 is as in (5.34).

PROOF We shall use the graph transformation method, and proceed in three steps.
1. Let $\varphi_0 \equiv 0$. Then, obviously,

$$\mathcal{M}_0 := \text{graph}(\varphi_0) = \mathcal{X}_1, \tag{5.46}$$

which is a trivial flat manifold. In accord to (5.44), we follow the evolution of this flat manifold. Proposition 5.20 shows that each $\mathcal{M}_t$ is again the graph of a function $\varphi_t \in \mathcal{G}_L$. Using the decay property and the uniform boundedness property, we will show that, as $t \to +\infty$, the functions $(\varphi_t)_{t \geq 0}$ converge to a function $m \in \mathcal{G}_L$, whose graph $\mathcal{M}$ is the desired inertial manifold for S.
2. Recalling (5.43) and (5.46), by proposition 5.20 we have that for all $t \geq 0$ there is $\varphi_t \in \mathcal{G}_L$ such that

$$\mathcal{M}_t = \text{graph}(\varphi_t) = S(t)\,\text{graph}(\varphi_0) = S(t)\mathcal{X}_1. \tag{5.47}$$

Since S satisfies the uniform boundedness property with respect to $\Phi = \{0\}$, and graph$(\varphi_0) = \mathcal{X}_1$, there is $R > 0$ such that (5.42) holds for all $t > 0$ and $y \in S(t)\mathcal{X}_1$. Given $\xi \in \mathcal{X}_1$ and $t > 0$, let $x = \kappa(t, \xi, 0)$. Then, $x \in S(t)\,\text{graph}(\varphi_0) = S(t)\mathcal{X}_1$, and $\pi_2 x = \varphi_t(\xi)$. Thus, by (5.42),

$$\|\varphi_t\|_{\mathcal{G}} = \sup_{\xi \in \mathcal{X}_1} \|\varphi_t(\xi)\|_{\mathcal{X}} = \sup_{\xi \in \mathcal{X}_1} \|\pi_2 \kappa(t, \xi, 0)\|_{\mathcal{X}} \leq R. \tag{5.48}$$

Fix t_1 and t_2, with $0 < t_1 < t_2$, and $\xi \in \mathcal{X}_1$. For $i = 1, 2$, let $x_i := \xi + \varphi_{t_i}(\xi) \in \mathcal{M}_{t_i}$. By (5.47), there are $\xi_1, \xi_2 \in \mathcal{X}_1$ such that $x_i = S(t_i)\xi_i$. Since

$$\pi_1 S(t_1)\xi_1 = \pi_1 S(t_2)\xi_2 = \xi,$$

(5.33) holds, and proposition 5.18 implies that, by (5.34),

$$\|\pi_2(S(t_1)\xi_1 - S(t_2)\xi_2)\|_{\mathcal{X}} \leq C_1 \|\pi_2(\xi_1 - S(t_2 - t_1)\xi_2)\|_{\mathcal{X}} e^{-\eta t_1}. \tag{5.49}$$

Now, since $\xi_1 \in \mathcal{X}_1$, $\pi_2 \xi_1 = 0$; since also

$$S(t_2 - t_1)\xi_2 \in S(t_2 - t_1)\mathcal{X}_1 = \mathcal{M}_{t_2 - t_1},$$

it follows that

$$\pi_2 S(t_2 - t_1)\xi_2 = \varphi_{t_2 - t_1}(\pi_1 S(t_2 - t_1)\xi_2).$$

Moreover,

$$\pi_2(S(t_1)\xi_1 - S(t_2)\xi_2) = \pi_2(x_1 - x_2) = \varphi_{t_1}(\xi) - \varphi_{t_2}(\xi).$$

Hence, we deduce from (5.49) and (5.48) that

$$\|\varphi_{t_1}(\xi) - \varphi_{t_2}(\xi)\|_{\mathcal{X}} \leq C_1 \|\varphi_{t_2 - t_1}(\pi_1 S(t_2 - t_1)\xi_2)\|_{\mathcal{X}} e^{-\eta t_1} \leq C_1 R e^{-\eta t_1}. \tag{5.50}$$

Since $\xi \in \mathcal{X}_1$ and t_1, t_2 are arbitrary, this implies that $(\varphi_t)_{t \geq 0}$ is a Cauchy set in $\mathcal{G}_L$: in fact, given $\varepsilon > 0$, by (5.50) we have that

$$\|\varphi_t - \varphi_\theta\|_{\mathcal{G}_L} = \sup_{\xi \in \mathcal{X}_1} \|\varphi_t(\xi) - \varphi_\theta(\xi)\| < \varepsilon$$

for all t and θ such that

$$\min\{t, \theta\} \geq T := \max\left\{0, \ln\left(\tfrac{C_1 R}{\varepsilon}\right)\right\}.$$

Since $\mathcal{G}_L$ is a complete metric space, we conclude that, as $t \to +\infty$, $(\varphi_t)_{t \geq 0}$ converges to a function $m \in \mathcal{G}_L$. As a consequence of (5.48), m is also bounded, with $\|m\|_{\mathcal{G}_L} \leq R$.

3. We proceed then to show that the set $\mathcal{M} := \text{graph}(m)$ is the desired inertial manifold of S. Thus, we need to show that $\mathcal{M}$ is positively invariant, and exponentially attracting.

3.1. The positive invariance of $\mathcal{M}$ is a consequence of the uniform convergence $\varphi_t \to m$. Indeed, let $x = \xi + m(\xi) \in \mathcal{M}$. For $t \geq 0$, let $x_t := \xi + \varphi_t(\xi) \in \mathcal{M}_t$. Then, as $t \to +\infty$, $x_t \to \xi + m(\xi) = x$. Fix $\tau > 0$: then, by (5.44),

$$S(\tau)x_t \in S(\tau)\mathcal{M}_t = S(\tau)S(t)\mathcal{M}_0 = S(\tau + t)\mathcal{M}_0 = \mathcal{M}_{\tau + t};$$

therefore,

$$S(\tau)x_t = \pi_1 S(\tau)x_t + \varphi_{\tau+t}(\pi_1 S(\tau)x_t). \tag{5.51}$$

Since $S(\tau)$ is continuous on $\mathcal{X}$, $S(\tau)x_t \to S(\tau)x$ as $t \to +\infty$; on the other hand, the right side of (5.51) converges to $\pi_1 S(\tau)x + m(\pi_1 S(\tau)x)$. It follows that

$$S(\tau)x = \pi_1 S(\tau)x + m(\pi_1 S(\tau)x) \in \operatorname{graph}(m) = \mathcal{M}.$$

Since $x \in \mathcal{M}$ and $\tau > 0$ are arbitrary, this means that $\mathcal{M}$ is positively invariant.

3.2. We now show that $\mathcal{M}$ is exponentially attracting, and (5.45) holds. Fix $x \in \mathcal{X} \setminus \mathcal{M}$, and $t > 0$. Let $\xi := \pi_1 S(t)x$ and, for $\theta > t$, $y := \xi + \varphi_\theta(\xi)$. Then $y \in \mathcal{M}_\theta = S(\theta)\mathcal{X}_1$, so there is $z \in \mathcal{X}_1$ such that $y = S(\theta)z$. Since $\pi_1 S(t)x = \pi_1 S(\theta)z = \xi$, by (5.34) of proposition 5.18 we have

$$\|\pi_2(S(t)x - S(\theta)z)\|_{\mathcal{X}} \le C_1 \|\pi_2(x - S(\theta - t)z)\|_{\mathcal{X}} e^{-\eta t}.$$

Let $w := S(\theta - t)z$. Then $w \in \mathcal{M}_{\theta-t}$, so $w = \alpha + \varphi_{\theta-t}(\alpha)$ for some $\alpha \in \mathcal{X}_1$. Since $\pi_2 S(\theta)z = \pi_2 y = \varphi_\theta(\xi)$, we can estimate

$$\begin{aligned}
d(S(t)x, \mathcal{M}) &\le \|S(t)x - (\xi + m(\xi))\|_{\mathcal{X}} = \|\pi_2 S(t)x - m(\xi)\|_{\mathcal{X}} \\
&= \lim_{\theta \to +\infty} \|\pi_2 S(t)x - \varphi_\theta(\xi)\|_{\mathcal{X}} = \lim_{\theta \to +\infty} \|\pi_2(S(t)x - S(\theta)z)\|_{\mathcal{X}} \\
&\le C_1 \lim_{\theta \to +\infty} \|\pi_2(x - S(\theta - t)z)\|_{\mathcal{X}} e^{-\eta t} \\
&= C_1 \lim_{\theta \to +\infty} \|\pi_2 x - \varphi_{\theta-t}(\alpha)\|_{\mathcal{X}} e^{-\eta t} \\
&= C_1 \|\pi_2 x - m(\alpha)\|_{\mathcal{X}} e^{-\eta t}.
\end{aligned}$$

Given then arbitrary $u = \beta + m(\beta) \in \mathcal{M}$, we deduce that

$$\begin{aligned}
d(S(t)x, \mathcal{M}) &\le C_1(\|\pi_2 x - m(\beta)\|_{\mathcal{X}} + \|m(\beta) - m(\alpha)\|_{\mathcal{X}})e^{-\eta t} \\
&= C_1(\|\pi_2(x - u)\|_{\mathcal{X}} + \|m(\beta) - m(\alpha)\|_{\mathcal{X}})e^{-\eta t} \\
&\le C_1(\|x - u\|_{\mathcal{X}} + 2\|m\|_{\mathcal{G}_L})e^{-\eta t}.
\end{aligned}$$

Taking inf with respect to $u \in \mathcal{M}$, we conclude that

$$d(S(t)x, \mathcal{M}) \le C_1(d(x, \mathcal{M}) + 2\|m\|_{\mathcal{G}_L})e^{-\eta t},$$

which is (5.45). This concludes the proof of theorem 5.21. $\quad\square$

We remark that the inertial manifold $\mathcal{M}$ so constructed is not compact. This is because $\mathcal{M}$ is the graph of a function m over $\mathcal{X}_1$, which is not bounded. To obtain a compact inertial manifold, we can apply proposition 5.9, and intersect $\mathcal{M}$ with a compact, positively invariant absorbing set. On the other hand, since m is Lipschitz continuous, proposition 2.61 implies that $\mathcal{M}$ has finite fractal dimension, with

$$\dim_{\mathrm{F}}(\mathcal{M}) = \dim \mathcal{X}_1.$$

We conclude this section with a property that characterizes the convergence of the sets $S(t)\mathcal{X}_1$ to the inertial manifold $\mathcal{M} = \mathrm{graph}(m)$ constructed in the previous section.

PROPOSITION 5.22

Let $(\varphi_t)_{t \geq 0}$ be the family of functions in $\mathcal{G}_L$ constructed in proposition 5.20 from the initial function $\varphi_0 \equiv 0$, as in theorem 5.21. Let

$$m := \lim_{t \to +\infty} \varphi_t \in \mathcal{G}_L .$$

Then the semiflow S satisfies the surjectivity property with respect to the set

$$\Phi_1 := (\varphi_t)_{t \geq 0} \cup \{m\} ,$$

and

$$\lim_{t \to +\infty} \mathrm{dist}(\mathrm{graph}(\varphi_t), \mathrm{graph}(m)) = 0 , \tag{5.52}$$

where dist *is the distance defined in (2.3).*

PROOF   1. Recalling (5.35), we need to prove that for all $t, \theta \geq 0$,

$$\pi_1 S(t) \, \mathrm{graph}(\varphi_\theta) = \mathcal{X}_1 , \quad \pi_1 S(t) \, \mathrm{graph}(m) = \mathcal{X}_1 . \tag{5.53}$$

Obviously, $\pi_1 S(t) \, \mathrm{graph}(\varphi_\theta) \subseteq \mathcal{X}_1$ and $\pi_1 S(t) \, \mathrm{graph}(m) \subseteq \mathcal{X}_1$. To prove the inverse inclusions, fix $\xi \in \mathcal{X}_1$ and $t, \theta > 0$. Let first $y := \xi + \varphi_{t+\theta}(\xi)$. Then, by (5.36) and (5.46),

$$y \in \mathrm{graph}(\varphi_{t+\theta}) = S(t+\theta)\mathcal{X}_1 = S(t)S(\theta)\mathcal{X}_1 ;$$

thus, there is $x_\theta \in S(\theta)\mathcal{X}_1$ such that $y = S(t)x_\theta$. But then $x_\theta \in \mathrm{graph}(\varphi_\theta)$, and $\pi_1 S(t)x_\theta = \xi$. This proves the first of (5.53). To prove the second, we proceed as before. Then, since $x_\theta \in \mathrm{graph}(\varphi_\theta)$, there is $\alpha \in \mathcal{X}_1$ such that $x_\theta = \alpha + \varphi_\theta(\alpha)$. Let $z := \alpha + m(\alpha)$. Since the convergence $\varphi_\theta \to m$ as $\theta \to +\infty$ is uniform, $x_\theta \to z$ as $\theta \to +\infty$. Thus, since both π_1 and $S(t)$ are continuous,

$$\|\xi - \pi_1 S(t)z\|_\mathcal{X} = \|\pi_1 (S(t)x_\theta - S(t)z)\|_\mathcal{X} \to 0$$

as well. It follows that $\pi_1 S(t)z = \xi$; since $z \in \mathcal{M}$, this proves the second of (5.53).

2. To prove (5.52), recalling (2.3), (5.36) and (5.46) we need to show that, as $t \to +\infty$,

$$\partial(S(t)\mathcal{X}_1, \mathcal{M}) \to 0, \quad \partial(\mathcal{M}, S(t)\mathcal{X}_1) \to 0 . \tag{5.54}$$

Given $a_t = \alpha + \varphi_t(\alpha) \in \mathrm{graph}(\varphi_t) = S(t)\mathcal{X}_1$, let $b := \alpha + m(\alpha) \in \mathcal{M}$. Then

$$0 \leq d(a_t, \mathcal{M}) \leq \|a_t - b\|_\mathcal{X} = \|\varphi_t(\alpha) - m(\alpha)\|_\mathcal{X} \to 0 . \tag{5.55}$$

Moreover, since the convergence $\varphi_t \to m$ is uniform with respect to $\alpha \in \mathcal{X}_1$, (5.55) implies that, in fact,

$$\sup_{a_t \in S(t)\mathcal{X}_1} d(a_t, \mathcal{M}) = \partial(S(t)\mathcal{X}_1, \mathcal{M}) \to 0.$$

This is the first of (5.54); the second is proven analogously.   ▯

REMARK 5.23   The second of (5.53) implies that for each $\xi \in \mathcal{X}_1$ and $t \geq 0$ we can determine $x \in \mathcal{M}$ such that

$$\pi_1 S(t)x = \xi. \tag{5.56}$$

The following result shows that this element x can be determined uniquely, if S satisfies the backward uniqueness property (recall definition 3.14). An instance when this property is satisfied is given in proposition 5.27 below.   ▯

PROPOSITION 5.24
In the same conditions of proposition 5.22, assume that S satisfies the backward uniqueness property. Then the element x in (5.56) is uniquely determined by ξ (and t).

PROOF   Assume that $y \in \mathcal{M}$ is also such that $\pi_1 S(t)y = \xi$. Then, since $m \in \mathcal{G}_L$, $x - y \in \mathcal{C}_L$. By the cone invariance property, also $S(t)x - S(t)y \in \mathcal{C}_L$. But since $\pi_1 S(t)x = \xi = \pi_1 S(t)y$, it follows that $S(t)x = S(t)y$. The backward uniqueness property implies then that $x = y$.   ▯

5.5 A Modification

In this section we present a second result, whereby the existence of an inertial manifold of the form (5.16) follows from a modified version of the strong squeezing property, related to a modification of the decay property. The main interest of this result lies in an improved estimate of the exponential convergence of the orbits to the manifold, which involves only the initial distance $d(x, \mathcal{M})$ of the orbit to the manifold.

5.5.1 The Modified Strong Squeezing Property

In analyzing the proof of theorem 5.21, we realize that the only point where the decay property was used in an essential way was in the proof of proposition 5.18. We further observe that its assumption (5.33) can be rewritten as

$$\pi_1 S(t_1)u = \pi_1 S(t_1)v, \tag{5.57}$$

for $u = x_1$ and $v = S(t_2 - t_1)x_2$. This is our starting point for the introduction of the modified strong squeezing property. Again, we assume for the moment that S does admit an inertial manifold $\mathcal{M}$ of the form (5.16), and proceed to deduce other natural geometric properties of the semiflow.

We go back to the last step of (5.31), with the difference that we now use the estimate

$$d(x_1, \mathcal{M}) \leq \|x_1 - z\|_{\mathcal{X}},$$

with $z := \pi_1 x_1 + m(\pi_1 x_1) \in \mathcal{M}$. Together with (5.31), this yields

$$\|S(t)x_1 - S(t)x_2\|_{\mathcal{X}} \leq CC_1 \|x_1 - z\|_{\mathcal{X}} \, e^{-\eta t}. \tag{5.58}$$

At this point, the right side of (5.58) still depends on the function m, which of course is the very function we want to determine, via the choice of z. To remove this dependency, we single out two properties of z that can be expressed without the explicit knowledge of m; namely, that $\pi_1 z = \pi_1 x_1$, and that $x_2 - z \in \mathcal{C}_L$. Of these, the first follows from the definition of z, and the second would follow from proposition 5.10, recalling that both z and $x_2 \in \mathcal{M}$ (the latter as assumed before (5.29)). Thus, we are led to introduce the set

$$\mathcal{Y}(x_1, x_2) := \{u \in \mathcal{X} : \pi_1 u = \pi_1 x_1, u - x_2 \in \mathcal{C}_L\}, \tag{5.59}$$

to which z would belong. We further note that if we also require that

$$\|S(t)x_1 - S(t)x_2\|_{\mathcal{X}} \leq CC_1 \inf_{u \in \mathcal{Y}(x_1, x_2)} \|x_1 - u\|_{\mathcal{X}} e^{-\eta t}, \tag{5.60}$$

then (5.58) would follow from (5.60), since $z \in \mathcal{Y}(x_1, x_2)$.

Finally, recalling (5.57), we further modify (5.60) by assuming that $\pi_1 S(t)x_1 = \pi_1 S(t)x_2$, so that the left side of (5.60) reduces to

$$\|\pi_2(S(t)x_1 - S(t)x_2)\|_{\mathcal{X}}.$$

In conclusion, renaming $t = T$, we have seen that if S admitted an inertial manifold $\mathcal{M}$ of the form (5.16), with $m \in \mathcal{G}_L$, then the conditions

$$\pi_1 S(T)x_1 = \pi_1 S(T)x_2, \qquad x_1, x_2 \in \mathcal{X}, \, T > 0, \tag{5.61}$$

would imply that there is $K > 0$ such that

$$\|\pi_2(S(T)x_1 - S(T)x_2)\|_{\mathcal{X}} \leq K \inf_{u \in \mathcal{Y}(x_1, x_2)} \|x_1 - u\|_{\mathcal{X}} e^{-\eta T}. \tag{5.62}$$

Our last step is to require that (5.62) holds not only for $t = T$, but also for all $t \in [0, T]$. That is, we give

DEFINITION 5.25   *Let L, $\eta > 0$, and $\mathcal{C}_L$ be as in (5.24). The semiflow S satisfies the* MODIFIED DECAY PROPERTY *with parameters L and η if there is $K > 0$ such that whenever $x_1, x_2 \in \mathcal{X}$ and $T > 0$ are such that (5.61) holds, and*

$$\pi_2 S(T)x_1 \neq \pi_2 S(T)x_2, \tag{5.63}$$

then for all $t \in [0, T]$,

$$\|\pi_2(S(t)x_1 - S(t)x_2)\|_{\mathcal{X}} \leq K \inf_{u \in \mathcal{Y}(x_1, x_2)} \|x_1 - u\|_{\mathcal{X}} e^{-\eta t}, \qquad (5.64)$$

where the set $\mathcal{Y}(x_1, x_2)$ is defined as in (5.59).

Combining the cone invariance property with the modified decay property we finally arrive at

DEFINITION 5.26   *In the same conditions of definition 5.25, the semiflow S satisfies the MODIFIED STRONG SQUEEZING PROPERTY with parameters L and η if it satisfies the cone invariance property with parameter L and the modified decay property with parameters L and η.*

5.5.2   Consequences of the Modified Strong Squeezing Property

In this section we present some immediate consequences of the modified strong squeezing property. We assume that L, η and $\mathcal{C}_L$ are as in the previous definitions, and that S satisfies the modified strong squeezing property with parameters L and η.

PROPOSITION 5.27 (Backward Uniqueness)
Assume that x_1, $x_2 \in \mathcal{X}$ and $T > 0$ are such that (5.61) and (5.63) hold. Then for all $t \in [0, T]$,

$$\|S(t)x_1 - S(t)x_2\|_{\mathcal{X}} \leq K\left(1 + \tfrac{1}{L}\right) \inf_{u \in \mathcal{Y}(x_1, x_2)} \|x_1 - u\|_{\mathcal{X}} e^{-\eta t}. \qquad (5.65)$$

In particular, if $x_1 - x_2 \in \mathcal{C}_L$, then $x_1 = x_2$.

PROOF   If (5.61) and (5.63) hold, it follows that for all $t \in [0, T]$, $S(t)x_1 - S(t)x_2 \notin \mathcal{C}_L$. Indeed, if there were $\theta \in [0, T]$ such that $S(\theta)x_1 - S(\theta)x_2 \in \mathcal{C}_L$, the cone invariance property would imply that $S(T)x_1 - S(T)x_2 \in \mathcal{C}_L$ as well. But then (5.61) would imply that $\pi_2 S(T)x_1 = \pi_2 S(T)x_2$, contradicting (5.63). Now, since the condition $S(t)x_1 - S(t)x_2 \notin \mathcal{C}_L$ is equivalent to the inequality

$$\|\pi_2(S(t)x_1 - S(t)x_2)\|_{\mathcal{X}} > L\|\pi_1(S(t)x_1 - S(t)x_2)\|_{\mathcal{X}}, \qquad (5.66)$$

(5.65) follows from (5.66) and (5.64). In particular, if $x_1 - x_2 \in \mathcal{C}_L$, then $x_1 \in \mathcal{Y}(x_1, x_2)$. Hence, if $x_1 - x_2 \in \mathcal{C}_L$, the right-hand side of (5.65), and therefore also the difference $S(t)x_1 - S(t)x_2$, vanishes for all $t \in [0, T]$. $\square$

In particular, from proposition 5.24 we deduce that if S satisfies the strong squeezing property, the element $x \in \mathcal{M}$ defined by (5.56) is uniquely determined by $\xi \in \mathcal{X}_1$.

PROPOSITION 5.28 (Contractivity)
Assume that φ_1, $\varphi_2 \in \mathcal{G}_L$, and $x_1 \in \text{graph}(\varphi_1)$, $x_2 \in \text{graph}(\varphi_2)$, $T > 0$ are such that (5.61) and (5.63) hold. There exists $C > 0$, independent of x_1, x_2, such that for all $t \in [0, T]$,

$$\|\pi_2(S(t)x_1 - S(t)x_2)\|_{\mathcal{X}} \leq C\|\varphi_1(\pi_1 x_1) - \varphi_2(\pi_1 x_2)\|_{\mathcal{X}} \, e^{-\eta t}. \qquad (5.67)$$

PROOF   Let $y := \pi_1 x_1 + \varphi_2(\pi_1 x_2)$. Then, since $x_2 \in \text{graph}(\varphi_2)$,

$$\pi_2(y - x_2) = \varphi_2(\pi_1 x_2) - \varphi_2(\pi_1 x_2) = 0.$$

Hence, $y - x_2 \in \mathcal{C}_L$ (trivially), and since $\pi_1 y = \pi_1 x_1$, it follows that $y \in \mathcal{Y}(x_1, x_2)$. Since also

$$x_1 - y = \pi_2 x_1 - \pi_2 y = \varphi_1(\pi_1 x_1) - \varphi_2(\pi_1 x_2),$$

(5.67) follows from (5.64).   ▯

PROPOSITION 5.29 (Asymptotic Phases)
Let $m \in \mathcal{G}_L$, and assume that $x_1 \in \mathcal{X}$, $x_2 \in \text{graph}(m)$ and $T > 0$ are such that (5.61) and (5.63) hold. Then for all $t \in [0, T]$,

$$\|\pi_2(S(t)x_1 - S(t)x_2)\|_{\mathcal{X}} \leq C\|\pi_2 x_1 - m(\pi_1 x_1)\|_{\mathcal{X}} e^{-\eta t}. \qquad (5.68)$$

PROOF   Let $y := \pi_1 x_1 + m(\pi_1 x_1)$. Then $y \in \text{graph}(m)$, and $y - x_2 \in \mathcal{C}_L$ by proposition 5.10. Hence, $y \in \mathcal{Y}(x_1, x_2)$. Since also $x_1 - y = \pi_2 x_1 - m(\pi_1 x_1)$, (5.68) follows from (5.64). We can then conclude, recalling definition 5.7.   ▯

5.5.3   Construction of the Inertial Manifold, 2

We can now show that if the semiflow S satisfies the modified strong squeezing property, it admits an inertial manifold of the form (5.16), which satisfies an exponential estimate involving only the initial distance of the orbits to the manifold. We claim:

THEOREM 5.30
Let S be a continuous semiflow on $\mathcal{X}$, which satisfies the modified strong squeezing property with parameters L and η, as well as the surjectivity and uniform boundedness properties with respect to the set $\Phi = \{0\}$. Then S admits an inertial manifold $\mathcal{M} \subseteq \mathcal{X}$ of the form (5.16). More precisely, there is $m \in \mathcal{G}_L$, such that the set $\mathcal{M} := \text{graph}(m)$ is an inertial manifold for S, having the exponential tracking property (5.5). That is, for all $x \in \mathcal{X} \setminus \mathcal{M}$, there is $z \in \mathcal{M}$ such that for all $t \geq 0$,

$$\|S(t)x - S(t)z\|_{\mathcal{X}} \leq C_2\|\pi_2 x - m(\pi_1 x)\|_{\mathcal{X}} e^{-\eta t}, \qquad (5.69)$$

with C_2 independent of x, z and t. Consequently, $\mathcal{M}$ is exponentially attracting.

PROOF We start as in the proof of theorem 5.21. Keeping the same notations, we see that we can proceed exactly in the same way up to the proof that $\mathcal{M}$ is positively invariant, except that now we cannot use proposition 5.18 to obtain (5.49). However, we can instead use proposition 5.28, with $\bar{x}_1 = \xi_1 \in \mathcal{X}_1 = \text{graph}(\varphi_0)$, and $\bar{x}_2 = S(t_2 - t_1)\xi_2 \in S(t_2 - t_1)\mathcal{X}_1 = \text{graph}(\varphi_{t_2 - t_1})$; note that $\pi_1 S(t_1)\bar{x}_1 = \pi_1 S(t_1)\bar{x}_2 = \xi$. Thus, by (5.67) we obtain

$$\begin{aligned}
\|\pi_2(x_1 - x_2)\|_{\mathcal{X}} &\leq \|\pi_2(S(t_1)\xi_1 - S(t_2)\xi_2)\|_{\mathcal{X}} \\
&= \|\pi_2(S(t_1)\xi_1 - S(t_1)S(t_2 - t_1)\xi_2)\|_{\mathcal{X}} \\
&\leq C_1 \|\varphi_0(\pi_1\xi_1) - \varphi_{t_2 - t_1}(\pi_1 S(t_2 - t_1)\xi_1)\|_{\mathcal{X}} e^{-\eta t_1}.
\end{aligned}$$

Since $\varphi_0(\pi_1\xi_1) = 0$, (5.49) follows.

To show that $\mathcal{M}$ has the exponential tracking property, given $x \in \mathcal{X} \setminus \mathcal{M}$ we consider the set

$$\mathcal{J}_x := \{t \geq 0 : (\exists z \in \mathcal{M} : \pi_1 S(t)z = \pi_1 S(t)x, \ \pi_2 S(t)z \neq \pi_2 S(t)x)\}.$$

This set is not empty, since $t = 0 \in \mathcal{J}_x$. To see this, let $z := \pi_1 x + m(\pi_1 x)$. Then $\pi_1 z = \pi_1 x$, but $\pi_2 z = m(\pi_1 x) \neq \pi_2 x$, since $z \in \mathcal{M}$ but $x \notin \mathcal{M}$.

Consider first the case when $\sup \mathcal{J}_x =: \bar{t} < +\infty$. Then for all $t \geq \bar{t}$ there is $z \in \mathcal{M}$ such that $S(t)z = S(t)x$. Then, (5.69) is trivial if $t \geq \bar{t}$, while if $t \leq \bar{t}$ it is a consequence of the modified decay property. Indeed, in this case the left side of (5.69) reduces to $\|\pi_2(S(t)x - S(t)z)\|_{\mathcal{X}}$, which we can estimate by (5.64). In particular, we can take $u = \pi_1 x + m(\pi_1 x)$ which is in $\mathcal{Y}(x, z)$, because both u and $z \in \mathcal{M}$. Hence, from (5.64),

$$\|\pi_2(S(t)x - S(t)z)\|_{\mathcal{X}} \leq K \|x - u\|_{\mathcal{X}} e^{-\eta t},$$

and (5.69) follows, because of the choice of u.

If instead $\sup \mathcal{J}_x = +\infty$, we can find a sequence $(t_k)_{k \in \mathbb{N}}$ such that $t_k \to +\infty$ monotonically, and a corresponding sequence $(z_k)_{k \in \mathbb{N}} \subset \mathcal{M}$ such that for all $k \in \mathbb{N}$,

$$\pi_1 S(t_k)z_k = \pi_1 S(t_k)x \quad \text{and} \quad \pi_2 S(t_k)z_k \neq \pi_2 S(t_k)x.$$

Since $m \in \mathcal{G}_L$, we can apply proposition 5.29 and (5.66), to deduce that for each $k \in \mathbb{N}$ and all $t \in [0, t_k]$,

$$L\|\pi_1(S(t)x - S(t)z_k)\|_{\mathcal{X}} \leq \|\pi_2(S(t)x - S(t)z_k)\|_{\mathcal{X}} \leq C\|\pi_2 x - m(\pi_1 x)\|_{\mathcal{X}} e^{-\eta t}. \tag{5.70}$$

In particular, for $t = 0$ this implies that for all $k \in \mathbb{N}$

$$\|\pi_1 x - \pi_1 z_k\|_{\mathcal{X}} \leq \frac{C}{L}\|\pi_2 x - m(\pi_1 x_1)\|_{\mathcal{X}} =: \rho.$$

This means that each $\zeta_k := \pi_1 z_k$ is in the closed ball of $\mathcal{X}_1$ of center $\pi_1 x$ and radius ρ. Since $\mathcal{X}_1$ has finite dimension, this ball is compact. Thus, there is a subsequence of $(t_k)_{k \in \mathbb{N}}$, still denoted $(t_k)_{k \in \mathbb{N}}$, such that the corresponding subsequence $(\zeta_k)_{k \in \mathbb{N}}$

converges to a limit $\zeta \in \mathcal{X}_1$, with $\|\pi_1 x - \zeta\|_{\mathcal{X}} \le \rho$. Let $z := \zeta + m(\zeta)$. Then for each $k \in \mathbb{N}$ and all $t \in [0, t_k]$, recalling (5.70) we can estimate

$$
\begin{aligned}
\|S(t)x - S(t)z\|_{\mathcal{X}} &\le \|S(t)x - S(t)z_k\|_{\mathcal{X}} + \|S(t)z_k - S(t)z\|_{\mathcal{X}} \\
&\le \left(1 + \tfrac{1}{L}\right)\|\pi_2(S(t)x - S(t)z_k)\|_{\mathcal{X}} + \|S(t)z_k - S(t)z\|_{\mathcal{X}} \quad (5.71) \\
&\le C\left(1 + \tfrac{1}{L}\right)\|\pi_2 x - m(\pi_1 x_1)\|_{\mathcal{X}} e^{-\eta t} + \|S(t)z_k - S(t)z\|_{\mathcal{X}}.
\end{aligned}
$$

Since m is continuous, and $z, z_k \in \mathcal{M}$ for all $k \in \mathbb{N}$, we have that $z_k \to z$. Since $S(t)$ is continuous, it follows that also $S(t)z_k \to S(t)z$. Thus, there is $k_0 \in \mathbb{N}$ (depending on t), such that if $k \ge k_0$,

$$
\|S(t)z_k - S(t)z\|_{\mathcal{X}} \le C\left(1 + \tfrac{1}{L}\right)\|\pi_2 x - m(\pi_1 x_1)\|_{\mathcal{X}} e^{-\eta t}. \quad (5.72)
$$

Consequently, (5.69) follows from (5.71) and (5.72) (recall that $z \in \mathcal{M}$). From (5.6) we know that the exponential tracking property implies that $\mathcal{M}$ is exponentially attracting; thus, the proof of theorem 5.30 is complete. $\qquad\square$

We remark that the inertial manifold so constructed enjoys all the properties described in section 5.4.2.

5.5.4   Comparison of the Squeezing Properties

We conclude this section with a comparison of the strong squeezing property and the discrete squeezing property introduced in definition 4.3, in the context of exponential attractors. We also compare the strong squeezing property with the modified strong squeezing property.

At first, we note that if the semiflow S satisfies the strong squeezing property of definition 5.16 with parameters η and L, and $L \le 1$, then S also satisfies the discrete squeezing property of definition 4.3, with $\mathcal{B} = \mathcal{X}$, $P_N = \pi_1$, $\mathcal{X}_N = \mathcal{X}_1$. Indeed, comparing (4.9) with (5.24) we have $\mathcal{C}_P = \mathcal{C}_{P_N} = \mathcal{C}_1 \subseteq \mathcal{C}_L$ for all $L \in \,]0,1]$. Given $\gamma \in \,]0, \tfrac{1}{2}[$, choose $t_* = T$ so large that $Ke^{-\eta T} \le \gamma$. Then, if $u, v \in \mathcal{X}$ are such that $S(T)u - S(T)v \notin \mathcal{C}_1$, and therefore also $S(T)u - S(T)v \notin \mathcal{C}_L$ for $L \le 1$, (5.32) implies that

$$
\|S(T)u - S(T)v\|_{\mathcal{X}} \le Ke^{-\eta T}\|u - v\|_{\mathcal{X}} \le \gamma\|u - v\|_{\mathcal{X}}.
$$

This means that the operator $S_* = S(T)$ satisfies (4.11); hence, the discrete squeezing property holds.

Next, to illustrate the relationship between the strong squeezing property and the modified strong squeezing property, we return to the observation we made at the beginning of section 5.3.2: namely, that we can often assume that the semiflow S also satisfies a second cone invariance property. In this case, however, we assume that this second cone invariance property involves a cone $\mathcal{C}_{L'}$, with parameter $L' > L$ (instead of $L' < L$, as in proposition 5.14). In these conditions, we claim:

PROPOSITION 5.31

Assume the continuous semiflow S satisfies the strong squeezing property with parameters L and η. Suppose further that there exists L' > L such that S also satisfies a second cone invariance property with parameter L'. Then S satisfies the modified strong squeezing property, with parameters L and η.

PROOF   Let $x_1, x_2 \in \mathcal{X}$, and $T > 0$ be such that

$$\pi_1 S(T)x_1 = \pi_1 S(T)x_2, \quad \pi_2 S(T)x_1 \neq \pi_2 S(T)x_2,$$

as required in (5.61) and (5.63). Then, we have that both

$$S(T)x_1 - S(T)x_2 \notin \mathcal{C}_L, \quad S(T)x_1 - S(T)x_2 \notin \mathcal{C}_{L'},$$

for, otherwise, definition (5.24) would imply that $\pi_2 S(T)x_1 = \pi_2 S(T)x_2$. Proposition 5.13 implies then that

$$S(t)x_1 - S(t)x_2 \notin \mathcal{C}_L, \quad S(t)x_1 - S(t)x_2 \notin \mathcal{C}_{L'}$$

for all $t \in [0, T]$. Consequently, for $t \in [0, T]$,

$$L\|\pi_1(S(t)x_1 - S(t)x_2)\|_{\mathcal{X}} < L'\|\pi_1(S(t)x_1 - S(t)x_2)\|_{\mathcal{X}}$$
$$< \|\pi_2(S(t)x_1 - S(t)x_2)\|_{\mathcal{X}}. \tag{5.73}$$

Let $x_3 \in \mathcal{Y}(x_1, x_2)$ (the set introduced in (5.59)). Then $\pi_1 x_3 = \pi_1 x_1$ and $x_3 - x_2 \in \mathcal{C}_L$, and (5.24) implies that

$$\|\pi_2(x_2 - x_3)\|_{\mathcal{X}} \leq L\|\pi_1(x_2 - x_1)\|_{\mathcal{X}}.$$

This, together with (5.73) for $t = 0$, implies

$$L'\|\pi_1(x_1 - x_2)\|_{\mathcal{X}} \leq \|\pi_2(x_1 - x_2)\|_{\mathcal{X}} \leq \|\pi_2(x_1 - x_3)\|_{\mathcal{X}} + \|\pi_2(x_2 - x_3)\|_{\mathcal{X}}$$
$$\leq \|\pi_2(x_1 - x_3)\|_{\mathcal{X}} + L\|\pi_1(x_1 - x_2)\|_{\mathcal{X}}. \tag{5.74}$$

Hence,

$$\|\pi_1(x_1 - x_2)\|_{\mathcal{X}} \leq \tfrac{1}{L'-L}\|\pi_2(x_1 - x_3)\|_{\mathcal{X}}, \tag{5.75}$$

and, replacing back into (5.74),

$$\|\pi_2(x_1 - x_2)\|_{\mathcal{X}} \leq \tfrac{L'}{L'-L}\|\pi_2(x_1 - x_3)\|_{\mathcal{X}}. \tag{5.76}$$

By the decay property (5.32), recalling (5.75) and (5.76),

$$\|S(t)x_1 - S(t)x_2\|_{\mathcal{X}} \leq K\|x_1 - x_2\|_{\mathcal{X}}e^{-\eta t} \leq K(\tfrac{1}{L'}+1)\|\pi_2(x_1 - x_2)\|_{\mathcal{X}}e^{-\eta t}$$
$$\leq \tfrac{K(1+L')}{L'-L}\|\pi_2(x_1 - x_3)\|_{\mathcal{X}}e^{-\eta t}. \tag{5.77}$$

Since $\pi_1 x_3 = \pi_1 x_1$, we also have

$$\|\pi_2 x_1 - \pi_1 x_3\|_{\mathcal{X}} = \|\pi_2 x_1 + \pi_1 x_1 - \pi_1 x_3 - \pi_2 x_3\|_{\mathcal{X}} = \|x_1 - x_3\|_{\mathcal{X}} ;$$

thus, from (5.77) we obtain

$$\|S(t)x_1 - S(t)x_2\|_{\mathcal{X}} \leq \frac{K(1+L')}{L'-L}\|x_1 - x_3\|_{\mathcal{X}} \mathrm{e}^{-\eta t}.$$

Taking inf of the right side of this inequality with respect to $x_3 \in \mathcal{Y}(x_1, x_2)$ we conclude that (5.64) holds, with K replaced by $\frac{K(1+L')}{L'-L}$. Thus, the modified strong squeezing property holds.   □

In conclusion, the strong squeezing property, together with the availability of a second cone invariance property with parameter $L' > L$, implies the modified strong squeezing property. In this sense, the modified strong squeezing property is a weaker assumption than the original strong squeezing property.

5.6   Inertial Manifolds for Evolution Equations

In this section we show how theorems 5.21 or 5.30 can be applied to construct an inertial manifold for the semiflow S generated by an autonomous evolution equation of the form (5.1) on a separable Hilbert space $\mathcal{X}$ with norm $\|\cdot\|_{\mathcal{X}}$ and scalar product $\langle \cdot, \cdot \rangle$.

5.6.1   The Evolution Problem

We consider an evolution equation of the general form (5.1), that is, again,

$$u_t + Au = F(u), \tag{5.78}$$

where the linear operator A is densely defined, with domain

$$\mathrm{dom}(A) := \{u \in \mathcal{X} : Au \in \mathcal{X}\},$$

and generates at least a C^0-SEMIGROUP $(\mathrm{e}^{-tA})_{t \geq 0}$ on $\mathcal{X}$ (see section A.3).

In the next section we introduce a condition on the point spectrum of A, known as a SPECTRAL GAP CONDITION. Together with the additional assumption that F be globally bounded, i.e. $F \in C_b(\mathcal{X}, \mathcal{X})$, and globally Lipschitz continuous on $\mathcal{X}$, this condition will allow us to show that the semiflow S defined in $\mathcal{X}$ by problem (5.78) verifies the cone invariance property defined in section 5.3.1. Other, less restrictive assumptions on the point spectrum of A will assure that S also satisfies one of the two versions of the decay properties described in section 5.4 and 5.5. Consequently, theorem 5.21 (respectively, 5.30) will imply that if the operator A satisfies the spectral gap condition, relative to the nonlinearity F, the corresponding semiflow generated by the evolution equation (5.78) admits an inertial manifold in $\mathcal{X}$.

Before proceeding, we remark that the conditions assumed on F, i.e. that F is globally bounded and globally Lipschitz continuous, are much stronger that those usually satisfied by the nonlinearities present in applications. For example, in the parabolic problem considered in sections 3.3, the nonlinearity $F(u) = u - u^3$ satisfies neither condition on $\mathcal{X}$ (in fact, this particular function F is bounded, and Lipschitz continuous, only from bounded subsets of the subspace $V \subset \mathcal{X} = \mathcal{H}$ into $\mathcal{X}$ (see proposition 3.15)). To overcome these problems, we shall suitably modify the nonlinearity outside of a bounded absorbing set, so that the nonlinearity of the modified system does satisfy the two desired conditions, and its long time dynamics coincides with that of the original one. In this sense, we take full advantage of the observation that the introduction of inertial manifolds is motivated by the desire to study the long time behavior of a system. We explain this procedure in detail in section 5.8.3, in the context of a problem concerning the Chafee-Infante equations.

5.6.2 The Spectral Gap Condition

1. Generally speaking, a spectral gap condition is a requirement on the point spectrum of A, whereby its eigenvalues should be spaced with sufficiently large gaps, so as to allow the linear part of equation (5.78), i.e. the term Au, to dominate, in a sense to be specified, the nonlinear one, i.e. the term $F(u)$. The effect of this condition is that the corresponding semiflow S satisfies the cone invariance property. Spectral gap conditions of various type were originally introduced in the context of Navier-Stokes equations in two dimensions of space. Here, we introduce a specific version of the spectral gap condition, which is sufficient for the purpose of showing that the semiflows generated by problems (P) and (H_ε) admit an inertial manifold, at least if the dimension of space is $n = 1$ and, in (3.4), if ε is sufficiently small. For simplicity, we shall refer to this particular version as "the" spectral gap condition.

Essentially, all versions of the spectral gap condition require that the point spectrum of the operator A can be divided into two parts

$$\sigma_{\mathrm{p}}(A) = \sigma_1 \cup \sigma_2, \tag{5.79}$$

of which σ_1 is finite, and such that, if

$$\Lambda_1 := \sup\{\mathfrak{Re}\,\lambda : \lambda \in \sigma_1\},$$
$$\Lambda_2 := \inf\{\mathfrak{Re}\,\lambda : \lambda \in \sigma_2\},$$

the difference $\Lambda_2 - \Lambda_1$ (the "spectral gap") is large enough, so as to satisfy an inequality of the form

$$\Lambda_2 - \Lambda_1 \geq C(\Lambda_1^\nu + \Lambda_2^\nu) \tag{5.80}$$

for certain nonnegative numbers C and ν related to the nonlinear term F of equation (5.78). Inequality (5.80) is usually interpreted in the sense that the spectral gap should be so large as to allow the linear part of the equation to dominate the nonlinear one.

To describe the particular version of the spectral gap condition we consider in the sequel, we assume that A admits a countable set of eigenvalues $(\lambda_j)_{j \in \mathbb{N}}$, with a corresponding orthonormal system of eigenfunctions $(w_j)_{j \in \mathbb{N}}$, spanning $\mathcal{X}$ (some or all of the functions w_j may be complex-valued), and that the eigenvalues are ordered in such a way that

$$\Lambda_2 - \Lambda_1 = \mathfrak{Re}(\lambda_{N+1} - \lambda_N) \tag{5.81}$$

for some $N \in \mathbb{N}$. Under these conditions, we give

DEFINITION 5.32   *Let $A \colon \mathcal{X} \to \mathcal{X}$ be as above, and assume that $F \in C_b(\mathcal{X}, \mathcal{X})$ satisfies the global Lipschitz condition*

$$\|F(u) - F(v)\|_{\mathcal{X}} \leq \ell_F \|u - v\|_{\mathcal{X}}, \tag{5.82}$$

for some ℓ_F independent of u and v. The operator A satisfies the SPECTRAL GAP CONDITION relative to F, if there is $N \in \mathbb{N}$ such that

$$\mathfrak{Re}(\lambda_{N+1} - \lambda_N) > 2\ell_F. \tag{5.83}$$

That is, recalling (5.81), we assume that (5.80) holds, with $v = 0$ and $C = \ell_F$.

2.   We now describe a condition under which the spectral gap condition (5.83) implies that the semiflow generated by equation (5.78) satisfies the cone invariance property. We fix $N \in \mathbb{N}$, as defined in (5.81), and set

$$\mathcal{X}_1 := \operatorname{span}\{w_1, \ldots, w_N\}.$$

We denote by P_1 the corresponding projection of $\mathcal{X}$ onto $\mathcal{X}_1$, and by $P_2 := I - P_1$ its complementary projection; note that A commutes with both P_1 and P_2. We assume that for all $u \in \mathcal{X}$,

$$\mathfrak{Re}\langle AP_1 u, P_1 u \rangle \leq \mathfrak{Re}\, \lambda_N \|P_1 u\|_{\mathcal{X}}^2, \tag{5.84}$$

$$\mathfrak{Re}\langle AP_2 u, P_2 u \rangle \geq \mathfrak{Re}\, \lambda_{N+1} \|P_2 u\|_{\mathcal{X}}^2. \tag{5.85}$$

Introducing then, for $L > 0$, the indefinite quadratic form $H_L \colon \mathcal{X} \to \mathbb{R}$ defined by

$$H_L(u) := \|P_2 u\|_{\mathcal{X}}^2 - L^2 \|P_1 u\|_{\mathcal{X}}^2, \qquad u \in \mathcal{X}, \tag{5.86}$$

we claim:

PROPOSITION 5.33
Assume $F \in C_b(\mathcal{X}, \mathcal{X})$ satisfies (5.82), that the spectral gap condition (5.83) holds, as well as conditions (5.84) and (5.85). There exist positive numbers L_1 and L_2, with $L_1 < 1 < L_2$, such that for all $t_0 \geq 0$, all $t \geq t_0$, x and $x' \in \mathcal{X}$, and $L \in [L_1, L_2]$,

$$H_L(S(t)x - S(t)x') \leq H_L(S(t_0)x - S(t_0)x')\, e^{-2(\mathfrak{Re}\, \lambda_{N+1} - \ell_F)(t - t_0)}. \tag{5.87}$$

PROOF Let $x, x' \in \mathcal{X}, t > 0$, and set

$$u_\Delta := S(t)x - S(t)x', \quad F_\Delta := F(S(t)x) - F(S(t)x'). \tag{5.88}$$

Then, recalling (5.84) and (5.85), and that, since the projections P_1 and P_2 are orthogonal, the identities

$$\langle P_2 x, P_1 y \rangle = 0, \quad \langle P_2 x, P_2 y \rangle = \langle P_2 x, y \rangle$$

hold for all $x, y \in \mathcal{X}$, we can estimate (omitting as usual the reference to the variable t):

$$
\begin{aligned}
\frac{1}{2} &\frac{\mathrm{d}}{\mathrm{d}t} H_L(u_\Delta) \\
&= \mathfrak{Re}\langle P_2 u_\Delta, -A u_\Delta + F_\Delta \rangle - L^2 \mathfrak{Re}\langle P_1 u_\Delta, -A u_\Delta + F_\Delta \rangle \\
&= -\mathfrak{Re}\langle P_2 u_\Delta, A u_\Delta \rangle + L^2 \mathfrak{Re}\langle P_1 u_\Delta, A u_\Delta \rangle + \mathfrak{Re}\langle P_2 u_\Delta - L^2 P_1 u_\Delta, F_\Delta \rangle \\
&\leq -\mathfrak{Re}\,\lambda_{N+1} \|P_2 u_\Delta\|_{\mathcal{X}}^2 + L^2 \mathfrak{Re}\,\lambda_N \|P_1 u_\Delta\|_{\mathcal{X}}^2 + \ell_F \|u_\Delta\|_{\mathcal{X}} \|P_2 u_\Delta - L^2 P_1 u_\Delta\|_{\mathcal{X}} \\
&\leq -\mathfrak{Re}\,\lambda_{N+1} \|P_2 u_\Delta\|_{\mathcal{X}}^2 + L^2 \mathfrak{Re}\,\lambda_N \|P_1 u_\Delta\|_{\mathcal{X}}^2 + \tfrac{1}{2}\ell_F (\|u_\Delta\|_{\mathcal{X}}^2 + \|P_2 u_\Delta - L^2 P_1 u_\Delta\|_{\mathcal{X}}^2) \\
&\leq -\mathfrak{Re}\,\lambda_{N+1} \|P_2 u_\Delta\|_{\mathcal{X}}^2 + L^2 \mathfrak{Re}\,\lambda_N \|P_1 u_\Delta\|_{\mathcal{X}}^2 \\
&\quad + \tfrac{1}{2}\ell_F (\|P_1 u_\Delta\|_{\mathcal{X}}^2 + \|P_2 u_\Delta\|_{\mathcal{X}}^2 + \|P_2 u_\Delta\|_{\mathcal{X}}^2 + L^4 \|P_1 u_\Delta\|_{\mathcal{X}}^2).
\end{aligned}
$$

From this, we deduce the estimate

$$\frac{\mathrm{d}}{\mathrm{d}t} H_L(u_\Delta) \leq -2\gamma H_L(u_\Delta), \tag{5.89}$$

where γ is any number satisfying the inequalities

$$\mathfrak{Re}\,\lambda_N + \tfrac{1}{2}\ell_F \left(L^2 + L^{-2}\right) \leq \gamma \leq \mathfrak{Re}\,\lambda_{N+1} - \ell_F. \tag{5.90}$$

For (5.90) to hold, it is necessary that

$$\mathfrak{Re}\,\lambda_N + \tfrac{1}{2}\ell_F \left(L^2 + L^{-2}\right) \leq \mathfrak{Re}\,\lambda_{N+1} - \ell_F,$$

an inequality that we can write as

$$\psi(L) := \mathfrak{Re}\,\lambda_{N+1} - \mathfrak{Re}\,\lambda_N - \ell_F \left(1 + \tfrac{1}{2}\left(L^2 + L^{-2}\right)\right) \geq 0. \tag{5.91}$$

The maximum value of ψ is attained when $L = 1$, and the spectral gap condition (5.83) implies that $\psi(1) > 0$. Since ψ is concave and $\psi(L) \to -\infty$ as $L \to 0^+$ and $L \to +\infty$, it follows that ψ has two positive zeros L_1, L_2, with $L_1 < 1 < L_2$ and, as we immediately verify, $L_1 L_2 = 1$. Thus, (5.91) holds for all $L \in [L_1, L_2]$. From (5.89) we deduce then that, with the choice $\gamma := \mathfrak{Re}\,\lambda_{N+1} - \ell_F$,

$$\frac{\mathrm{d}}{\mathrm{d}t} H_L(S(t)x - S(t)x') \leq -2(\mathfrak{Re}\,\lambda_{N+1} - \ell_F) H_L(S(t)x - S(t)x').$$

Integrating this inequality on any interval $[t_0, t]$, with $0 \leq t_0 < t$, we obtain (5.87). □

5.6.3 The Strong Squeezing Properties

In this section we show how the squeezing properties follow from the spectral gap condition (5.83). We first show that, as a consequence of proposition 5.33, the spectral gap condition implies the cone invariance property.

PROPOSITION 5.34

In the same conditions of proposition 5.33, the semiflow S generated by the evolution equation (5.78) satisfies the cone invariance property in $\mathcal{X}$, with parameter L, for all $L \in [L_1, L_2]$, where L_1 and L_2 are the positive zeros of the function ψ defined in (5.91).

PROOF   Let $x_1, x_2 \in \mathcal{X}$, and $L \in [L_1, L_2]$. We have to show that if $x_1 - x_2 \in \mathcal{C}_L$, then $S(t)x_1 - S(t)x_2 \in \mathcal{C}_L$ for all $t \geq 0$. But, recalling (5.24), the condition $x_1 - x_2 \in \mathcal{C}_L$ is equivalent to $H_L(x_1 - x_2) \leq 0$; hence, from (5.87) with $t_0 = 0$ we deduce that $H_L(S(t)x_1 - S(t)x_2) \leq 0$. This means that $S(t)x_1 - S(t)x_2 \in \mathcal{C}_L$. $\qquad\Box$

We next show that if $\mathfrak{Re}\,\lambda_{N+1}$ is sufficiently large, the decay properties hold.

PROPOSITION 5.35

In the same conditions of proposition 5.34, assume that $N \in \mathbb{N}$ is such that

$$\eta := \mathfrak{Re}\,\lambda_{N+1} - \ell_F \sqrt{1 + L^{-2}} > 0, \qquad (5.92)$$

with $L \in [L_1, L_2]$. Then S satisfies the decay property with parameters L and η.

PROOF   Let $x_1, x_2 \in \mathcal{X}$, and $T > 0$ be such that $S(T)x_1 - S(T)x_2 \notin \mathcal{C}_L$. We have to show that there is $K > 0$ such that for all $t \in [0, T]$, (5.32) holds, with η defined by (5.92) and $\pi_1 = P_1$, $\pi_2 = P_2$. By proposition 5.34, the semiflow S satisfies the cone invariance property; hence, by proposition 5.13, $S(t)x_1 - S(t)x_2 \notin \mathcal{C}_L$ for all $t \in [0, T]$. With the same definitions (5.88) of u_Δ and F_Δ, we have then

$$L \|P_1 u_\Delta(t)\|_{\mathcal{X}} \leq \|P_2 u_\Delta(t)\|_{\mathcal{X}}.$$

Thus, as in the proof of proposition 5.33 (omitting again the reference to the variable t),

$$\begin{aligned}
\frac{\mathrm{d}}{\mathrm{d}t}\|P_2 u_\Delta\|_{\mathcal{X}}^2 &= 2\mathfrak{Re}\langle P_2 u_\Delta, F_\Delta - A u_\Delta\rangle \\
&\leq -2\mathfrak{Re}\,\lambda_{N+1}\|P_2 u_\Delta\|_{\mathcal{X}}^2 + 2\ell_F\|P_2 u_\Delta\|_{\mathcal{X}}\|u_\Delta\|_{\mathcal{X}} \\
&\leq -2\mathfrak{Re}\,\lambda_{N+1}\|P_2 u_\Delta\|_{\mathcal{X}}^2 + 2\ell_F\|P_2 u_\Delta\|_{\mathcal{X}}(\|P_1 u_\Delta\|_{\mathcal{X}}^2 + \|P_2 u_\Delta\|_{\mathcal{X}}^2)^{1/2} \\
&\leq -2\mathfrak{Re}\,\lambda_{N+1}\|P_2 u_\Delta\|_{\mathcal{X}}^2 + 2\ell_F\left(1 + L^{-2}\right)^{1/2}\|P_2 u_\Delta\|_{\mathcal{X}}^2 \\
&= -2\eta\|P_2 u_\Delta\|_{\mathcal{X}}^2.
\end{aligned}$$

From this, we conclude that for $t \in [0, T]$,

$$\|P_2(S(t)x_1 - S(t)x_2)\|_{\mathcal{X}} \leq e^{-\eta t}\|P_2(x_1 - x_2)\|_{\mathcal{X}},$$

and (5.32) follows, with $K = 1$. ⬜

PROPOSITION 5.36
In the same conditions of proposition 5.34, assume that $L \in [L_1, L_2[$, and that $N \in \mathbb{N}$ is such that

$$\eta_1 := \mathfrak{Re}\,\lambda_{N+1} - \ell_F > 0. \tag{5.93}$$

Then S satisfies the modified decay property, with parameters L and η.

PROOF   To prove the modified decay property, recalling (5.64) we consider x_1, $x_2 \in \mathcal{X}$ and $T > 0$, such that

$$P_1 S(T)x_1 = P_1 S(T)x_2, \tag{5.94}$$

and $y \in \mathcal{Y}(x_1, x_2)$, i.e. such that $P_1 y = P_1 x_1$ and $y - x_2 \in \mathcal{C}_L$. We have to show that there is $K > 0$, independent of x_1 and x_2, such that for all $t \in [0, T]$,

$$\|P_2(S(t)x_1 - S(t)x_2)\|_{\mathcal{X}} \leq K\|P_2(x_1 - y)\|_{\mathcal{X}}e^{-\eta t}. \tag{5.95}$$

Setting $u_\Delta(t) := S(t)x_1 - S(t)x_2$ as in the proof of proposition 5.33, (5.94) implies that for all $L \in [L_1, L_2]$,

$$H_L(u_\Delta(T)) = \|P_2(u_\Delta(T))\|_{\mathcal{X}}^2 \geq 0.$$

We claim that this implies that $H_L(u_\Delta(t)) \geq 0$ for all $t \in [0, T]$ and $L \in [L_1, L_2]$. Indeed, if otherwise $H_{L'}(u_\Delta(t')) < 0$ for some $L' \in [L_1, L_2]$ and $t' \in [0, T]$, proposition 5.33 with $t = T$ and $t_0 = t'$ would yield a contradiction to the inequality $0 \leq H_{L'}(u_\Delta(T))$. Thus, by proposition 5.33 with $t_0 = 0$ we obtain that

$$0 \leq H_L(u_\Delta(t)) \leq e^{-2(\mathfrak{Re}\,\lambda_{N+1} - \ell_F)t}H_L(x_1 - x_2) \tag{5.96}$$

for all $t \in [0, T]$ and $L \in [L_1, L_2]$. If now $L \in [L_1, L_2[$, choosing $\Lambda \in]L, L_2[$ we find

$$\begin{aligned}
H_\Lambda(u_\Delta(t)) &= \|P_2 u_\Delta(t)\|_{\mathcal{X}}^2 - \Lambda^2\|P_1 u_\Delta(t)\|_{\mathcal{X}}^2 \\
&= \left(1 - \Lambda^2 L_2^{-2}\right)\|P_2 u_\Delta(t)\|_{\mathcal{X}}^2 + \Lambda^2 L_2^{-2}(\|P_2 u_\Delta(t)\|_{\mathcal{X}}^2 - L_2^2\|P_1 u_\Delta(t)\|_{\mathcal{X}}^2) \\
&= \left(1 - \Lambda^2 L_2^{-2}\right)\|P_2 u_\Delta(t)\|_{\mathcal{X}}^2 + \Lambda^2 L_2^{-2}H_{L_2}(u_\Delta(t)) \\
&\geq \left(1 - \Lambda^2 L_2^{-2}\right)\|P_2 u_\Delta(t)\|_{\mathcal{X}}^2.
\end{aligned}$$

From this, and (5.96) with $L = \Lambda$, it follows that

$$\|P_2 u_\Delta(t)\|_{\mathcal{X}}^2 \leq \frac{L_2^2}{L_2^2 - \Lambda^2}e^{-2(\mathfrak{Re}\,\lambda_{N+1} - \ell_F)t}H_\Lambda(x_1 - x_2). \tag{5.97}$$

Since $y - x_2 \in \mathcal{C}_L$,

$$\|P_2(y - x_2)\|_{\mathcal{X}} \le L \|P_1(y - x_2)\|_{\mathcal{X}} \, ;$$

thus, recalling that $P_1 y = P_1 x_1$, we deduce that for all $r > 0$

$$
\begin{aligned}
H_\Lambda(x_1 - x_2) &= \|P_2(x_1 - x_2)\|_{\mathcal{X}}^2 - \Lambda^2 \|P_1(x_1 - x_2)\|_{\mathcal{X}}^2 \\
&\le (\|P_2(x_1 - y)\|_{\mathcal{X}} + \|P_2(y - x_2)\|_{\mathcal{X}})^2 - \Lambda^2 \|P_1(x_1 - x_2)\|_{\mathcal{X}}^2 \\
&\le (\|P_2(x_1 - y)\|_{\mathcal{X}} + L \|P_1(x_2 - y)\|_{\mathcal{X}})^2 - \Lambda^2 \|P_1(x_1 - x_2)\|_{\mathcal{X}}^2 \\
&\le (1 + r^{-1}) \|P_2(x_1 - y)\|_{\mathcal{X}}^2 + ((1+r)L^2 - \Lambda^2) \|P_1(x_1 - x_2)\|_{\mathcal{X}}^2 \, .
\end{aligned}
$$

Choosing $r = \Lambda^2 L^{-2} - 1$, we obtain

$$H_\Lambda(x_1 - x_2) \le \tfrac{\Lambda^2}{\Lambda^2 - L^2} \|P_2(x_1 - y)\|_{\mathcal{X}}^2 \, ;$$

replacing this into (5.97), we deduce that, for $t \in [0, T]$,

$$\|P_2 u_\Delta(t)\|_{\mathcal{X}}^2 \le \tfrac{L_2^2}{L_2^2 - \Lambda^2} \tfrac{\Lambda^2}{\Lambda^2 - L^2} \, e^{-2(\mathfrak{Re}\,\lambda_{N+1} - \ell_F)t} \|P_2(x_1 - y)\|_{\mathcal{X}}^2 \, .$$

This is (5.95), with

$$K := L_2 \Lambda ((L_2^2 - \Lambda^2)(\Lambda^2 - L^2))^{-1/2} \, .$$

Since $y \in \mathcal{Y}(x_1, x_2)$ is arbitrary, (5.64) follows, and the proof of proposition 5.36 is complete. $\qquad\Box$

From propositions 5.22, 5.35 and 5.36 we immediately derive

COROLLARY 5.37

Assume that N is such that the spectral gap condition (5.83) holds, and $\mathfrak{Re}\,\lambda_{N+1} > \ell_F$. Then for any $L \in [L_1, L_2[$, the semiflow S satisfies the modified strong squeezing property, with parameters L and $\eta := \mathfrak{Re}\,\lambda_{N+1} - \ell_F$. If in addition $\mathfrak{Re}\,\lambda_{N+1} > \ell_F \sqrt{1 + L^{-2}}$, with $L \in [L_1, L_2]$, the semiflow S satisfies the strong squeezing property, with parameters L and $\eta := \mathfrak{Re}\,\lambda_{N+1} - \ell_F \sqrt{1 + L^{-2}}$.

Propositions 5.35 and 5.36 illustrate our previous remark, at the end of section 5.5.4, that the modified strong squeezing property is weaker than the strong squeezing property. Indeed, in proposition 5.36 we cannot take $L = L_2$, and the constant K in (5.95) is such that $K \to +\infty$ as $L \to L_2^-$. On the other hand, since $L < L_2$, we can in fact choose $L' \in]L, L_2]$, and proposition 5.34 guarantees that the semiflow S does satisfy the cone invariance property, with parameter $L' > L$ as required in proposition 5.31. Moreover, in the modified strong squeezing property we can take smaller parameters L than for the strong squeezing property, and since (5.92) implies (5.93), the rate η_1 of the exponential decay is larger.

5.6.4 Uniform Boundedness and Surjectivity

We now proceed to show that the semiflow S generated by the evolution equation (5.78) satisfies the uniform boundedness property with respect to bounded subsets $\Phi \subseteq \mathcal{G}_L$, as well as the surjectivity property with respect to arbitrary subsets $\Phi \subseteq \mathcal{G}_L$. Here, $L > 0$ is arbitrary, and π_1, π_2 are, of course, the projections P_1 and P_2 of the previous sections. As usual, we set $\mathcal{X}_1 := P_1 \mathcal{X}$ and $\mathcal{X}_2 := P_2 \mathcal{X}$. Recalling that we assume the nonlinearity $F \colon \mathcal{X} \to \mathcal{X}$ to be bounded, we also set

$$F_0 := \sup_{u \in \mathcal{X}} \|F(u)\| .$$

1. We start with the uniform boundedness of the semiflow.

PROPOSITION 5.38 (Uniform Boundedness)
Assume that there is $N \in \mathbb{N}$ such that (5.85) holds. Let $\Phi \subseteq \mathcal{G}_L$ be a bounded set of Lipschitz continuous functions $\varphi \colon \mathcal{X}_1 \to \mathcal{X}_2$. Then for all $t \geq 0$ and all $\varphi \in \Phi$, the set $P_2 S(t) \operatorname{graph}(\varphi)$ is bounded in $\mathcal{X}$.

PROOF Given $\varphi \in \Phi$ and $u_0 \in \operatorname{graph}(\varphi)$, fix $t \geq 0$ and let $u(t) := S(t)u_0$. Then u solves (5.78) and, by (5.85), we can estimate

$$\frac{\mathrm{d}}{\mathrm{d}t} \|P_2 u\|_{\mathcal{X}}^2 = 2 \Re\mathfrak{e}\, \lambda_{N+1} \langle P_2 u, P_2 u_t \rangle = 2 \Re\mathfrak{e}\, \lambda_{N+1} \langle P_2 u, f(u) - Au \rangle$$

$$\leq -2 \Re\mathfrak{e}\, \lambda_{N+1} \|P_2 u\|_{\mathcal{X}}^2 + 2 F_0 \|P_2 u\|_{\mathcal{X}} . \tag{5.98}$$

From this differential inequality we easily derive that for all $t \geq 0$,

$$\|P_2 u(t)\|_{\mathcal{X}} \leq \left(\|P_2 u_0\|_{\mathcal{X}} - \frac{F_0}{\Re\mathfrak{e}\, \lambda_{N+1}} \right) \mathrm{e}^{-\Re\mathfrak{e}\, \lambda_{N+1} t} + \frac{F_0}{\Re\mathfrak{e}\, \lambda_{N+1}} . \tag{5.99}$$

Since $u_0 \in \operatorname{graph}(\varphi)$, and Φ is a bounded set of functions,

$$\|P_2 u_0\|_{\mathcal{X}} = \|\varphi(P_1 u_0)\|_{\mathcal{X}} \leq R_1$$

for some R_1 independent of u_0 and φ. Thus, we deduce from (5.99) that

$$\|P_2 u(t)\|_{\mathcal{X}} \leq R_1 + \frac{F_0}{|\Re\mathfrak{e}\, \lambda_{N+1}|} ,$$

that is, $\|P_2 S(t)u_0\|_{\mathcal{X}}$ can be bounded independently of $t \geq 0$ and $\varphi \in \Phi$. This means that S satisfies the uniform boundedness property. □

We remark that (5.99) implies that for all $t \geq 0$,

$$\|P_2 S(t)u_0\|_{\mathcal{X}} \leq \max \left\{ \|P_2 u_0\|_{\mathcal{X}}, \frac{K_0}{\Re\mathfrak{e}\, \lambda_{N+1}} \right\} .$$

2.   We now prove two intermediate results, which we need in order to show the surjectivity property.

PROPOSITION 5.39 (Backward Uniqueness)

Assume S satisfies the cone invariance property with parameter $L \in [L_1, L_2]$. Assume that (5.84) holds for some $N \in \mathbb{N}$, and let u_0, $v_0 \in \mathcal{X}$ and $T > 0$ be such that $P_1 S(T) u_0 = P_1 S(T) v_0$ and $H_L(u_0 - v_0) \le 0$. Then $u_0 = v_0$.

PROOF   Let, as before,

$$u_\Delta(t) := S(t) u_0 - S(t) v_0, \quad F_\Delta(t) := F(S(t) u_0) - F(S(t) v_0).$$

The condition $H_L(u_0 - v_0) \le 0$ implies that $u_0 - v_0 \in \mathcal{C}_L$; then, by the cone invariance property, $u_\Delta(t) \in \mathcal{C}_L$ for all $t \ge 0$, i.e.

$$\|P_2 u_\Delta(t)\|_{\mathcal{X}} \le L \|P_1 u_\Delta(t)\|_{\mathcal{X}}. \tag{5.100}$$

Recalling (5.84), we can therefore estimate (omitting as usual the argument t)

$$\begin{aligned}
\frac{d}{dt} \|P_1 u_\Delta\|_{\mathcal{X}}^2 &= 2 \mathfrak{Re} \, \lambda_N \langle P_1 u_\Delta, F_\Delta - A u_\Delta \rangle \\
&\ge -2 \mathfrak{Re} \, \lambda_N \|P_1 u_\Delta\|_{\mathcal{X}}^2 - 2\ell_F \|u_\Delta\|_{\mathcal{X}} \|P_1 u_\Delta\|_{\mathcal{X}} \\
&\ge -2 \mathfrak{Re} \, \lambda_N \|P_1 u_\Delta\|_{\mathcal{X}}^2 - 2\ell_F \sqrt{1 + L^2} \|P_1 u_\Delta\|_{\mathcal{X}}^2.
\end{aligned}$$

Integrating this inequality on $[t, T]$, $0 \le t < T$, we find

$$\|P_1 u_\Delta(t)\|_{\mathcal{X}} \le e^{\left(\mathfrak{Re} \, \lambda_N + \ell_F \sqrt{1 + L^2} \right)(T - t)} \|P_1 u_\Delta(T)\|_{\mathcal{X}};$$

and since

$$P_1 u_\Delta(T) = P_1(S(T) u_0 - S(T) v_0) = 0,$$

we conclude that $P_1 u_\Delta(t) = 0$ for all $t \in [0, T]$. Together with (5.100), this implies that $u_\Delta(t) \equiv 0$, i.e. that $S(t) u_0 \equiv S(t) v_0$ in $[0, T]$.   □

PROPOSITION 5.40 (Coercivity)

Under the same conditions of proposition 5.39, the semiflow S is COERCIVE, i.e. for all $x \in \mathcal{X}$ and $t \ge 0$,

$$\|P_1 S(t) x\|_{\mathcal{X}} \to +\infty \quad \text{as} \quad \|P_1 x\|_{\mathcal{X}} \to +\infty. \tag{5.101}$$

PROOF   Proceeding as in (5.98), but considering $P_1 u$ instead of $P_2 u$, we obtain the estimate

$$\frac{d}{dt} \|P_1 u\|_{\mathcal{X}}^2 \ge -2 \mathfrak{Re} \, \lambda_N \|P_1 u\|_{\mathcal{X}}^2 - 2F_0 \|P_1 u\|_{\mathcal{X}},$$

from which we deduce that for all $t \geq 0$,

$$\|P_1 u(t)\|_{\mathcal{X}} \geq \|P_1 u_0\|_{\mathcal{X}} e^{-\Re e \lambda_N t} - \frac{F_0}{\Re e \lambda_N}(1 - e^{-\Re e \lambda_N t}).$$

Keeping in mind that $t > 0$ is kept fixed, this inequality implies (5.101). □

3. We can now show that S satisfies the surjectivity property, with respect to all of $\mathcal{G}_L$.

PROPOSITION 5.41 (Surjectivity)
Assume S satisfies the cone invariance property with parameter $L \in [L_1, L_2]$. Then for all $\varphi \in \mathcal{G}_L$ and $t \geq 0$,

$$P_1 S(t) \operatorname{graph}(\varphi) = P_1 \mathcal{X}.$$

PROOF The inclusion $P_1 S(t) \operatorname{graph}(\varphi) \subseteq P_1 \mathcal{X}$ is obvious. For the inverse inclusion, fix $\varphi \in \mathcal{G}_L$ and $t > 0$, and set

$$\Phi := P_1 S(t) \operatorname{graph}(\varphi) \subseteq P_1 \mathcal{X}.$$

We define a function $h \colon \mathcal{X}_1 \to \mathcal{X}_1$ by

$$h(\zeta) := P_1 S(t)(\zeta + \varphi(\zeta)), \qquad \zeta \in \mathcal{X}_1. \tag{5.102}$$

Since φ, $S(t)$ and P_1 are continuous, h is also continuous. We now show that h is also injective. Indeed, let ζ_0 and $\zeta_1 \in \mathcal{X}_1$ be such that $h(\zeta_0) = h(\zeta_1)$, and set $u_0 = \zeta_0 + \varphi(\zeta_0)$, $u_1 = \zeta_1 + \varphi(\zeta_1)$. Then, u_0 and $u_1 \in \operatorname{graph}(\varphi)$. Since $\varphi \in \mathcal{G}_L$,

$$\|P_2(u_0 - u_1)\|_{\mathcal{X}} = \|\varphi(\zeta_0) - \varphi(\zeta_1)\|_{\mathcal{X}} \leq L\|\zeta_0 - \zeta_1\|_{\mathcal{X}} = L\|P_1(u_0 - u_1)\|_{\mathcal{X}};$$

thus, recalling (5.86), $H_L(u_0 - u_1) \leq 0$. By (5.102), the condition $h(\zeta_0) = h(\zeta_1)$ means that $P_1 S(t)u_0 = P_1 S(t)u_1$; hence, by proposition 5.39, $u_0 = u_1$. Therefore, $\zeta_0 = \zeta_1$, which means that h is injective.

We can then define the inverse function $h^{-1} \colon \Phi \to \mathcal{X}_1$. We now show that h^{-1} is also continuous. Arguing by contradiction, assume that there is $\xi \in \Phi$ such that h^{-1} is not continuous at ξ. We can then find a number $\varepsilon_0 > 0$, and a sequence $(\xi_k)_{k \in \mathbb{N}} \subset \Phi$, such that $\xi_k \to \xi$ as $k \to \infty$, but for all $k \in \mathbb{N}$

$$\|h^{-1}(\xi_k) - h^{-1}(\xi)\|_{\mathcal{X}} \geq \varepsilon_0. \tag{5.103}$$

Let $\zeta := h^{-1}(\xi)$, $\zeta_k := h^{-1}(\xi_k)$, $z_k := \zeta_k + \varphi(\zeta_k)$, and $z := \zeta + \varphi(\zeta)$. Then $\xi_k = h(\zeta_k) = P_1 S(t)z_k$, and $\xi = h(\zeta) = P_1 S(t)z$. The coercivity property implies that the sequence $(\zeta_k)_{k \in \mathbb{N}}$ is bounded in $\mathcal{X}_1$: in fact, if otherwise $\|\zeta_k\|_{\mathcal{X}} = \|P_1 z_k\|_{\mathcal{X}} \to +\infty$ as $k \to +\infty$, then by (5.101) also $\|P_1 S(t)z_k\|_{\mathcal{X}} = \|\xi_k\|_{\mathcal{X}} \to +\infty$, contradicting the fact that $\xi_k \to \xi$. Since $P_1 \mathcal{X}$ is finite dimensional, there is a subsequence $(\zeta_{k_m})_{m \in \mathbb{N}}$, converging to some limit $\overline{\zeta} \in P_1 \mathcal{X}$. The continuity of h implies that $h(\zeta_{k_m}) \to h(\overline{\zeta})$; but since $h(\zeta_{k_m}) = \xi_{k_m}$, we conclude that $h(\overline{\zeta}) = \xi$, i.e. that $\overline{\zeta} = h^{-1}(\xi)$. Thus, $h^{-1}(\xi_{k_m}) \to h^{-1}(\xi)$, contradicting (5.103). It follows that h^{-1} is continuous.

By proposition 5.40, h is coercive on $P_1 \mathcal{X}$; since this space is finite dimensional, h is a sequentially continuous mapping. Then, by theorem A.44, h is a homeomorphism from $\mathcal{X}_1$ into itself. This concludes the proof of proposition 5.41. ☐

4.   In conclusion, we have seen that if operator A of the evolution equation (5.78) satisfies the spectral gap condition (5.83), with N sufficiently large so that (5.93) or the stronger inequality (5.92) holds, then the corresponding semiflow S admits an inertial manifold $\mathcal{M} \subseteq \mathcal{X}$, of the form (5.16). Since we also know that S admits a compact, positively invariant absorbing set, we conclude that S admits a compact inertial manifold. We summarize this result in

THEOREM 5.42
Let A be a densely defined operator generating a C^0-semigroup on a separable Hilbert space $\mathcal{X}$. Let $F \in C_b(\mathcal{X}, \mathcal{X})$ satisfy the Lipschitz condition (5.82). Assume A satisfies the spectral gap condition (5.83), with N so large that

$$\mathfrak{Re}\, \lambda_{N+1} > \ell_F. \qquad (5.104)$$

Then the semiflow S generated by the autonomous evolution equation (5.78) admits an inertial manifold $\mathcal{M}$ in $\mathcal{X}$, such that $\dim_F(\mathcal{M}) = \dim \mathcal{X}_1$.

PROOF   By corollary 5.37, S satisfies at least the modified strong squeezing property, with parameters $L \in [L_1, L_2]$ and $\eta = \mathfrak{Re}\, \lambda_{N+1} - \ell_F$. By propositions 5.38 and 5.41, S also satisfies the uniform boundedness and surjectivity properties, with respect to any bounded subset $\Phi \subseteq \mathcal{G}_L$. The existence of an inertial manifold $\mathcal{M}$ follows then from theorem 5.30. Since $\mathcal{M}$ is the graph of a Lipschitz continuous function over $\mathcal{X}_1$, its fractal dimension is the same as that of $\mathcal{X}_1$ (i.e., N). ☐

REMARK 5.43   1. If the stronger condition $\mathfrak{Re}\, \lambda_{N+1} > \ell_F \sqrt{1 + L^{-2}}$ holds, the existence of $\mathcal{M}$ would also follow from theorem 5.21.

2. If the point spectrum of A is such that $\mathfrak{Re}\, \lambda_N \geq 0$ for all $N \in \mathbb{N}$, then condition (5.104) in theorem 5.42 is redundant, since it follows from the spectral gap condition (5.83). Indeed, in this case we have that $\mathfrak{Re}\, \lambda_{N+1} > \mathfrak{Re}\, \lambda_N + 2\ell_F > \ell_F$. ☐

5.7 Applications

In this section we briefly describe how the spectral gap condition can be formulated for parabolic and hyperbolic evolution problems of the type (P) and (H$_\varepsilon$). In the next section, we shall see that the spectral gap condition is actually satisfied for the specific situation when, in these problems, the space dimension is $n = 1$ and, for problem (H_ε), ε is sufficiently small.

5.7.1 Semilinear Heat Equations

We consider a semilinear heat equation of the form

$$u_t - \Delta u = f(x) - g(u), \tag{5.105}$$

in the space $\mathcal{X} := L^2(\Omega)$. As in (3.1), we subject u to homogeneous Dirichlet boundary conditions, but we assume that $f : \mathcal{X} \to \mathcal{X}$ is globally bounded and globally Lipschitz continuous, with Lipschitz constant ℓ. In this case, equation (5.105) is directly in the form (5.78), with $A = -\Delta$, $\text{dom}(A) = H^2(\Omega) \cap H_0^1(\Omega)$, and $F(u) := f - g(u)$. By theorem A.79, A generates an analytic semigroup of operators in $\mathcal{X}$. Moreover, as seen in section 3.2.2, A admits a complete system of eigenfunctions $\{w_j\}_{j \in \mathbb{N}}$, corresponding to all real eigenvalues $0 < \lambda_1 \le \lambda_2 \le \cdots \to +\infty$ (see theorem A.76). Then, each $u \in \mathcal{X}$ admits the uniquely determined Fourier series expansion (3.21), that is,

$$u = \sum_{j=1}^\infty \langle u, w_j \rangle w_j \tag{5.106}$$

(the series converging in $\mathcal{X}$), and Parseval's formula (3.22), i.e.

$$\|u\|_{\mathcal{X}}^2 = \sum_{j=1}^\infty \langle u, w_j \rangle^2, \tag{5.107}$$

holds for all $u \in \mathcal{X}$. Then,

$$\text{dom}(A) = \{ u \in \mathcal{X} : \sum_{j=1}^\infty \lambda_j \langle u, w_j \rangle w_j \quad \text{converges in } \mathcal{X} \}.$$

Let $u \in \text{dom}(A)$. Then, $\langle Au, u \rangle$ is real and nonnegative, and, by (5.106),

$$\langle Au, u \rangle = \left\langle \sum_{j=1}^\infty \langle u, w_j \rangle A w_j, \sum_{k=1}^\infty \langle u, w_k \rangle w_k \right\rangle = \left\langle \sum_{j=1}^\infty \langle u, w_j \rangle \lambda_j w_j, \sum_{k=1}^\infty \langle u, w_k \rangle w_k \right\rangle$$

$$= \sum_{j=0}^\infty \lambda_j \langle u, w_j \rangle^2, \tag{5.108}$$

the last step being a consequence of the orthogonality of the eigenfunctions w_j. As in section 3.2.3, for $N \in \mathbb{N}_{>0}$ we set

$$\mathcal{X}_1 := \text{span}\{w_1, \ldots, w_N\},$$

and define the corresponding projection $P_1 : \mathcal{X} \to \mathcal{X}_1$ by (3.27). Setting as usual $P_2 := I - P_1$, we see that (5.108) implies that A satisfies conditions (5.84) and (5.85).

In fact, if $v := P_1 u$, then $\langle v, w_j \rangle = 0$ for all $j \geq N+1$, so (5.108) and (5.107) imply that

$$\langle Av, v \rangle = \sum_{j=1}^{N} \lambda_j \langle v, w_j \rangle^2 \leq \lambda_N \sum_{j=1}^{N} \langle v, w_j \rangle^2 = \lambda_N \|v\|_{\mathcal{X}}^2 . \tag{5.109}$$

Likewise, if $w := P_2 u$,

$$\langle Aw, w \rangle = \sum_{j=N+1}^{\infty} \lambda_j \langle w, w_j \rangle^2 \geq \lambda_{N+1} \|w\|_{\mathcal{X}}^2 . \tag{5.110}$$

It follows that proposition 5.33 can be applied, and the semiflow S generated by equation (5.105) satisfies the cone invariance property, with parameter L, provided that N is so large that the spectral gap condition

$$\lambda_{N+1} - \lambda_N \geq 2\ell \tag{5.111}$$

holds. As we observed in remark 5.43, condition (5.104) holds, since $\lambda_N \geq 0$ for all $N \in \mathbb{N}_{>0}$. In conclusion, if N is so large that (5.111) holds, the semiflow S generated by the parabolic PDE (5.105) admits an inertial manifold in $\mathcal{X}$. In section 5.8, we shall see that the spectral gap condition (5.111) can certainly be satisfied if the space dimension is $n = 1$.

5.7.2   Semilinear Wave Equations

We consider a semilinear wave equation of the form

$$\varepsilon u_{tt} + u_t - \Delta u = f(x) - g(u), \qquad \varepsilon > 0. \tag{5.112}$$

As in (3.4), we subject u to homogeneous Dirichlet boundary conditions, but, as in (5.105), we assume $g: L^2(\Omega) \to L^2(\Omega)$ to be globally bounded and globally Lipschitz continuous, with Lipschitz constant ℓ. Equation (5.112) is formally equivalent to a first order system of the form (3.14), i.e.

$$U_t + AU = F(U) \tag{5.113}$$

for $U = (u, v) \in \mathcal{X} := H^1(\Omega) \times L^2(\Omega)$, where

$$A = \begin{pmatrix} 0 & -\frac{1}{\varepsilon} \\ -\Delta & \frac{1}{\varepsilon} \end{pmatrix}, \quad F(U) = \begin{pmatrix} 0 \\ f - g(u) \end{pmatrix}, \tag{5.114}$$

and $\mathrm{dom}(A) = \left(H^2(\Omega) \cap H^1_0(\Omega) \right) \times H^1(\Omega)$ (we identify for convenience pairs (u, v) of functions with the corresponding column vector $(u, v)^\top$).

The operator A defined in (5.114) generates a C^0-semigroup in $\mathcal{X}$ (see theorem A.53). However, this semigroup is not analytic, and A is neither self-adjoint nor positive. In particular, conditions (5.84) and (5.85) do not hold. Still, by theorem

3.20 we know that (5.113) generates a semiflow S in $\mathcal{X}$ (for an alternative proof directly concerning system (5.113), see e.g. Sell and You, [SY02, sct. 5.2]).

Our goal is to show that, if ε is sufficiently small, A satisfies the spectral gap condition, and, consequently, S admits an inertial manifold in $\mathcal{X}$. To this end, we will consider in $\mathcal{X}$ an equivalent norm, introduced by Mora in [Mor87].

In order to conform to Mora's presentation, we introduce the time rescaling $t \mapsto \sqrt{\varepsilon t}$; that is, we consider the new unknown

$$w(x,t) := u(x, \sqrt{\varepsilon t}),$$

which solves the equation

$$w_{tt} + \tfrac{1}{\sqrt{\varepsilon}} w_t - \Delta w = f(x) - g(w).$$

Setting then $\alpha := \tfrac{1}{2\sqrt{\varepsilon}}$, and renaming w, we finally consider the equation

$$u_{tt} + 2\alpha u_t - \Delta u = f(x) - g(u).$$

This equation is formally equivalent to system (5.113), where now

$$A = \begin{pmatrix} 0 & -1 \\ -\Delta & 2\alpha \end{pmatrix}. \tag{5.115}$$

Moreover, we can now consider in $\mathcal{X}$ the standard graph norm, i.e.

$$\|(u,v)\|_{\mathcal{X}}^2 = \|\nabla u\|^2 + \|v\|^2, \qquad (u,v) \in \mathcal{X}. \tag{5.116}$$

To determine the spectrum of A, we observe that the eigenvalue equation

$$AU = \mu U, \qquad U = (u,v) \in \mathcal{X},$$

is equivalent to the system

$$\begin{cases} v = -\mu u \\ -\Delta u + 2\alpha v = \mu v. \end{cases} \tag{5.117}$$

Thus, u must satisfy the conditions

$$u \in H_0^1(\Omega), \qquad -\Delta u = (2\alpha\mu - \mu^2)u.$$

It follows that $2\alpha\mu - \mu^2$ must be an eigenvalue of $-\Delta$, with respect to homogeneous Dirichlet boundary conditions; that is, if $(\lambda_j)_{j \in \mathbb{N}}$ denotes the sequence of these eigenvalues, we must have

$$2\alpha\mu - \mu^2 = \lambda_j, \qquad j \geq 1. \tag{5.118}$$

For each $j \geq 1$, this equation has the two complex solutions

$$\mu_j^{\pm} := \alpha \pm \sqrt{\alpha^2 - \lambda_j}; \tag{5.119}$$

thus, A does have a countable set of eigenvalues. Since $\min_{j\geq 1} \lambda_j = \lambda_1 > 0$, we see that if ε is so large that

$$\alpha^2 = \tfrac{1}{4\varepsilon} < \lambda_1,$$

then no eigenvalue of A is real. Moreover, since in this case all eigenvalues have the same real part α, it is clearly impossible to find any decomposition of the point spectrum $\sigma_p(A)$, as in (5.79), such that the spectral gap condition (5.83) holds. Hence, the existence of an inertial manifold for equation (5.112) is, in general, open. Indeed, as we have mentioned, in chapter 7 we present an explicit *non*-existence result for a similar problem, which holds precisely when ε is sufficiently large. On the other hand, when ε is small (i.e., when α is large), some eigenvalues of A will be real, positive and distinct, so that it makes sense to investigate whether the spectral gap condition may hold. In fact, we have the following result.

THEOREM 5.44
Let ℓ be the Lipschitz constant of f. Assume there is $m \in \mathbb{N}$ such that

$$\lambda_{m+1} - \lambda_m > 4\ell, \tag{5.120}$$

and let N be the smallest integer such that (5.120) holds. Assume that α is so large that

$$\alpha^2 > \lambda_{N+1}. \tag{5.121}$$

There is then an equivalent norm in $\mathcal{X}$, such that the operator A defined in (5.115) satisfies the spectral gap condition relative to F. As a consequence, the semiflow S generated by system (5.113) admits an inertial manifold in $\mathcal{X}$, whose fractal dimension does not exceed N.

PROOF   1. We set

$$\mathbb{N}_0 := \{j \in \mathbb{N}: \lambda_j \leq \alpha^2\}, \quad \mathbb{N}_1 := \{j \in \mathbb{N}: \lambda_j > \alpha^2\}.$$

Then, $\mathbb{N}_0$ is not empty, because $\lambda_{N+1} \in \mathbb{N}_0$ by (5.121). The eigenvalues $\mu_j^\pm$ of A are such that $\mu_j^\pm \in \mathbb{R}$ if $j \in \mathbb{N}_0$, while $\mu_j^\pm \in \mathbb{C} \setminus \mathbb{R}$ if $j \in \mathbb{N}_1$; in the latter case,

$$\Re \mu_j^\pm = \alpha. \tag{5.122}$$

We decompose the point spectrum of A as in (5.79), with

$$\sigma_1 := \{\mu_1^-, \ldots, \mu_N^-\};$$

thus, $\Lambda_1 = \mu_N^- = \alpha - \sqrt{\alpha^2 - \lambda_N}$. To determine Λ_2, observe first that if $k \in \mathbb{N}_0$, then $\mu_k^+ \in \mathbb{R}$, and

$$\mu_k^+ = \alpha + \sqrt{\alpha^2 - \lambda_k} \geq \alpha - \sqrt{\alpha^2 - \lambda_{N+1}} = \mu_{N+1}^-; \tag{5.123}$$

also, if $k \in \mathbb{N}_0$ and $k \geq N+2$, then $\mu_k^- \in \mathbb{R}$, and

$$\mu_k^- = \alpha - \sqrt{\alpha^2 - \lambda_k} \geq \alpha - \sqrt{\alpha^2 - \lambda_{N+1}} = \mu_{N+1}^-. \tag{5.124}$$

Together with (5.123) and (5.122), (5.124) implies that $\Lambda_2 = \mu_{N+1}^-$; consequently,

$$\Lambda_2 - \Lambda_1 = \sqrt{\alpha^2 - \lambda_N} - \sqrt{\alpha^2 - \lambda_{N+1}}. \tag{5.125}$$

Corresponding to the eigenvalues (5.119), A has eigenfunctions $U_j^\pm$, which are complex-valued if $j \in \mathbb{N}_1$. By (5.117), these eigenfunctions have the form

$$U_j^\pm = (u_j, -\mu_j^\pm u_j), \quad \text{with } -\Delta u_j = \lambda_j u_j. \tag{5.126}$$

As we immediately verify, the system $(U_j^\pm)_{j \in \mathbb{N}}$ is linearly independent and complete in $\mathcal{X}$; that is,

$$\overline{\operatorname{span}\{U_j^\pm : j \in \mathbb{N}\}} = \mathcal{X}$$

(closure in $\mathcal{X}$). Thus, for $N \in \mathbb{N}_{>0}$ we can set

$$\mathcal{X}_1 := \operatorname{span}\{U_1^-, \ldots, U_N^-\}. \tag{5.127}$$

This subspace of $\mathcal{X}$ is finite dimensional, but is not orthogonal to the subspace

$$\mathcal{X}_2 := \operatorname{span}\{U_1^+, \ldots, U_N^+, U_{N+1}^-, U_{N+1}^+, U_{N+2}^-, U_{N+2}^+, \cdots\} \tag{5.128}$$

spanned by the other eigenfunctions of A. To see this, it is sufficient to note that, because of (5.126), for $1 \leq j \leq N$

$$\langle U_j^-, U_j^+ \rangle_{\mathcal{X}} = \|\nabla u_j\|^2 + \langle -\mu_j^- u_j, -\mu_j^+ \bar{u}_j \rangle = \lambda_j \|u_j\|^2 + \mu_j^- \mu_j^+ \|u_j\|^2$$
$$= 2\lambda_j \neq 0,$$

having recalled (5.116), and that $\|u_j\|^2 = 1$, $\|\nabla u_j\|^2 = \lambda_j$ (by (3.24)), and $\mu_j^- \mu_j^+ = \lambda_j$.

2. To overcome this difficulty, we define a new norm in $\mathcal{X}$, equivalent to the norm (5.116), with the purpose of making $\mathcal{X}_1$ and $\mathcal{X}_2$ orthogonal with respect to this new norm. Then, we will see that, as a consequence of (5.120) and (5.121), the operator A satisfies the spectral gap condition relative to F, expressed in term of the new norm.

Adapting Mora's presentation of [Mor87], we further decompose σ_2 into the sets

$$\sigma_{21} := \{\mu_1^+, \ldots, \mu_N^+\}, \quad \sigma_{2\infty} := (\mu_j^\pm)_{j \geq N+1}$$

and define corresponding subspaces of $\mathcal{X}$

$$\mathcal{X}_{21} := \operatorname{span}\{U_1^+, \ldots, U_N^+\}, \quad \mathcal{X}_{2\infty} := \operatorname{span}\{U_j^\pm, j \geq N+1\};$$

we also set

$$\mathcal{X}_{11} := \operatorname{span}\{U_j^\pm : j \leq N\} = \mathcal{X}_1 \oplus \mathcal{X}_{21}.$$

Let $U = (u,v) \in \mathcal{X}$. We immediately note that if $U \in \mathcal{X}_1$, or $U \in \mathcal{X}_{21}$, then, as a consequence of (5.126),

$$
\langle v, u_j \rangle = \begin{cases} -\mu_j^- \langle u, u_j \rangle & \text{if } U \in \mathcal{X}_1, \\ -\mu_j^+ \langle u, u_j \rangle & \text{if } U \in \mathcal{X}_{21}, \end{cases}
$$

for $1 \leq j \leq N$. On the other hand, if $U \in \mathcal{X}_{11}$ or $U \in \mathcal{X}_{2\infty}$, there is no special relations between the Fourier coefficients of the components of U, and we have the standard decompositions

$$
\begin{cases} U = \sum_{j=1}^N (\langle u, u_j \rangle u_j, \langle v, u_j \rangle u_j) & \text{if } U \in \mathcal{X}_{11}, \\ U = \sum_{j=N+1}^\infty (\langle u, u_j \rangle u_j, \langle v, u_j \rangle u_j) & \text{if } U \in \mathcal{X}_{2\infty}. \end{cases}
$$

To see this, let e.g. $U \in \mathcal{X}_{2\infty}$. Then

$$
U = \sum_{j=N+1}^\infty (\xi_j^- U_j^- + \xi_j^+ U_j^+), \tag{5.129}
$$

for suitable sequences $(\xi_j^\pm)_{j \geq N+1} \subset \mathbb{C}$ (note that the second component of $U_j^\pm$ is complex-valued, because $\mu_j^\pm \notin \mathbb{R}$ if $j \in \mathbb{N}_1$). Recalling (5.126), (5.129) can be written as

$$
U = \sum_{j=N+1}^\infty ((\xi_j^- + \xi_j^+) u_j, (\xi_j^- \mu_j^- + \xi_j^+ \mu_j^+) u_j);
$$

thus, we deduce that the Fourier coefficients of u and v must satisfy the system

$$
\begin{cases} \langle u, u_j \rangle = \xi_j^- + \xi_j^+, \\ \langle v, u_j \rangle = -(\mu_j^- \xi_j^- + \mu_j^+ \xi_j^+). \end{cases} \tag{5.130}
$$

Since the determinant of the matrix

$$
\begin{pmatrix} 1 & 1 \\ \mu_j^- & \mu_j^+ \end{pmatrix}
$$

equals $\mu_j^+ - \mu_j^- = 2\sqrt{\alpha^2 - \lambda_j} \neq 0$, system (5.130) uniquely determines each pair (ξ_j^-, ξ_j^+), and vice versa.

Finally, we remark that, in analogy with (5.109) and (5.110), we have the implications

$$
(u,v) \in \mathcal{X}_{11} \implies \|\nabla u\|^2 \leq \lambda_N \|u\|^2, \tag{5.131}
$$

$$
(u,v) \in \mathcal{X}_{2\infty} \implies \|\nabla u\|^2 \geq \lambda_{N+1} \|u\|^2, \tag{5.132}
$$

which follow from the Fourier series expansion of u in $L^2(\Omega)$.

3. We now define a function $\Phi\colon \mathcal{X}_{11} \times \mathcal{X}_{11} \to \mathbb{R}$, by

$$\Phi(U,V) := \alpha^2 \langle u, \bar{y} \rangle - \langle \nabla u, \nabla \bar{y} \rangle + \langle \alpha u + v, \alpha \bar{y} + \bar{z} \rangle,$$

for $U = (u,v)$ and $V = (y,z) \in \mathcal{X}_{11}$; the bar denotes complex conjugation. Because of (5.131), and recalling that

$$\lambda_N < \lambda_{N+1} < \alpha^2 = \tfrac{1}{4\varepsilon},$$

we have that for all $U \in \mathcal{X}_{11}$,

$$\Phi(U,U) = \alpha^2 \|u\|^2 - \|\nabla u\|^2 + \|\alpha u + v\|^2 \geq (\alpha^2 - \lambda_N)\|u\|^2 \geq 0. \qquad (5.133)$$

Thus, Φ is positive definite, and therefore defines a scalar product on $\mathcal{X}_{11}$; we denote by $\| \cdot \|_{11}$ the corresponding norm. We also denote by $\| \cdot \|_{2\infty}$ the norm induced in the orthogonal space $\mathcal{X}_{2\infty}$ by the standard norm (5.116); an analogous notation denotes the corresponding scalar product in $\mathcal{X}_{2\infty}$.

We are now ready to redefine the scalar product and the norm in $\mathcal{X}$; to avoid introducing more symbols than necessary, for the remainder of this section we agree to keep the notations $\langle \cdot, \cdot \rangle_{\mathcal{X}}$ and $\| \cdot \|_{\mathcal{X}}$ to denote these new scalar product and norm.

We fix $R > 0$ such that

$$R \geq \frac{\alpha^2}{\lambda_{N+1}}, \qquad (5.134)$$

and define the new scalar product in $\mathcal{X}$ by

$$\langle U,V \rangle_{\mathcal{X}} := \Phi(P_{11}U, P_{11}V) + R \langle P_{2\infty}U, P_{2\infty}V \rangle_{2\infty}, \qquad (5.135)$$

where P_{11} and $P_{2\infty}$ are the projections from $\mathcal{X}$ into, respectively, $\mathcal{X}_{11}$ and $\mathcal{X}_{2\infty}$, defined by means of the Fourier series expansion of $U \in \mathcal{X}$ (for instance, $P_{2\infty}U$ is defined by (5.129)). With some abuse of notation, we shall abbreviate (5.135) by

$$\langle U,V \rangle_{\mathcal{X}} := \Phi(U,V) + R \langle U,V \rangle_{2\infty}. \qquad (5.136)$$

It is then easy to see that, as a consequence of (5.131), (5.136) does define an equivalent norm in $\mathcal{X}$. We now show that the spaces $\mathcal{X}_1$ and $\mathcal{X}_2$, defined in (5.127) and (5.128), are now orthogonal with respect to this new norm.

PROPOSITION 5.45

Let $\mathcal{X}$ be endowed with the scalar product (5.136). Let $j, k \geq 1$. Then:

$$\langle U_j^{\pm}, U_k^{\pm} \rangle_{\mathcal{X}} = \langle U_j^{\pm}, U_k^{\mp} \rangle_{\mathcal{X}} = 0 \quad \text{if } j \neq k; \qquad (5.137)$$

$$\|U_j^{\pm}\|_{11}^2 = 2(\alpha^2 - \lambda_j) \quad \text{if } 1 \leq j \leq N; \qquad (5.138)$$

$$\|U_j^{\pm}\|_{2\infty}^2 = \begin{cases} 2\alpha \left(\alpha \pm \sqrt{\alpha^2 - \lambda_{N+1}} \right) & \text{if } j \geq N+1, \; j \in \mathbb{N}_0, \\ 2\lambda_j & \text{if } j \geq N+1, \; j \in \mathbb{N}_1. \end{cases} \qquad (5.139)$$

In particular, $\mathcal{X}_1 \perp \mathcal{X}_{21}$ and $\mathcal{X}_{21} \perp \mathcal{X}_{2\infty}$. Therefore, since $\mathcal{X}_2 = \mathcal{X}_{21} \oplus \mathcal{X}_{2\infty}$,

$$\mathcal{X}_1 \perp \mathcal{X}_2. \tag{5.140}$$

PROOF   All the identities in (5.137) are an immediate consequence of (5.126) and the orthogonality of the system $(u_j)_{j \geq 1}$, both in $L^2(\Omega)$ and $H_0^1(\Omega)$. In particular, this implies the orthogonality of $\mathcal{X}_{11}$ and $\mathcal{X}_{2\infty}$. To see that $\mathcal{X}_1 \perp \mathcal{X}_{21}$, we still have to show that, in addition to (5.137) for $1 \leq j, k \leq N$, also

$$\langle U_j^-, U_j^+ \rangle_{\mathcal{X}} = 0, \quad 1 \leq j \leq N. \tag{5.141}$$

To see this, recalling (5.133) and (5.126) we compute

$$\begin{aligned}
\langle U_j^-, U_j^+ \rangle_{\mathcal{X}} &= \Phi(U_j^-, U_j^+) \\
&= \alpha^2 \|u_j\|^2 - \|\nabla u_j\|^2 + \langle (\alpha - \mu_j^-)u_j, (\alpha - \mu_j^+)\bar{u}_j \rangle \\
&= 2\alpha^2 \|u_j\|^2 - \lambda_j \|u_j\|^2 - \alpha(\mu_j^+ + \mu_j^-)\|u_j\|^2 + \mu_j^- \mu_j^+ \|u_j\|^2.
\end{aligned}$$

By (5.118), $\mu_j^+ + \mu_j^- = 2\alpha$, and $\mu_j^- \mu_j^+ = \lambda_j$; hence, (5.141) follows.
   Identity (5.138) is a consequence of (5.133), by which we have

$$\begin{aligned}
\|U_j^\pm\|_{11}^2 &= \Phi\left(U_j^\pm, U_j^\pm\right) = \alpha^2 \|u_j\|^2 - \|\nabla u_j\|^2 + \|(\alpha - \mu_j^\pm)u_j\|^2 \\
&= (\alpha^2 - \lambda_j)\|u_j\|^2 + \left|\mp\sqrt{\alpha^2 - \lambda_j}\right|^2 \|u_j\|^2,
\end{aligned}$$

recalling that $\|u_j\|^2 = 1$. As for (5.139), acting likewise we find that

$$\|U_j^\pm\|_{2\infty}^2 = \left(\lambda_j + |\mu_j^\pm|^2\right) \|u_j\|^2.$$

If $j \in \mathbb{N}_0$, $\mu_j^\pm = \alpha \pm \sqrt{\alpha^2 - \lambda_j} \in \mathbb{R}$, and

$$\lambda_j + |\mu_j^\pm|^2 = \lambda_j + \alpha^2 + \alpha^2 - \lambda_j \pm 2\alpha\sqrt{\alpha^2 - \lambda_j} = 2\alpha\left(\alpha \pm \sqrt{\alpha^2 - \lambda_j}\right),$$

while if $j \in \mathbb{N}_1$, $\mu_j^\pm = \alpha \pm i\sqrt{\lambda_j - \alpha^2} \in \mathbb{C} \setminus \mathbb{R}$, and

$$\lambda_j + |\mu_j^\pm|^2 = \lambda_j + \alpha^2 + \lambda_j - \alpha^2 = 2\lambda_j.$$

This concludes the proof of proposition 5.45.   □

4. Having thus established the desired orthogonal decomposition (5.140), we call $P_1 : \mathcal{X} \to \mathcal{X}_1$ and $P_2 : \mathcal{X} \to \mathcal{X}_2$ the corresponding orthogonal projections, and proceed to show that A, P_1 and P_2 satisfy inequalities (5.84) and (5.85). We actually show that

$$\forall U \in \mathcal{X}_1: \quad \langle AU, U \rangle_{\mathcal{X}} = \langle U, AU \rangle_{\mathcal{X}} \leq \mu_N^- \|U\|_{\mathcal{X}}^2, \tag{5.142}$$

$$\forall U \in \mathcal{X}_2: \quad \mathfrak{Re}\langle AU, U \rangle_\mathcal{X} = \mathfrak{Re}\langle U, AU \rangle_\mathcal{X} \geq \mu_{N+1}^- \|U\|_\mathcal{X}^2. \tag{5.143}$$

In fact, these inequalities are immediate. To show (5.142), let $U \in \mathcal{X}_1$. Then,

$$U = \sum_{j=1}^N \gamma_j U_j^-, \quad \gamma_j := \langle u, u_j \rangle;$$

thus,

$$AU = \sum_{j=1}^N \gamma_j AU_j^- = \sum_{j=1}^N \gamma_j \mu_j^- U_j^-,$$

and, therefore, by (5.137)

$$\langle AU, U \rangle_\mathcal{X} = \Phi(AU, U) = \Phi\left(\sum_{j=1}^N \gamma_j \mu_j^- U_j^-, \sum_{k=1}^N \gamma_k U_k^-\right)$$

$$= \sum_{i=1}^N \gamma_i^2 \mu_i^- \|U_i^-\|_{11}^2 = \Phi(U, AU) = \langle U, AU \rangle_\mathcal{X}.$$

This shows the first half of (5.142). The second half follows from the inequality $\mu_j^- \leq \mu_N^-$, $1 \leq j \leq N$, and recalling that if $U \in \mathcal{X}_1$, again by (5.137)

$$\|U\|_\mathcal{X}^2 = \|U\|_{11}^2 = \Phi\left(\sum_{j=1}^N \gamma_j U_j^-, \sum_{k=1}^N \gamma_k U_k^-\right) = \sum_{i=1}^N \gamma_i^2 \|U_i^-\|_{11}^2. \tag{5.144}$$

To show (5.143), let $U \in \mathcal{X}_2$. Then,

$$U = \sum_{j=1}^N \gamma_j U_j^+ + \sum_{j=N+1}^\infty (\xi_j^- U_j^- + \xi_j^+ U_j^+) =: U_{21} + U_{2\infty},$$

with $\gamma_j := \langle u, u_j \rangle \in \mathbb{R}$ and $\xi_j^\pm \in \mathbb{C}$, as in (5.129). Again by (5.137),

$$\langle AU, U \rangle_\mathcal{X} = \Phi(AU_{21}, U_{21}) + \langle AU_{2\infty}, U_{2\infty} \rangle_{2\infty}$$

$$= \sum_{i=1}^N \gamma_i^2 \mu_i^+ \|U_i^+\|_{11}^2 + \sum_{i=N+1}^\infty (|\gamma_i^-|^2 \mu_i^- \|U_i^-\|_{2\infty}^2 + |\gamma_i^+|^2 \mu_i^+ \|U_i^+\|_{2\infty}^2),$$

$$\langle U, AU \rangle_\mathcal{X} = \Phi(U_{21}, AU_{21}) + \langle U_{2\infty}, AU_{2\infty} \rangle_{2\infty}$$

$$= \sum_{i=1}^N \gamma_i^2 \mu_i^+ \|U_i^+\|_{11}^2 + \sum_{i=N+1}^\infty (|\gamma_i^-|^2 \overline{\mu_i^-} \|U_i^-\|_{2\infty}^2 + |\gamma_i^+|^2 \overline{\mu_i^+} \|U_i^+\|_{2\infty}^2).$$

Thus,

$$\mathfrak{Re}\langle AU, U \rangle_\mathcal{X} = \mathfrak{Re}\langle U, AU \rangle_\mathcal{X}.$$

The second half of (5.143) follows then from (5.122), and from the obvious inequalities

$$\mu_j^+ = \alpha + \sqrt{\alpha^2 - \lambda_j} > \alpha - \sqrt{\alpha^2 - \lambda_{N+1}} \quad \text{if } j \in \mathbb{N}_0, j \leq N,$$

$$\mu_j^- = \alpha - \sqrt{\alpha^2 - \lambda_j} \geq \alpha - \sqrt{\alpha^2 - \lambda_{N+1}} \quad \text{if } j \in \mathbb{N}_0, j \geq N+1.$$

Indeed, these inequalities imply that

$$\Re\langle AU, U\rangle_{\mathcal{X}} \geq \mu_{N+1}^- \left(\sum_{i=1}^{N} \gamma_i^2 \|U_i^+\|_{11}^2 + \sum_{i=N+1}^{\infty} \left(|\gamma_i^-|^2 \|U_i^-\|_{2\infty}^2 + |\gamma_i^+|^2 \|U_i^+\|_{2\infty}^2 \right) \right)$$

$$= \mu_{N+1}^- \|U\|_{\mathcal{X}}^2 ,$$

the last identity being established as in (5.144).

5. Our last step is to estimate the Lipschitz constant of F in (5.113); recall that $F(U) := (0, f - g(u))$. To this end, we first remark that the orthogonal projections P_{11} and $P_{2\infty}$ introduced in (5.135) induce corresponding orthogonal projections p_{11} and $p_{2\infty}$ in $H_0^1(\Omega)$ and $L^2(\Omega)$, naturally defined as follows. If $U = (u,v) \in \mathcal{X}$ and $U_{ij} := (u_{ij}, v_{ij}) = P_{ij}U$, $ij = 11$ or $ij = 2\infty$, then

$$p_{ij}u := u_{ij}, \quad p_{ij}v := v_{ij}.$$

Thus, from (5.133) and (5.132) it follows that, for any $U = (u,v) \in \mathcal{X}$,

$$\|U\|_{\mathcal{X}}^2 = \|P_{11}U\|_{11}^2 + R\|P_{2\infty}U\|_{2\infty}^2$$
$$\geq (\alpha^2 - \lambda_N)\|p_{11}u\|^2 + R\|\nabla p_{2\infty}u\|^2 + \|p_{2\infty}v\|^2$$
$$\geq (\alpha^2 - \lambda_N)\|p_{11}u\|^2 + R\lambda_{N+1}\|p_{2\infty}u\|^2 . \tag{5.145}$$

From (5.134) we deduce that

$$R\lambda_{N+1} \geq \alpha^2 - \lambda_N ;$$

hence, we obtain from (5.145) that

$$\|U\|_{\mathcal{X}}^2 \geq (\alpha^2 - \lambda_N)\left(\|p_{11}u\|^2 + \|p_{2\infty}u\|^2\right) = (\alpha^2 - \lambda_N)\|u\|^2 . \tag{5.146}$$

Given then $U = (u,v)$ and $V = (y,z) \in \mathcal{X}$, we compute

$$\|F(U) - F(V)\|_{\mathcal{X}}^2 = \|P_{11}(F(U) - F(V))\|_{\mathcal{X}}^2 + \|P_{2\infty}(F(U) - F(V))\|_{\mathcal{X}}^2$$
$$= \|p_{11}(f(u) - f(v))\|^2 + \|p_{2\infty}(f(u) - f(v))\|^2$$
$$= \|f(u) - f(v)\|^2 \leq \ell^2 \|u - v\|^2$$
$$\leq \frac{\ell^2}{\alpha^2 - \lambda_N}\|U - V\|_{\mathcal{X}}^2 ,$$

the last step following from (5.146). Thus,

$$\ell_F \leq \frac{\ell}{\sqrt{\alpha^2 - \lambda_N}}. \tag{5.147}$$

6. We can now conclude the proof of theorem 5.44. From (5.125) and (5.147) we deduce that the operator A defined by (5.115) satisfies the spectral gap condition (5.83), relative to F, if there exists $N \in \mathbb{N}$ such that $\lambda_{N+1} < \alpha^2$ and

$$\sqrt{\alpha^2 - \lambda_N} - \sqrt{\alpha^2 - \lambda_{N+1}} > \frac{2\ell}{\sqrt{\alpha^2 - \lambda_N}}. \tag{5.148}$$

Multiplying by $\sqrt{\alpha^2 - \lambda_N}$, and then squaring, we see that (5.148) is equivalent to the inequality

$$(\alpha^2 - \lambda_N)(\lambda_{N+1} - \lambda_N - 4\ell) + 4\ell^2 > 0,$$

which in turn is implied by (5.120) (with the lowest value $m = N$). As observed in remark 5.43, condition (5.104) also holds, since $\mathfrak{Re}\,\mu_N \geq 0$ for all $N \in \mathbb{N}_{>0}$. Thus, theorem 5.42 can be applied, and theorem 5.44 follows. $\Box$

In conclusion, let N be the lowest possible value of the integers m such that condition (5.120) holds. Then, if α is so large that (5.121) holds, then the semiflow S generated by the semilinear damped wave equation (5.112) admits an inertial manifold in $\mathcal{X}$. In section 5.8.5 we shall see that conditions (5.120) and (5.121) can certainly be satisfied if, again, the dimension of space is $n = 1$.

5.8 Semilinear Evolution Equations in One Space Dimension

In this section we show that, when the dimension of space is $n = 1$, theorem 5.42 can be applied to problems (P) and (H$_\varepsilon$) (the latter, at least if ε is sufficiently small). As a consequence, in these cases we can deduce the existence of an inertial manifold for the corresponding semiflows. In the parabolic problem, the estimates we obtain will also allow us to slightly improve our results of section 3.3.1, concerning the absorbing sets of the related semiflow.

5.8.1 The Parabolic Problem

We consider the Chafee-Infante reaction-diffusion equation (3.1) in one space dimension, i.e. the IBVP on $]0, +\infty[\times]0, \pi[$

$$\begin{cases} u_t - u_{xx} + k(u^3 - u) = f(x) \\ u(0, x) = u_0(x) \\ u(t, 0) = u(t, \pi) = 0, \end{cases} \tag{5.149}$$

where $k > 0$. From the results of chapter 3, we know that problem (5.149) generates a semiflow S, both in $\mathcal{X} = L^2(0, \pi)$ and in $\mathcal{V} = H_0^1(0, \pi)$. We also know that S admits a bounded, positively invariant absorbing set in both these spaces. In particular, since

$H_0^1(0,\pi) \hookrightarrow C([0,\pi])$, S has a bounded absorbing set in $L^\infty(0,\pi)$. S also admits a compact attractor $\mathcal{A} \subseteq \mathcal{X}$, which is contained and bounded in $\mathcal{V}$. When $f \equiv 0$, the structure of this attractor is known in large detail, at least when k is not the square of an integer; indeed, the following results hold (see e.g. Jolly, [Jol89]).

THEOREM 5.46

Assume that $m^2 < k < (m+1)^2$, $m \in \mathbb{N}$. Then, there are m pairs $(\varphi_j^+, \varphi_j^-)$, $j = 0,\dots,m-1$, of nontrivial equilibria for (5.149), having the following properties:

1. *Each φ_j^+ and φ_j^-, $j = 0,\dots,m-1$, is an hyperbolic equilibrium (in the sense of definition 2.21), is connected to the origin (i.e., to the zero solution) by a heteroclinic orbit (see definition 2.24), and has exactly j simple zeroes in $]0,\pi[$.*

2. *If $0 \le j \le i \le m-1$, there also exist connecting heteroclinic orbits from $\varphi_j^\pm$ to $\varphi_i^\pm$.*

3. *The global attractor is the union of the unstable manifolds of these equilibria.*

The result on the zeroes of the equilibria $\varphi_j^\pm$ is shown in Chafee-Infante, [CI74]. A proof of the other results can be found e.g. in Henry, [Hen85], where additional results on the dimension of the global attractor are also given. The existence of an inertial manifold for problem (5.149) is also known: see e.g. Jolly, [Jol89], where an inertial manifold of the form (5.16) is constructed, resorting to a local extension of the unstable manifolds along suitable stable directions. Consequently, S also has an exponential attractor $\mathcal{E}$, which is obtained as the intersection of the inertial manifold and a compact absorbing, positively invariant ball $\mathcal{B}$ (whose existence we mentioned above).

In the sequel, we will assume that $f \equiv 0$ for simplicity, and show how the inertial manifold for S can be obtained as an application of theorem 5.42. Thus, we have to show that the semiflow S satisfies the spectral gap condition (5.83), with respect to the nonlinearity $F(u) := k(u - u^3)$. This requires that we overcome the problem that, as we have already remarked, F is *not* Lipschitz continuous from $\mathcal{X}$ into $\mathcal{X}$, but only from bounded sets of $\mathcal{X}_1$ into $\mathcal{X}$, as seen in proposition 3.15. (Actually, the proof of that proposition shows that F is Lipschitz continuous from bounded sets of $L^\infty(0,\pi)$ into $\mathcal{X}$.) We shall take care of this problem by suitably "adjusting" F outside a fixed absorbing set $\mathcal{B}$ in $L^\infty(0,\pi)$, so as to make F globally bounded and Lipschitz continuous. Since both the attractor and the inertial manifold are contained in $\mathcal{B}$, and $\mathcal{B}$ is absorbing, this adjustment of F will not affect the long-time dynamics of the system.

5.8.2   Absorbing Sets

In this section we prove the existence of bounded, positively invariant absorbing sets for S in the space $L^\infty(0,\pi)$, without resorting to the imbedding of $\mathcal{V}$ into this space. Rather, as in Eden, Foias, Nicolaenko and Temam, [EFNT94], we establish estimates on the norms $\|u(\cdot,t)\|_{L^{2p}(0,\pi)}$, $1 \le p \le \infty$, which allow us not only to deduce the existence of this absorbing set, but also to obtain a specific control on the Lipschitz constant ℓ of the nonlinearity F. We will use this new result to show that, when we introduce the stated modification of F, the modified equation satisfies the spectral gap condition (5.83).

In the sequel, we denote as usual by $|\cdot|_p$ the norm in $L^p(0,\pi)$, for $1 \le p \le +\infty$, and, for $a > 0$, we set

$$\mathcal{B}_p(a) := \{u \in H_0^1(0,\pi) : |u|_p \le a\}.$$

We claim:

LEMMA 5.47
For all $a > 0$,

$$\mathcal{B}_\infty(a) = \bigcap_{p \ge 1} \mathcal{B}_{2p}(\pi^{1/2p}a).$$

PROOF The inclusion

$$\mathcal{B}_\infty(a) \subseteq \bigcap_{p \ge 1} \mathcal{B}_{2p}(\pi^{1/2p}a)$$

follows from the estimate

$$|u|_{2p} \le \pi^{1/2p}|u|_\infty.$$

As for the opposite inclusion, since

$$|u|_\infty = \lim_{p \to \infty} |u|_{2p}, \quad \lim_{p \to \infty} \pi^{1/2p} = 1, \tag{5.150}$$

given arbitrary $\varepsilon > 0$ there is $p_0 \ge 1$ such that for all $p \ge p_0$,

$$|u|_\infty \le |u|_{2p} + \varepsilon, \quad \pi^{1/2p}a \le a + \varepsilon.$$

Consequently, for $u \in \mathcal{B}_{2p}(\pi^{1/2p}a)$, $p \ge p_0$,

$$|u|_\infty \le \pi^{1/2p}a + \varepsilon \le a + 2\varepsilon.$$

Letting $\varepsilon \to 0$, we deduce that $|u|_\infty \le a$, that is, $u \in \mathcal{B}_\infty(a)$. ☐

We proceed then to prove

PROPOSITION 5.48
If $a \ge \pi^{1/2p}$, $1 \le p < +\infty$, $\mathcal{B}_{2p}(a)$ is positively invariant with respect to S. The same is true for $\mathcal{B}_\infty(a)$ if $a \ge 1$.

PROOF Let u be a solution of (5.149) and, for $1 \le r \le \infty$, set

$$v_r(t) := |u(t, \cdot)|_r.$$

Our first step is an estimate of $v_{2p}(t)$; for convenience, we sometimes omit the dependence of the functions from some or all of their variables. For each $p \in [1, +\infty]$, we compute

$$\frac{1}{2}\frac{d}{dt}\left(v_{2p}^{2p}(t)\right) = \frac{1}{2}\frac{d}{dt}\int_0^\pi |u(t,x)|^{2p}\,dx = p\int_0^\pi u^{2p-1}u_t\,dx$$

$$= p\int_0^\pi u^{2p-1}(k(u-u^3)+u_{xx})\,dx$$

$$= pk\int_0^\pi u^{2p}\,dx - pk\int_0^\pi u^{2p+2}\,dx + p\int_0^\pi u^{2p-1}u_{xx}\,dx$$

$$= pk(v_{2p})^{2p} - pk(v_{2p+2})^{2p+2} + p[u^{2p-1}u_x]_{x=0}^{x=\pi}$$

$$- p(2p-1)\int_0^\pi u^{2p-2}(u_x)^2\,dx.$$

Because of the boundary conditions $u(t,0) = u(t,\pi) = 0$, the terms within the square brackets equal zero; hence, neglecting the last term, which is negative, we obtain the inequality

$$\frac{d}{dt}\left((v_{2p})^{2p}\right) \le 2pk(v_{2p})^{2p} - 2pk(v_{2p+2})^{2p+2}. \tag{5.151}$$

By Hölder's inequality,

$$(v_{2p}(t))^{2p} = \int_0^\pi |u(t,x)|^{2p}\,dx \le \pi^{1/(p+1)}\left(\int_0^\pi |u(t,x)|^{2p+2}\,dx\right)^{p/(p+1)}$$

$$= \pi^{1/(p+1)}(v_{2p+2}(t))^{2p};$$

thus,

$$v_{2p+2}(t) \ge \pi^{-1/2p(p+1)}v_{2p}(t),$$

and from (5.151) we obtain the estimate

$$\frac{d}{dt}\left((v_{2p})^{2p}\right) \le 2pk(v_{2p})^{2p} - 2pk\pi^{-1/p}(v_{2p})^{2p+2}. \tag{5.152}$$

We can now show that the set $\mathcal{B}_{2p}(a)$ is positively invariant if $a \ge \pi^{1/2p}$. Proceeding by contradiction, assume that for some $g \in \mathcal{B}_{2p}(a)$ there is $t_1 > 0$ such that $u(t_1, \cdot) = S(t_1)g \notin \mathcal{B}_{2p}(a)$. From (5.152), written as

$$\frac{d}{dt}\left((v_{2p})^{2p}\right) \le 2pk(v_{2p})^{2p}\left(1 - \pi^{-1/p}(v_{2p})^2\right),$$

we deduce that v_{2p} is decreasing on any interval where $v_{2p}(t) \ge \pi^{1/2p}$. Since

$$v_{2p}(t_1) = |u(t_1, \cdot)|_{2p} > a, \quad v_{2p}(0) = |u(0, \cdot)|_{2p} = |g|_{2p} \le a,$$

there is an interval $[t_1 - r, t_1 + r]$ such that $v_{2p}(t) \ge a$ for $|t - t_1| \le r$, and $v_{2p}(t_1 - r) = a$. But since $a \ge \pi^{1/2p}$, this implies the contradiction

$$a < v_{2p}(t_1) \le v_{2p}(t_1 - r) = a.$$

Hence, each set $\mathcal{B}_{2p}(a)$, with $a \geq \pi^{1/2p}$, is positively invariant as claimed. We can then conclude. Since $\pi^{1/2p}a \geq \pi^{1/2p}$ if $a \geq 1$, each set $\mathcal{B}_{2p}(\pi^{1/2p}a)$ is also invariant. As a consequence of lemma 5.47, the set $\mathcal{B}_\infty(a)$, being the intersection of positive invariant sets, is positively invariant if $a \geq 1$.   ☐

Finally, we have the following result.

PROPOSITION 5.49
If $a > 1$, $\mathcal{B}_\infty(a)$ is absorbing with respect to S.

PROOF   We rewrite (5.152) as

$$\frac{\mathrm{d}}{\mathrm{d}t}\left((v_{2p})^{2p}\right) \leq 2pk(v_{2p})^{2p} - 2pk\pi^{-1/p}\left((v_{2p})^{2p}\right)^{1+1/p}.$$

This is an inequality of the form

$$y' \leq \alpha y - \beta y^\gamma, \tag{5.153}$$

with $\alpha = 2pk$, $\beta = 2pk\pi^{-1/p}$, and $\gamma = 1 + \frac{1}{p} > 1$. From (5.153) we deduce

$$y^{-\gamma}y' = \frac{1}{1-\gamma}\frac{\mathrm{d}}{\mathrm{d}t}y^{1-\gamma} \leq \alpha y^{1-\gamma} - \beta,$$

$$\frac{\mathrm{d}}{\mathrm{d}t}y^{1-\gamma} \geq \alpha(1-\gamma)y^{1-\gamma} - \beta(1-\gamma);$$

thus, the function $w := y^{1-\gamma}$ satisfies the inequality

$$w' + \alpha(\gamma-1)w \geq \beta(\gamma-1).$$

Consequently, for $t > 0$,

$$w(t) \geq w(0)e^{\alpha(1-\gamma)t} + \frac{\beta}{\alpha}\left(1 - e^{\alpha(1-\gamma)t}\right);$$

from this, recalling that $1 - \gamma < 0$, we deduce that

$$y(t) \leq \left(y(0)^{1-\gamma}e^{\alpha(1-\gamma)t} + \frac{\beta}{\alpha}(1 - e^{\alpha(1-\gamma)t})\right)^{1/(1-\gamma)}.$$

Since $1 - \gamma = -\frac{1}{p}$, $\alpha(1-\gamma) = -2k$, and $\frac{\beta}{\alpha} = \pi^{-1/p}$, we conclude that

$$v_{2p}(t) \leq \left((v_{2p}(0))^{-2}e^{-2kt} + \pi^{-1/p}\left(1 - e^{-2kt}\right)\right)^{-1/2}$$

$$= \left(\pi^{-1/p} + e^{-2kt}\left((v_{2p}(0))^{-2} - \pi^{-1/p}\right)\right)^{-1/2}.$$

Thus, if $u(0, \cdot) = u_0$,

$$|u(t, \cdot)|_{2p} \leq \left(\pi^{-1/p} + e^{-2kt} \left(|u_0|_{2p}^{-2} - \pi^{-1/p} \right) \right)^{-1/2} =: b_{2p}(t). \qquad (5.154)$$

We have to prove that if $|u_0|_\infty > a$, there is $T_a > 0$ such that $|u(t, \cdot)|_\infty \leq a$ for all $t \geq T_a$. To obtain this, we first show that there is $p_0 \geq 1$ such that for all $p \geq p_0$ and $t \geq 0$,

$$b_{2p}(t) \geq \pi^{1/2p}. \qquad (5.155)$$

Indeed, it is immediate to see that, for each $t \geq 0$, $b_{2p}(t) \geq \pi^{1/2p}$ if and only if $|u_0|_{2p} \geq \pi^{1/2p}$. But by (5.150), recalling that $a > 1$ we deduce that there is $p_0 \geq 1$ such that for all $p \geq p_0$,

$$|u_0|_{2p} \geq a \geq \pi^{1/2p};$$

hence, (5.155) holds. Consequently, by the first part of proposition 5.48, the set $\mathcal{B}_{2p}(b_{2p}(t))$ is positively invariant. Now, (5.150) also implies that

$$\lim_{p \to \infty} b_{2p}(t) = \left(1 + e^{-2kt} \left(|u_0|_\infty^{-2} - 1 \right) \right)^{-1/2} =: b_\infty(t),$$

pointwise in t. As in the second part of the proof of lemma 5.47, we can then show that for all $t \geq 0$,

$$\bigcap_{p \geq 1} \mathcal{B}_{2p}(b_{2p}(t)) \subseteq \mathcal{B}_\infty(b_\infty(t)). \qquad (5.156)$$

Consequently, (5.156) and (5.154) imply that for all $t \geq 0$ and $p \geq 1$,

$$u(t, \cdot) \in \mathcal{B}_{2p}(b_{2p}(t)) \subseteq \mathcal{B}_\infty(b_\infty(t)),$$

that is,

$$|u(t, \cdot)|_\infty \leq b_\infty(t). \qquad (5.157)$$

Since $b_\infty(t)$ decreases monotonically to 1 as $t \to +\infty$, and $b_\infty(0) = |u_0|_\infty > a > 1$, there is $T_a > 0$ such that $b_\infty(t) \leq a$ for all $t \geq T_a$. Hence, from (5.157) we conclude that $|u(t, \cdot)|_\infty \leq a$ if $t \geq T_a$. This proves that $\mathcal{B}_\infty(a)$ is absorbing if $a > 1$. □

5.8.3   Adjusting the Nonlinearity

In this section we show how to modify the nonlinearity $F(u) = k(u - u^3)$ so as to make it globally bounded and Lipschitz continuous from $\mathcal{X}$ into $\mathcal{X}$. Before doing so, we show that F is locally Lipschitz continuous, with respect to the L^2 norm, from L$^\infty(0, \pi)$ into $\mathcal{X}$.

PROPOSITION 5.50
Let $a > 0$. For all $u, v \in \mathcal{B}_\infty(a)$,

$$\|F(u) - F(v)\| \leq k(1 + 3a^2)\|u - v\|.$$

PROOF The result is an immediate consequence of the estimate

$$\|F(u) - F(v)\|^2 = k^2 \int_0^\pi |(u-v) - (u^3 - v^3)|^2 \, dx$$
$$= k^2 \int_0^\pi |u-v|^2 |1 - (u^2 + uv + v^2)|^2 \, dx$$
$$\leq k^2 (1 + 3a^2)^2 \int_0^\pi |u-v|^2 \, dx. \tag{5.158}$$

☐

To modify F, we fix $a > 1$ and choose a function $b \in C_b^\infty(\mathbb{R})$, such that $|b(r)| \leq 2a - 1$, $|b'(r)| \leq 1$ for all $r \in \mathbb{R}$, and

$$b(r) = r \quad \text{if} \quad |r| \leq a. \tag{5.159}$$

Then, we set $F_a(u) := k(b(u) - (b(u))^3)$, and claim:

PROPOSITION 5.51
Let $a > 1$. Then F_a is globally bounded and Lipschitz continuous from $\mathcal{X}$ into $\mathcal{X}$, with Lipschitz constant $\ell_a := k(1 + 3(2a - 1)^2)$.

PROOF The boundedness of F_a in $\mathcal{X} = L^2(0, \pi)$ follows from the boundedness of b on $\mathbb{R}$. Similarly (omitting the dependence of u and v on x), as in (5.158)

$$\int_0^\pi |F_a(u) - F_a(v)|^2 \, dx$$
$$= k^2 \int_0^\pi |b(u) - b(v) - (b(u))^3 - (b(v))^3|^2 \, dx$$
$$= k^2 \int_0^\pi |b(u) - b(v)|^2 |1 - ((b(u))^2 + b(u)b(v) + (b(v))^2)|^2 \, dx$$
$$\leq k^2 (1 + 3(2a - 1)^2)^2 \int_0^\pi |b(u) - b(v)|^2 \, dx$$
$$\leq k^2 (1 + 3(2a - 1)^2)^2 \int_0^\pi |u - v|^2 \, dx,$$

having recalled that $|b'(r)| \leq 1$ for all $r \in \mathbb{R}$. ☐

5.8.4 The Inertial Manifold

We are now in a position to apply theorem 5.42 to the modified equation

$$u_t - u_{xx} = F_a(u). \tag{5.160}$$

To this end, we recall that the eigenvalues of the operator

$$A := -\frac{d^2}{dx^2}$$

in $L^2(0, \pi)$, with domain $H^2(0, \pi) \cap H_0^1(0, \pi)$, are $\lambda_n = n^2$, $n \in \mathbb{N}_{>0}$, and the corresponding eigenfunctions are the functions $w_n(x) = \sin(nx)$. Thus, the spectral gap condition (5.83) reads

$$(N+1)^2 - N^2 = 2N + 1 > 2\ell_a = 2k\left(1 + 3(2a-1)^2\right), \qquad (5.161)$$

and it can obviously be satisfied for sufficiently large N. Moreover, $\mathfrak{Re}\, \lambda_N = N^2 \geq 0$ for all N, so that, as observed in remark 5.43, condition (5.104) also holds. It follows that the semiflow S_a generated by the modified equation (5.160) admits an inertial manifold of the form

$$\mathcal{M}_a = \mathrm{graph}(m_a) \cap \mathcal{B}_\infty(a),$$

where $m_a \in \mathcal{G}_{\ell_a}$. Since, by (5.159), $F_a(u) = F(u)$ if $|u| \leq a$, it follows that the restriction of the semiflow S_a to $\mathcal{M}_a$ coincides with that of the "original" semiflow S on $\mathcal{M}_a$. Hence, $\mathcal{M}_a$ is also the desired inertial manifold for S. To see this, let $u_0 \in \mathcal{X}$. By the smoothing effect of the parabolic operator (see section 3.3.3), we have that $S(t)u_0 \in H^1(0, \pi) \hookrightarrow L^\infty(0, \pi)$ for all $t > 0$. Since the set $\mathcal{B}_\infty(a)$ is absorbing for S_a, there is $T_a > 0$ such that $u_a(t, \cdot) := S_a(t)u_0 \in \mathcal{B}_\infty(a)$ for all $t \geq T_a$. Then, since $|u_a(t, x)| \leq a$, (5.159) implies that u_a solves the equation of (5.149). By the uniqueness of solutions of this initial-boundary value problem (with initial values at $t = T_a$), it follows that $S(t) = S_a(t)$ on $[T_a, +\infty[$. In particular, for $t \geq T_a$,

$$d(S(t)u_0, \mathcal{M}_a) = d(S_a(t)u_0, \mathcal{M}_a) \leq C_a e^{-\eta t}.$$

Finally, we recall that the dimension of $\mathcal{M}_a$ is N, as determined by (5.161); thus, N depends on a, via ℓ_a. Since

$$\inf_{a>1} \ell_a = \ell_1 = 4k,$$

condition (5.161) yields the lower bound

$$N_0 = \lfloor 4k + \tfrac{1}{2} \rfloor \qquad (5.162)$$

for the dimension of the inertial manifold, i.e., if $N \geq N_0$ then theorem 5.42 yields the existence of an N-dimensional inertial manifold for the modified equation (5.160).

In fact, it is possible to obtain a better lower bound, in the following way. For $r > 0$ to be determined, we rewrite the differential equation in (5.149), with $f = 0$, as

$$u_t - u_{xx} + kru = k((1+r)u - u^3).$$

We set $F_{ra}(u) := k(b((1+r)u) - (b(u))^3)$ and consider the modified equation

$$u_t - u_{xx} + kru = F_{ar}(u). \qquad (5.163)$$

Then, as in proposition 5.51, F_{ar} is again globally bounded and globally Lipschitz continuous, but now the Lipschitz constant of F_{ar} is

$$\ell_{ar} = k \max\{1 + r, |3a^2 - r - 1|\} =: k\ell(a, r). \tag{5.164}$$

To see this, we proceed as in the proof of proposition 5.51; in particular, it is sufficient to estimate $\|F_{ar}(u) - F_{ar}(v)\|$ when both $|u|_\infty \le a$, $|v|_\infty \le a$. Let $h := u^2 + uv + v^2$. Then, $h(x) \ge 0$ for all $x \in [0, \pi]$. Moreover: if $h(x) \le 1 + r$,

$$|1 + r - h(x)| = 1 + r - h(x) \le 1 + r,$$

while if $h(x) \ge 1 + r$,

$$|1 + r - h(x)| = h(x) - 1 - r \le 3a^2 - 1 - r.$$

Consequently, for all $x \in [0, \pi]$

$$|1 + r - h(x)| \le \ell(a, r).$$

For $u, v \in \mathcal{B}_\infty(a)$ we compute then that

$$\int_0^\pi |F_{ar}(u) - F_{ar}(v)|^2 \, dx = k^2 \int_0^\pi |u - v|^2 |1 + r - h|^2 \, dx \le (k\ell(a,r))^2 \int_0^\pi |u - v|^2 \, dx,$$

from which (5.164) follows. We turn then to the spectral gap condition for the operator

$$A_r := -\frac{d^2}{dx^2} + kr$$

which appears in (5.163). Since the eigenvalues of A_r are the numbers $\lambda_j = j^2 + kr$, $j \ge 1$, the spectral gap condition (5.161) now reads

$$2N + 1 > k\ell(a, r). \tag{5.165}$$

As before, we wish to minimize the right side of (5.165). Letting $a \searrow 1$, we have a first lower bound, given by $k\ell(1, r)$; in turn, this is minimized by $r = \frac{1}{2}$, with $\ell(1, \frac{1}{2}) = \frac{3}{2}$. Thus, from (5.165) we deduce the lower bound for the dimension of the inertial manifold

$$N_1 := \lfloor \tfrac{3}{2}k + \tfrac{1}{2} \rfloor,$$

i.e., if $N \ge N_1$ then theorem 5.42 yields the existence of an N-dimensional inertial manifold for the modified equation (5.163). This bound is obviously lower than the bound N_0 obtained in (5.162).

5.8.5   The Hyperbolic Perturbation

In this section we briefly consider the hyperbolic perturbation of problem (5.149), i.e. the one-dimensional IBVP

$$
\begin{cases}
\varepsilon u_{tt} + u_t - u_{xx} + k(u^3 - u) = f(x) \\
u(0,x) = u_0(x), \quad u_t(0,x) = u_1(x) \\
u(t,0) = u(t,\pi) = 0.
\end{cases}
\tag{5.166}
$$

Problem (5.166) is a version of (3.4), with, as in the parabolic case (5.149), $n = 1$. Thus, we know that (5.166) generates a semiflow S, both in $\mathcal{X} = H_0^1(0\pi) \times L^2(0,\pi)$ and in

$$
\mathcal{X}_1 = \left(H^2(0,\pi) \cap H_0^1(0,\pi)\right) \times H_0^1(0,\pi).
$$

We also know that S admits a bounded, positively invariant absorbing set in both these spaces, as well as a global attractor $\mathcal{A} \subseteq \mathcal{X}$, which is contained and bounded in $\mathcal{X}_1$; S also admits an exponential attractor $\mathcal{E} \subseteq \mathcal{X}$. We now proceed to show that theorem 5.44 can be applied to problem (5.166); therefore, S also admits an inertial manifold in $\mathcal{X}$. To this end, as in section 5.7.2 we first rescale the time variable and then transform the equation into a first order system of the form (5.113), with $\mathcal{A}$ given by (5.115), and $F(U) = (0, f - k(u - u^3))$ if $U = (u,v) \in \mathcal{X}$. Thus, we need to verify conditions (5.120) and (5.121), after overcoming the difficulty that F is neither globally bounded nor globally Lipschitz continuous on $\mathcal{X}$. To this end, we proceed exactly as in section 5.8.3, by modifying F outside of a bounded, positively invariant absorbing set $\mathcal{B}$ (which we know to exist). To do so, we note that the definition of F only involves the first component u of $U = (u,v)$; now, since the projection $\tilde{\mathcal{B}}$ of $\mathcal{B}$ into $H_0^1(0,\pi)$ is bounded, and $H_0^1(0,\pi) \hookrightarrow L^\infty(0,\pi)$, it follows that $\tilde{\mathcal{B}}$ is contained in a ball $\mathcal{B}_\infty(a)$ of $L^\infty(0,\pi)$. Let a be the radius of this ball and, as in section 5.8.3, set $\ell_a := k(1 + 3(2a - 1)^2)$, $g_a(u) := b(f + k(u - u^3))$, and $F_a(U) := (0, g_a(u))$, for $U = (u,v) \in \mathcal{X}$.

Recalling that $\lambda_N = N^2$, conditions (5.120) and (5.121) read, respectively,

$$
2m + 1 > 4\ell_a, \quad \alpha^2 > (N+1)^2.
$$

The first of these is certainly satisfied for all m such that

$$
m \geq \lfloor 2\ell_a + \tfrac{1}{2} \rfloor =: N;
\tag{5.167}
$$

thus, if α is so large that

$$
\alpha^2 > (N+1)^2,
$$

the semiflow S_a generated by the modified equation

$$
\varepsilon u_{tt} + u_t - u_{xx} = g_a(u)
$$

admits an inertial manifold $\mathcal{M}_a \subseteq \mathcal{X}$. Since, by (5.159), $F_a(U) = F(U)$ if $U = (u,v)$ with $|u| \leq a$, it follows that the restriction of the semiflow S_a to $\mathcal{M}_a$ coincides with that of the "original" semiflow S on $\mathcal{M}_a$. Hence, $\mathcal{M}_a$ is also the desired inertial manifold for S. To see this, let $U_0 = (u_0, v_0) \in \mathcal{X}$, and $U_a(t) = (u_a(t), v_a(t)) := S_a(t)U_0$. Since $\mathcal{B}$ is absorbing, there is $T_a > 0$ such that $u_a(t, \cdot) \in \mathcal{B}_\infty(a)$ for all $t \geq T_a$. Then, since $|u_a(t, x)| \leq a$, (5.159) implies that u_a solves the equation of (5.166). By the uniqueness of solutions of this initial-boundary value problem (with initial values at $t = T_a$), it follows that $S(t) = S_a(t)$ on $[T_a, +\infty[$. In particular, for $t \geq T_a$,

$$d(S(t)U_0, \mathcal{M}_a) = d(S_a(t)U_0, \mathcal{M}_a) \leq C_a \, e^{-\eta t}.$$

Finally, we recall that the dimension of $\mathcal{M}_a$ is N_0, as determined by (5.167). Thus, N_0 depends on a, via ℓ_a, and we can obtain lower bounds of the dimension by minimizing ℓ_a, with $a > 1$. Since ℓ_a is minimized by $a = 1$, with $\ell_1 = 4k$, (5.167) implies the lower bound $N_0 = 8k$ for the dimension of the inertial manifold.

5.8.6   Concluding Remarks

1.   The results of this section 5.8 show that when the perturbation parameter ε is small, the asymptotic properties of the semiflow generated by problems (5.166) are qualitatively similar to those of the semiflow generated by the limit problem when $\varepsilon = 0$, i.e. by (5.149). It would in fact be interesting to further investigate this question, for example in the framework of the upper and lower semicontinuity of these inertial manifolds as $\varepsilon \to 0$, as we did in section 3.6 for the attractors. As regards to this, a first difficulty is, of course, that, in general, the inertial manifolds are, in contrast to the global attractors, only positively invariant, and not necessarily invariant.

2.   The construction of an inertial manifold for the Chafee-Infante equations with the method described in section 5.6 requires the verification of the spectral gap condition (5.83). In one dimension of space, this is a consequence of (5.161), which guarantees that the eigenvalues of the Laplacian admit arbitrarily large gaps. When $n > 1$, however, one only knows that, for general domains $\Omega \subset \mathbb{R}^n$, $\lambda_N = O(N^{2/n})$ as $N \to +\infty$ (see e.g. Robinson, [Rob93, Rob96]); therefore, it is not known if the point spectrum of the operator contains large gaps. For example, already in the two-dimensional case the condition $\lambda_N = O(N)$ is clearly not enough to guarantee the existence of large gaps. On the other hand, the situation may be better, if Ω has some particular geometric properties. For instance, the spectral gap condition is satisfied for rectangular domains (see Robinson, [Rob01, sct. 15.4.2]). Similarly, in [MPS88] Mallet-Paret and Sell prove the strong squeezing property for parallelepipeds in $\mathbb{R}^3$, using a technique of spatial averaging. This means that, in this case, one has some type of degenerate spectral gap condition, which can be satisfied without large gaps. Since the method of spatial averaging is a consequence of general properties of the Laplace operator on a domain, the strong squeezing property also holds on some nonrectangular domains in $\mathbb{R}^3$.

3. Once again, these difficulties in constructing inertial manifolds may explain the interest in exploring, at least in some cases, the existence of other types of attracting sets which, as in the case of the exponential attractor, retain most of the properties of an inertial manifold; in particular, the finite dimensionality, and the exponential convergence of the orbits.

4. The notion of inertial manifold can be translated and extended to more general classes of differential equations, such as nonautonomous differential equations (see e.g. [GV97, WF97, LL99]), retarded parabolic differential equations (see e.g. [TY94, BdMCR98]), or differential equations with random or stochastic perturbations (see e.g. [Chu95, BF95, CL99, CS01, DLS03]). An extension to abstract nonautonomous dynamical systems, which includes the systems generated by all the equations mentioned above, can be found in [KS02b, KS02a, KS03].

Chapter 6

Examples

In this chapter we present four examples of semiflows, generated by some specific initial-boundary value problems from mathematical physics. These semiflows admit a global attractor in a suitable phase space $\mathcal{X}$; except for one case, we are also able to show that they also admit an exponential attractor, or even an inertial manifold, in $\mathcal{X}$. The models we consider are:

1. The hyperbolic perturbations of the viscous and nonviscous CAHN-HILLIARD equations, whose corresponding semiflows admit a global attractor, an exponential attractor and an inertial manifold;

2. EXTENSIBLE BEAM equation, whose corresponding semiflow admits a global attractor, an exponential attractor and, for special types of forcing terms, an inertial manifold; analogous results are valid for a model of the VON KÁRMÁN equation in one dimension of space;

3. The NAVIER-STOKES equations in two dimension of space, whose corresponding semiflow admits a global attractor and an exponential attractor, but the existence of an inertial manifold is open;

4. MAXWELL's equations in a ferromagnetic medium, for which we can only show the existence of a global attractor.

The PDEEs in the first two models are dissipative hyperbolic; the Navier-Stokes equations are essentially parabolic, due to the presence of a viscosity term, while the quasi-stationary Maxwell equations we consider are transformed into a system of quasilinear parabolic equations. All these models can be treated with a suitable application of the methods described in chapters 3, 4 and 5.

For other examples of semiflows generated by nonlinear PDEEs, we refer to the books by Temam, [Tem83]; Sell-You, [SY02]; Eden, Foias, Nicolaenko and Temam, [EFNT94], and Constantin, Foias, Nicolaenko and Temam, [CFNT89].

6.1 Cahn-Hilliard Equations

In this section we consider the singular perturbations of two IBV problems, concerning respectively the viscous and the nonviscous CAHN-HILLIARD equations, in one dimension of space. Resorting to the techniques presented in the previous chapters, we will show that, at least when the perturbation parameter is sufficiently small, the semiflows generated by these two problems admit global attractors, exponential attractors and inertial manifolds in a suitable phase space. The material we present is mostly taken from Zheng-Milani, [ZM03, ZM05].

6.1.1 Introduction

1. The equations we consider have the unified form

$$\varepsilon u_{tt} + u_t + \Delta(\Delta u - u^3 + u - \delta u_t) = 0, \qquad (6.1)$$

where $\varepsilon > 0$, $\delta \geq 0$, $u = u(t,x)$, with $t > 0$ and $x \in \Omega :=]0,\pi[$, and $\Delta := \partial^2/\partial x^2$.

We assume that u satisfies the homogeneous Dirichlet boundary conditions

$$u(t,0) = u(t,\pi) = 0, \quad \Delta u(t,0) = \Delta u(t,\pi) = 0, \qquad t \geq 0. \qquad (6.2)$$

2. When $\varepsilon = 0$, (6.1) yields the classical viscous and nonviscous Cahn-Hilliard equations, corresponding respectively to the cases $\delta > 0$ and $\delta = 0$; these equations are parabolic, and are considered together with the initial condition $u(0,x) = u_0(x)$, $x \in]0,\pi[$. This case has been extensively studied; in particular, we refer to Temam, [Tem83, ch. III.4.2], or Sell-You, [SY02, ch. 5.5], and the references cited therein. Other references can be found in Zheng-Milani, [ZM03].

In summary, it is known that the Cahn-Hilliard equations generate a semiflow in the space $\mathcal{H} := L^2(0,\pi)$, and that these semiflows admit a compact global attractor and an inertial manifold in $\mathcal{H}$ (for the viscous case, this was proven in Zheng-Milani, [ZM05]). Moreover, the global attractors are lower- and upper-semicontinuous as $\delta \to 0$. We are not aware of any result on exponential attractors for either problem; nevertheless, since both semiflows admit a compact absorbing set (because they admit a global attractor), as well as a closed inertial manifold, they also admit an exponential attractor. As we mentioned in the introduction to chapter 5, this exponential attractor is given by the intersection of the absorbing ball with the inertial manifold.

3. When $\varepsilon > 0$ (which is the case we present here), equation (6.1) is hyperbolic; indeed, it is an example of a nonlinear beam equation with viscous damping, of the type we consider in the next section 6.2. Thus, we impose the initial conditions

$$u(0,x) = u_0(x), \quad u_t(0,x) = u_1(x) \qquad x \in]0,\pi[. \qquad (6.3)$$

In the sequel, we shall refer to the IBV problem (6.1)+ (6.3)+(6.2) as

$$\text{problem} \quad (\text{CH}_{\varepsilon\delta}),$$

with, for simplicity, $\varepsilon \in]0,1]$ and $\delta \in [0,1]$.

4.   To define the phase spaces in which we study problem $(\mathrm{CH}_{\varepsilon\delta})$, we introduce the following notations. We set $\mathcal{H}^0 := \mathrm{L}^2(0,\pi)$, with the usual notations $\langle\,\cdot\,,\cdot\,\rangle$ and $\|\cdot\|$ for the scalar product and norm. For integer $m \geq 1$, we set

$$\mathcal{H}^m := \mathrm{H}^m(0,\pi)\cap\mathrm{H}_0^1(0,\pi)\,, \quad \mathcal{H}_0^m := \mathrm{H}_0^m(0,\pi)\,, \quad \mathcal{H}^{-m} := (\mathrm{H}_0^m)'\,,$$

and denote by $\|\cdot\|_m$ the norm in $\mathcal{H}^m$, $m \in \mathbb{Z}$. Because of Poincaré's inequality, we can choose in $\mathcal{H}^1$ the norm $\|u\|_1 = \|\nabla u\|$. Finally, for $1 \leq p \leq +\infty$ we set $L^p := \mathrm{L}^p(0,\pi)$, and denote by $|\cdot|_p$ its norm.

The phase spaces we consider are then the following:

$$\mathcal{X}_0 := \mathcal{H}^1 \times \mathcal{H}^{-1}\,, \quad \mathcal{X}_1 := \mathcal{H}^2 \times L^2\,, \quad \mathcal{X}_2 := \mathcal{Y}\times\mathcal{H}^1\,, \quad \mathcal{X}_{-1} := \mathcal{H}^{-1}\times\mathcal{Y}'\,, \quad (6.4)$$

where $\mathcal{Y} := \{u \in \mathcal{H}^1 : -\Delta u \in \mathcal{H}^1\}$. Note that $\mathcal{H}_0^3 \hookrightarrow \mathcal{Y} \hookrightarrow \mathcal{H}^3$ (the second inclusion being a consequence of standard elliptic theory); thus, $\mathcal{Y}' \hookrightarrow \mathcal{H}^{-3}$. We will consider in $\mathcal{X}_0$ and $\mathcal{X}_1$ two equivalent norms, which we define in (6.8) and (6.11), (6.12) below.

5.   We conclude this introductory part by mentioning some results that we will need in the sequel.

At first, we recall (see section A.5.5) that $-\Delta$ is an unbounded operator in $\mathrm{L}^2(0,\pi)$, with domain $\mathcal{H}^2$. In particular, $-\Delta$ is an isomorphism between $\mathcal{H}^m$ and $\mathcal{H}^{m-2}$, $m \in \mathbb{N}$; we denote its inverse by $(-\Delta)^{-1}$. We consider then the equation formally obtained from (6.1) by taking $(-\Delta)^{-1}$, that is, the equation

$$\varepsilon(-\Delta)^{-1}u_{tt} + (-\Delta)^{-1}u_t - \Delta u + u^3 - u + \delta u_t = 0\,. \quad (6.5)$$

For u and $v \in \mathcal{H}^{-1}$ we set

$$[u,v] := \langle v, (-\Delta)^{-1}u\rangle_{\mathcal{H}^{-1}\times\mathcal{H}^1}\,; \quad (6.6)$$

then, $[u,u] = \|u\|_{-1}^2$, and, letting $\varphi := (-\Delta)^{-1}u$ and $\psi := (-\Delta)^{-1}v$,

$$[u,v] = \langle -\Delta\psi, \varphi\rangle = \langle\nabla\psi,\nabla\varphi\rangle \leq \|\nabla\psi\|\cdot\|\nabla\varphi\| = \|\psi\|_1\cdot\|\varphi\|_1 = \|u\|_{-1}\cdot\|v\|_{-1}\,.$$

Finally, we recall that, since $N = 1$, the continuous imbedding $\mathrm{H}^1(0,\pi)\hookrightarrow\mathrm{L}^\infty(0,\pi)$ holds; we reserve the letter K to denote a constant such that the inequalities

$$\|u\|_{-1} \leq K\|u\| \leq K^2\|\nabla u\|\,, \quad |u|_p \leq K\|u\|_1 = K\|\nabla u\|\,, \quad (6.7)$$

hold for all $u \in \mathcal{H}^1$, and $1 \leq p \leq +\infty$. Without loss of generality, we can assume that $K \geq 1$.

6.   We now introduce in $\mathcal{X}_0$ an equivalent norm, whose square is defined by

$$E_0(u,v) := \varepsilon\|v\|_{-1}^2 + \varepsilon[u,v] + \tfrac{1}{2}\|u\|_{-1}^2 + \|\nabla u\|^2\,, \quad (u,v) \in \mathcal{X}\,. \quad (6.8)$$

Since we assume that $\varepsilon \leq 1$, E_0 does define a norm: indeed, by (6.7) we immediately derive that for all $(u,v) \in \mathcal{X}$,

$$\tfrac{1}{2}\left(\varepsilon\|v\|_{-1}^2 + \|\nabla u\|^2\right) \leq E_0(u,v) \leq \alpha\left(\varepsilon\|v\|_{-1}^2 + \|\nabla u\|^2\right), \qquad (6.9)$$

with

$$\alpha := \max\{\tfrac{3}{2}, K^4 + 1\}. \qquad (6.10)$$

Similarly, we immediately verify that we can consider in $\mathcal{X}_1$ the equivalent norms whose squares are defined by

$$E_1(u,v) := \varepsilon\|v\|^2 + \varepsilon\langle u,v\rangle + \tfrac{1}{2}\|u\|^2 + \|\Delta u\|^2, \qquad (6.11)$$

if $\delta > 0$, or, if $\delta = 0$, by

$$E_{1\varepsilon}(u,v) := \varepsilon\|v\|^2 + \langle u,v\rangle + \tfrac{1}{2\varepsilon}\|u\|^2 + \|\Delta u\|^2. \qquad (6.12)$$

6.1.2   The Cahn-Hilliard Semiflows

In this section, we show that problem $(CH_{\varepsilon\delta})$ generates a semiflow S, both in $\mathcal{X}_0$ and in $\mathcal{X}_1$. In some cases, the situation is slightly different in the viscous and the nonviscous cases; for example, when $\delta = 0$ some results are valid only under some limitation on the size of ε. To reduce the length of our presentation, in these cases we will only give a proof for the nonviscous problem, which is more difficult, and refer to Zheng-Milani, [ZM03, ZM05], for the viscous one.

1.   At first, we recall the following global existence, uniqueness and regularity result.

THEOREM 6.1
For all $\varepsilon \in \,]0,1]$, $\delta \in [0,1]$ and $(u_0,u_1) \in \mathcal{X}_0$, there exists a unique function $u \in C_b([0,+\infty[;\mathcal{H}^1) \cap C_b^1([0,+\infty[;\mathcal{H}^{-1})$, which is a weak solution of problem $(CH_{\varepsilon\delta})$ (i.e. with (6.5) satisfied in $\mathcal{H}^{-1}$, almost everywhere in t). If in addition $(u_0,u_1) \in \mathcal{X}_1$, then

$$u \in C_b([0,+\infty[;\mathcal{H}^2) \cap C_b^1([0,+\infty[;\mathcal{H}^0). \qquad (6.13)$$

PROOF   As usual, we limit ourselves to establish formal a priori estimates on weak solutions of problem $(CH_{\varepsilon\delta})$.

At first, we consider the function $\Phi_0 \colon \mathcal{X}_0 \to \mathbb{R}$ defined by

$$\Phi_0(u,v) := E_0(u,v) + \tfrac{1}{2}|u|_4^4 - \|u\|^2 + \tfrac{1}{2}\delta\|u\|^2, \qquad (6.14)$$

and note that Φ_0 is bounded from below. Indeed, by Minkowski's inequality, for all $\eta > 0$ there is $C_\eta > 0$ such that for all $u \in L^4(0,\pi)$,

$$\|u\|^2 \leq C_\eta + \eta|u|_4^4; \qquad (6.15)$$

hence, taking e.g. $\eta = \frac{1}{2}$ and recalling (6.9),

$$\Phi_0(u,v) \geq E_0(u,v) - C_{1/2} \geq -C_{1/2}. \qquad (6.16)$$

Let now $M_1 := C_{2/3}$, as defined in (6.15). We claim that for all $t \geq 0$,

$$\Phi_0(u(t),u_t(t)) \leq (\Phi_0(u_0,u_1) - \alpha M_1)\,e^{-t/\alpha} + \alpha M_1, \qquad (6.17)$$

where α is as in (6.10). To show this, we begin by multiplying equation (6.5) in $\mathcal{H}$ by $2u_t$ and u, and adding the resulting identities. Recalling (6.14), we obtain

$$\frac{\mathrm{d}}{\mathrm{d}t}\Phi_0(u,u_t) + (2-\varepsilon)\|u_t\|_{-1}^2 + \|\nabla u\|^2 + |u|_4^4 - \|u\|^2 + 2\delta\|u_t\|^2 = 0. \qquad (6.18)$$

From this, recalling that $\varepsilon \leq 1$, we deduce that

$$\frac{\mathrm{d}}{\mathrm{d}t}\Phi_0(u,u_t) + \varepsilon\|u_t\|_{-1}^2 + \|\nabla u\|^2 + |u|_4^4 \leq \|u\|^2.$$

From (6.14) and (6.9), since also $\delta \leq 1$,

$$\Phi_0(u,u_t) \leq \alpha\left(\varepsilon\|u_t\|_{-1}^2 + \|\nabla u\|^2\right) + \tfrac{1}{2}|u|_4^4 - \tfrac{1}{2}\|u\|^2; \qquad (6.19)$$

thus, from (6.18) and (6.19), recalling (6.15) and that $\alpha \geq \frac{3}{2}$, we obtain

$$\frac{\mathrm{d}}{\mathrm{d}t}\Phi_0(u,u_t) + \tfrac{1}{\alpha}\Phi_0(u,u_t) + \tfrac{1}{2\alpha}\|u\|^2 + \tfrac{2}{3}|u|_4^4 \leq \|u\|^2 \leq C_{2/3} + \tfrac{2}{3}|u|_4^4.$$

From this, we conclude that

$$\frac{\mathrm{d}}{\mathrm{d}t}\Phi_0(u,u_t) + \tfrac{1}{\alpha}\Phi_0(u,u_t) \leq M_1,$$

and (6.17) follows by integration.

As a consequence of (6.17), we deduce the boundedness of the function $t \mapsto (u(t),u_t(t))$ in $\mathcal{X}_0$. Indeed, we immediately deduce that there exists $M_2 > 0$ such that for all $t \geq 0$,

$$E_0(u(t),u_t(t)) \leq M_2. \qquad (6.20)$$

Finally, the proof of the regularity result (6.13) follows from additional a priori estimates, similar to those we establish in proposition 6.4 below. $\qquad\Box$

2. Theorem 6.1 allows us to define the solution operator $S_{\varepsilon\delta} = (S_{\varepsilon\delta}(t))_{t\geq 0}$ associated to problem $(\mathrm{CH}_{\varepsilon\delta})$, with $\varepsilon \in\,]0,1]$ and $\delta \in [0,1]$. We now show that $S_{\varepsilon\delta}$ is in fact a semiflow.

PROPOSITION 6.2
Let ε, $\delta \in\,]0,1]$, or, if $\delta = 0$, $\varepsilon \in\,]0,\frac{1}{3}]$. Then, for each $t > 0$, the operator $S_{\varepsilon\delta}$ is Lipschitz continuous on $\mathcal{X}_0$.

PROOF   We consider only the case $\delta = 0$. Let $z := u - \tilde{u}$ denote the difference of two solutions of problem (CH$_{\varepsilon 0}$). Then, z solves the equations

$$\varepsilon z_{tt} + z_t + \Delta\left(\Delta z - (u^3 - \tilde{u}^3) + z\right) = 0,$$
$$\varepsilon(-\Delta)^{-1} z_{tt} + (-\Delta)^{-1} z_t - \Delta z + (u^3 - \tilde{u}^3) - z = 0. \tag{6.21}$$

As usual, we multiply (6.21) in $\mathcal{H}$ by $2z_t$ and z, and add the resulting identities. Setting $h := u^2 + u\tilde{u} + \tilde{u}^2$, we obtain

$$\frac{\mathrm{d}}{\mathrm{d}t}\left(\varepsilon\|z_t\|_{-1}^2 + \varepsilon[z_t, z] + \tfrac{1}{2}\|z\|_{-1}^2 + \|\nabla z\|^2 + \langle u^3 - \tilde{u}^3, z\rangle\right)$$
$$+ (2 - \varepsilon)\|z_t\|_{-1}^2 + \|\nabla z\|^2 + \langle u^3 - \tilde{u}^3, z\rangle = \langle h_t z, z\rangle + \langle z, 2z_t + z\rangle. \tag{6.22}$$

By (6.20), we can start the estimate of the first term of the right side of (6.22) by

$$2\langle u u_t z, z\rangle \le \|u_t\|_{-1}\|u z^2\|_1 \le C\left(\|\nabla u\| |z|_\infty^2 + 2|u|_\infty |z|_\infty \|\nabla z\|\right), \tag{6.23}$$

where C depends only on M_2 of (6.20). Resorting then to the Gagliardo-Nirenberg inequality

$$|z|_\infty \le C\|\nabla z\|^{1/2}\|z\|^{1/2} + C\|z\|$$

(see theorem A.70), we obtain from (6.23)

$$2\langle u u_t z, z\rangle \le C\left(\|\nabla z\|\,\|z\| + \|\nabla z\|^{3/2}\|z\|^{1/2} + \|z\|^2\right) \le \tfrac{1}{24}\|\nabla z\|^2 + C_2\|z\|^2.$$

The other three terms of $\langle h_t z, z\rangle$ can be treated in the same way, leading to the estimate

$$\langle h_t z, z\rangle \le \tfrac{1}{6}\|\nabla z\|^2 + C\|z\|^2. \tag{6.24}$$

Next, we estimate

$$\langle z, 2z_t + z\rangle \le \tfrac{8}{5}\|z_t\|_{-1}^2 + \tfrac{5}{8}\|\nabla z\|^2 + \|z\|^2. \tag{6.25}$$

Calling $\Psi(z, z_t)$ the differentiated term of the left side of (6.22), i.e.

$$\Psi(z, z_t) := \varepsilon\|z_t\|_{-1}^2 + \varepsilon[z_t, z] + \tfrac{1}{2}\|z\|_{-1}^2 + \|\nabla z\|^2 + \langle u^3 - \tilde{u}^3, z\rangle, \tag{6.26}$$

by (6.24) and (6.25) we obtain from (6.22) that

$$\frac{\mathrm{d}}{\mathrm{d}t}\Psi(z, z_t) + (\tfrac{2}{5} - \varepsilon)\|z_t\|_{-1}^2 + \tfrac{5}{24}\|\nabla z\|^2 + \langle u^3 - \tilde{u}^3, z\rangle \le C\|z\|^2. \tag{6.27}$$

Assume now e.g. that $\varepsilon \le \tfrac{1}{3}$. Then, $\tfrac{2}{5} - \varepsilon > \tfrac{1}{6}\varepsilon$; and since the function $u \mapsto u^3$ is monotone, we deduce from (6.27) that

$$\frac{\mathrm{d}}{\mathrm{d}t}\Psi(z, z_t) + \tfrac{1}{6}\left(\|z_t\|_{-1}^2 + \|\nabla z\|^2 + \langle u^3 - \tilde{u}^3, z\rangle\right) \le C\|z\|^2. \tag{6.28}$$

It is now easy to verify that, since $\varepsilon \leq 1$,

$$\Psi(z, z_t) \leq \alpha \left(\varepsilon \|z_t\|_{-1}^2 + \|\nabla z\|^2 + \langle u^3 - \tilde{u}^3, z \rangle \right),$$

with α as in (6.10); consequently, we obtain from (6.28) that

$$\frac{d}{dt} \Psi(z, z_t) + \frac{1}{6\alpha} \Psi(z, z_t) \leq C \|z\|^2. \tag{6.29}$$

Integrating (6.29), we obtain that for all $t \geq 0$

$$\Psi(z(t), z_t(t)) \leq \Psi(z(0), z_t(0)) e^{-t/6\alpha} + C \int_0^t \|z\|^2 \, ds. \tag{6.30}$$

From (6.26) and (6.9), we deduce that

$$\Psi(z, z_t) \geq \tfrac{1}{2} \left(\varepsilon \|z_t\|_{-1}^2 + \|\nabla z\|^2 \right) \geq \tfrac{1}{2\alpha} E_0(z, z_t); \tag{6.31}$$

moreover, we also have that

$$0 \leq \langle u^3 - \tilde{u}^3, z \rangle = \langle hz, z \rangle \leq \left(|u|_\infty^2 + |u|_\infty |\tilde{u}|_\infty + |\tilde{u}|_\infty^2 \right) \|z\|^2 \leq C_3 \|\nabla z\|^2,$$

where C_3 depends only on M_2 and K. Hence, recalling that the sum of the first three terms of Ψ is positive definite, and that $\varepsilon \leq 1$, we also have

$$\Psi(z, z_t) \leq \left(\varepsilon \|z_t\|_{-1}^2 + \varepsilon [z_t, z] + \tfrac{1}{2} \|z\|_{-1}^2 + (1 + C_3 + K) \|\nabla z\|^2 \right) \leq C_4 E_0(z, z_t). \tag{6.32}$$

Finally, since $\|u\|^2 \leq 4 E_0(u, v)$ for all $(u, v) \in \mathcal{X}_0$, we deduce from (6.30), (6.31) and (6.32) that z satisfies the estimate

$$E_0(z(t), z_t(t)) \leq M \alpha E_0(z(0), z_t(0)) e^{-t/6\alpha} + M \alpha \int_0^t E_0(z, z_t) \, ds, \tag{6.33}$$

with $M := \max\{2C_4, 8C\}$. Applying Gronwall's inequality, we deduce from (6.33) that, for all $t \geq 0$,

$$E_0(z(t), z_t(t)) \leq M \alpha E_0(z(0), z_t(0)) e^{M\alpha t}. \tag{6.34}$$

Thus, each operator $S_{\varepsilon 0}(t)$ is Lipschitz continuous in $\mathcal{X}_0$, as claimed. The proof in the case $\delta > 0$ is similar (and actually simpler). $\quad\square$

6.1.3   Absorbing Sets

1.   The existence of a bounded, positively invariant absorbing set for the semiflow $S_{\varepsilon\delta}$ in $\mathcal{X}_0$ is an immediate consequence of estimate (6.17).

PROPOSITION 6.3
Let $\varepsilon \in {]}0, 1]$ and $\delta \in [0, 1]$. For any $R_0 > \alpha M_1 + C_1$, the ball

$$\mathcal{B}_0 := \{(u, v) \in \mathcal{X}_0 \colon E_0(u, v) \leq R_0\}$$

is absorbing for $S_{\varepsilon\delta}$. Moreover, for any $R > \alpha M_1$, the set

$$\mathcal{B}_0 := \{(u,v) \in \mathcal{X} : \Phi_0(u,v) \leq R\} \tag{6.35}$$

is bounded, positively invariant and absorbing for $S_{\varepsilon\delta}$ in $\mathcal{X}_0$.

PROOF   The first claim follows from (6.16) and (6.17). In particular, for all $t \geq 0$,

$$E_0(u(t), u_t(t)) \leq (\Phi_0(u_0, u_1) - \alpha M_1)e^{-t/\alpha} + \alpha M_1 + C_1. \tag{6.36}$$

Assume now that (u_0, u_1) is in a bounded set $\mathcal{G} \subseteq \mathcal{X}_0$. There exists then $\Gamma \geq 1$ such that $E_0(u_0, u_1) \leq \Gamma$. Now, from (6.14) and (6.7), recalling also (6.10),

$$\Phi_0(u_0, u_1) \leq E_0(u_0, u_1) + \tfrac{1}{4}|u_0|_4^4 \leq \Gamma + \tfrac{1}{4}K^4\Gamma^2 \leq \alpha\Gamma^2;$$

thus, from (6.36) we deduce that for all $t \geq 0$,

$$E_0(u(t), u_t(t)) \leq \alpha(\Gamma^2 - M_1)e^{-t/\alpha} + \alpha M_1 + C_1.$$

From this it follows that if $\alpha(\Gamma^2 - M_1) \leq R_0 - (\alpha M_1 + C_1)$, then $E_0(u(t), u_t(t)) \leq R_0$ for all $t \geq 0$, while if $\alpha(\Gamma^2 - M_1) > R_0 - (\alpha M_1 + C_1)$, then $E_0(u(t), u_t(t)) \leq R_0$ for all $t \geq T_{\mathcal{G}}$, with

$$T_{\mathcal{G}} := \alpha \ln \frac{\alpha(\Gamma^2 - M_1)}{R_0 - (\alpha M_1 + C_1)}.$$

This proves that the ball $\mathcal{B}_0$ is absorbing. The boundedness of the set $\mathcal{B}_0$ follows from (6.16) and (6.17), and its positive invariance is a direct consequence of (6.17). In fact, if $\Phi_0(u_0, u_1) \leq R$, then for all $t \geq 0$

$$\Phi_0(u(t), u_t(t)) \leq (R - \alpha M_1)e^{-t/\alpha} + \alpha M_1 \leq (R - \alpha M_1) + \alpha M_1 \leq R.$$

Finally, we prove that $\mathcal{B}_0$ is absorbing exactly in the same way as we did for $\mathcal{B}_0$; we find that $\Phi_0(u(t), u_t(t)) \leq R$ for all $t \geq \tilde{T}_{\mathcal{G}}$, where now

$$\tilde{T}_{\mathcal{G}} := \begin{cases} 0 & \text{if} \quad \alpha(\Gamma^2 - M_1) \leq R - \alpha M_1; \\ \dfrac{\alpha(\Gamma^2 - M_1)}{R - \alpha M_1} & \text{if} \quad \alpha(\Gamma^2 - M_1) > R - \alpha M_1. \end{cases}$$

This concludes the proof of proposition 6.3; we remark that the set $\mathcal{B}_0$ is not a ball of $\mathcal{X}_0$.   □

2.   We now show that the semiflow $S_{\varepsilon\delta}$ also admits an absorbing set in $\mathcal{X}_1$.

PROPOSITION 6.4
For each $\delta \in \,]0,1]$, there exists $R_{1\delta} > 0$, depending on δ but not on ε, such that for all $\varepsilon \in \,]0,1]$, the set

$$\mathcal{B}_{1\delta} := \{(u,v) \in \mathcal{X}_1 \cap \mathcal{B} : E_1(u,v) \leq R_{1\delta}\},$$

where E_1 is defined in (6.11), is positively invariant, bounded and absorbing in $\mathcal{X}_1$ for the semiflow $S_{\varepsilon\delta}$. Analogously, if $\delta = 0$, there exist $\varepsilon_0 \in \left]0, \frac{1}{3}\right]$, with the property that for all $\varepsilon \in \left]0, \varepsilon_0\right]$, there is $R_{1\varepsilon}$ such that the set

$$\mathcal{B}_{1\varepsilon} := \{(u,v) \in \mathcal{X}_1 \cap \mathcal{B} \colon E_{1\varepsilon}(u,v) \leq R_{1\varepsilon}\}, \tag{6.37}$$

where $E_{1\varepsilon}$ is defined in (6.12), is positively invariant, bounded and absorbing in $\mathcal{X}_1$ for the semiflow $S_{\varepsilon 0}$.

PROOF   We only consider the case $\delta = 0$. We multiply equation (6.1) by $2u_t$ and $\frac{1}{\varepsilon}u$ in $\mathcal{H}$. We obtain

$$\frac{\mathrm{d}}{\mathrm{d}t}E_{1\varepsilon}(u,u_t) + \|u_t\|^2 + \frac{1}{\varepsilon}\|\Delta u\|^2 = -\frac{1}{\varepsilon}\langle \nabla(u^3 - u), \nabla u\rangle + 2\langle \Delta(u^3 - u), u_t\rangle.$$

Denoting by C_b a generic positive constant, depending on the uniform bound on u in $H^1(0,\pi) \hookrightarrow L^\infty(0,\pi)$, we can estimate

$$-\frac{1}{\varepsilon}\langle \nabla(u^3 - u), \nabla u\rangle = -\frac{1}{\varepsilon}\langle (3u^2 - 1)\nabla u, \nabla u\rangle \leq \frac{1}{\varepsilon}\|\nabla u\|^2 \leq \frac{1}{\varepsilon}C_b. \tag{6.38}$$

Since

$$\Delta(u^3 - u) = \mathrm{div}((3u^2 - 1)\nabla u) = 6u\nabla u \cdot \nabla u + (3u^2 - 1)\Delta u,$$

resorting to the Gagliardo-Nirenberg inequality

$$|\nabla u|_4 \leq C\|\Delta^2 u\|^{1/4}\|\nabla u\|^{3/4} + C\|\nabla u\|,$$

and to the elliptic estimate

$$\|u\|_2 \leq C\|\Delta u\| + C\|u\|$$

(see theorems A.70 and A.77), we have

$$\|u\,\nabla u \cdot \nabla u\| \leq |u|_\infty |\nabla u|_4^2 \leq C_b(1 + \|u\|_2) \leq C_b(1 + \|\Delta u\|),$$
$$\|(3u^2 - 1)\Delta u\| \leq (3|u|_\infty^2 + 1)\|\Delta u\| \leq C_b\|\Delta u\|.$$

Consequently (for different C_b),

$$2\langle \Delta(u^3 - u), u_t\rangle \leq \frac{1}{2}\|u_t\|^2 + C_b + C_b\|\Delta u\|^2. \tag{6.39}$$

Choose now ε_0 so small that $2\varepsilon_0 C_b \leq 1$. Then, if $\varepsilon \leq \varepsilon_0$, by (6.39) and (6.38) we obtain from (6.38)

$$\frac{\mathrm{d}}{\mathrm{d}t}E_{1\varepsilon}(u,u_t) + \frac{1}{2\varepsilon}\left(\varepsilon\|u_t\|^2 + \|\Delta u\|^2\right) \leq \left(1 + \frac{1}{\varepsilon}\right)C_b. \tag{6.40}$$

Since, as we easily verify,

$$E_{1\varepsilon}(u,u_t) \leq \frac{3}{2}\left(\varepsilon\|u_t\|^2 + \|\Delta u\|^2\right) + \frac{1}{\varepsilon}\|u\|^2,$$

we deduce from (6.40) that, if also $\varepsilon \leq \frac{1}{2}$,

$$\frac{d}{dt}E_{1\varepsilon}(u,u_t) + \frac{1}{3\varepsilon}E_{1\varepsilon}(u,u_t) \leq \frac{2}{\varepsilon}C_b + \frac{1}{\varepsilon^2}\|u\|^2 \leq \frac{2}{\varepsilon^2}C_b. \tag{6.41}$$

Let $C_\varepsilon := \frac{6}{\varepsilon}C_b$. Integrating (6.41), we deduce that for all $t \geq 0$,

$$E_{1\varepsilon}(u(t),u_t(t)) \leq (E_{1\varepsilon}(u_0,u_1) - C_\varepsilon)e^{-t/3\varepsilon} + C_\varepsilon.$$

Thus, it follows that if $R_{1\varepsilon} > C_\varepsilon$, the set $\mathcal{B}_{2\varepsilon}$ defined in (6.37) is positively invariant and absorbing for $S_{\varepsilon 0}$. This concludes the proof of proposition 6.4 (when $\delta = 0$); note that, in general, $R_{1\varepsilon}$ is unbounded as $\varepsilon \to 0$. ⬜

6.1.4   The Global Attractor

In this section we resort to theorem 2.56 of chapter 2 to show that the semiflow S generated by problem (CH$_{\varepsilon\delta}$) admits a global attractor $\mathcal{A}_{\varepsilon\delta}$ in $\mathcal{X}_0$. More precisely, we resort to the α-contraction method described in section 3.4.5, to show that the ω-limit set $\omega_{\varepsilon\delta}(\mathcal{B}_0)$, where $\mathcal{B}_0$ is the absorbing set determined in proposition 6.3, is the global attractor for $S_{\varepsilon\delta}$ in $\mathcal{X}_0$. Note that this set is not empty, since it contains the stationary solutions of problem (CH$_{\varepsilon\delta}$). Recalling proposition 2.59, to apply theorem 2.56 it is sufficient to find $t_* > 0$ such that the operator $S_{\varepsilon\delta}(t_*)$ is an α-contraction in $\mathcal{X}_0$, up to a precompact pseudometric.

THEOREM 6.5
Let ε, $\delta \in]0,1]$, or, if $\delta = 0$, $\varepsilon \in]0,\frac{1}{3}]$. The set $\mathcal{A}_{\varepsilon\delta} := \omega_{\varepsilon\delta}(\mathcal{B}_0)$ is a global attractor in $\mathcal{X}_0$ for the semiflow $S_{\varepsilon\delta}$ generated by problem (CH$_{\varepsilon\delta}$).

PROOF   It is sufficient to note that, as a second consequence of estimate (6.33), we can apply theorem 2.56 to the semiflow $S_{\varepsilon\delta}$. Indeed, if e.g. we choose $t_* > 0$ such that $q_* := M\,\alpha e^{-t_*/6\alpha} < 1$, the operator $S_{\varepsilon\delta}(t_*)$ is a strict contraction in $\mathcal{X}_0$, up to the pseudometric ψ_* defined by

$$\psi_*\big((u,v),(\tilde{u},\tilde{v})\big) := \left(\alpha M \int_0^{t_*} \|z(s)\|^2\,ds\right)^{1/2},$$

where, for (u,v), $(\tilde{u},\tilde{v}) \in \mathcal{X}_0$, $z := u - \tilde{u}$ is the difference of the solutions to problem (CH$_{\varepsilon\delta}$), corresponding to initial values (u,v) and $(\tilde{u},\tilde{v})$. This pseudometric is clearly precompact, because of the compactness of the injection

$$\{u \in L^2(0,t_*;\mathcal{H}^1): u_t \in L^2(0,t_*;\mathcal{H}^{-1})\} \hookrightarrow L^2(0,t_*;L^2(0,\pi)).$$

Thus, by proposition 2.59, the map $S_{\varepsilon\delta}(t_*)$ is an α-contraction. In turn, theorem 2.56 implies that the ω-limit set of the set $\mathcal{B}_0$ defined in (6.35) is the global attractor for the semiflow $S_{\varepsilon\delta}$. ⬜

REMARK 6.6   We can say a lot more on the dependence of the attractors $\mathcal{A}_{\varepsilon\delta}$ on the parameters ε and δ. In particular, we can show that $\mathcal{A}_{\varepsilon\delta}$ is uniformly bounded, in $\mathcal{X}_0$, with respect to both parameters ε and δ. Moreover, if $\delta > 0$, the attractor $\mathcal{A}_{\varepsilon\delta}$ is contained in a bounded set of $\mathcal{X}_2$, which is independent of ε. The additional regularity of the attractor when $\delta > 0$ is due to the presence of the term $-\delta\Delta u_t$, which has a regularizing effect on the solution. In contrast, the question of the validity of the analogous result for $\mathcal{A}_{\varepsilon 0}$ is open; in fact, we do not even know if the inclusion $\mathcal{A}_{\varepsilon 0} \subset \mathcal{X}_1$ holds. Finally, the attractors $\mathcal{A}_{\varepsilon\delta}$ are also upper-semicontinuous with respect to both ε and δ. More precisely, let $\mathcal{A}_{0\delta} \subset \mathcal{H}^3$ be the global attractor of the semiflow $S_{0\delta}$ generated by the parabolic problem (CH$_{0\delta}$), and define the set

$$\mathcal{A}_{0\delta} := \{(u,v) \in \mathcal{X}_0 : u \in \mathcal{A}_{0\delta}, v = -\Delta(I - \delta\Delta)^{-1}(u - u^3 + \Delta u)\}.$$

We interpret $\mathcal{A}_{0\delta}$ as the "natural" imbedding of $\mathcal{A}_{0\delta}$ in $\mathcal{X}_0$. Then, we have the commutative diagram

$$\begin{array}{ccc} \mathcal{A}_{\varepsilon\delta} & \longrightarrow & \mathcal{A}_{\varepsilon 0} \\ \downarrow & & \downarrow \\ \mathcal{A}_{0\delta} & \longrightarrow & \mathcal{A}_{00} \end{array} \ ,$$

where the vertical arrows mean convergence as $\varepsilon \to 0$, and the horizontal arrows mean convergence as $\delta \to 0$, in the sense of the semidistance ∂ of (2.2), in the topology of $\mathcal{X}_1$. For a proof of these results, based on techniques analogous of those we used in sections 3.5 and 3.6, we refer to Zheng-Milani, [ZM03].   □

6.1.5   The Exponential Attractor

In this section we show that the semiflow $S_{\varepsilon\delta}$ also admits an exponential attractor $\mathcal{E}_{\varepsilon\delta}$ in $\mathcal{X}_0$, which contains the global attractor $\mathcal{A}_{\varepsilon\delta}$.

THEOREM 6.7
In the same conditions of proposition 6.4, the semiflow $S_{\varepsilon\delta}$ generated by problem (CH$_{\varepsilon\delta}$) admits an exponential attractor $\mathcal{E}_{\varepsilon\delta}$ in $\mathcal{X}_0$.

PROOF   As before, we prove theorem 6.7 in detail only in the case $\delta = 0$. We apply theorem 4.5 of chapter 4. To this end, we consider the absorbing sets $\mathcal{B}_{1\delta}$ or $\mathcal{B}_{1\varepsilon}$ for $S_{\varepsilon\delta}$ in $\mathcal{X}_1$, determined in proposition 6.4. Since the injection $\mathcal{X}_1 \hookrightarrow \mathcal{X}_0$ is compact, these sets are compact in $\mathcal{X}_0$. We propose then to show that $S_{\varepsilon\delta}$ satisfies the discrete squeezing property (see definition 4.3), relative to the set $\mathcal{B}_1$, where $\mathcal{B}_1 = \mathcal{B}_{1\delta}$ if $\delta > 0$, and $\mathcal{B}_1 = \mathcal{B}_{1\varepsilon}$ if $\delta = 0$.

1.   We proceed almost exactly as in the proof of theorem 4.10 of section 4.4.2, of which we keep the same choices of N and $\mathcal{X}_N$, and denote by $\| \cdot \|_0$ the norm induced on $\mathcal{X}_0$ by E_0. Thus, we show that, given any $t_* > 0$ and $\gamma \in]0, \frac{1}{2}[$, there exists an integer N_*, with the property that if u_0, $\bar{u}_0 \in \mathcal{B}_1$ are such that $S(t_*)u_0 - S(t_*)\bar{u}_0 \notin \mathcal{C}_{N_*}$, i.e. if

$$\|P_{N_*}(S(t_*)u_0 - S(t_*)\bar{u}_0)\|_0 < \|Q_{N_*}(S(t_*)u_0 - S(t_*)\bar{u}_0)\|_0 \tag{6.42}$$

(that is, if (4.13) holds for the operator $S(t_*)$), then (4.11) must hold, i.e.

$$\|S(t_*)u_0 - S(t_*)\bar{u}_0\|_0 \leq \gamma \|u_0 - v_0\|_0. \tag{6.43}$$

To this end, we define on $\mathcal{X}_0$ the function

$$M_0(u,v) := \varepsilon \|v\|_{-1}^2 + [u,v] + \|\nabla u\|^2 + \tfrac{\delta}{2\varepsilon}\|u\|^2.$$

It is then easy to verify that, if $\varepsilon\lambda_{N+1} \geq K^2 + 1$, with K as in (6.7), M_0 is the square of an equivalent norm in the subspace $\mathcal{Q}_N(\mathcal{X})$ (the projection $\mathcal{Q}_N$ being defined as in (4.38)), with

$$\tfrac{1}{2\alpha}E_0(u,v) \leq M_0(u,v) \leq 3E_0(u,v). \tag{6.44}$$

2.   We now estimate the difference of solutions of (6.1), whose orbits are in $\mathcal{B}_1$. If u and $\tilde{u}$ are two such solutions, corresponding to initial values $U_0 := (u_0, u_1)$, $\tilde{U}_0 := (\tilde{u}_0, \tilde{u}_1) \in \mathcal{B}_1$, we set $z(t) := u(\cdot,t) - \tilde{u}(\cdot,t)$, and $Z(t) := (z(t), z_t(t))$. At first, we recall estimate (6.34), which provides a control of the growth of Z on bounded time intervals. We can rewrite (6.34) as

$$E_0(z(t), z_t(t)) \leq C E_0(z(0), z_t(0))e^{ct},$$

with C and c independent of u, $\tilde{u}$ and t.   Next, we establish a linear differential inequality on Z in $\mathcal{Q}_N\mathcal{X}_0$, for $N \in \mathbb{N}$ so large that $\varepsilon\lambda_{N+1} \geq \max\{K^2 + 1, 4\}$. More precisely, we set $q := \mathcal{Q}_N(z)$, and claim that the function $t \mapsto M_0(q(t), q_t(t))$ satisfies the linear differential inequality

$$\frac{d}{dt}M_0(q,q_t) + \tfrac{1}{3\varepsilon}M_0(q,q_t) \leq K_\delta \|\nabla z(t)\|^2, \tag{6.45}$$

where, when $\delta = 0$, the constant K_0 is independent of t, N and ε, and, if $\delta > 0$, $K_\delta := \frac{K_0}{\delta\lambda_{N+1}}$. As stated above, we show (6.45) only for the case $\delta = 0$. Applying $\mathcal{Q}_N$ to equation (6.21), and noting that $\mathcal{Q}_N$ commutes with $-\Delta$, we see that q satisfies the equation

$$\varepsilon(-\Delta)^{-1}q_{tt} + (-\Delta)^{-1}q_t - \Delta q = q - \mathcal{Q}_N(u^3 - \tilde{u}^3) =: g_N. \tag{6.46}$$

Multiplying (6.46) in $L^2(0,\pi)$ by $2q_t$ and $\tfrac{1}{\varepsilon}q$, adding the resulting identities, we obtain

$$\frac{d}{dt}M_0(q,q_t) + \|q_t\|_{-1}^2 + \tfrac{1}{\varepsilon}[q_t,q] + \tfrac{1}{\varepsilon}\|\nabla q\|^2 = \tfrac{1}{\varepsilon}\langle g_N, 2\varepsilon q_t + q\rangle. \tag{6.47}$$

Splitting the term $\frac{1}{\varepsilon}[q_t,q]$ in two, we deduce from (6.47)

$$\frac{d}{dt}M_0(q,q_t)+\|q_t\|_{-1}^2+\frac{1}{2\varepsilon}[q_t,q]+\frac{1}{\varepsilon}\|\nabla q\|^2 \le \frac{1}{\varepsilon}\langle g_N,2\varepsilon q_t+q\rangle -\frac{1}{2\varepsilon}[q_t,q]. \quad (6.48)$$

Since $\varepsilon \le 1$, we have that

$$-\frac{1}{2\varepsilon}[q_t,q] \le \frac{1}{4}\|q_t\|_{-1}^2+\frac{K^2+1}{4\varepsilon^2\lambda_{N+1}}\|\nabla q\|^2 \le \frac{1}{4}\|q_t\|_{-1}^2+\frac{1}{4\varepsilon}\|\nabla q\|^2. \quad (6.49)$$

Next, we estimate

$$\frac{1}{\varepsilon}\langle g_N,2\varepsilon q_t+q\rangle \le C\frac{2}{\varepsilon}\|\nabla q-\nabla(u^3-\tilde{u}^3)\|\,(\varepsilon\|q_t\|_{-1}+\|q\|). \quad (6.50)$$

Writing

$$\nabla(u^3-\tilde{u}^3)=3u^2(\nabla u-\nabla\tilde{u})+3(u^2-\tilde{u}^2)\nabla\tilde{u},$$

we have

$$\|\nabla q-\nabla(u^3-\tilde{u}^3)\| \le |3u^2-1|_\infty\|\nabla z\|+3|u+\tilde{u}|_\infty\|\nabla\tilde{u}\|\,|z|_\infty \le C\|\nabla z\|, \quad (6.51)$$

where C depends on the uniform bounds on u and $\tilde{u}$ in $\mathcal{H}^1$. Inserting (6.51) into (6.50), and recalling that $\varepsilon\lambda_{N+1} \ge 4$, we obtain (for different constants C)

$$\frac{1}{\varepsilon}\langle g_N,2\varepsilon q_t+q\rangle \le C\|\nabla z\|\left(\|q_t\|_{-1}+\frac{1}{\varepsilon\sqrt{\lambda_{N+1}}}\|\nabla q\|\right)$$

$$\le C\|\nabla z\|^2+\frac{1}{4}\|q_t\|_{-1}^2+\frac{1}{\varepsilon^2\lambda_{N+1}}\|\nabla q\|^2 \quad (6.52)$$

$$\le C\|\nabla z\|^2+\frac{1}{4}\|q_t\|_{-1}^2+\frac{1}{4\varepsilon}\|\nabla q\|^2.$$

From (6.52), (6.49) and (6.48) we deduce then that, for suitable $C>0$,

$$\frac{d}{dt}M_0(q,q_t)+\frac{1}{2}\left(\|q_t\|_{-1}^2+\frac{1}{\varepsilon}[q_t,q]+\frac{1}{\varepsilon}\|\nabla q\|^2\right) \le C\|\nabla z\|^2,$$

which yields (6.45) when $\delta=0$.

3. We now recall (6.34), which provides an estimate on the difference of solutions on bounded intervals. Setting $\beta := M\alpha$, (6.34) yields

$$\frac{1}{2}\|\nabla z(t)\|^2 \le E_0(z(t),z_t(t)) \le \beta E_0(z(0),z_t(0))e^{\beta t},$$

for all $t \ge 0$. Replacing this into (6.45), then integrating, and recalling (6.44), we obtain

$$M_0(q(t),q_t(t)) \le 3E_0(z(0),z_t(0))e^{-t/3\varepsilon}+6\beta\varepsilon M_0\,E_0(z(0),z_t(0))e^{\beta t} \quad (6.53)$$

(for $\delta = 0$). Given $\gamma \in \,]0, \frac{1}{2}[$, we first choose $t_* > 0$ so large that $8\alpha\,e^{-t_*/3\varepsilon_0} \le 1$, and then $\varepsilon_1 \in \,]0, \varepsilon_0]$ so small that $16\alpha\beta\varepsilon_1 C_5 e^{\beta t_*} \le \gamma^2$. With these choices, we deduce from (6.53), that, if $\varepsilon \le \varepsilon_1$,

$$M_0(q(t_*), q_t(t_*)) \le \tfrac{1}{4\alpha}\gamma^2 E_0(z(0), z_t(0)).$$

Thus, if (6.42) holds, then, by (6.44),

$$E_0(z(t_*), z_t(t_*)) \le 2E_0\left(\mathcal{Q}_{N_*}(z(t_*), z_t(t_*))\right)$$
$$\le 4\alpha M_0(q(t_*), q_t(*)) \le \gamma^2 E_0(z(0), z_t(0)).$$

This means that (6.43) holds, as desired. Consequently, we conclude that if $\delta = 0$ and $\varepsilon \le \varepsilon_1$, the semiflow $S_{\varepsilon 0}$ satisfies the discrete squeezing property. We can then apply theorem 4.5, and conclude the proof of theorem 6.7 when $\delta = 0$. The proof when $\delta > 0$ is similar (and actually simpler).   □

6.1.6   The Inertial Manifold

In this section we show that the semiflow $S_{\varepsilon\delta}$ also admits an inertial manifold in $\mathcal{X}_0$, constructed with the techniques of section 5.7.2 of chapter 5. That is, the inertial manifold is the graph of a Lipschitz continuous function defined over a finite dimensional subspace of $\mathcal{X}_0$.

1.   As in section 5.7.2, we introduce the time rescaling $t \mapsto \sqrt{\varepsilon}t$, which transforms equation (6.1) into

$$u_{tt} + 2\alpha u_t + \Delta\left(\Delta u - u^3 + u - 2\alpha\delta u_t\right) = 0, \tag{6.54}$$

with $\alpha := \frac{1}{2\sqrt{\varepsilon}}$. Then, we transform equation (6.54) into the first order system

$$U_t + AU = F(U) \tag{6.55}$$

for $U = (u, v) \in \mathcal{X}_0$, where

$$A = \begin{pmatrix} 0 & -1 \\ \Delta^2 & 2\alpha(1 - \delta\Delta) \end{pmatrix}, \quad F(U) := \begin{pmatrix} 0 \\ g(u) \end{pmatrix}, \tag{6.56}$$

with $g(u) := \Delta(u^3 - u)$. Note that, if $\delta > 0$,

$$\mathrm{dom}(A) = \{u \in \mathrm{H}^3(0, \pi) \colon u \in \mathcal{H}, u_{xx} \in \mathcal{H}^2\} \times \mathcal{H}^1,$$

while if $\delta = 0$, the second factor is to be replaced by $\mathcal{H}^{-1}$.

System (6.55) generates a semiflow in $\mathcal{X}$, which we still denote by $S_{\varepsilon\delta}$: our goal is to show that the semiflow generated by system (6.55) satisfies the strong squeezing property. In turn, this will be a consequence of the fact that the operator A in (6.55) satisfies the spectral gap condition (5.83) of definition 5.32, either with respect to the standard graph norm in $\mathcal{X}_0$, defined in (6.57) below, or to an equivalent one.

2.   We consider in $\mathcal{X}_0$ the usual graph norm, induced by the scalar product

$$\langle U,V\rangle_0 := \langle \nabla u, \nabla \bar{y}\rangle + [\bar{z},v]\,, \quad U=(u,v)\,, \quad V=(y,z)\in \mathcal{X}_0\,, \tag{6.57}$$

where the bar denotes complex conjugation. Note that the last term of (6.57), defined in (6.6), makes sense, because $z \in \mathcal{H}^{-1}$, and $(-\Delta)^{-1}v \in \mathcal{H}^1$ since $v \in \mathcal{H}^{-1}$. Moreover, the operator A defined in (6.56) is monotone. Indeed, for $U \in \mathrm{dom}(A)$, $\langle AU,U\rangle_0$ is real and nonnegative, since

$$\begin{aligned}
\langle AU,U\rangle_0 &= -\langle \nabla v, \nabla \bar{u}\rangle + \langle \bar{v},(-\Delta)^{-1}(\Delta^2 u + 2\alpha(1-\delta\Delta)v)\rangle\\
&= -\langle \nabla v, \nabla \bar{u}\rangle + \langle \nabla \bar{v}, \nabla u\rangle + 2\alpha\|v\|_{-1}^2 + 2\alpha\delta\|v\|^2\\
&= 2\alpha\left(\|v\|_{-1}^2 + \delta\|v\|^2\right).
\end{aligned}$$

To determine the eigenvalues of A, we observe that the eigenvalue equation

$$AU = \mu U\,, \quad U=(u,v)\in \mathcal{X}_0\,,$$

is equivalent to the system

$$-v = \mu u\,, \quad \Delta^2 u + 2\alpha(1-\delta\Delta)v = \mu v\,. \tag{6.58}$$

Thus, u must solve the eigenvalue problem

$$\begin{cases} \Delta^2 u + 2\alpha\delta\mu\Delta u = (2\alpha\mu - \mu^2)u\,,\\ u(0) = u(\pi) = 0\,,\\ \Delta u(0) = \Delta u(\pi) = 0\,. \end{cases} \tag{6.59}$$

We easily see that (6.59) has, for each positive integer j, the pair of eigenvalues

$$\mu_j^\pm := \alpha(1+\delta\,j^2)\pm\sqrt{\alpha^2(1+\delta\,j^2)^2 - j^4} \in \mathbb{C}\,;$$

thus, A does have a countable set of eigenvalues, with $\mathfrak{Re}\,\mu_j^\pm > 0$ for all j. Because of the first of (6.58), the corresponding eigenfunctions have the form $U_j^\pm = (u_j, -\mu_j^\pm u_j)$, with $u_j(x) = \sqrt{\frac{2}{\pi}}\sin(jx)$. For future reference, we note that for all $j \ge 1$

$$\|\nabla u_j\| = j\,, \quad \|u_j\|_{-1} = j^{-1}\,. \tag{6.60}$$

3.   A remarkable feature of system (6.55), with A and F defined in (6.56), is a difference in the distribution of the eigenvalues $\mu_j^\pm$, and the consequent possibility of satisfying the spectral gap condition, according to whether $\delta > 0$ or $\delta = 0$. Indeed, if $\delta = 0$ the eigenvalues reduce to

$$\mu_j^\pm = \alpha \pm \sqrt{\alpha^2 - j^4}\,;$$

thus, if $\alpha < 1$, i.e. if ε is so large that $4\varepsilon > 1$, all the eigenvalues of A are complex, nonreal, and have the same real part α. This situation is similar to the semilinear dissipative wave equation considered in section 5.7.2; as in that example, in this case it is impossible to find any decomposition of the eigenvalues of A such that the spectral gap condition (5.83) holds, and the existence of an inertial manifold for equation (6.55) cannot be guaranteed. On the other hand, when $\alpha > 1$ some eigenvalues of A will be real, positive and distinct, so that the spectral gap condition may hold. In contrast, we will see that the spectral gap condition always holds if $\delta > 0$, even if all eigenvalues $\mu_j^{\pm}$ are complex, nonreal.

4.   In the remainder of this section, we will assume that the nonlinearity $g \colon \mathcal{H}^1 \to \mathcal{H}^{-1}$ in (6.56) is globally bounded and globally Lipschitz continuous, with Lipschitz constant ℓ. To construct the inertial manifold for the original semiflow $S_{\varepsilon\delta}$, we would then need to adjust the nonlinearity, in a way similar to section 5.8.3; for this technical part, we refer to Zheng-Milani, [ZM05]. Under this assumption, in the viscous case $\delta > 0$ we have

THEOREM 6.8

Let $N_1 \in \mathbb{N}$ be so large that if $N \geq N_1$, then:

1. *If $0 < \alpha\delta < 1$, $\alpha\delta(2N+1) > 2\ell$;*

2. *If $\alpha\delta = 1$, $2N \geq 2\ell + \sqrt{2\alpha}$ and*

$$\sqrt{2\alpha N^2 + \alpha^2} - \sqrt{2\alpha(N+1)^2 + \alpha^2} + \sqrt{2\alpha} \geq -1 .$$

3. *If $\alpha\delta > 1$, the inequalities*

$$(2N+1)\left(\alpha\delta - \sqrt{\alpha^2\delta^2 - 1}\right) \geq \frac{2\ell}{\sqrt{\alpha\delta - 1}} + 1, \quad (6.61)$$

$$\left|\sqrt{R(N)} - \sqrt{R(N+1)} + (2N+1)\sqrt{\alpha^2\delta^2 - 1}\right| \leq 1 \quad (6.62)$$

hold, where

$$R(N) := (\alpha^2\delta^2 - 1)N^4 + 2\alpha^2\delta N^2 + \alpha^2 . \quad (6.63)$$

In each of these cases, the operator A satisfies the spectral gap condition (5.83) in $\mathcal{X}_0$, with respect to either the graph norm (6.57), or an equivalent one. Consequently, the semiflow $S_{\varepsilon\delta}$ generated by (6.55) admits an inertial manifold in $\mathcal{X}_0$, of the form (5.16).

Similarly, in the nonviscous case $\delta = 0$ we have

THEOREM 6.9

Let $N \in \mathbb{N}$ be the smallest integer such that $(N+1)^4 - N^4 > 4\ell$. Assume that $\alpha = \frac{1}{2\sqrt{\varepsilon}}$ is so large that

$$\alpha^2 > (N+1)^4 + \frac{4\ell^2}{(N+1)^4 - 4\ell} \, .$$

There is then an equivalent norm in $\mathcal{X}$ such that the operator A defined in (6.56), with $\delta = 0$, satisfies the spectral gap condition. Consequently, the semiflow S generated by (6.55) admits an inertial manifold in $\mathcal{X}_0$, of the form (5.16).

5. We will only prove theorem 6.8 in the case $\alpha\delta > 1$, which is the most difficult; for all other cases, we refer to Zheng-Milani, [ZM05].

Since $\alpha\delta > 1$, all eigenvalues $\mu_j^{\pm}$ of A are real and positive, and we easily see that both sequences $(\mu_j^-)_{j\geq 1}$ and $(\mu_j^+)_{j\geq 1}$ are increasing. We will proceed in four steps.

5.1. Setting $\gamma := \alpha\delta + \sqrt{\alpha^2\delta^2 - 1}$, we easily check that, as $j \to +\infty$

$$\mu_j^{\pm} = \gamma j^2 + \alpha \pm \frac{\alpha^2\delta}{\sqrt{\alpha^2\delta^2 - 1}} + O(j^{-2}). \tag{6.64}$$

Since $\gamma > 1$, (6.64) implies that it is impossible to decompose the point spectrum of A in such a way that the corresponding subspaces $\mathcal{X}_1$ and $\mathcal{X}_2$ are orthogonal. Indeed, for any such decomposition $\sigma_1 \cup \sigma_2$ there is at least one index j such that $U_j^- \in \mathcal{X}_1$ and $U_j^+ \in \mathcal{X}_2$; for this j, recalling (6.57) and (6.60), and noting that $\mu_j^- \mu_j^+ = j^4$, we compute that

$$\left\langle U_j^-, U_j^+ \right\rangle_{\mathcal{X}} = \|\nabla u_j\|^2 + [-\mu_j^- u_j, -\mu_j^+ u_j] = \|\nabla u_j\|^2 + \mu_j^- \mu_j^+ \|u_j\|_{-1}^2 = 2j^2 \, .$$

To overcome this difficulty, we will define a new scalar product in $\mathcal{X}_0$, equivalent to (6.57), with respect to which the subspaces $\mathcal{X}_1$ and $\mathcal{X}_2$ will be orthogonal. To this end, we note that the same asymptotic distribution (6.64) assures that, if the eigenvalues $\mu_j^{\pm}$ are listed in nondecreasing order, then for arbitrarily large N there are consecutive eigenvalues μ_N^- and μ_{N+1}^-. More precisely,

PROPOSITION 6.10

Let the eigenvalues $\mu_j^{\pm}$, $j \geq 1$, be arranged in nondecreasing order. For all $m \in \mathbb{N}$, there is $N \geq m$ such that μ_N^- and μ_{N+1}^- are consecutive.

PROOF For $m \geq 1$, let q_m denote the number of indices n such that $\mu_m^+ < \mu_n^- \leq \mu_{m+1}^+$. We have to show that for all $m \in \mathbb{N}$ there is $N \geq m$ such that $q_N \geq 2$. Assume otherwise, i.e., that there is $m_0 \geq 1$ such that for all $m \geq m_0$, $q_m \leq 1$. This means that for each $m \geq m_0$ there is at most one $\mu_n^- \in \,]\mu_m^+, \mu_{m+1}^+]$. In turn, this defines a function

$m \mapsto n(m)$, $m \geq m_0$. For $r \geq 0$, let $m_r := m_0 + r$ and $n_r := n(m_r)$. Then, after possibly a finite number of them, the eigenvalues are ordered as

$$\mu_{m_0}^+ < q_{m_0}\mu_{n_0}^- + (1 - q_{m_0})\mu_{m_1}^+ \leq \mu_{n_1}^- < q_{m_1}\mu_{n_1}^- + (1 - q_{m_1})\mu_{m_2}^+$$

$$\leq \cdots \leq \mu_{m_r}^+ < q_{m_r}\mu_{n_r}^- + (1 - q_{m_r})\mu_{m_{r+1}}^+ \leq \mu_{m_{r+1}}^- < \cdots . \qquad (6.65)$$

Since the sequence $(\mu_j^-)_{j \geq 1}$ is increasing, (6.65) implies that $n_r = n_0 + r$, and, therefore, $\mu_{n_0+r}^- > \mu_{m_0+r}^+$ for all $r \geq 0$. Now, from (6.64) we have that

$$\tfrac{1}{\gamma}(n_0 + r)^2 > \gamma(m_0 + r)^2 + \frac{2\alpha^2\delta}{\sqrt{\alpha^2\delta^2 - 1}} + O\left(r^{-2}\right)$$

as $r \to +\infty$. However, this is impossible, since $\gamma > 1$. $\qquad\qquad\Box$

5.2.   Given then N such that μ_N^- and μ_{N+1}^- are consecutive, we separate the eigenvalues of A as follows. Denoting by $\sigma_{\mathrm{p}}(A)$ the point spectrum of A, i.e. the sequence of its eigenvalues $(\mu_j^\pm)_{j \in \mathbb{N}_{>0}}$, and by $\mathcal{U}$ the set of the corresponding sequence of eigenvectors $(U_j^\pm)_{j \in \mathbb{N}_{>0}}$, we set

$$I_0 = \{j \in \mathbb{N}: \mu_j^- \leq \mu_j^+ \leq \mu_N^-\}, \quad I_1 = \{j \in \mathbb{N}: \mu_j^- \leq \mu_N^- < \mu_j^+\}$$

$$\sigma_1 := \{\mu_j^\pm : j \in I_0\} \cup \{\mu_j^- : j \in I_1\}, \quad \sigma_2 := \sigma_{\mathrm{p}}(A) \setminus \sigma_1,$$

$$\mathcal{U}_1 = \{U_j^\pm : j \in I_0\} \cup \{U_j^- : j \in I_1\}, \quad \mathcal{U}_2 = \mathcal{U} \setminus \mathcal{U}_1, \qquad (6.66)$$

and consider the corresponding decomposition of

$$\mathcal{X}_1 := \mathrm{span}\,\mathcal{U}_1, \quad \mathcal{X}_2 := \mathrm{span}\,\mathcal{U}_2. \qquad (6.67)$$

We explicitly note that, in this section, $\mathcal{X}_1$ and $\mathcal{X}_2$ denote the subspaces of $\mathcal{X}_0$ defined in (6.67), and not those defined in (6.4) and (6.7). Our goal is to make these two subspaces orthogonal, and to show that the spectral inequality (5.83) holds, with $\Lambda_1 = \mu_N^-$ and $\Lambda_2 = \mu_{N+1}^-$, in accord with (6.66). We further decompose $\mathcal{X}_2 := \mathcal{X}_C \oplus \mathcal{X}_R$, with

$$\mathcal{X}_C := \mathrm{span}\,\mathcal{U}_C, \quad \mathcal{U}_C = \{U_j^+ : j \in I_1\},$$

$$\mathcal{X}_R := \mathrm{span}\,\mathcal{U}_R, \quad \mathcal{U}_R = \mathcal{U}_2 \setminus \mathcal{U}_C, \qquad (6.68)$$

and set $\mathcal{X}_N := \mathcal{X}_1 \oplus \mathcal{X}_C$. Note that $\mathcal{X}_1$ and $\mathcal{X}_C$ are finite dimensional, that $U_N^- \in \mathcal{X}_1$, $U_{N+1}^- \in \mathcal{X}_R$, and that the reason why $\mathcal{X}_1$ is not orthogonal to $\mathcal{X}_2$ is that, while it is orthogonal to $\mathcal{X}_R$, $\mathcal{X}_1$ is not orthogonal to $\mathcal{X}_C$.

We now introduce two functions $\Phi: \mathcal{X}_N \to \mathbb{R}$ and $\Psi: \mathcal{X}_R \to \mathbb{R}$, defined by

$$\Phi(U,V) := 2\alpha\langle u, \bar{y}\rangle + (2\alpha\delta - 1)\langle \nabla u, \nabla \bar{y}\rangle + \left\langle (-\Delta)^{-1/2}\bar{z}, (-\Delta)^{1/2}u\right\rangle$$

$$+ \left\langle (-\Delta)^{-1/2}\bar{v}, (-\Delta)^{1/2}y\right\rangle + [\bar{z}, v], \qquad (6.69)$$

$$\Psi(U,V) := \alpha\delta\langle\nabla u, \nabla\bar{y}\rangle + \left\langle(-\Delta)^{-1/2}\bar{z}, (-\Delta)^{1/2}u\right\rangle$$

$$+ \left\langle(-\Delta)^{-1/2}\bar{v}, (-\Delta)^{1/2}y\right\rangle + [\bar{z},v], \tag{6.70}$$

with $U = (u,v)$, $V = (y,z) \in \mathcal{X}_N$ or, respectively, $\mathcal{X}_R$. These functions are well defined: Indeed, since $u \in \mathrm{H}^1(0,\pi)$ and $z \in \mathrm{H}^{-1}(0,\pi)$, then both $(-\Delta)^{-1/2}z$ and $(-\Delta)^{1/2}u$ are in $\mathrm{L}^2(0,\pi)$, and analogously for y and v. We now show that Φ and Ψ are positive definite. Let first $U = (u,v) \in \mathcal{X}_N$: then,

$$\Phi(U,U) = 2\alpha\|u\|^2 + (2\alpha\delta - 1)\|\nabla u\|^2 + 2\left\langle(-\Delta)^{-1/2}\bar{v}, (-\Delta)^{1/2}u\right\rangle + \|v\|_{-1}^2$$

$$\geq 2\alpha\|u\|^2 + (2\alpha\delta - 1)\|\nabla u\|^2 - 2\|v\|_{-1}\|\nabla u\| + \|v\|_{-1}^2$$

$$\geq 2\alpha\|u\|^2 + (2\alpha\delta - 1)\|\nabla u\|^2 - \|v\|_{-1}^2 - \|\nabla u\|^2 + \|v\|_{-1}^2$$

$$\geq 2\alpha\|u\|^2 + 2(\alpha\delta - 1)\|\nabla u\|^2. \tag{6.71}$$

Since $\alpha\delta > 1$, we conclude that $\Phi(U,U) \geq 0$ for all $U \in \mathcal{X}_N$. Analogously, for $U \in \mathcal{X}_R$:

$$\Psi(U,U) = \alpha\delta\|\nabla u\|^2 + 2\left\langle(-\Delta)^{-1/2}\bar{v}, (-\Delta)^{1/2}u\right\rangle + \|v\|_{-1}^2$$

$$\geq (\alpha\delta - 1)\|\nabla u\|^2 + \|\nabla u\|^2 - 2\|v\|_{-1}\|\nabla u\| + \|v\|_{-1}^2 \tag{6.72}$$

$$\geq (\alpha\delta - 1)\|\nabla u\|^2, \tag{6.73}$$

from which we conclude that also $\Psi(U,U) \geq 0$ for all $U \in \mathcal{X}_R$. Thus, Φ and Ψ define a scalar product, respectively on $\mathcal{X}_N$ and $\mathcal{X}_R$, and we can define an equivalent scalar product in $\mathcal{X}_0$, by

$$\langle\!\langle U,V\rangle\!\rangle := \Phi(P_N U, P_N V) + \Psi(P_R U, P_R V), \tag{6.74}$$

where P_N and P_R are, respectively, the projections of $\mathcal{X}$ onto $\mathcal{X}_N$ and $\mathcal{X}_R$. For simplicity, with a slight abuse of notation we shall write (6.74) simply as

$$\langle\!\langle U,V\rangle\!\rangle := \Phi(U,V) + \Psi(U,V). \tag{6.75}$$

We proceed then to show that the subspaces $\mathcal{X}_1$ and $\mathcal{X}_2$ defined in (6.67) are orthogonal with respect to the scalar product (6.75). In fact, it is sufficient to show that $\mathcal{X}_1$ is orthogonal to $\mathcal{X}_C$; in turn, this reduces to showing that $\langle\!\langle U_j^-, U_j^+\rangle\!\rangle_{\mathcal{X}} = 0$ if $U_j^- \in \mathcal{X}_N$ and $U_j^+ \in \mathcal{X}_C$. Recalling (6.69) and (6.70), we immediately compute that

$$\langle\!\langle U_j^-, U_j^+\rangle\!\rangle = \Phi(U_j^-, U_j^+)$$

$$= 2\alpha\|u_j\|^2 + (\alpha\delta - 1)\|\nabla u_j\|^2 - (\mu_j^+ + \mu_j^-)\|u_j\|^2$$

$$+ \mu_j^+ \mu_j^- \|u_j\|_{-1}^2. \tag{6.76}$$

Recalling (6.60), and noting that $\mu_j^- \mu_j^+ = j^4$ and $\mu_j^+ + \mu_j^- = 2\alpha(1 + \delta j^2)$, we conclude from (6.76) that $\langle\!\langle U_j^-, U_j^+\rangle\!\rangle = 0$, as claimed.

5.3. Having thus established the desired orthogonal decomposition, we proceed to show that A satisfies the spectral gap inequality (5.83), with respect to the equivalent norm $\||\cdot\||$ in $\mathcal{X}_0$ defined by the scalar product (6.75). For this, we first need to estimate the Lipschitz constant ℓ_F of F; recall that $F(U) := (0, g(u))$, and that we are assuming that g is globally Lipschitz continuous. Let $P_1 : \mathcal{X} \to \mathcal{X}_1$ and $P_2 : \mathcal{X} \to \mathcal{X}_2$ be the orthogonal projections corresponding to the decomposition $\mathcal{X}_0 = \mathcal{X}_1 \oplus \mathcal{X}_2$. P_1 and P_2 induce corresponding projections p_1 and p_2 in $\mathcal{H}^1$ and $\mathcal{H}^{-1}$ in a natural way. Recalling (6.71), (6.73), it follows that, for $U = (u, v) \in \mathcal{X}_0$,

$$\||U\||^2 = \Phi(P_1 U, P_1 U) + \Psi(P_2 U, P_2 U) \geq 2\alpha \|p_1 u\|^2 + (\alpha\delta - 1)\|p_2 u\|^2$$
$$\geq (\alpha\delta - 1)\left(\|p_N u\|^2 + \|p_2 u\|^2\right) = (\alpha\delta - 1)\|u\|^2. \tag{6.77}$$

Given then $U = (u, \tilde{u})$ and $V = (v, \tilde{v}) \in \mathcal{X}$, we compute

$$\||F(U) - F(V)\|| = \|g(u) - g(v)\|_{-1} \leq \ell \|\nabla u - \nabla v\| \leq \frac{\ell}{\sqrt{\alpha\delta - 1}} \||U - V\||,$$

the last step following form (6.77) (recall that we are assuming $\alpha\delta > 1$). Thus,

$$\ell_F \leq \frac{\ell}{\sqrt{\alpha\delta - 1}}. \tag{6.78}$$

5.4.   We are now ready to conclude. By (6.78), the spectral gap inequality is satisfied if

$$\bar{\mu}_{N+1} - \bar{\mu}_N > \frac{\ell}{\sqrt{\alpha\delta - 1}}. \tag{6.79}$$

Recalling (6.63), we compute that

$$\bar{\mu}_{N+1} - \bar{\mu}_N = \alpha\delta(2N + 1) + \sqrt{R(N)} - \sqrt{R(N + 1)}. \tag{6.80}$$

We shall prove below that

$$\lim_{N \to +\infty} \left(\sqrt{R(N)} - \sqrt{R(N + 1)} + (2N + 1)\sqrt{\alpha^2\delta^2 - 1}\right) = 0; \tag{6.81}$$

assuming this for the moment, we can determine $N_1 > 0$ such that for all $N \geq N_1$, (6.62) holds. Then, from (6.80) we deduce that if $N \geq N_1$,

$$\bar{\mu}_{N+1} - \bar{\mu}_N \geq (2N + 1)(\alpha\delta - \sqrt{\alpha^2\delta^2 - 1}) - 1. \tag{6.82}$$

This means that if (6.81) holds and $N \geq N_1$ satisfies (6.61), the spectral gap inequality (6.79) follows from (6.80) and (6.82). To prove (6.81), setting

$$R_1(N) := 1 + \frac{2\alpha^2\delta}{(\alpha^2\delta^2 - 1)N^2} + \frac{\alpha^2}{(\alpha^2\delta^2 - 1)N^4},$$

we compute that

$$\sqrt{R(N)} - \sqrt{R(N+1)} + (2N+1)\sqrt{\alpha^2\delta^2 - 1}$$
$$= \sqrt{\alpha^2\delta^2 - 1}\left((N+1)^2\left(1 - \sqrt{R_1(N+1)}\right) - N^2\left(1 - \sqrt{R_1(N)}\right)\right). \quad (6.83)$$

We easily see that

$$\lim_{N \to +\infty} N^2\left(1 - \sqrt{R_1(N)}\right) = -\frac{\alpha^2\delta}{\alpha^2\delta^2 - 1};$$

consequently, (6.83) yields (6.81). This concludes the proof of theorem 6.9 if $\alpha\delta > 1$.

6.2 Beam and von Kármán Equation

In this section we consider the generalized BEAM EQUATION

$$\varepsilon u_{tt} + u_t + \Delta^2 u = \left(\int_\Omega |\nabla u|^2 \, dx - \beta\right)\Delta u + f, \quad (6.84)$$

with $\varepsilon > 0$ and $\beta \in \mathbb{R}$. Equation (6.84) describes the displacement of a solid beam, filling a bounded domain $\Omega \subseteq \mathbb{R}^N$, subject to an external load force f. As in the Cahn-Hilliard equations (6.1), the principal part of equation (6.84) is of the fourth order in the space variables; thus, the equation can be regarded as semilinear. Note, however, that the nonlinearity is nonlocal in character, due to the coefficient $\|\nabla u\|^2$ at the right side of (6.84).

In the sequel, we assume for simplicity that $0 < \varepsilon \leq 1$. Most of the material we present is taken from Eden-Milani, [EM93]; for more information on general types of beam equations, we refer to the references therein, and, in particular, to Ball, [Bal73].

6.2.1 Functional Framework and Notations

We assume that Ω has a Lipschitz continuous boundary $\partial\Omega$, on which we impose the so-called "hinged" boundary conditions

$$u_{|\partial\Omega} = 0, \quad \Delta u_{|\partial\Omega} = 0. \quad (6.85)$$

Finally, we supplement (6.84) with the initial conditions

$$u(0,\cdot) = u_0, \quad u_t(0,\cdot) = u_1, \quad (6.86)$$

and refer to the initial-boundary value problem (6.84)+(6.85)+(6.86) as problem   (BE) .

We set $\mathcal{H} := L^2(\Omega)$, with the usual norm $\|\cdot\|$ and scalar product $\langle\cdot,\cdot\rangle$, and define

$$\mathcal{V} := H^2(\Omega)\cap H_0^1(\Omega), \quad \mathcal{D} := \{u \in H^4(\Omega): u, \Delta u \in H_0^1(\Omega)\}.$$

Then, $\mathcal{V} \hookrightarrow \mathcal{H} \hookrightarrow \mathcal{V}'$ is a Gelfand triple, with compact injection $\mathcal{V} \hookrightarrow \mathcal{H}$. We also consider the product spaces

$$\mathcal{X}_0 := \mathcal{V}\times\mathcal{H}, \quad \mathcal{X}_1 := \mathcal{D}\times\mathcal{V},$$

on which we define the functions

$$E_0(u,v) := \varepsilon\|v\|^2 + \langle u,v\rangle + \tfrac{1}{2\varepsilon}\|u\|^2 + \|\Delta u\|^2,$$
$$\Phi_0(u,v) := E_0(u,v) + \|\nabla u\|^4 - \beta\|\nabla u\|^2.$$

By Schwarz' inequality, we immediately see that E_0 is the square of an equivalent norm in $\mathcal{X}_0$. Likewise, we define the following functions on the space $\mathcal{X}_1$:

$$E_1(u,v) := \varepsilon\|\Delta v\|^2 + \varepsilon\langle\Delta u,\Delta v\rangle + \tfrac{1}{2}\|\Delta u\|^2 + \|\Delta^2 u\|^2,$$
$$\Phi_1(u,v) := E_1(u,v) + \|\nabla u\|^2\,\|\nabla\Delta u\|^2 - \beta\|\nabla\Delta u\|^2 - 2\langle f(t,\cdot),\Delta^2 u\rangle.$$

Again, we easily check that, since we are assuming $\varepsilon \in\,]0,1]$, E_1 is the square of an equivalent norm in $\mathcal{X}_1$.

6.2.2   The Beam Equation Semiflow

As for equation (6.9) of section 3.1, we will show that problem (BE) generates a semiflow S, both in $\mathcal{X}_0$ and in $\mathcal{X}_1$. At first, we recall the following global existence, uniqueness and regularity result.

THEOREM 6.11
For all $\beta \in \mathbb{R}$, $(u_0,u_1) \in \mathcal{X}_0$ and $f \in C([0,+\infty[;\mathcal{H})$, there exists a unique

$$u \in C([0,+\infty[;\mathcal{V})\cap C^1([0,+\infty[;\mathcal{H}),$$

which is a weak solution of problem (BE) (i.e. with (6.84) satisfied in $\mathcal{V}'$, almost everywhere in t). If in addition $(u_0,u_1) \in \mathcal{X}_1$ and $f \in C^1([0,+\infty[;\mathcal{H})$, then

$$u \in C([0,+\infty[;\mathcal{D})\cap C^1([0,+\infty[;\mathcal{V})\cap C^2([0,+\infty[;\mathcal{H}).$$

PROOF   See e.g. Ball, [Bal73].   ⬚

Theorem 6.11 allows us to define the solution operator $S = (S(t))_{t\geq 0}$ in $\mathcal{X}_0$, associated to problem (BE). When f is independent of t, S is a semigroup; to show that S is also a semiflow, it is sufficient to prove the following

PROPOSITION 6.12

For each $t \geq 0$, $S(t)$ *is locally Lipschitz continuous in* $\mathcal{X}_0$.

PROOF   Assume that u and $\bar{u}$ are two solutions of (6.84), and let $z := u - \bar{u}$. Then, z satisfies the equation

$$\varepsilon z_{tt} + z_t + \Delta^2 z - \left(\|\nabla u\|^2 - \beta \right) \Delta z = \left(\|\nabla u\|^2 - \|\nabla \bar{u}\|^2 \right) \Delta \bar{u} =: g. \qquad (6.87)$$

Multiplying this by $2z_t$ and $\frac{1}{\varepsilon} z$ in $\mathcal{H}$, and adding the resulting identities, we obtain

$$\frac{\mathrm{d}}{\mathrm{d}t} \left(E_0(z, z_t) + \|\nabla u\|^2 \|\nabla z\|^2 \right) + \|z_t\|^2 + \frac{1}{\varepsilon} \|\Delta z\|^2 + \frac{1}{\varepsilon} \|\nabla u\|^2 \|\nabla z\|^2$$
$$= 2 \langle \nabla u, \nabla u_t \rangle \|\nabla z\|^2 + 2\beta \langle \nabla z, \nabla z_t \rangle + \beta \frac{1}{\varepsilon} \|\nabla z\|^2 + \frac{1}{\varepsilon} \langle g, 2\varepsilon z_t + z \rangle. \qquad (6.88)$$

We now use integration by parts, to write

$$\langle \nabla u, \nabla u_t \rangle = \langle -\Delta u, u_t \rangle, \quad \|\nabla z\|^2 = \langle -\Delta z, z \rangle,$$

and note that

$$\|\nabla u\|^2 - \|\nabla \bar{u}\|^2 = -\langle \Delta u + \Delta \bar{u}, u - \bar{u} \rangle.$$

Consequently, we obtain from (6.88)

$$\frac{\mathrm{d}}{\mathrm{d}t} \left(E_0(z, z_t) + \|\nabla u\|^2 \|\nabla z\|^2 \right) + \|z_t\|^2 + \frac{1}{\varepsilon} \|\Delta z\|^2 + \frac{1}{\varepsilon} \|\nabla u\|^2 \|\nabla z\|^2$$
$$\leq 2\|\Delta u\| \, \|u_t\| \, \|\Delta z\| \, \|z\| + |\beta| \, \|\Delta z\| \, \|z_t\| + |\beta| \frac{1}{\varepsilon} \|\Delta z\| \, \|z\|$$
$$+ \frac{1}{\varepsilon} \|\Delta u + \Delta \bar{u}\| \, \|\Delta \bar{u}\| \, \|z\| \left(\varepsilon \|z_t\| + \|z\| \right). \qquad (6.89)$$

Since the map $[0, +\infty[\ni t \mapsto (u(t, \cdot), \sqrt{\varepsilon} u_t(t, \cdot)) \in \mathcal{V}$ is locally bounded, and analogously for $\bar{u}$, recalling that we are assuming that $\varepsilon \leq 1$ we deduce from (6.89) that

$$\frac{\mathrm{d}}{\mathrm{d}t} \left(E_0(z, z_t) + \|\nabla u\|^2 \|\nabla z\|^2 \right) + \|z_t\|^2 + \frac{1}{\varepsilon} \|\Delta z\|^2 + \frac{1}{\varepsilon} \|\nabla u\|^2 \|\nabla z\|^2$$
$$\leq \frac{1}{2\varepsilon} \|\Delta z\|^2 + \left(\frac{1}{4} + C\varepsilon \right) \|z_t\|^2 + C \frac{1}{\varepsilon} \|z\|^2, \qquad (6.90)$$

where C depends on $|\beta|$, u and $\bar{u}$. Integrating (6.90) and applying Gronwall's inequality (2.62), we can then easily conclude the proof of proposition 6.12. $\square$

6.2.3   Absorbing Sets

We now show the existence of absorbing sets for the semiflow S generated by problem (BE).

1.   We first show that S admits a bounded, positively invariant absorbing set in $\mathcal{X}_0$.

PROPOSITION 6.13

Assume that $f \in C_b([0, +\infty[; \mathcal{H})$. There exists $R_0 > 0$, dependent on ε, such that the set

$$\mathcal{B}_0 := \{(u, v) \in \mathcal{X}_0 : \Phi_0(u, v) \le R_0^2\} \tag{6.91}$$

is bounded, positively invariant and absorbing for the solution operator S generated by problem (BE) in $\mathcal{X}_0$ (that is, $\mathcal{B}_0$ absorbs all bounded sets of $\mathcal{X}_0$).

PROOF   To show that $\mathcal{B}_0$ is bounded, we see that, if $(u, v) \in \mathcal{B}_0$,

$$E_0(u, v) = \Phi_0(u, v) + \beta \|\nabla u\|^2 - \|\nabla u\|^4 \le \Phi_0(u, v) + \tfrac{1}{4}\beta^2 \le R_0^2 + \tfrac{1}{4}\beta^2. \tag{6.92}$$

To show that $\mathcal{B}_0$ is positively invariant and absorbing, we establish an exponential inequality on $\Phi_0(u, u_t)$. Multiplying equation (6.84) in $\mathcal{H}$ by $2u_t$ and $\frac{1}{\varepsilon}u$, and adding the resulting identities, we obtain

$$\frac{d}{dt}\Phi_0(u, u_t) + \|u_t\|^2 + \tfrac{1}{\varepsilon}\|\Delta u\|^2 + \tfrac{1}{\varepsilon}\left(\|\nabla u\|^2 - \beta\right)\|\nabla u\|^2 = \tfrac{1}{\varepsilon}\langle f, 2\varepsilon u_t + u\rangle. \tag{6.93}$$

Let λ_1 denote the first eigenvalue of the operator $(-\Delta)^2$, relative to the boundary conditions (6.85), so that

$$\lambda_1 \|u\|^2 \le \|\Delta u\|^2.$$

Then, we can estimate the right side of (6.93) by

$$2\|f\|^2 + \tfrac{1}{2}\|u_t\|^2 + \tfrac{1}{2\varepsilon\lambda_1}\|f\|^2 + \tfrac{1}{2\varepsilon}\|\Delta u\|^2,$$

and, therefore, obtain from (6.93)

$$\frac{d}{dt}\Phi_0(u, u_t) + \tfrac{1}{2}\|u_t\|^2 + \tfrac{1}{2\varepsilon}\|\Delta u\|^2 + \tfrac{1}{\varepsilon}\left(\|\nabla u\|^2 - \beta\right)\|\nabla u\|^2 \le \tfrac{1}{\varepsilon}C_f, \tag{6.94}$$

where the constant C_f depends on $\sup_{t \ge 0}\|f(t)\|$. We now set

$$\alpha := \max\{3, 2(\lambda_1^{-1} + 1)\}, \tag{6.95}$$

and note that, since $\varepsilon \le 1$,

$$\Phi_0(u, u_t) \le \tfrac{3}{2}\varepsilon\|u_t\|^2 + (\tfrac{1}{\lambda_1\varepsilon} + 1)\|\Delta u\|^2 + \|\nabla u\|^4 - \beta\|\nabla u\|^2 \tag{6.96}$$

$$\le \alpha\left(\tfrac{1}{2}\|u_t\|^2 + \tfrac{1}{2\varepsilon}\|\Delta u\|^2 + \tfrac{1}{2\varepsilon}\|\nabla u\|^4 - \tfrac{1}{\varepsilon}\beta\|\nabla u\|^2\right) + \beta(\tfrac{\alpha}{\varepsilon} - 1)\|\nabla u\|^2.$$

Consequently, we deduce from (6.94) that

$$\frac{d}{dt}\Phi_0(u, u_t) + \tfrac{1}{\alpha}\Phi_0(u, u_t) + \tfrac{1}{2\varepsilon}\|\nabla u\|^4 \le \tfrac{1}{\varepsilon}C_f + \tfrac{1}{\varepsilon}|\beta(1 - \alpha\varepsilon)| \cdot \|\nabla u\|^2$$

$$\le \tfrac{1}{\varepsilon}C_f + \tfrac{1}{2\varepsilon}|\beta(1 - \alpha\varepsilon)|^2 + \tfrac{1}{2\varepsilon}\|\nabla u\|^4.$$

Thus, $\Phi_0(u, u_t)$ satisfies the exponential inequality

$$\frac{d}{dt}\Phi_0(u, u_t) + \tfrac{1}{\alpha}\Phi_0(u, u_t) \le M_{1\varepsilon} := \tfrac{1}{\varepsilon}C_f + \tfrac{1}{2\varepsilon}|\beta(1 - \alpha\varepsilon)|^2,$$

so that we can conclude, as usual, the existence of a bounded, positively invariant absorbing set for S in $\mathcal{X}_0$. Note that $M_{1\varepsilon}$ is unbounded as $\varepsilon \to 0$.   □

2.   We now proceed to show that S also admits a bounded, positively invariant absorbing set in $\mathcal{X}_1$.

PROPOSITION 6.14
Assume that $f \in C_b^1([0, +\infty[; \mathcal{H})$, and let $\mathcal{B}_0$ be the set defined in (6.91). There exists $R_1 > 0$, dependent on ε, such that the set

$$\mathcal{B}_1 := \{(u,v) \in \mathcal{X}_1 : \Phi_1(u,v) \leq R_1^2\} \cap \mathcal{B}_0 \tag{6.97}$$

is bounded, positively invariant and absorbing for the solution operator S generated by problem (BE) in $\mathcal{X}_1$ (that is, $\mathcal{B}_0$ absorbs all bounded sets of $\mathcal{X}_1$).

PROOF   To show that $\mathcal{B}_1$ is bounded, let $(u,v) \in \mathcal{B}_1$, and set $F_0 := \sup_{t \geq 0} \|f(t)\|$. If $\|\nabla u\|^2 \geq \beta$, we have that

$$E_1(u,v) \leq \Phi_1(u,v) + 2\langle f, \Delta^2 u\rangle \leq R_1^2 + 2\|f\|^2 + \tfrac{1}{2}\|\Delta^2 u\|^2\,.$$

Thus, since clearly $\|\Delta^2 u\|^2 \leq E_1(u,v)$ for all $(u,v) \in \mathcal{X}_1$,

$$\tfrac{1}{2}E_1(u,v) \leq R_1^2 + 2F_0^2\,. \tag{6.98}$$

If instead $\|\nabla u\|^2 \leq \beta$, then $\beta > 0$, and

$$E_1(u,v) \leq \Phi_1(u,v) + \beta\|\nabla \Delta u\|^2 + 2\langle f, \Delta^2\rangle\,. \tag{6.99}$$

By the Gagliardo-Nirenberg inequalities (see theorem A.70), and elliptic estimates similar to those of theorem A.77, we can estimate

$$\|\nabla \Delta u\| \leq C\|\Delta^2 u\|^{2/3}\|\nabla u\|^{1/3}\,; \tag{6.100}$$

thus, we obtain from (6.99) that

$$E_1(u,v) \leq R_1^2 + C\beta^{5/3}\|\Delta^2 u\|^{4/3} + 2\|f\|\,\|\Delta^2 u\| \leq R_1^2 + C^3\beta^5 + 4\|f\|^2 + \tfrac{1}{2}\|\Delta^2 u\|\,.$$

Consequently,

$$\tfrac{1}{2}E_1(u,v) \leq R_1^2 + C^3\beta^5 + 4F_0^2\,. \tag{6.101}$$

Together with (6.98), (6.101) shows that $\mathcal{B}_1$ is bounded in $\mathcal{X}_1$.

To show that $\mathcal{B}_1$ is positively invariant and absorbing, we establish an exponential inequality on $\Phi_1(u, u_t)$. Multiplying equation (6.84) in $\mathcal{H}$ by $2\Delta^2 u_t$ and $\Delta^2 u$, and adding the resulting identities, we obtain

$$\frac{\mathrm{d}}{\mathrm{d}t}\Phi_1(u,u_t) + (2-\varepsilon)\|\Delta u_t\|^2 + \|\Delta^2 u\|^2 + (\|\nabla u\|^2 - \beta)\|\nabla \Delta u\|^2 - \langle f, \Delta^2 u\rangle$$
$$= 2\langle \nabla u, \nabla u_t\rangle\|\nabla \Delta u\| - 2\langle f_t, \Delta^2 u\rangle$$

$$= 2\langle \Delta u, u_t \rangle \langle \Delta^2 u, \Delta u \rangle - 2\langle f_t, \Delta^2 u \rangle$$
$$\leq 2\|\Delta u\|^2 \|u_t\| \, \|\Delta^2 u\| + 2\|f_t\| \, \|\Delta^2 u\|$$
$$\leq 8\|\Delta u\|^4 \|u_t\|^2 + 8\|f_t\|^2 + \tfrac{1}{4} \|\Delta^2 u\|^2 .$$

Therefore, setting $F_1 := \sup_{t \geq 0} \|f_t(t)\|$ and, for $(u,v) \in \mathcal{X}_1$,

$$\Psi_1(u,v) := \|\Delta v\|^2 + \tfrac{1}{2}\|\Delta^2 u\|^2 + \left(\|\nabla u\|^2 - \beta\right)\|\nabla \Delta u\|^2 - \langle f, \Delta^2 u \rangle ,$$

and recalling that $\varepsilon \leq 1$, we deduce that

$$\frac{d}{dt}\Phi_1(u,u_t) + \Psi_1(u,u_t) + \tfrac{1}{4}\|\Delta^2 u\|^2 \leq 8F_1^2 + 8\|\Delta u\|^4 \|u_t\|^2 . \tag{6.102}$$

We now easily verify that, with α defined in (6.95),

$$\Phi_1(u,v) \leq \alpha \Psi_1(u,v) + (1-\alpha)\left(\|\nabla u\|^2 - \beta\right)\|\nabla \Delta u\|^2 + (\alpha - 2)\langle f, \Delta^2 u \rangle .$$

Consequently, we obtain from (6.102) that

$$\frac{d}{dt}\Phi_1(u,u_t) + \tfrac{1}{\alpha}\Phi_1(u,u_t) + \tfrac{1}{4}\|\Delta^2 u\|^2 \leq 8F_1^2 + 8\|\Delta u\|^4 \|u_t\|^2$$
$$+ \tfrac{1-\alpha}{\alpha}\left(\|\nabla u\|^2 - \beta\right)\|\nabla \Delta u\|^2 + \tfrac{\alpha-2}{\alpha}\|f\| \, \|\Delta^2 u\| . \tag{6.103}$$

Recalling that $\alpha > 1$, we have then that, if $\|\nabla u\|^2 \geq \beta$,

$$\frac{d}{dt}\Phi_1(u,u_t) + \tfrac{1}{\alpha}\Phi_1(u,u_t) \leq 8F_1^2 + 8\|\Delta u\|^4 \|u_t\|^2 + F_0^2 . \tag{6.104}$$

If instead $\|\nabla u\|^2 \leq \beta$, resorting again to estimate (6.100) we proceed from (6.103) with

$$\frac{d}{dt}\Phi_1(u,u_t) + \tfrac{1}{\alpha}\Phi_1(u,u_t) + \tfrac{1}{4}\|\Delta^2 u\|^2$$
$$\leq 8F_1^2 + 8\|\Delta u\|^4 \|u_t\|^2 + C\beta^{5/3}\|\Delta^2 u\|^{4/3} + \|f\| \, \|\Delta^2 u\|$$
$$\leq 8F_1^2 + 8\|\Delta u\|^4 \|u_t\|^2 + C_1\beta^5 + 2\|f\|^2 + \tfrac{1}{4}\|\Delta^2 u\|^2 .$$

Together with (6.104), this shows that, in either case, there exists a constant $K > 0$, depending on F_0, F_1 and β, such that

$$\frac{d}{dt}\Phi_1(u,u_t) + \tfrac{1}{\alpha}\Phi_1(u,u_t) \leq K + 8\|\Delta u\|^4 \|u_t\|^2 . \tag{6.105}$$

In the sequel, we denote by $M_{r,\varepsilon}$, $r \geq 1$, various positive constants, independent of u (but unbounded as $\varepsilon \to 0$). Assume now that (u_0, u_1) is in a bounded set of $\mathcal{X}_1$. Then, (u_0, u_1) is also in a bounded set of $\mathcal{X}_0$, and since $\mathcal{B}_0$ is absorbing, there is $T_0 \geq 0$ such that $\Phi_0(u(t), u_t(t)) \leq R_0^2$ for all $t \geq T_0$ (with $T_0 = 0$ if $(u_0, u_1) \in \mathcal{B}_1$, since $\mathcal{B}_1$ is positively invariant). We easily verify that, for each $(u,v) \in \mathcal{X}_0$,

$$\|\Delta u\|^2 \leq E_0(u,v), \quad \|v\|^2 \leq \tfrac{2}{\varepsilon}E_0(u,v) ; \tag{6.106}$$

hence, recalling (6.92), we deduce that

$$8\|\Delta u\|^4 \|u_t\|^2 \le \tfrac{16}{\varepsilon}\left(R_0^2 + \tfrac{1}{4}\beta\right)^3 =: M_{2\varepsilon}.$$

Inserting this into (6.105), we obtain the exponential inequality

$$\frac{d}{dt}\Phi_1(u, u_t) + \tfrac{1}{\alpha}\Phi_1(u, u_t) \le K + M_{2\varepsilon} =: M_{3\varepsilon}, \tag{6.107}$$

from which we can conclude that if $R_1^2 > \alpha M_{3\varepsilon}$, the set $\mathcal{B}_1$ defined in (6.97) is positively invariant and absorbing for S in $\mathcal{X}_1$. Indeed, from (6.107) we obtain that, for $t \ge T_0$,

$$\Phi_1(u(t), u_t(t)) \le \left(\Phi_1(u(T_0), u_t(T_0)) - \alpha M_{3\varepsilon}\right)e^{-(t-T_0)/\alpha} + \alpha M_{3\varepsilon}.$$

Consequently, we deduce that $(u(t), u_t(t)) \in \mathcal{B}_1$ for all $t \ge T_1$, with $T_1 > T_0$ defined by the identity

$$\left(\Phi_1(u(T_0), u_t(T_0)) - \alpha M_{3\varepsilon}\right)e^{-(T_1-T_0)/\alpha} + \alpha M_{3\varepsilon} = R_1^2,$$

if $\Phi_1(u(T_0), u_t(T_0)) > R_1^2$, or $T_1 = T_0$ if instead $\Phi_1(u(T_0), u_t(T_0)) \le R_1^2$. This concludes the proof of proposition 6.14; note that R_1 depends on R_0, via $M_{2\varepsilon}$. $\quad\Box$

6.2.4 The Global Attractor

In this section we resort to theorem 2.56 of chapter 2 to show that the semiflow S generated by problem (BE) admits a global attractor $\mathcal{A}$ in $\mathcal{X}_0$. More precisely, we resort to the α-contraction method described in section 3.4.5, to show that $\mathcal{A} = \omega(\mathcal{B}_0)$, where $\mathcal{B}_0$ is the absorbing set determined in proposition 6.13. Note that $\omega(\mathcal{B}_0)$ is not empty, since it contains the stationary solutions of problem (BE).

Recalling proposition 2.59, to apply theorem 2.56 it is sufficient to find an appropriate pseudometric δ on $\mathcal{X}_0$, and a number $t_* > 0$, such that condition (2.47) of chapter 3 holds, with $T = S(t_*)$.

For fixed $\tau > 0$ we define in $\mathcal{X}_0 \times \mathcal{X}_0$ the function

$$\delta_\tau((u,v),(\bar{u},\bar{v})) := \left(\int_0^\tau \|P_1(S(t)(u,v)) - P_1(S(t)(\bar{u},\bar{v}))\|^2\,dt\right)^{1/2}, \tag{6.108}$$

where P_1 is the projection from $\mathcal{X}_0$ onto $\mathcal{V}$. With exactly the same proof of proposition 3.25, we have that, for each $\tau > 0$, δ_τ is a pseudometric on $\mathcal{X}$, precompact on $\mathcal{B}_0$ with respect to the norm of $\mathcal{X}_0$ defined by E_0.

We proceed then to establish an estimate of the difference of two solutions of problem (BE), that allows us to apply proposition 2.59.

PROPOSITION 6.15

There are positive constants γ_i, $i = 1, 2, 3$, depending on ε, β and R_0, but not on t, such that for all $U_0 := (u_0, u_1)$, $\overline{U}_0 := (\bar{u}_0, \bar{u}_1) \in \mathcal{B}_0$, and $t \ge 0$,

$$E_0(S(t)U_0 - S(t)\overline{U}_0) \le \gamma_1 e^{-\gamma_2 t/2} E_0(U_0 - \overline{U}_0) + \gamma_3\left(\delta_t(U_0, \overline{U}_0)\right)^2, \tag{6.109}$$

with δ_t defined in (6.108).

PROOF   We start from (6.89), which we rewrite as

$$\frac{d}{dt}\left(E_0(z,z_t)+\left(\|\nabla u\|^2-\beta\right)\|\nabla z\|^2\right)+\|z_t\|^2+\tfrac{1}{\varepsilon}\|\Delta z\|^2+\tfrac{1}{\varepsilon}\left(\|\nabla u\|^2-\beta\right)\|\nabla z\|^2$$
$$\leq 2\|\Delta u\|\,\|u_t\|\,\|\Delta z\|\,\|z\|+\tfrac{1}{\varepsilon}\|\Delta u+\Delta\bar{u}\|\,\|\Delta\bar{u}\|\,\|z\|\,(\varepsilon\|z_t\|+\|z\|)=:\rho_1\,.\quad(6.110)$$

Recalling (6.106), and that $U_0,\overline{U}_0\in\mathcal{B}_0$, we have that

$$\|\Delta u\|^2\leq R_0^2+\tfrac{1}{4}\beta\,,\quad \varepsilon\|u_t\|^2\leq 2\left(R_0^2+\tfrac{1}{4}\beta\right)\,.$$

Thus, the right side of (6.110) can be estimated by

$$\rho_1\leq 2\sqrt{2}\frac{1}{\sqrt{\varepsilon}}\left(R_0^2+\tfrac{1}{4}\beta\right)\|\Delta z\|\,\|z\|+\tfrac{2}{\varepsilon}\left(R_0^2+\tfrac{1}{4}\beta\right)\|z\|\,(\varepsilon\|z_t\|+\|z\|)$$
$$\leq C\tfrac{1}{\varepsilon}\|z\|^2+\tfrac{1}{4\varepsilon}\|\Delta z\|^2+\tfrac{1}{2}\|z_t\|^2\,.$$

Inserting this into (6.110), and setting

$$Z:=E_0(z,z_t)+\left(\|\nabla u\|^2-\beta\right)\|\nabla z\|^2\,,$$

we obtain

$$\frac{dZ}{dt}+\left(\tfrac{1}{2}\|z_t\|^2+\tfrac{1}{2\varepsilon}\|\Delta z\|^2+\tfrac{1}{\varepsilon}\left(\|\nabla u\|^2-\beta\right)\|\nabla z\|^2\right)+\tfrac{1}{4\varepsilon}\|\Delta z\|^2\leq C\tfrac{1}{\varepsilon}\|z\|^2\,.\quad(6.111)$$

Acting as in (6.96), we easily see that

$$Z\leq\alpha\left(\tfrac{1}{2}\|z_t\|^2+\tfrac{1}{2\varepsilon}\|\Delta z\|^2+\tfrac{1}{\varepsilon}\left(\|\nabla u\|^2-\beta\right)\|\nabla z\|^2\right)+\tfrac{1}{\varepsilon}\beta(\alpha-\varepsilon)\|\nabla z\|^2\,;$$

consequently, we obtain from (6.111) that (for different C)

$$\frac{dZ}{dt}+\tfrac{1}{\alpha}Z+\tfrac{1}{4\varepsilon}\|\Delta z\|^2\leq C\tfrac{1}{\varepsilon}\|z\|^2+C\|\nabla z\|^2\leq C\tfrac{1}{\varepsilon}\|z\|^2+\tfrac{1}{4\varepsilon}\|\Delta z\|^2\,.\quad(6.112)$$

From (6.112) we immediately obtain that, for all $t\geq 0$,

$$Z(t)\leq Z(0)e^{-t/\alpha}+C\tfrac{1}{\varepsilon}\int_0^t\|z(s)\|^2\,ds\,;\quad(6.113)$$

since we obviously have that, for suitable constant C, depending on R_0, and all $t\geq 0$,

$$E_0(z(t),z_t(t))\leq Z(t)\leq C\,E_0(z(t),z_t(t))\,,$$

we obtain from (6.113) that, for all $t\geq 0$,

$$E_0(z(t),z_t(t))\leq C\,E_0(z(0),z_t(0))e^{-t/\alpha}+C\tfrac{1}{\varepsilon}\int_0^t\|z(s)\|^2\,ds\,,\quad(6.114)$$

from which (6.109) follows. ◻

The existence of a global attractor for S follows now theorem 2.56 and proposition 2.59. Indeed, choosing $t_* > 0$ such that $q := e^{-t_*/2\alpha} < 1$, from (6.109) we see that the operator $T = S(t_*)$ and the pseudometric δ_{t_*} satisfy condition (2.47) of proposition 2.59. Hence, T is an α-contraction, and the set $\mathcal{A} = \omega(\mathcal{B}_0)$ is the desired attractor for S.

REMARK 6.16 If ε is sufficiently small, it is possible to show, with techniques analogous to those of section 3.5, that the attractor $\mathcal{A}$ is bounded in $\mathcal{X}_1$. For an alternative proof, see Eden-Milani, [EM93]. ◻

6.2.5 The Exponential Attractor

In this section we show that the semiflow S also admits an exponential attractor $\mathcal{E}$ in $\mathcal{X}_0$, which contains the global attractor $\mathcal{A}$.

THEOREM 6.17
In the same conditions of proposition 6.14, with $f_t \equiv 0$, the semiflow S generated by problem (BE) admits an exponential attractor $\mathcal{E}$ in $\mathcal{X}_0$.

PROOF We apply theorem 4.5 of chapter 4. To this end, we consider the absorbing set $\mathcal{B}_1$ for S in $\mathcal{X}_1$, determined in proposition 6.14; since the injection $\mathcal{X}_1 \hookrightarrow \mathcal{X}_0$ is compact (because so is the injection $\mathcal{V} \hookrightarrow \mathcal{H}$), $\mathcal{B}_1$ is compact in $\mathcal{X}_0$. We propose then to show that S satisfies the discrete squeezing property (see definition 4.3), relative to $\mathcal{B}_1$. We proceed almost exactly as in the proof of theorem 4.10 of section 4.4.2, of which we keep the same choices of N and $\mathcal{X}_N$, and denote by $\| \cdot \|_0$ the norm induced on $\mathcal{X}_0$ by E_0. Thus, we show that, given any $t_* > 0$ and $\gamma \in]0, \frac{1}{2}[$, there exists an integer N_*, with the property that if $u_0, \bar{u}_0 \in \mathcal{B}_1$ are such that $S(t_*)u_0 - S(t_*)\bar{u}_0 \notin \mathcal{C}_{N_*}$ (i.e. if (6.42) holds for the operator $S(t_*)$), then (6.43) must hold. To this end, we first note that, since $\|z\|^2 \leq 4\varepsilon E_0(z, z_t)$, using Gronwall's inequality we obtain from (6.114) that

$$E_0(z(t), z_t(t)) \leq C E_0(z(0), z_t(0)) e^{4Ct}. \tag{6.115}$$

Next, we apply the projection q_N to the equation (6.87) satisfied by z, to obtain that the function $q := q_N z$ satisfies the equation

$$\varepsilon q_{tt} + q_t + \Delta^2 q - \left(\|\nabla u\|^2 - \beta \right) \Delta q = -\langle \Delta u + \Delta \bar{u}, z \rangle \Delta q_N \bar{u}. \tag{6.116}$$

For $(q, p) \in Q_N(\mathcal{X}_0)$, we define

$$M(q, p) := \varepsilon \|p\|^2 + \langle p, q \rangle + \|\Delta q\|^2 + \frac{1}{2\varepsilon} \left(\|\nabla u\|^2 - \beta \right) \|\nabla q\|^2.$$

Multiplying (6.116) in $\mathcal{H}$ by $2q_t$ and $\frac{1}{\varepsilon}q$, and adding the resulting identities, we obtain

$$\frac{\mathrm{d}}{\mathrm{d}t}M(q,q_t) + \|q_t\|^2 + \frac{1}{2\varepsilon}\langle q,q_t\rangle + \frac{1}{\varepsilon}\|\Delta q\|^2 + \frac{1}{2\varepsilon}\left(\|\nabla u\|^2 - \beta\right)\|\nabla q\|^2$$

$$= 2\langle -\Delta u, u_t\rangle \|\nabla q\|^2 + \frac{1}{\varepsilon}\langle \Delta u + \Delta\bar{u}, z\rangle \langle \Delta q_N \bar{u}, 2\varepsilon q_t + q\rangle$$

$$- \frac{1}{2\varepsilon}\langle q,q_t\rangle - \frac{1}{2\varepsilon}\left(\|\nabla u\|^2 - \beta\right)\|\nabla q\|^2 =: \rho_2 . \qquad (6.117)$$

We now proceed as in the proof of proposition 6.15: Recalling that U_0 and $\overline{U}_0 \in \mathcal{B}_0$, that $\varepsilon \leq 1$, and that, in analogy to (3.28) of proposition 3.6,

$$\|q\| \leq \frac{1}{\sqrt{\lambda_{N+1}}}\|\Delta q\| , \qquad (6.118)$$

we can estimate the right side of (6.117) by

$$\rho_2 \leq \frac{2C_0}{\sqrt{\varepsilon\lambda_{N+1}}}\|\Delta q\|^2 + \frac{2C_0}{\sqrt{\lambda_{N+1}}}\|\Delta\bar{u}\|\,\|z\|\,\|q_t\|$$

$$+ \frac{C_0}{\varepsilon\sqrt{\lambda_{N+1}}}\|z\|\,\|\Delta q\| + \frac{1}{2\varepsilon\sqrt{\lambda_{N+1}}}\|\Delta q\|\,\|q_t\| + \frac{|\beta|}{2\varepsilon\sqrt{\lambda_{N+1}}}\|\Delta q\|^2$$

$$\leq \frac{C_0}{\varepsilon^2\lambda_{N+1}}\|\Delta q\|^2 + \frac{C_0}{\varepsilon\lambda_{N+1}}\|z\|^2 + \frac{1}{2}\|q_t\|^2 + \frac{1}{4\varepsilon}\|\Delta q\|^2 , \qquad (6.119)$$

where the constant C_0 depends on R_0 and β. Since $\lambda_N \to +\infty$ as $N \to +\infty$, we can choose $N_1 \in \mathbb{N}$ so large that if $N \geq N_1$, $4C_0 \leq \varepsilon\lambda_{N+1}$. For such N, we obtain from (6.117) and (6.119), together with (6.115), that

$$\frac{\mathrm{d}}{\mathrm{d}t}M(q,q_t) + \frac{1}{2}M(q,q_t) \leq \frac{4C_0}{\lambda_{N+1}}E_0(z(0),z_t(0))\,\mathrm{e}^{4Ct} .$$

Integrating this inequality, we obtain that, for all $t \geq 0$,

$$M(q(t),q_t(t)) \leq M(q(0),q_t(0))\mathrm{e}^{-t/2\varepsilon} + \frac{8\varepsilon C_0}{(1+8C\varepsilon)\lambda_{N+1}}\,\mathrm{e}^{4Ct} . \qquad (6.120)$$

This estimate is the analogous of estimate (4.48) of section 4.4.2; it follows that, to conclude the proof of (6.43), it is sufficient to choose t_* and N_* so that the right side of (6.120) is (arbitrarily) small. Thus, given any $\eta > 0$, we first choose t_* so large that $M(q(0),q_t(0))\mathrm{e}^{-t_*/2\varepsilon} \leq \eta$, and then $N_* \geq N_1$ so large that

$$\frac{8\varepsilon C_0}{(1+8C\varepsilon)\lambda_{N_*+1}}\,\mathrm{e}^{4Ct_*} \leq \eta .$$

With these choices, we obtain from (6.120) that

$$M(q(t_*),q_t(t_*)) \leq 2\eta .$$

Proceeding then as in the proof of theorem 4.10, we can then deduce that the discrete squeezing property holds, relative to the set $\mathcal{B}_1$. By theorem 4.5, this is sufficient to conclude the proof of theorem 6.17.   □

6.2.6 Inertial Manifold

In this last section, we show that if the source term f in (6.84) has the special form $f = p_N f$, for some $N \in \mathbb{N}$, where $p_N = I_{\mathcal{H}} - q_N$ is the projection on $\mathcal{H}$ considered in the last section, then the semiflow S admits a (trivial) inertial manifold. More precisely, we have

THEOREM 6.18
Let $N \in \mathbb{N}$ and $f \in \mathcal{H}$ be such that $f \in p_N \mathcal{H}$. There exists $M \geq N$ such that the flat manifold $\mathcal{M} := P_M \mathcal{X}_0$ is an inertial manifold for S.

PROOF Let u be a solution of problem (BE), corresponding to initial values $U_0 = (u_0, u_1)$ in a bounded set $\mathcal{G}$ of $\mathcal{X}_0$, and set $U(t) := (u(t), u_t(t)) = S(t)U_0$. Fix $M \geq N$, to be determined. Since

$$\partial(S(t)U_0, \mathcal{M}) \leq E_0(U(t) - P_M U(t)),$$

setting $q := q_M u$ it is sufficient to show that

$$E_0(U(t) - P_M U(t)) = E_0(q(t), q_t(t)) \leq k_1 e^{-k_2 t}, \qquad (6.121)$$

for suitable positive constants k_1, k_2, depending only on $\mathcal{G}$. To this end, we see that, since $q_M f = 0$ because $M \geq N$, q satisfies the equation

$$\varepsilon q_{tt} + q_t + \Delta^2 q - \left(\|\nabla u\|^2 - \beta \right) \Delta q = 0.$$

This equation is similar to (6.116); acting as in (6.117), we arrive at the estimate

$$\frac{d}{dt} M(q, q_t) + \tfrac{1}{2\varepsilon} M(q, q_t) + \tfrac{1}{2}\|q_t\|^2 + \tfrac{1}{2\varepsilon}\|\Delta q\|^2$$
$$= 2\langle -\Delta u, u_t \rangle \|\nabla q\|^2 - \tfrac{1}{2\varepsilon}\langle q, q_t \rangle - \tfrac{1}{2\varepsilon}\left(\|\nabla u\|^2 - \beta \right) \|\nabla q\|^2$$
$$\leq C_0 \left(\frac{1}{\sqrt{\varepsilon \lambda_{M+1}}} + \frac{1}{\varepsilon^2 \lambda_{M+1}} + \frac{1}{\varepsilon \sqrt{\lambda_{M+1}}} \right) \|\Delta q\|^2 + \tfrac{1}{2}\|q_t\|^2. \quad (6.122)$$

Taking $M \geq N$ so large that

$$C_0 \left(\frac{1}{\sqrt{\varepsilon \lambda_{M+1}}} + \frac{1}{\varepsilon^2 \lambda_{M+1}} + \frac{1}{\varepsilon \sqrt{\lambda_{M+1}}} \right) \leq \frac{1}{2\varepsilon},$$

we obtain from (6.122) that

$$\frac{d}{dt} M(q, q_t) + \tfrac{1}{2\varepsilon} M(q, q_t) \leq 0.$$

This estimate yields the exponential decay of $M(q, q_t)$; in turn, because of (6.118), this yields the exponential decay of $E_0(q, q_t)$. Thus, we can deduce that (6.121) holds, and complete the proof of theorem 6.18. ☐

6.2.7   von Kármán Equations

We conclude this section with the remark that we can proceed with almost exactly the same techniques we have used for problem (BE), to study the long-time behavior of weak solutions to the equation

$$u_{tt} + \sqrt{\delta\rho}u_t + \partial_x^4 u + \left(1 - \kappa\|u_x\|^2 - \sigma\langle u_x, u_{xt}\rangle\right)u_{xx} + \alpha\partial_x^4 u_t + \rho u_x = 0, \quad (6.123)$$

in one-dimension of space (e.g., with $\Omega = \,]0,1[$). This equation represents a one-dimensional version of the so-called VON KÁRMÁN equations for a thin plate; more precisely, it describes the displacement of a thin elastic plate subject to an axial force load and to the flow of a fluid along its surface. In (6.123), the positive parameters α, σ, δ represent various damping parameters associated to the plate and the fluid; κ is a measure of the elastic properties of the plate, and ρ is the flow rate of the fluid flowing along its surface. We refer to Guckenheimer-Holmes, [GH83, sct. VII.7.6], for a detailed study of various IBV problems associated to (6.123), and for the consequent definition of the associated semiflow. In Eden-Milani, [EM93], we briefly outline the argument leading to the existence of a global and an exponential attractor for this semiflow, in a phase space analogous to the space $\mathcal{X}_0$ considered for problem (BE).

6.3   Navier-Stokes Equations

In this section we consider the so-called NAVIER-STOKES equations in two dimensions of space. These equations describe the motion of a two-dimensional viscous, incompressible fluid in a bounded set $\Omega \subset \mathbb{R}^2$. For more information on the Navier-Stokes equations, also in three space dimensions, we refer e.g. to Lions, [Lio69, sct. 1.6], Temam, [Tem83], Constantin-Foias, [CF88], and Sell-You, [SY02, ch. 6].

For $j = 1, 2$, we set $\partial_j := \partial/\partial x_j$. Given two smooth vector fields $\vec{a} = (a_1, a_2)$ and $\vec{b} = (b_1, b_2)$, we define a third field $(\vec{a} \cdot \nabla)\vec{b}$ by

$$(\vec{a} \cdot \nabla)\vec{b} := (a_1\partial_1 b_1 + a_2\partial_2 b_1, a_1\partial_1 b_2 + a_2\partial_2 b_2). \quad (6.124)$$

6.3.1   The Equations and their Functional Framework

Denoting by $\vec{u}$ the velocity of the fluid, and by p its pressure, the Navier-Stokes equations we consider have the form

$$\rho_0\left(\vec{u}_t + (\vec{u} \cdot \nabla)\vec{u}\right) - \nu\Delta\vec{u} = \vec{f} - \nabla p, \quad (6.125)$$

$$\operatorname{div}\vec{u} = 0, \quad (6.126)$$

where $\rho_0 > 0$ is the density of the fluid (which we assume to be constant; in the sequel, we take $\rho_0 = 1$), $\nu > 0$ is its kinematic viscosity, and $\vec{f}$ is a measure of the external forces applied to the fluid. ν is proportional to the reciprocal of the Reynolds number; in many fluids, v is small. Equation (6.126) translates the requirement that the fluid be incompressible. We supplement (6.125) and (6.126) with the initial and boundary conditions

$$\begin{aligned} \vec{u}(0,x) &= \vec{u}_0(x), & x \in \Omega, \\ \vec{u}(t,x) &= 0, & (t,x) \in [0,+\infty[\,\times\partial\Omega. \end{aligned} \qquad (6.127)$$

The homogeneous boundary condition in (6.127) means that we assume the fluid to be at rest at the boundary of Ω; other type of boundary conditions can be considered, such as space-periodic ones, or so-called "nonpenetrating" boundary conditions, of the form

$$\vec{n} \cdot \vec{u} = 0, \quad \vec{n} \times \operatorname{curl} \vec{u} = 0,$$

where $\vec{n}$ is the outward unit normal to the boundary $\partial\Omega$. We refer to the evolution problem (6.125) + (6.126) + (6.127) as

$$\text{problem} \quad (NS).$$

The function spaces in which we consider problem (NS) are suitable subspaces of the spaces $H(\operatorname{div},\Omega)$ and $H(\operatorname{curl},\Omega)$ introduced in section A.7, to which we refer. In particular, we adopt its convention A.83, whereby if $\mathcal{X}$ is a space of scalar functions on Ω, such as $L^2(\Omega)$, then, with abuse of notation, we write $\vec{u} \in \mathcal{X}$ to mean that the components of $\vec{u}$ are in $\mathcal{X}$. Thus, we rely on the context to know when the notation $\mathcal{X}$ denotes a space of scalar valued or of vector valued functions (i.e. when $\mathcal{X}$ is actually an abbreviation for the space $\mathcal{X} \times \mathcal{X}$). Also, when there is no danger of confusion, we denote vectors simply by u, instead of $\vec{u}$. Finally, we denote as usual the norm and scalar product in $L^2(\Omega)$ by $\| \cdot \|$ and $\langle \cdot, \cdot \rangle$, and by $|\cdot|_p$ the norm in $L^p(\Omega)$, $1 \le p \le +\infty$.

Setting then

$$\begin{aligned} \mathcal{V} &:= \{\vec{u} \in H_0^1(\Omega): \operatorname{div}\vec{u} = 0\}, \\ \mathcal{H} &:= H^0(\operatorname{div},\Omega) = \{\vec{u} \in L^2(\Omega): \operatorname{div}\vec{u} = 0\}, \end{aligned}$$

theorem A.85 allows us to consider $\mathcal{V} \hookrightarrow \mathcal{H} \hookrightarrow \mathcal{V}'$ as a Gelfand triple. Next, recalling (6.124), we define a trilinear form b on $\mathcal{V}$, by

$$b(u,v,w) := \langle (u \cdot \nabla)v, w \rangle = \sum_{i,j=1}^{2} \int_{\Omega} u_i(\partial_i v_j)w_j \, dx. \qquad (6.128)$$

The following proposition describes the main properties of b that we need in the sequel.

PROPOSITION 6.19

The trilinear form b defined in (6.128) is a continuous map from $\mathcal{V} \times \mathcal{V} \times \mathcal{V}$ to $\mathbb{R}$, which satisfies the estimate

$$|b(u,v,w)| \le |u|_4 |\nabla v|_2 |w|_4, \qquad (6.129)$$

for all u, v and w ∈ V. Moreover, for all u,v ∈ V,

$$b(u,u,v) = -b(u,v,u). \tag{6.130}$$

In particular,

$$b(u,v,v) = 0, \quad b(u,u,u) = 0. \tag{6.131}$$

PROOF Estimate (6.129) follows from theorem A.58 (recall that, since $N = 2$, $V \hookrightarrow H_0^1(\Omega) \hookrightarrow L^p(\Omega)$ for all $p \in [2, +\infty[$). To prove identity (6.130), we use the integration by parts formula

$$\int_{\partial\Omega} (\vec{n} \cdot \vec{f}) g \, ds = \int_{\Omega} (\operatorname{div} \vec{f}) g \, ds + \int_{\Omega} \vec{f} \cdot \nabla g \, dx, \tag{6.132}$$

where $\vec{n}$ is the outward unit normal to $\partial\Omega$; the validity of this formula for functions in V is justified in section A.7. Indeed, we easily compute that

$$b(u,u,v) + b(u,v,u) = \int_{\Omega} \vec{u} \cdot \nabla(\vec{u} \cdot \vec{v}) \, dx;$$

hence, by (6.132) we have that

$$b(u,u,v) + b(u,v,u) = -\int_{\Omega} (\operatorname{div} \vec{u})(\vec{u} \cdot \vec{v}) \, dx + \int_{\partial\Omega} (\vec{n} \cdot \vec{u})(\vec{u} \cdot \vec{v}) \, ds = 0,$$

keeping in mind that $u \in V$.   ☐

PROPOSITION 6.20
Let $T > 0$, and $u \in L^2(0,T;V) \cap L^\infty(0,T;\mathcal{H})$. Then, $(u \cdot \nabla)u \in L^2(0,T;V')$.

PROOF Let $v \in L^2(0,T;V)$. By the Gagliardo-Nirenberg inequality

$$|u|_4 \le C\|\nabla u\|^{1/2}\|u\|^{1/2}, \tag{6.133}$$

which holds because u vanishes at $\partial\Omega$ (see theorem A.70), recalling (6.130) and (6.129), we can estimate

$$\begin{aligned}
\int_0^T \int_{\Omega} |(u \cdot \nabla)u \cdot v| \, dx \, dt &= \int_0^T |b(u,u,v)| \, dt = \int_0^T |b(u,v,u)| \, dt \\
&\le \int_0^T |u|_4^2 \|\nabla v\| \, dt \le C \int_0^T \|\nabla u\| \|u\| \|\nabla v\| \, dt \\
&\le C\|u\|_{L^\infty(0,T;\mathcal{H})} \|u\|_{L^2(0,T;V)} \|v\|_{L^2(0,T;V)},
\end{aligned}$$

from which the conclusion follows.   ☐

6.3.2   The 2-Dimensional Navier-Stokes Semiflow

Global existence of weak solutions to problem (NS) is provided by the following result, a proof of which can be found e.g. in Lions, [Lio69, sct. 1.6] (see also Temam, [Tem83]).

THEOREM 6.21
Let $T > 0$. For all $f \in L^2(0,T;\mathcal{V}')$ and $u_0 \in \mathcal{H}$, there exists a unique $u \in L^2(0,T;\mathcal{V})$, with $u_t \in L^2(0,T;\mathcal{V}')$, which is a weak solution of problem (NS), in the sense that $u(0,\cdot) = u_0$ (this makes sense because, by theorem A.80, $u \in C([0,T];\mathcal{H})$), and for all $v \in \mathcal{V}$,

$$\langle u_t + (u \cdot \nabla)u - \nu \Delta u - f, v \rangle_{\mathcal{V}' \times \mathcal{V}} = 0, \tag{6.134}$$

almost everywhere in $t \in [0,T]$. If in addition $f \in L^2(0,T;\mathcal{H})$, $f_t \in L^2(0,T;\mathcal{V}')$, then

$$u \in C([\tau,T];\mathcal{V}) \cap L^2(\tau,T;H^2(\Omega) \cap \mathcal{V}) \tag{6.135}$$

for all $\tau \in {]}0,T{[}$. If $u_0 \in \mathcal{V}$, we can take $\tau = 0$ in (6.135).

REMARK 6.22   In the weak formulation (6.134), the unknown p is not present. This is because, by (6.132),
$$\langle \nabla p, v \rangle_{\mathcal{V}' \times \mathcal{V}} = 0$$
for all $v \in \mathcal{V}$. On the other hand, once a weak solution u of problem (NS) has been found, we can formally determine p in the following way (this procedure can be justified rigorously by the results of section A.7). By proposition 6.20, we know that

$$w := u_t + (u \cdot \nabla)u - \nu \Delta u - f \in \mathcal{V}'$$

for almost all $t \in [0,T]$. Let now $\psi \in C_0^\infty(\Omega)$. Then, curl $\psi \in \mathcal{V}$, so that (6.134) implies that
$$\langle \operatorname{curl} w, \psi \rangle_{\mathcal{V}' \times \mathcal{V}} = \langle w, \operatorname{curl} \psi \rangle_{\mathcal{V}' \times \mathcal{V}} = 0.$$
The arbitrariness of ψ implies then that curl $w = 0$; thus, there is a scalar function p such that $w = -\nabla p$, as desired. $\Box$

Theorem 6.21 allows us to define the solution operator $S = (S(t))_{t \geq 0}$ in $\mathcal{H}$, associated to problem (NS). When f is independent of t, S is a semigroup; to show that S is also a semiflow, it is sufficient to prove the following

PROPOSITION 6.23
For each $t \geq 0$, $S(t)$ is locally Lipschitz continuous in $\mathcal{H}$.

PROOF   Assume that u and $\bar{u}$ are two solutions of (6.134), and let $z := u - \bar{u}$. Then, z satisfies the equations

$$\langle z_t - \nu \Delta z, v \rangle_{\mathcal{V}' \times \mathcal{V}} + b(u,z,v) + b(z,\bar{u},v) = 0, \tag{6.136}$$

for all $v \in \mathcal{V}$. Take now $v = z$ in (6.136) (which is legitimate), and integrate in $[0,t]$: recalling (6.131) and (6.133), we obtain

$$\tfrac{1}{2}\|z(t)\|^2 + v \int_0^t \|\nabla z\|^2\,dt = \tfrac{1}{2}\|z(0)\|^2 - \int_0^t b(z,\bar{u},z)\,dt$$

$$\leq \tfrac{1}{2}\|z(0)\|^2 + C \int_0^t |z|_4^2 \|\nabla \bar{u}\|\,dt$$

$$\leq \tfrac{1}{2}\|z(0)\|^2 + C \int_0^t \|z\|\,\|\nabla z\|\,\|\nabla \bar{u}\|\,dt$$

$$\leq \tfrac{1}{2}\|z(0)\|^2 + v \int_0^t \|\nabla z\|^2\,dt + C \int_0^t \|z\|^2 \|\nabla \bar{u}\|^2\,dt .$$

Consequently, we obtain that

$$\|z(t)\|^2 \leq \|z(0)\|^2 + C \int_0^t \|z\|^2 \|\nabla \bar{u}\|^2\,dt\,;$$

recalling that $\bar{u} \in L^2(0,T;\mathcal{V})$, the conclusion follows by Gronwall's inequality (2.62).
$\square$

We remark that the validity of theorem 6.21 and proposition 6.23 is strictly limited to the two-dimensional case ($N = 2$). For example, as the proof of proposition 6.23 shows, if $N = 3$ the question of uniqueness of weak solutions is open.

6.3.3 Absorbing Sets and Attractor

We now show the existence of a bounded, positively invariant absorbing set for S, first in $\mathcal{H}$ and then in $\mathcal{V}$. As a consequence of the asymptotic smoothness of the semiflow, we deduce that S also admits a global attractor in $\mathcal{H}$.

PROPOSITION 6.24
Assume that $f \in L^\infty(0,+\infty;\mathcal{V}')$. There exists a positively invariant, absorbing ball $\mathcal{B} \subseteq \mathcal{H}$ for the solution operator S generated by problem (NS) (that is, $\mathcal{B}$ absorbs all bounded sets of $\mathcal{H}$). If in addition $f \in L^\infty(0,+\infty;\mathcal{H})$, S also admits a positively invariant absorbing ball $\mathcal{B}_1$ in $\mathcal{V}$ (that is, $\mathcal{B}_1$ absorbs all bounded sets of $\mathcal{V}$).

PROOF   In the sequel, we set

$$F_0 := \sup_{t \geq 0} \|f(t)\|_{\mathcal{V}'}^2 , \quad F_1 := \sup_{t \geq 0} \|f(t)\|_{\mathcal{H}}^2 .$$

1. Multiplying equation (6.125) in $\mathcal{H}$ by $2u$, and recalling the second of (6.131), we obtain that

$$\frac{d}{dt}\|u\|^2 + 2v\|\nabla u\|^2 = 2\langle f,u\rangle_{\mathcal{V}'\times\mathcal{V}} \leq C\|f\|_{\mathcal{V}'}^2 + v\|\nabla u\|^2 . \tag{6.137}$$

By Poincaré's inequality (3.16), we deduce from (6.137) the exponential inequality

$$\frac{d}{dt}\|u\|^2 + v\lambda_1\|u\|^2 \le CF_0.$$

The existence of a positively invariant absorbing ball for S in $\mathcal{H}$ follows then by proposition 2.64.

2. To prove the existence of an absorbing ball in $\mathcal{V}$, we formally multiply equation (6.125) in $\mathcal{H}$ by $-2\Delta u$. Since we do not know that $-\Delta u \in \mathcal{H}$, this procedure is formal; to fully justify it, we should consider, as usual, the Galerkin approximations of u, constructed on subspaces of the eigenvalues of $-\Delta$ (with respect to the homogeneous Dirichlet boundary conditions). We obtain that

$$\frac{d}{dt}\|\nabla u\|^2 + 2v\|\Delta u\|^2 + 2b(u,u,-\Delta u) = 2\langle f, -\Delta u\rangle$$

$$\le C\|f\|^2 + \tfrac{1}{2}v\|\Delta u\|^2. \tag{6.138}$$

Recalling (6.130), and resorting to the Gagliardo-Nirenberg and elliptic estimates

$$|u|_4 \le C\|\Delta^2 u\|^{1/4}\|u\|^{3/4} \le C\|\Delta u\|^{1/4}\|u\|^{3/4}$$

(see theorems A.70 and A.77), we estimate

$$b(u,u,-\Delta u) = b(u,\Delta u,u) \le C|u|_4^2\|\Delta u\|$$

$$\le C\|u\|^{3/2}\|\Delta u\|^{3/2} \le C\|u\|^6 + \tfrac{1}{2}v\|\Delta u\|^2.$$

Inserting this into (6.138), we obtain that

$$\frac{d}{dt}\|\nabla u\|^2 + v\|\Delta u\|^2 \le CF_1 + C\|u\|^6. \tag{6.139}$$

Since the first part of this proof provides an estimate on $\|u\|$ independent of $t \ge 0$, recalling the second Poincaré inequality (3.17) we finally deduce from (6.139) the exponential inequality

$$\frac{d}{dt}\|\nabla u\|^2 + v\lambda_1\|\nabla u\|^2 \le CF_1 + C_1. \tag{6.140}$$

The existence of a positively invariant absorbing ball for S in $\mathcal{V}$ follows then again by proposition 2.64. This concludes the proof of proposition 6.24. $\square$

For future reference, we remark that from (6.139) we also deduce that, for all t, $s \in [0,+\infty[$, with $s < t$,

$$v\int_s^t \|\Delta u\|^2\, dt \le \|\nabla u(s)\|^2 + C(t-s)$$

and, since (6.140) implies that the function $s \mapsto \|\nabla u(s)\|$ is bounded, we conclude that there is $C > 0$, independent of s and t, such that

$$v\int_s^t \|\Delta u\|^2\, dt \le C(1+t-s). \tag{6.141}$$

We can now deduce the existence of a global attractor for the semiflow S in $\mathcal{H}$.

THEOREM 6.25
Let $f(t, \cdot) \equiv f \in \mathcal{H}$. The semiflow S generated by problem (NS) admits a compact, global attractor $\mathcal{A}$ in $\mathcal{H}$.

PROOF   It is sufficient to apply theorem 2.46 of chapter 2. Indeed, the last claim of theorem 6.21 implies that the semiflow S is uniformly compact for large t; hence, the desired attractor is the set $\mathcal{A} = \omega(\mathcal{B})$, where $\mathcal{B}$ is the absorbing ball $\mathcal{B}$ constructed in proposition 6.24.   ☐

6.3.4   The Exponential Attractor

In this section we show that the semiflow S also admits an exponential attractor $\mathcal{E}$ in $\mathcal{H}$, which contains the global attractor $\mathcal{A}$.

THEOREM 6.26
In the same conditions of theorem 6.25, the semiflow S generated by problem (NS) admits an exponential attractor $\mathcal{E}$ in $\mathcal{H}$.

PROOF   We apply theorem 4.5 of chapter 4. To this end, we consider the absorbing ball $\mathcal{B}_1$ for S in $\mathcal{V}$, determined in proposition 6.24: since the injection $\mathcal{V} \hookrightarrow \mathcal{H}$ is compact (because the map $\mathcal{V} \in u \mapsto \operatorname{div} u \in \mathcal{H}$ is continuous), $\mathcal{B}_1$ is compact in $\mathcal{H}$. We propose then to show that S satisfies the discrete squeezing property (see definition 4.3), relative to $\mathcal{B}_1$. We proceed almost exactly as in the proof of theorem 4.7 of section 4.3.2, of which we keep the same choices of N and $\mathcal{X}_N$. In particular, we show that, given any $t_* > 0$ and $\gamma \in \,]0, \frac{1}{2}[$, there exists an integer N_*, with the property that if u_0, $v_0 \in \mathcal{B}_1$ are such that $S(t_*)u_0 - S(t_*)v_0 \notin C_{N_*}$, i.e. if

$$\|P_{N_*}(S(t_*)u_0 - S(t_*)v_0)\| < \|Q_{N_*}(S(t_*)u_0 - S(t_*)v_0)\| \tag{6.142}$$

(that is, if (4.13) holds for the operator $S(t_*)$), then (4.11) must hold, i.e.

$$\|S(t_*)u_0 - S(t_*)v_0\| \le \gamma \|u_0 - v_0\|. \tag{6.143}$$

Thus, we follow the evolution of the quotient norm Λ introduced in (3.58), i.e.

$$\Lambda(t) := \frac{\|\nabla z(t)\|^2}{\|z(t)\|^2},$$

where $z(t) := S(t)u_0 - S(t)v_0$ is the difference of the two solutions of problem (NS), with initial data u_0 and v_0. Acting as in the proof of (3.61), we arrive at the identity

$$\frac{1}{2}\frac{d\Lambda}{dt} = \frac{1}{\|z\|^2}\langle -\Delta z - \Lambda z, v(\Delta z + \Lambda z) - (u \cdot \nabla)z - (z \cdot \nabla v)\rangle. \tag{6.144}$$

Setting $w := \frac{z}{\|z\|}$, we deduce from (6.144) that

$$\frac{d\Lambda}{dt} = -2v\|\Delta w + \Lambda w\|^2 + 2v\langle \Delta w + \Lambda w, (u \cdot \nabla)w + (w \cdot \nabla v)\rangle$$

$$\leq v\left(\|(u \cdot \nabla)w\|^2 + \|(w \cdot \nabla)u\|^2\right) =: v(R_1 + R_2). \tag{6.145}$$

We first have that

$$R_1 \leq C|u|_\infty^2 \|\nabla w\|^2 ; \tag{6.146}$$

thus, resorting to the Gagliardo-Nirenberg and elliptic inequalities

$$|u|_\infty \leq C\|\Delta^2 u\|^{1/2}\|u\|^{1/2} \leq C\|\Delta u\|^{1/2}\|u\|^{1/2}$$

(see theorems A.70 and A.77), and recalling that $u \in \mathcal{B}_1$, so that its norm in V, and therefore in $\mathcal{H}$, is uniformly bounded in t, we deduce from (6.146) that

$$R_1 \leq C_1\|\Delta u\|\Lambda . \tag{6.147}$$

Similarly, since $\|w\| = 1$ and $v \in \mathcal{B}_1$, recalling (6.133) we can estimate

$$R_2 \leq C|w|_4^2|\nabla v|_4^2 \leq C\|\nabla w\|\,\|w\|\,\|\Delta v\|\,\|\nabla v\| \leq C\|\Delta v\|\sqrt{\Lambda}. \tag{6.148}$$

Inserting (6.147) and (6.148) into (6.145), we obtain that (for different constants C)

$$\frac{d\Lambda}{dt} \leq C\|\Delta v\|^2 + C(1 + \|\Delta u\|)\Lambda .$$

Integrating this inequality for $0 < s < t$ yields

$$\Lambda(t) \leq \Lambda(s) + C\int_s^t \|\Delta v\|^2 \, d\theta + C\int_s^t (1 + \|\Delta u\|)\Lambda(s) \, d\theta .$$

By Gronwall's inequality, and recalling estimate (6.141), which also holds for v, we have then

$$\Lambda(t) \leq \left(\Lambda(s) + C\int_s^t \|\Delta v\|^2 \, d\theta\right) \exp\left(C\int_s^t (1 + \|\Delta u\|) \, d\theta\right)$$

$$\leq (\Lambda(s) + C(1 + t - s)) \exp\left(C(t - s + \sqrt{t - s}\sqrt{1 + t - s})\right)$$

$$\leq (\Lambda(s) + C(1 + t - s)) e^{C(1 + t - s)},$$

from which we obtain that

$$\Lambda(s) \geq e^{-C(1 + t - s)}\Lambda(t) - C(1 + t - s).$$

Integrating this inequality with respect to s in the interval $[0, t]$, we finally deduce the estimate

$$\int_0^t \Lambda(s) \, ds \geq C_1\Lambda(t)(1 - e^{-Ct}) - C\left(1 + t + \tfrac{1}{2}t^2\right), \tag{6.149}$$

with $C_1 := (Ce^C)^{-1}$. Our next step is to recall (6.136) (with $\bar{u}$ replaced by v): Since $v \in \mathcal{B}_1$, which is bounded in V, we easily obtain that, as in (4.32),

$$\|z(t)\|^2 \le \|z(0)\|^2 \exp\left(C_2 t - v \int_0^t \Lambda(s)\,ds\right),$$

where C_2 depends on the bound on $\|\nabla v\|$ provided by $\mathcal{B}_1$. Consequently, setting

$$\varphi(t) := C_2 t + C\left(1 + t + \tfrac{1}{2}t^2\right),$$

we deduce from (6.149) that

$$\|z(t)\|^2 \le \|z(0)\|^2 \exp\left(\varphi(t) - C_1 \Lambda(t)(1 - e^{-Ct})\right). \tag{6.150}$$

Let now $t_* > 0$ be such that (6.142) holds: then, as in (4.29), $\Lambda(t_*) \ge \tfrac{1}{2}\lambda_{N+1}$. Consequently, (6.150) yields that

$$\|z(t_*)\|^2 \le \|z(0)\|^2 \exp\left(\varphi(t_*) - C_1 \lambda_{N+1}(1 - e^{-Ct_*})\right).$$

Given then $\gamma \in \,]0, \tfrac{1}{2}[$, we can make

$$\varphi(t_*) - C\lambda_{N+1}(1 - e^{-ct_*}) \le 2\ln\gamma$$

by choosing N so large that

$$\varphi(t_*) - 2\ln\gamma \le C\lambda_{N+1}(1 - e^{-ct_*}).$$

Thus, (6.143) holds. The rest of the proof of theorem 6.26 proceeds then as that of theorem 4.7.   ☐

We conclude this section with the remark that, in contrast to theorems 6.25 and 6.26, the existence of an inertial manifold for the semiflow generated by the Navier-Stokes equations is an open question, even in two dimensions of space, and for different types of boundary conditions, such as periodic ones.

6.4 Maxwell's Equations

In this section we consider a model for the quasi-stationary MAXWELL'S equations, which describe the evolution of the electromagnetic fields and inductions in a ferromagnetic medium. This situation is characterized by a nonlinear dependence between the magnetic field and induction, and by the fact that the displacement currents are negligible with respect to the eddy ones. The former feature gives rise to a quasilinear system, while the latter allows us to consider a reduced problem, which is of parabolic type. For more information on Maxwell's equations, we refer e.g. to Duvaut-Lions, [DL69, ch. 7].

6.4.1   The Equations and their Functional Framework

1.   The complete system of Maxwell's equations is the first order linear system, essentially derived from the so-called AMPÈRE'S theorem and FARADAY'S LAW,

$$D_t - \operatorname{curl} H = G - J, \tag{6.151}$$

$$B_t + \operatorname{curl} E = 0, \tag{6.152}$$

$$\operatorname{div} D = \rho, \tag{6.153}$$

$$\operatorname{div} B = 0, \tag{6.154}$$

where E and H denote, respectively, the electric and the magnetic fields, and D, B, the corresponding inductions. The vector functions G and J in (6.151) represent, respectively, an external source, and the so-called eddy currents, while the scalar function ρ in (6.153) is a measure of the total electric charge. We refer to equations (6.151), ..., (6.154) collectively as

$$\text{system   (ME)}.$$

This system is considered in a bounded domain $\Omega \subset \mathbb{R}^3$, with a Lipschitz boundary $\partial\Omega$. We supplement system (ME) with the initial conditions

$$\overline{B}(0,\cdot) = D_0, \quad B(0,\cdot) = B_0, \tag{6.155}$$

and impose the boundary conditions

$$\nu \times E = 0, \quad \nu \cdot B = 0, \tag{6.156}$$

where ν is the unit outward normal to $\partial\Omega$. The first of conditions (6.156) translates the assumption that the boundary of Ω be a so-called "perfect" conductor (i.e., an ideal metal). As we remark below, the second boundary condition in (6.156) is in general redundant, since it is a consequence of the first condition, and of equation (6.152).

REMARK 6.27   The second Maxwell's equation (6.152) implies that

$$\operatorname{div} B_t = 0, \quad \nu \cdot B_t = 0$$

(for the latter, see e.g. proposition A.89). Thus, if we assume, as it is customary, that the initial value B_0 satisfies the conditions

$$\operatorname{div} B_0 = 0, \quad \nu \cdot B_0 = 0, \tag{6.157}$$

both equation (6.154) and the second boundary condition in (6.156) are a consequence of (6.152).   ☐

System (ME) consists of eight conditions on twelve unknowns (the components of D, E, B and H). To make (ME) a determined system, we assume the constituent relations

$$D = \varepsilon E, \quad H \in \zeta(B),$$

where ε is, for simplicity, a positive constant (known as the dielectric constant, which measures the effects of the displacement currents), and ζ is a monotone map, in general multivalued because of the presence of hysteresis. Here, again for simplicity, we consider an idealized model, in which the hysteresis phenomena are neglected; thus, we assume that ζ is a monotone function. Finally, we assume that the eddy currents are everywhere present, and caused entirely by the conductivity of the medium; that is, that

$$J = \sigma E,$$

where σ is, for simplicity, a positive constant measuring conductivity. With these assumptions, Maxwell's equations (ME) can be written as

$$\varepsilon E_t - \operatorname{curl} \zeta(B) = G - \sigma E, \tag{6.158}$$
$$B_t + \operatorname{curl} E = 0,$$
$$\operatorname{div} E = \tfrac{1}{\varepsilon}\rho, \tag{6.159}$$
$$\operatorname{div} B = 0, \tag{6.160}$$

which we again refer to as

$$\text{system (ME)}.$$

Again, equation (6.160) is redundant if (6.157) holds. To make the system determined, we further assume that G and ρ satisfy the compatibility condition

$$\rho_t + \tfrac{1}{\varepsilon}\sigma\rho = \operatorname{div} G,$$

which is derived by taking the divergence of (6.158).

In ferromagnetic media, the effect of displacement currents is usually negligible in comparison to those of the eddy currents; that is, $\varepsilon \ll \sigma$. It is then common, in applications, to neglect the term εE_t in equation (6.158), and to consider instead the reduced equations

$$\sigma E - \operatorname{curl} \zeta(B) = G, \tag{6.161}$$
$$B_t + \operatorname{curl} E = 0. \tag{6.162}$$

These equations are known as the QUASI-STATIONARY Maxwell's equations; we refer to them as

$$\text{system (QS)}.$$

We remark that, now, equation (6.159) loses sense, and the divergence of E is determined by equation (6.161), that is

$$\operatorname{div} E = \frac{1}{\sigma} \operatorname{div} G. \tag{6.163}$$

The first initial condition on D in (6.155) is also lost; indeed, system (QS) must be considered as a singular limit problem for the complete system (ME).

2.   We now transform the first order system (QS) into a formally parabolic evolution equation, by the introduction of suitable electromagnetic potentials. To describe this process, assume that system (QS) has a solution (B, E), with $B(t, \cdot) \in H_0^0(\mathrm{div}, \Omega)$ and $E(t, \cdot) \in H(\mathrm{div}, \Omega)$ for almost all $t > 0$ (see section A.7 for the definition of these and related spaces). By proposition A.91, we can then determine a vector function $A(t, \cdot) \in H_0(\mathrm{curl}, \Omega) \cap H^0(\mathrm{div}, \Omega)$, such that

$$\begin{cases} \mathrm{curl}\, A = B, \\ \mathrm{div}\, A = 0, \\ v \times A = 0. \end{cases} \tag{6.164}$$

Then, equation (6.162) implies that $A_t + E \in H^0(\mathrm{curl}, \Omega) \cap H(\mathrm{div}, \Omega)$ for almost all t; hence, by theorem A.93, there is $\varphi \in H_0^1(\Omega)$, such that

$$A_t + E = -\nabla \varphi. \tag{6.165}$$

Recalling (6.163) and the second of the equations in (6.164), we see φ is determined as the solution of the boundary value problem

$$\begin{cases} -\Delta \varphi = \frac{1}{\sigma} \mathrm{div}\, G, \\ \varphi \big|_{\partial \Omega} = 0. \end{cases} \tag{6.166}$$

Replacing the expressions of B and E obtained in (6.164) and (6.165) into system (QS), we finally obtain the IBV problem

$$\begin{cases} \sigma A_t + \mathrm{curl}\, \zeta\,(\mathrm{curl}\, A) = -G - \sigma \nabla \varphi =: F, \\ A(0, \cdot) = A_0, \\ v \times A = 0, \end{cases} \tag{6.167}$$

where the initial value A_0 is determined from B_0 by means of the boundary value problem (6.164), at $t = 0$.

REMARK 6.28   The equation in problem (6.167) is quasilinear, and has the same form as equation (3.11). As such, it does not immediately fit within the framework of the semilinear equations we have considered in chapter 3; however, if the function ζ is monotone, problem (6.167) is of parabolic type, and a suitable weak solution theory can be established, by means of classical results on evolution equations with monotone operators (see e.g. Brezis, [Bre73]). In contrast, for the corresponding IBV problem for the complete system of Maxwell's equation (i.e. when $\varepsilon > 0$), which is

$$\begin{cases} \varepsilon A_{tt} + \sigma A_t + \mathrm{curl}\, \zeta\,(\mathrm{curl}\, A) = -G - \varepsilon \nabla \varphi_t - \sigma \nabla \varphi =: F, \\ A(0, \cdot) = A_0, \\ A_t(0, \cdot) = A_1 := -\nabla \varphi(0, \cdot) - E_0, \\ v \times A = 0, \end{cases}$$

an analogous weak solution theory is not yet available. This is the principal reason why we consider only the reduced problem (QS).   ▯

3.   In the sequel, we take for simplicity $\sigma = 1$. In accord with (6.157), we assume that $B_0 \in H_0^0(\text{div}, \Omega)$, and that the source term G satisfies

$$G \in L^2(0, +\infty; L^2(\Omega)) \cap L^\infty(0, +\infty; L^2(\Omega)), \quad G_t \in L^2(0, +\infty; L^2(\Omega)).$$
$$(6.168)$$

Then, $A_0 \in H_0(\text{curl}, \Omega) \cap H^0(\text{div}, \Omega)$, and the function φ defined in (6.166) is such that φ and $\varphi_t \in L^2(0, +\infty; H_0^1(\Omega))$. Therefore, the source F in (6.167) has the same regularity (6.168) as G. We set then

$$\mathcal{H} := H^0(\text{div} \, \Omega), \quad \mathcal{V} := H_0(\text{curl}, \Omega) \cap H^0(\text{div}, \Omega).$$

By theorem A.85, we have that $\mathcal{V} \hookrightarrow \mathcal{H} \hookrightarrow \mathcal{V}'$ is a Gelfand triple. Moreover, by Friedrichs' inequality (A.80), we can choose in $\mathcal{V}$ the norm

$$\|u\|_{\mathcal{V}} := \|\text{curl} \, u\|.$$

In fact, since $\text{div} \, u = 0$ if $u \in \mathcal{V}$, we have that

$$\|u\| \le \lambda_F \|\text{curl} \, u\|,$$
$$(6.169)$$

for all $u \in \mathcal{V}$, with λ_F independent of u. Finally, we also set

$$\mathcal{D} := \{u \in \mathcal{V}: \text{curl} \, \zeta(\text{curl} \, u) \in H^0(\text{div} \, \Omega)\}.$$
$$(6.170)$$

As we have stated above, we assume that ζ is a monotone function. More precisely, we assume that $\zeta \colon \mathbb{R}^3 \to \mathbb{R}^3$ is a globally Lipschitz continuous function, and the derivative of a convex function $Z \in C^{1,1}(\mathbb{R}^3, \mathbb{R})$. Without loss of generality, we can choose Z so that $Z(0) = 0$. We also assume that the derivative $\zeta'(p)$, which is defined for almost all $p \in \mathbb{R}^3$, is a uniformly strictly positive matrix, that is

$$\exists \gamma > 0 \, \forall q \in \mathbb{R}^3: \quad \langle \zeta'(p)q, q \rangle_{\mathbb{R}^3} \ge \gamma |q|^2,$$
$$(6.171)$$

with γ independent of p. We have then the following estimates:

PROPOSITION 6.29

1. *Let ζ satisfy the assumptions stated above, and L be the Lipschitz constant of ζ. Then, for all u and $v \in \mathcal{H}$,*

$$\gamma \|u\|^2 + \langle \zeta(0), u \rangle \le \langle \zeta(u), u \rangle \le L \|u\|^2 + \langle \zeta(0), u \rangle,$$
$$(6.172)$$
$$\langle \zeta'(u)v, v \rangle \ge \gamma \|v\|^2.$$
$$(6.173)$$

In particular, for all $u \in \mathcal{V}$,

$$\gamma \|u\|_{\mathcal{V}}^2 \le \langle \zeta(\text{curl} \, u), \text{curl} \, u \rangle \le L \|u\|_{\mathcal{V}}^2.$$
$$(6.174)$$

2. *Given Z as above, define a function* $N: L^2(\Omega) \to [0, +\infty[$ *by*

$$N(u) := \int_\Omega Z(u(x))\, dx.$$

Then, for all $u \in L^2(\Omega)$,

$$\tfrac{1}{2}\gamma\|u\|^2 + \langle \zeta(0), u\rangle \le N(u) \le \tfrac{1}{2}L\|u\|^2 + \langle \zeta(0), u\rangle. \tag{6.175}$$

In particular, for all $u \in V$,

$$\tfrac{1}{2}\gamma\|u\|_V^2 \le N(\operatorname{curl} u) \le \tfrac{1}{2}L\|u\|_V^2. \tag{6.176}$$

PROOF 1. Since

$$\langle \zeta(u), u\rangle = \langle \zeta(u) - \zeta(0), u - 0\rangle + \langle \zeta(0), u\rangle,$$

(6.172) is an immediate consequence of the Lipschitz continuity of ζ and of (6.171), which also implies (6.173). Then, (6.174) follows from (6.172), since $\langle \zeta(0), \operatorname{curl} u\rangle = 0$, as follows by integration by parts.

2. Since $\zeta = Z'$, and $Z(0) = 0$, recalling (6.172) we compute that, for $u \in L^2(\Omega)$,

$$N(u) = \int_\Omega (Z(u(x)) - Z(0))\, dx = \int_\Omega \int_0^1 \zeta(r u(x)) \cdot u(x)\, dr\, dx$$
$$= \int_0^1 \tfrac{1}{r}\langle \zeta(ru), ru\rangle\, dr \ge \gamma\|u\|^2 \int_0^1 r\, dr + \int_0^1 \langle \zeta(0), u\rangle.$$

This implies the first half of (6.175); the second half follows similarly. Finally, (6.176) follows from (6.175). □

REMARK 6.30 Under the stated assumptions on ζ, it can be shown, by means of elliptic regularity results similar to theorem A.77, that if $\mathcal{D}$ is the space defined in (6.170), then $\mathcal{D} = H^2(\Omega) \cap V$. □

6.4.2 The Quasi-Stationary Maxwell Semiflow

1.   The following theorem provides the global existence and regularity of weak solutions to problem (6.167).

THEOREM 6.31
Let $T > 0$. *For all* $F \in L^2(0,T;V')$ *and* $A_0 \in \mathcal{H}$, *there exists a unique* $A \in C([0,T];\mathcal{H}) \cap L^2(0,T;V)$, *with* $A_t \in L^2(0,T;V')$, *which is a weak solution of problem* (6.167), *in the sense that* $A(0,\cdot) = A_0$ *(this makes sense because, by theorem A.80,* $A \in C([0,T];\mathcal{H})$*), and the equation in* (6.167) *is satisfied in* V' *for almost all* $t \in]0,T[$. *If in addition* $F \in L^2(0,T;\mathcal{H})$, *then*

$$A \in C([\tau,T];V) \cap L^2(\tau,T;\mathcal{D}), \quad A_t \in L^2(\tau,T;\mathcal{H}), \tag{6.177}$$

*for all $\tau \in]0,T[$. If $A_0 \in \mathcal{V}$, we can take $\tau = 0$ in (6.177). Furthermore, if $F_t \in$
$L^2(0,T;\mathcal{V}')$, then*

$$A_t \in C([\tau,T];\mathcal{H}) \cap L^2(\tau,T;\mathcal{V}), \quad A \in C([\tau,T];\mathcal{D}), \qquad (6.178)$$

with the choice $\tau = 0$ admissible if $A_0 \in \mathcal{D}$ (so that $A_1 := \text{curl}\,\zeta\,(\text{curl}\,A_0) + F(0,\cdot) \in \mathcal{H}$).

PROOF   A proof of this theorem, based on the Galerkin method, can be given
following the same procedure of the proof of theorem 3.9 of chapter 3, where we
considered the semilinear parabolic problem. Here, we limit ourselves to establish
the necessary a priori estimates that we shall also need in the sequel for the existence
of the absorbing balls for the semiflow generated by (6.167). At first, multiplying the
equation in (6.167) in $\mathcal{H}$ by $2A$ we obtain

$$\frac{d}{dt}\|A\|^2 + 2\langle \zeta(\text{curl}\,A), \text{curl}\,A\rangle = 2\langle F, \text{curl}\,A\rangle_{\mathcal{V}' \times \mathcal{V}};$$

recalling (6.174), we obtain then that

$$\frac{d}{dt}\|A\|^2 + \gamma\|A\|_{\mathcal{V}}^2 \leq \tfrac{1}{2\gamma}\|F\|_{\mathcal{V}'}^2 + \tfrac{1}{2}\gamma\|A\|_{\mathcal{V}}^2, \qquad (6.179)$$

from which, integrating in $[0,T[$, we obtain that $A \in L^\infty(0,T;\mathcal{H}) \cap L^2(0,T;\mathcal{V})$.
Hence, $A_t \in L^2(0,T;\mathcal{V}')$, so that $A \in C([0,T];\mathcal{H})$. Next, we multiply the equation in
(6.167) in $\mathcal{H}$ by $(e^t - 1)A_t$, obtaining

$$(e^t - 1)\|A_t\|^2 + \frac{d}{dt}\left((e^t - 1)N(\text{curl}\,A)\right)$$
$$\leq \tfrac{1}{2}(e^t - 1)\|F\|^2 + \tfrac{1}{2}(e^t - 1)\|A_t\|^2 + (e^t - 1)N(\text{curl}\,A) + N(\text{curl}\,A).$$

Integrating this, and recalling (6.176), we obtain

$$\int_0^t \left((e^\theta - 1)\|A_t\|^2\right)\,d\theta + (e^t - 1)N(\text{curl}\,A)$$
$$\leq \int_0^t \left((e^\theta - 1)\|F\|^2\right)\,d\theta + L\int_0^t \|A\|_{\mathcal{V}}^2\,d\theta$$
$$+ \int_0^t \left((e^\theta - 1)N(\text{curl}\,A)\right)\,d\theta. \qquad (6.180)$$

Recalling that $A \in L^2(0,T;\mathcal{V})$, we can then deduce, by means of Gronwall's inequal-
ity, that $A \in C([\tau,T];\mathcal{V})$ and $A_t \in L^2(\tau,T;\mathcal{H})$, $\tau \in]0,T[$. Thus, $A \in L^2(\tau,T;\mathcal{D})$, and
(6.177) holds. Note that if $A_0 \in \mathcal{V}$, we do not need to multiply by the factor $(e^t - 1)$,
so that we can take $\tau = 0$ in (6.180). In fact, for future reference we note that, in this
case, we have the estimate

$$\int_0^t \|A_t\|^2\,d\theta + \gamma\|A(t)\|_{\mathcal{V}}^2 \leq L\|A_0\|_{\mathcal{V}}^2 + \int_0^T \|F\|^2\,d\theta. \qquad (6.181)$$

Finally, we differentiate the equation in (6.167) with respect to t, and multiply the resulting equation in $\mathcal{H}$ by $2(e^t - 1)A_t$, obtaining

$$\frac{d}{dt}\left((e^t - 1)\|A_t\|^2\right) + 2(e^t - 1)\langle \zeta'(\operatorname{curl}A)\operatorname{curl}A_t, \operatorname{curl}A_t\rangle$$
$$= 2(e^t - 1)\langle F_t, A_t\rangle + e^t\|A_t\|^2.$$

From this, recalling (6.173), we obtain

$$\frac{d}{dt}\left((e^t - 1)\|A_t\|^2\right) + 2\gamma(e^t - 1)\|A_t\|_{\mathcal{V}}^2 \leq (e^t - 1)\|F_t\|^2 + (2e^t - 1)\|A_t\|^2,$$

and, integrating,

$$(e^t - 1)\|A_t\|^2 + 2\gamma\int_0^t (e^\theta - 1)\|A_t\|_{\mathcal{V}}^2\, d\theta$$
$$\leq \int_0^T (e^t - 1)\|F_t\|^2\, d\theta + \int_0^t (2e^\theta - 1)\|A_t\|^2\, d\theta. \tag{6.182}$$

Recalling that $A_t \in L^2(0, T; \mathcal{H})$ if $A_0 \in \mathcal{V}$, we conclude that (6.178) holds. $\square$

2.   If $F \in L^2(0, +\infty; \mathcal{V}')$, theorem 6.31 shows that problem (6.167) generates a solution operator $S = (S(t))_{t\geq 0}$, with $S(t): \mathcal{H} \to \mathcal{H}$ for all $t \geq 0$, and $S(t): \mathcal{V} \to \mathcal{V}$ if $F \in L^2(0, +\infty; \mathcal{H})$. We now show that, if F is independent of t, these solution operators are actually semiflows, both in $\mathcal{H}$ and in $\mathcal{V}$. To this end, it is sufficient to estimate the difference $\alpha(t) := A(t) - \overline{A}(t) = S(t)A_0 - S(t)\overline{A}_0$, in $\mathcal{H}$ and $\mathcal{V}$. At first, we immediately have that

$$\frac{d}{dt}\|\alpha\|^2 + 2\langle \zeta(\operatorname{curl}A) - \zeta(\operatorname{curl}\overline{A}), \operatorname{curl}\alpha\rangle = 0,$$

from which, recalling (6.173),

$$\frac{d}{dt}\|\alpha\|^2 + \gamma\|\alpha\|_{\mathcal{V}}^2 \leq 0. \tag{6.183}$$

Integrating this inequality we deduce that $S(t)$ is Lipschitz continuous in $\mathcal{H}$, for all $t \geq 0$.

We now prove that each operator $S(t)$ is also continuous in $\mathcal{V}$; that is, that for all $t > 0$, $A_0 \in \mathcal{V}$, and $\eta > 0$, there is $\delta > 0$ such that

$$\|S(t)A_0 - S(t)\overline{A}_0\|_{\mathcal{V}} = \|A(t) - \overline{A}(t)\|_{\mathcal{V}} \leq \eta \quad \text{if} \quad \|A_0 - \overline{A}_0\|_{\mathcal{V}} \leq \delta.$$

Proceeding by contradiction, we assume that there are $t_0 > 0$, $A_0 \in \mathcal{V}$ and $\eta_0 > 0$ such that for all $\delta > 0$, there is $\overline{A}_0 \in \mathcal{V}$ with

$$\|A_0 - \overline{A}_0\|_{\mathcal{V}} \leq \delta \quad \text{but} \quad \|A(t_0) - \overline{A}(t_0)\|_{\mathcal{V}} \geq \eta_0. \tag{6.184}$$

From (6.183) and (6.169) we deduce that

$$\gamma \|\alpha(t_0)\|_{\mathcal{V}}^2 \le 2\|\alpha_t(t_0)\| \, \|\alpha(t_0)\| \le 2\|\alpha_t(t_0)\| \, \|\alpha(0)\| \le 2\lambda_F \|\alpha_t(t_0)\| \, \|\alpha(0)\|_{\mathcal{V}} . \tag{6.185}$$

Consider now any $\overline{A}_0 \in \mathcal{V}$, with $\|A_0 - \overline{A}_0\|_{\mathcal{V}} \le 1$. From (6.182) (with $F_t \equiv 0$) and (6.181) we deduce that

$$\|A_t(t_0)\|^2 \le \frac{2e^{t_0} - 1}{e^{t_0} - 1} \int_0^{t_0} \|A_t\|^2 \, d\theta \le \frac{2e^{t_0} - 1}{e^{t_0} - 1} \left(L\|A_0\|_{\mathcal{V}}^2 + t_0\|F\|^2 \right) ;$$

similarly,

$$\|\overline{A}_t(t_0)\|^2 \le \frac{2e^{t_0} - 1}{e^{t_0} - 1} \left(2L(\|A_0\|_{\mathcal{V}}^2 + 1) + t_0\|F\|^2 \right) .$$

Therefore, we deduce that

$$\|\alpha_t(t_0)\|^2 \le \frac{6(2e^{t_0} - 1)}{e^{t_0} - 1} \left(L(1 + \|A_0\|_{\mathcal{V}}^2) + t_0\|F\|^2 \right) . \tag{6.186}$$

Inserting this into (6.185) we obtain

$$\|\alpha(t_0)\|_{\mathcal{V}}^2 \le C_0 \|A_0 - \overline{A}_0\|_{\mathcal{V}} , \tag{6.187}$$

where the constant C_0 depends on t_0 and $\|A_0\|_{\mathcal{V}}$, as per (6.186). Let then

$$\delta := \min\{\tfrac{1}{2}\eta_0^2 C_0^{-1}, 1\},$$

and $\overline{A}_0 \in \mathcal{V}$ be determined as in (6.184). Then, $\|A_0 - \overline{A}_0\|_{\mathcal{V}} \le 1$; thus, (6.187) holds, and this implies the contradiction

$$\eta_0^2 \le \|\alpha(t_0)\|_{\mathcal{V}}^2 \le C_0\delta \le \tfrac{1}{2}\eta_0^2 .$$

This allows us to conclude that each solution operator $S(t)$ is also continuous in $\mathcal{V}$. Hence, S is a semiflow in both $\mathcal{H}$ and $\mathcal{V}$.

6.4.3 Absorbing Sets and Attractors

1. We now proceed to show the existence of bounded, positively invariant absorbing sets for S, both in $\mathcal{H}$ and in $\mathcal{V}$.

PROPOSITION 6.32
Assume that F is independent of t. If $F \in \mathcal{V}'$, there exists a ball $\mathcal{B}_0 \subset \mathcal{H}$, which is positively invariant and absorbing for the semiflow S in $\mathcal{H}$. If $F \in \mathcal{H}$, there exists a ball $\mathcal{B}_1 \subset \mathcal{V}$, which is positively invariant and absorbing for the semiflow S in $\mathcal{V}$.

PROOF   From (6.179) we first obtain, recalling (6.169), the exponential inequality

$$\frac{d}{dt}\|A\|^2 + \frac{\gamma}{2\lambda_F}\|A\|^2 \le \frac{1}{2\gamma}\|F\|_{\mathcal{V}'}^2 ,$$

from which we deduce the existence of a bounded, positively invariant absorbing set B_0 for S in $\mathcal{H}$ in the usual way. In particular, for any bounded set $\mathcal{G} \subset \mathcal{H}$, there is $M > 0$ such that for all $A_0 \in \mathcal{G}$ and all $t \geq 0$,

$$\|S(t)A_0\| =: \|A(t)\| \leq M. \tag{6.188}$$

Next, we multiply the equation in (6.167) in $\mathcal{H}$ by A_t, to obtain

$$\|A_t\|^2 + \frac{d}{dt} N(\mathrm{curl}\, A) \leq \tfrac{1}{2}\|F\|^2 + \tfrac{1}{2}\|A_t\|^2. \tag{6.189}$$

From (6.179) we also obtain, recalling (6.176)

$$\tfrac{2\gamma}{L} N(\mathrm{curl}\, A) \leq \tfrac{1}{2\gamma}\|F\|_{V'}^2 + 2\|A\|\,\|A_t\| \leq \tfrac{C}{2\gamma}\|F\|^2 + 2\|A\|^2 + \tfrac{1}{2}\|A_t\|^2. \tag{6.190}$$

Summing (6.189) and (6.190), and recalling (6.188), we obtain the exponential inequality

$$\frac{d}{dt} N(\mathrm{curl}\, A) + \tfrac{2\gamma}{L} N(\mathrm{curl}\, A) \leq \tfrac{1}{2}\left(1 + \tfrac{C}{\gamma}\right)\|F\|^2 + 2M^2 =: M_1. \tag{6.191}$$

Because of (6.176), we can deduce from (6.191) the existence of a bounded, positively invariant absorbing set B_1 for S in $\mathcal{V}$. ▯

2.   Since the injections $\mathcal{D} \hookrightarrow \mathcal{V} \hookrightarrow \mathcal{H}$ are compact, (6.178) implies that S is asymptotically uniformly compact, both with respect to the topologies of $\mathcal{V}$ and $\mathcal{H}$; hence, theorem 2.46 implies that the semiflow S defined by problem (6.167) admits the sets $\mathcal{A}_0 = \omega(B_0)$ and $\mathcal{A}_1 = \omega(B_1)$ as global attractors, respectively in $\mathcal{H}$ and $\mathcal{V}$.

REMARK 6.33   The importance of showing the existence of an attractor for S in $\mathcal{V}$, and not just in $\mathcal{H}$, resides in the fact that we can then deduce the existence of an attractor, in $\mathcal{H}$, for the semiflow $\tilde{S}$ defined by the original first order system (QS), that is, recalling (6.164), by

$$\tilde{S}(t)B_0 = B(t, \cdot) := \mathrm{curl}\, A(t, \cdot) = \mathrm{curl}(S(t)A_0),$$

where $B_0 \in \mathcal{H}$, and $A_0 \in \mathcal{V}$ is determined from B_0 as the solution of problem (6.164). As for E, note that, by (6.165), $E(t, \cdot) := -A_t(t, \cdot) - \nabla\varphi$, with $\nabla\varphi \in L^2(\Omega)$ (recall that in the autonomous case, the source term G is independent of t), and $A_t(t, \cdot) \in \mathcal{H}$ for all $t > 0$, as stated in (6.178). Hence, $E(t, \cdot)$ remains in a bounded set of $L^2(\Omega)$, for all $t \geq 0$. ▯

We conclude our presentation of the quasi-stationary Maxwell's equations with the remark that, since problem (6.167) is quasilinear, the techniques we have developed in chapters 4 and 5 are not directly applicable. Hence, the existence of an exponential attractor for S, even only in $\mathcal{H}$, not to mention that of an inertial manifold, is an open problem.

Chapter 7

A Nonexistence Result for Inertial Manifolds

In this chapter we present a result, due to Mora and Solà-Morales, [MSM87], concerning the nonexistence of inertial manifolds for the semiflow generated by a semilinear dissipative wave equation of the form (3.4).

7.1 The Initial-Boundary Value Problem

1. We consider again the hyperbolic perturbation, for $\varepsilon > 0$, of the Chafee-Infante equation of (5.149) in one space dimension, with $f = 0$, i.e.

$$\varepsilon u_{tt} + u_t - u_{xx} = k(u - u^3) =: g(u), \tag{7.1}$$

with $k > 0$ and $t > 0$, $x \in \,]0, \pi[$. We supplement (7.1) with the initial conditions

$$u(0,x) = u_0(x), \quad u_t(0,x) = u_1(x), \quad x \in \,]0, \pi[, \tag{7.2}$$

but, in contrast to problem (5.166), we impose homogeneous boundary conditions of Neumann type, i.e.

$$u_x(t,0) = u_x(t,\pi) = 0, \quad t > 0. \tag{7.3}$$

We refer to the IBVP (7.1)+(7.2)+(7.3) as

$$\text{problem} \quad (H_N).$$

In section 5.8.5 of chapter 5 we have seen that if ε is sufficiently small, the semiflow generated by the corresponding IBVP with Dirichlet boundary conditions admits an inertial manifold. In contrast, Mora and Solà-Morales have shown in [MSM87] that, if instead ε is sufficiently large, the semiflow generated by problem (H_N) cannot admit an inertial manifold of class C^1, containing the global attractor of this semiflow.

2. In order to conform to Mora and Solà-Morales' presentation, we introduce again the time rescaling $t \mapsto \sqrt{\varepsilon} t$, which transforms (7.1) into the equation

$$u_{tt} + 2\alpha u_t - u_{xx} = k(u - u^3), \quad \alpha := \frac{1}{2\sqrt{\varepsilon}}. \tag{7.4}$$

Exactly as in section 3.4, we can prove that problem (H$_N$) generates a flow in the product space $\mathcal{X} := H^1(0,\pi) \times L^2(0,\pi)$. In fact, arguing as in section 3.4.1 we have the following result.

THEOREM 7.1

Problem (H$_N$) generates a continuous flow $\tilde{S} = (\tilde{S}(t))_{t \in \mathbb{R}}$ on $\mathcal{X}$, which admits a global attractor $\tilde{A}$ in $\mathcal{X}$. Moreover, for each compact interval $[a,b] \subset \mathbb{R}$, the map

$$\mathcal{X} \ni U_0 \mapsto \tilde{S}(\cdot)U_0 \in C([a,b];\mathcal{X}) \tag{7.5}$$

is of class C^2 and bounded.

The proof of this theorem can be established along the same lines as that of theorem 3.20, with the exception of the C^2 dependence of the orbits from their initial values, expressed in (7.5). This part can be proven as in Henry, [Hen81, thm. 3.4.4], even if the "space" operator A of the first order formulation of equation (7.4) is not sectorial (see section A.3). Here, we limit ourselves to prove the following forward uniqueness and continuous dependence result in $\mathcal{X}$, which we will need in the sequel.

PROPOSITION 7.2

Problem (H$_N$) is well posed in $\mathcal{X}$, on any compact interval $[T_0, T] \subset \mathbb{R}$ (see definition 1.5).

PROOF   Without loss of generality, we can take $T_0 = 0$. Let

$$\mathcal{Z} := C(\mathbb{R};H^1(0,\pi)) \cap C^1(\mathbb{R};L^2(0,\pi)),$$

and assume that $u, \bar{u} \in \mathcal{Z}$ are two solutions of the initial boundary value problem (7.4)+(7.3). Then, their difference z solves the equation

$$z_{tt} + 2\alpha z_t - z_{xx} = k(z - u^3 + \bar{u}^3), \tag{7.6}$$

with homogeneous initial and boundary conditions. Multiplying (7.6) in $L^2(0,\pi)$ by $2z_t$, we obtain, as in section 3.4.1:

$$\frac{d}{dt}(\|z_t\|^2 + \|z_x\|^2) + 4\alpha\|z_t\|^2 = 2k((1 + (u^2 + u\bar{u} + \bar{u}^2))z, z_t)$$

$$\leq 2k(1 + |u^2 + u\bar{u} + \bar{u}^2|_\infty)\|z\| \, \|z_t\|, \tag{7.7}$$

where $\|\cdot\|$ denotes, as usual, the norm in $L^2(0,\pi)$. Since $H^1(0,\pi) \hookrightarrow L^\infty(0,\pi)$, and $u, \bar{u} \in \mathcal{Z}$, given $T > 0$ there is $C_0 > 0$, depending on u and $\bar{u}$, such that

$$|u^2 + u\bar{u} + \bar{u}^2|_\infty \leq C(\|u\|_1^2 + \|\bar{u}\|_1^2) =: \leq C_0.$$

Thus, integrating (7.7) in $[0,t]$, $0 < t < T$, we obtain

$$\varphi(t) := \|z_t(t,\cdot)\|^2 + \|z_x(t,\cdot)\|^2 + 2\alpha \int_0^t \|z_t(\theta,\cdot)\|^2 \, d\theta$$

$$\leq \|z_t(0,\cdot)\|^2 + \|z_x(0,\cdot)\|^2 + C_1 \int_0^t \|z(\theta,\cdot)\|^2 \, d\theta \,, \tag{7.8}$$

with C_1 depending on C_0 and α. Estimating

$$\|z(\theta,\cdot)\|^2 \leq 2\|z(0,\cdot)\|^2 + 2\left(\int_0^\theta \|z_t(\tau,\cdot)\| \, d\tau\right)^2$$

$$\leq 2\|z(0,\cdot)\|^2 + 2\theta \int_0^\theta \|z_t(\tau,\cdot)\|^2 \, d\tau \,, \tag{7.9}$$

we obtain from (7.8) that

$$\varphi(t) \leq \|z_t(0,\cdot)\|^2 + \|z_x(0,\cdot)\|^2 + 2C_1\|z(0,\cdot)\|^2 t + \frac{C_1}{\alpha}\int_0^t \theta\varphi(\theta) \, d\theta \,.$$

By Gronwall's inequality, we deduce then that

$$\varphi(t) \leq (\|z_t(0,\cdot)\|^2 + \|z_x(0,\cdot)\|^2 + 2C\|z(0,\cdot)\|^2 t)e^{C_1 t^2/(2\alpha)} =: \tilde{\varphi}(t) \,. \tag{7.10}$$

Replacing this into (7.9), we have

$$\|z(t,\cdot)\|^2 \leq 2\|z(0,\cdot)\|^2 + \tfrac{1}{\alpha}t\,\tilde{\varphi}(t) \,.$$

Together with (7.10), this implies the conclusion of proposition 7.2. In particular, solutions of problem (H_N) in $\mathcal{X}$ are uniquely determined by their initial values. ∎

7.2   Overview of the Argument

Mora and Solà-Morales' nonexistence result is proven by contradiction; in its general lines, it runs as follows. The second order equation (7.4) is transformed into the first order system in $\mathcal{X}$

$$U_t + \alpha U = AU + G(U) \,, \tag{7.11}$$

where A is the linear operator, unbounded on $\mathcal{X}$, formally defined by

$$A := \begin{pmatrix} 0 & 1 \\ \alpha^2 + \partial_{xx} & 0 \end{pmatrix} \,, \tag{7.12}$$

and, for $U = (u, \tilde{u}) \in \mathcal{X}$, $G(U) := (0, k(u - u^3))$ (compare to (5.114). We define $S = (S(t))_{t\in\mathbb{R}}$ to be the flow generated by (7.11) in $\mathcal{X}$. This flow has, in the phase space $\mathcal{X}$, three stationary states P_{-1}, P_0 and P_1, which are the constant functions

$$P_{-1}(x) \equiv (-1,0), \quad P_0(x) \equiv (0,0), \quad P_1(x) \equiv (1,0); \tag{7.13}$$

the stationary states P_0 and P_1 are joined by a heteroclinic orbit γ, which is the graph of a C^1 function. By theorem 7.1, translated to the equivalent first order system (7.11), the flow S admits a global attractor $\mathcal{A}$ in $\mathcal{X}$; thus, $\gamma \subseteq \mathcal{A}$. In particular, denoting by $\mathcal{F}_S$ the class of the finite dimensional manifolds, which are locally invariant under the flow S, are of class C^1, contain P_1, and are differentiable at P_1, then, for any neighborhood $\mathcal{V}_\eta$ of P_1 in $\mathcal{X}$ we have that $\gamma \cap \mathcal{V}_\eta \in \mathcal{F}_S$.

Suppose now that S admitted a closed inertial manifold $\mathcal{M} \subseteq \mathcal{X}$. Then, $\gamma \subseteq \mathcal{A} \subseteq \mathcal{M}$; therefore, $\mathcal{M} \cap \mathcal{V}_\eta \in \mathcal{F}_S$. The central point of Mora and Solà-Morales' argument is to show that this cannot happen. They achieve this, by proving that the set $\mathcal{F}_S$ is at most countable, and, therefore, it contains "too few" elements for $\mathcal{M} \cap \mathcal{V}_\eta$ to be in $\mathcal{F}_S$, for a suitable choice of $\mathcal{V}_\eta$ sufficiently small.

To show that $\mathcal{F}_S$ is countable, Mora and Solà-Morales' first prove that the same is true for the analogous set $\mathcal{F}_R$, corresponding to the linear flow R, obtained by linearizing (7.11) at P_1 (see (A.5)). Then, they resort to an equivalence result, analogous to the Hartman-Grobman theorem (see theorem A.8), to "transfer" the result back to the original flow S.

Linearizing (7.11) at P_1, we obtain the system

$$U_t + \alpha U = A_k U, \tag{7.14}$$

where A_k is the unbounded linear operator in $\mathcal{X}$, formally defined by

$$A_k := \begin{pmatrix} 0 & 1 \\ \alpha^2 - 2k + \partial_{xx} & 0 \end{pmatrix}. \tag{7.15}$$

The linear system (7.14) generates a flow in $\mathcal{X}$, which we denote by $R = (R(t))_{t \in \mathbb{R}}$. Acting as in (5.117), we see that if α is so small, i.e. ε is so large, that

$$\alpha^2 < 2k, \tag{7.16}$$

then A_k admits a sequence of eigenvalues, which are all purely imaginary. In section 7.4 we show that, as a consequence, the set $\mathcal{F}_R$ consists of at most one countable family; that is, $\mathcal{F}_R = (\mathcal{M}_{Rm})_{m \in \mathbb{N}}$, each manifold $\mathcal{M}_{Rm}$ being finite dimensional, locally invariant under R, of class C^1, containing P_1, and differentiable at P_1.

By a C^1 linearization equivalence result, shown in section 7.10, the same holds in a neighborhood $\mathcal{V}_\eta$ of P_1 in $\mathcal{X}$, for the original nonlinear problem (7.11). That is, also the set $\mathcal{F}_S$ consists of at most one countable family $\mathcal{F}_S = (\mathcal{M}_m)_{m \in \mathbb{N}}$. As we discussed above, from this it follows that, if S admitted a closed inertial manifold $\mathcal{M}$, then there must exist $m \in \mathbb{N}$ such that $\mathcal{M} \cap \mathcal{V}_\eta = \mathcal{M}_m \cap \mathcal{V}_\eta$.

Mora and Solà-Morales' proceed then to show, by means of a perturbation argument, that this cannot happen. To describe this step of their argument, in section 7.6 we show that system (7.11) can be perturbed in such a way that the corresponding flow S_f still admits P_{-1}, P_0 and P_1 as stationary states, with P_0 and P_1 joined by a heteroclinic orbit γ_f, contained in the global attractor $\mathcal{A}_f$ of S_f. This perturbation can be made so, that if $\mathcal{V}_\eta$ is a sufficiently small neighborhood of P_1 in $\mathcal{X}$, then the restriction of S_f to V_η coincides with that of S. Thus, $\mathcal{A}_f \cap V_\eta = \mathcal{A} \cap V_\eta$ and, consequently, $\gamma_f \cap V_\eta = \gamma \cap V_\eta \in \mathcal{F}_S$. That is, γ_f must converge to P_1 along one of the countable manifolds $\mathcal{M}_m$ of the family $\mathcal{F}_S$. In the final part of their argument, Mora and Solà-Morales' show then explicitly that this *cannot* happen; that is, γ_f does not converge to P_1 along any of the manifolds $\mathcal{M}_m$. From this contradiction, they can then deduce that the flow generated by the original system (7.11) cannot admit an inertial manifold.

7.3   The Linearized Problem

1.   We consider the linearized problem (7.14) in $\mathcal{X} = H^1(0, \pi) \times L^2(0, \pi)$. The domain of the operator A_k defined by (7.15) is the space

$$\mathcal{Y} := \mathcal{H}_N^2(0, \pi) \times H^1(0, \pi), \tag{7.17}$$

where

$$\mathcal{H}_N^2(0, \pi) := \{u \in H^2(0, \pi): u_x(0) = u_x(\pi) = 0\}.$$

System (7.14) is equivalent to the second order equation

$$u_{tt} + 2\alpha u_t + L_k u = 0,$$

where L_k is the linear operator on $L^2(0, \pi)$, with domain $H^2(0, \pi)$, defined by

$$u \mapsto L_k u := -u_{xx} + 2ku.$$

The eigenvalues of L_k, and the corresponding eigenfunctions, are

$$\lambda_m = m^2 + 2k, \quad w_m(x) = \cos(mx), \quad m \in \mathbb{N}, \tag{7.18}$$

and the sequence $(w_m)_{m \in \mathbb{N}}$ forms an orthogonal system in both $L^2(0, \pi)$ and $H^1(0, \pi)$. This means that, if for $m \in \mathbb{N}$ we set

$$\mathcal{H}_m := \text{span}\{w_m\}, \quad \mathcal{X}_m := \mathcal{H}_m \times \mathcal{H}_m, \tag{7.19}$$

then the orthogonal decompositions

$$L^2(0, \pi) = \bigoplus_{m=0}^{\infty} \mathcal{H}_m, \quad \mathcal{X} = \bigoplus_{m=0}^{\infty} \mathcal{X}_m \tag{7.20}$$

hold, the first being invariant with respect to L_k. Finally, we recall that, since the domain of L_k, i.e. $\mathcal{H}_N^2(0, \pi)$, is compactly imbedded into $L^2(0, \pi)$, the operator L_k has compact resolvent (see theorem A.24).

2.   We now turn our attention to the operator A_k defined by (7.15). Our goal is to choose in the product space $\mathcal{X}$ a norm, with respect to which A_k generates a group of unitary operators (see section A.2). More precisely, we assume that (7.16) holds, that is, $\alpha^2 < 2k$, set $\alpha_1 := 2k - \alpha^2$, and endow $\mathcal{X}$ with the equivalent scalar product defined, for $U = (u, \tilde{u})$ and $V = (v, \tilde{v}) \in \mathcal{X}$, by

$$\langle U, V \rangle_{\mathcal{X}} := \langle u_x, \bar{v}_x \rangle + \alpha_1 \langle u, \bar{v} \rangle + \langle \tilde{u}, \bar{\tilde{v}} \rangle. \qquad (7.21)$$

In (7.21), $\langle \cdot, \cdot \rangle$ denotes, as usual, the scalar product in $L^2(0, \pi)$, and the bar denotes complex conjugation (recall that all the eigenfunctions of A_k are complex valued). We claim:

PROPOSITION 7.3
Let $\mathcal{Y}$ be defined as in (7.17). For all $U \in \mathcal{Y}$, $\mathfrak{Re} \langle A_k U, U \rangle_{\mathcal{X}} = 0$. Consequently, the operator iA_k (where i is the imaginary unit) is self-adjoint.

PROOF   1. Let $U = (u, \tilde{u}) \in \mathcal{Y}$. Recalling (7.21), we compute

$$\begin{aligned}
\langle A_k U, U \rangle_{\mathcal{X}} &= \langle \tilde{u}_x, \bar{u}_x \rangle + \alpha_1 \langle \tilde{u}, \bar{u} \rangle + \langle -\alpha_1 u + u_{xx}, \bar{\tilde{u}} \rangle \\
&= (\langle \tilde{u}_x, \bar{u}_x \rangle - \langle u_x, \bar{\tilde{u}}_x \rangle) - \alpha_1 (\langle u, \bar{\tilde{u}} \rangle - \langle \tilde{u}, \bar{u} \rangle) \\
&= (\overline{\langle u_x, \bar{\tilde{u}}_x \rangle} - \langle u_x, \bar{\tilde{u}}_x \rangle) - \alpha_1 (\overline{\langle \tilde{u}, \bar{u} \rangle} - \langle \tilde{u}, \bar{u} \rangle).
\end{aligned}$$

Since $\mathfrak{Re}(z - \bar{z}) = 0$ for all complex numbers z, we conclude that $\mathfrak{Re} \langle A_k U, U \rangle_{\mathcal{X}} = 0$ as claimed.
    2. Analogously, for $U = (u, \tilde{u})$ and $V = (v, \tilde{v}) \in \mathcal{Y}$, we compute

$$\begin{aligned}
\langle iA_k U, V \rangle_{\mathcal{X}} &= \langle i\tilde{u}_x, \bar{v}_x \rangle + \alpha_1 \langle i\tilde{u}, \bar{v} \rangle + \langle i(-\alpha_1 u + u_{xx}), \bar{v} \rangle \\
&= -\langle \tilde{u}_x, \overline{iv}_x \rangle - \alpha_1 \langle \tilde{u}, \overline{iv} \rangle + \alpha_1 \langle u, \overline{iv} \rangle + \langle u_x, \overline{iv}_x \rangle, \\
\langle U, iA_k V \rangle_{\mathcal{X}} &= \langle u_x, \overline{iv}_x \rangle + \alpha_1 \langle u, \overline{iv} \rangle + \langle \tilde{u}, \overline{i(-\alpha_1 v + v_{xx})} \rangle \\
&= \langle u_x, \overline{iv}_x \rangle + \alpha_1 \langle u, \overline{iv} \rangle - \alpha_1 \langle \tilde{u}, \overline{iv} \rangle - \langle \tilde{u}_x, \overline{iv}_x \rangle.
\end{aligned}$$

Consequently, $\langle iA_k U, V \rangle_{\mathcal{X}} = \langle U, iA_k V \rangle_{\mathcal{X}}$, as claimed.   □

3.   By Stone's theorem (see theorem A.50), A_k generates a C_0 group $Z = (Z(t))_{t \in \mathbb{R}}$ of unitary operators in $\mathcal{X}$. In particular, each $Z(t)$ is an isometry; that is, for all $U_0 \in \mathcal{X}$ and all $t \in \mathbb{R}$,

$$\|Z(t)U_0\|_{\mathcal{X}} = \|U_0\|_{\mathcal{X}}, \qquad (7.22)$$

where now $\|\cdot\|_{\mathcal{X}}$ denotes the norm in $\mathcal{X}$ corresponding to the scalar product (7.21), i.e.

$$\|U\|_{\mathcal{X}}^2 := \|u_x\|^2 + \alpha_1 \|u\|^2 + \|\tilde{u}\|^2$$

(recall that $\alpha_1 = 2k - \alpha^2 > 0$, by (7.16)). We claim then:

PROPOSITION 7.4
For all $t \in \mathbb{R}$, $R(t) = e^{-\alpha t} Z(t)$.

PROOF   Since Z is generated by A_k, for all $U_0 \in \mathcal{X}$ the function $t \mapsto V(t) := Z(t)U_0$ solves the initial value problem

$$V' = A_k V, \quad V(0) = U_0.$$

Then, the function $t \mapsto U(t) := e^{-\alpha t} V(t)$ solves the initial value problem

$$U' = e^{-\alpha t}(V' - \alpha V) = e^{-\alpha t}(A_k V - \alpha V) = A_k U - \alpha U,$$
$$U(0) = V(0) = U_0.$$

Thus, (7.14) holds and, therefore, $U(t) = R(t)U_0$. It follows that

$$R(t)U_0 = e^{-\alpha t} V(t) = e^{-\alpha t} Z(t)U_0;$$

since U_0 is arbitrary in $\mathcal{X}$, the conclusion follows.   ◻

PROPOSITION 7.5
Let $t \in \mathbb{R}$ and $m \in \mathbb{N}$. The restriction of $Z(t)$ to the finite dimensional subspace $\mathcal{X}_m$ defined in (7.19) is the rotation matrix

$$Z_m(t) := \begin{pmatrix} \cos(\omega_m t) & \omega_m^{-1}\sin(\omega_m t) \\ -\omega_m \sin(\omega_m t) & \cos(\omega_m t) \end{pmatrix},$$

where

$$\omega_m := \sqrt{m^2 + \alpha_1} = \sqrt{m^2 + 2k - \alpha^2}.$$

PROOF   Recalling (7.19) and (7.18), a generic element $U_0 \in \mathcal{X}_m$ has the form

$$U_0(x) = (\alpha\cos(mx), \beta\cos(mx)), \quad \alpha, \beta \in \mathbb{R}.$$

Then, the components of the function $t \mapsto U(t) := Z(t)U_0 := (u(t), \tilde{u}(t))$ solve the initial value problem

$$\begin{cases} u_t = \tilde{u} \\ \tilde{u}_t = -\alpha_1 u + u_{xx} \end{cases}, \quad \begin{cases} u(x,0) = \alpha\cos(mx) \\ \tilde{u}(x,0) = \beta\cos(mx) \end{cases}.$$

Equivalently, u solves the second order initial value problem

$$\begin{cases} u_{tt} - u_{xx} + \alpha_1 u = 0 \\ u(x,0) = \alpha\cos(mx) \\ u_t(x,0) = \beta\cos(mx). \end{cases}$$

By separation of variables, we easily obtain that

$$\begin{cases} u(x,t) = (\alpha\cos(\omega_m t) + \omega_m^{-1}\beta\sin(\omega_m t))\cos(mx), \\ \tilde{u}(x,t) = (-\alpha\omega_m\sin(\omega_m t) + \beta\cos(\omega_m t))\cos(mx). \end{cases} \tag{7.23}$$

Since this can be written as $U(t) = Z_m(t)U_0$, the conclusion follows.   ⬚

4. Setting now

$$a_m := \frac{1}{2\omega_m}(\alpha\omega_m - i\beta), \quad b_m := \overline{a_m}, \quad c_m := i\omega_m a_m, \quad d_m := -i\omega_m b_m,$$

we can write (7.23) in the complex form

$$\begin{cases} u(x,t) = (a_m e^{i\omega_m t} + b_m e^{-i\omega_m t})\cos(mx), \\ \tilde{u}(x,t) = (c_m e^{i\omega_m t} + d_m e^{-i\omega_m t})\cos(mx). \end{cases} \tag{7.24}$$

Consequently, if $U_0 \in \mathcal{X}$, by (7.24) and the second of (7.20) we have the series expansions

$$\begin{cases} u(x,t) = \sum_{m=0}^{\infty}(a_m e^{i\omega_m t} + b_m e^{-i\omega_m t})\cos(mx), \\ \tilde{u}(x,t) = \sum_{m=0}^{\infty}(c_m e^{i\omega_m t} + d_m e^{-i\omega_m t})\cos(mx). \end{cases} \tag{7.25}$$

Since $\omega_m \in \mathbb{R}$ for each $m \in \mathbb{N}$, it follows from (7.25) and theorem A.98 that for all $U_0 \in \mathcal{X}$ the function

$$\mathbb{R} \ni t \mapsto U(\cdot,t) = Z(t)U_0 \in \mathcal{X}$$

is almost periodic in t. Thus, by Bochner's theorem A.97, the set of translations $\{U(\cdot + \tau): \tau \in \mathbb{R}\}$ is relatively compact in $C_b(\mathbb{R};\mathcal{X})$, with respect to the sup norm.

7.4   Inertial Manifolds for the Linearized Problem

1.   We are now in a position to prove the basic result of this section, which will allow us to show that the flow $R = (R(t))_{t\in\mathbb{R}}$ generated by the linearized problem (7.14) has only at most countably many finite dimensional, locally invariant submanifolds containing the origin, and differentiable there. We start with

DEFINITION 7.6   *Let $\mathcal{M}$ be a global, trivial C^1 manifold in a Banach space $\mathcal{X}$; that is, $\mathcal{M}$ is the graph of a function $\varphi \in C^1(\mathcal{X}_1;\mathcal{X})$, with $\mathcal{X}_1$ a finite dimensional subspace of $\mathcal{X}$ (see definition 5.2). Let $x_0 = \varphi(\xi_0) \in \mathcal{M}$, $\xi_0 \in \mathcal{X}_1$. Consider the affine map $L_{x_0}: \mathcal{X}_1 \to \mathcal{X}$, defined by*

$$L_{x_0}(\xi) := x_0 + \varphi'(\xi_0)(\xi - \xi_0), \quad \xi \in \mathcal{X}_1.$$

The image of L_{x_0} is called the TANGENT SPACE to $\mathcal{M}$ at x_0, and denoted $\mathrm{T}_{x_0}\mathcal{M}$, if

$$\lim_{\xi \to \xi_0} \frac{\|\varphi(\xi) - L_{x_0}(\xi)\|_{\mathcal{X}}}{\|\xi - \xi_0\|_{\mathcal{X}}} = 0.$$

Example 7.7

The set

$$\mathcal{M} := \{(x,y) \in \mathbb{R}^2 : y = \arctan x\}$$

is a one dimensional, trivial Lipschitz submanifold of $\mathbb{R}^2$, whose tangent space at $x_0 = \frac{1}{4}$ is the set

$$\mathrm{T}_{1/4}\mathcal{M} := \{(\xi,\eta) \in \mathbb{R}^2 : 16\xi - 17\eta + 17\arctan \tfrac{1}{4} - 4 = 0\}.$$

$\square$

We have then the following result.

PROPOSITION 7.8

Let $\mathcal{M} \subseteq \mathcal{X}$ be a submanifold of $\mathcal{X}$, such that $0 \in \mathcal{M}$, M is differentiable at 0, and M is invariant with respect to the flow R. Let $\mathrm{T}\mathcal{M}$ denote the tangent space to $\mathcal{M}$ at 0. Then $\mathcal{M} \subseteq \mathrm{T}\mathcal{M}$.

PROOF 1. Let $P \colon \mathcal{X} \to \mathrm{T}\mathcal{M}$ be the projection onto $\mathrm{T}\mathcal{M}$, and $Q = I_{\mathcal{X}} - P$. Since $\mathrm{T}\mathcal{M}$ is tangent to $\mathcal{M}$ at 0,

$$\lim_{\substack{U \in \mathcal{M} \\ U \to 0}} \frac{\|QU\|_{\mathcal{X}}}{\|PU\|_{\mathcal{X}}} = 0. \tag{7.26}$$

Let $U_0 \in \mathcal{M}$, and $V(t) := R(t)U_0$. By proposition 7.4 and (7.22),

$$\|V(t)\|_{\mathcal{X}} = \mathrm{e}^{-\alpha t}\|Z(t)U_0\|_{\mathcal{X}} = \mathrm{e}^{-\alpha t}\|U_0\|_{\mathcal{X}} \to 0$$

as $t \to +\infty$. Since $V(t) \in \mathcal{M}$ for all $t \geq 0$, by (7.26)

$$\lim_{t \to +\infty} \frac{\|QV(t)\|_{\mathcal{X}}}{\|PV(t)\|_{\mathcal{X}}} = 0. \tag{7.27}$$

Let $U(t) := Z(t)U_0$. Again by proposition 7.4, (7.27) implies

$$\lim_{t \to +\infty} \frac{\|QU(t)\|_{\mathcal{X}}}{\|PU(t)\|_{\mathcal{X}}} = \lim_{t \to +\infty} \frac{\|QV(t)\|_{\mathcal{X}}}{\|PV(t)\|_{\mathcal{X}}} = 0. \tag{7.28}$$

Since P and Q are orthogonal, and $Z(t)$ is an isomorphism, we have that

$$0 \leq \|U_0\|^2 - \|PU(t)\|_{\mathcal{X}}^2 = \|QU(t)\|_{\mathcal{X}}^2 = \frac{\|QU(t)\|_{\mathcal{X}}^2}{\|PU(t)\|_{\mathcal{X}}^2}\|PU(t)\|_{\mathcal{X}}^2$$

$$\leq \frac{\|QU(t)\|_{\mathcal{X}}^2}{\|PU(t)\|_{\mathcal{X}}^2} \|U(t)\|_{\mathcal{X}}^2 = \frac{\|QU(t)\|_{\mathcal{X}}^2}{\|PU(t)\|_{\mathcal{X}}^2} \|U_0\|_{\mathcal{X}}^2.$$

Therefore, by (7.28),

$$\lim_{t\to+\infty} \|PU(t)\|_{\mathcal{X}} = \|U_0\|_{\mathcal{X}}, \quad \lim_{t\to+\infty} \|QU(t)\|_{\mathcal{X}} = 0,$$

from which

$$\lim_{t\to+\infty} QU(t) = 0, \quad \lim_{t\to+\infty} PU(t) = U_0. \tag{7.29}$$

Let $(\tau_m)_{m\in\mathbb{N}} \subset \mathbb{R}_{>0}$ be such that $\tau_m \to +\infty$. As we have recalled at the end of the previous section, the sequence $(U(\cdot+\tau_m))_{m\in\mathbb{N}}$ is relatively compact, by Bochner's theorem; hence, we can extract a subsequence $(U(\cdot+\tau_{m_k}))_{k\in\mathbb{N}}$, converging uniformly to a function $W \in C_b(\mathbb{R};\mathcal{X})$. Since P is continuous, we deduce that $PU(\tau_{m_k}) \to PW(0)$ as $k \to \infty$. But then, the second of (7.29) implies that $U_0 = PW(0) \in T\mathcal{M}$. Thus, $\mathcal{M} \subseteq T\mathcal{M}$, as claimed. ∎

As an immediate consequence, we have that if $\mathcal{M}$ is an invariant submanifold of $\mathcal{X}$, differentiable at 0, then M is in fact a closed, *linear* subspace of $\mathcal{X}$. Indeed, proposition 7.8 implies that $\mathcal{M} \subseteq T\mathcal{M}$; on the other hand, $\mathcal{M}$ and $T\mathcal{M}$ are homeomorphic, via the mapping $\mathcal{M} \ni u \mapsto \varphi'(0)\varphi^{-1}u \in T\mathcal{M}$, where φ is the map whose graph is $\mathcal{M}$, as per definition 5.2 (keep in mind that since φ is differentiable at 0, and φ^{-1} is Lipschitz continuous, $\varphi'(0)$ is invertible). Hence, $\mathcal{M} = T\mathcal{M}$, and the tangent space $T\mathcal{M}$ is linear.

2.   We now give a further characterization of the linear subspaces of $\mathcal{X}$ which are invariant with respect to the flow R.

THEOREM 7.9
Assume $\mathcal{F} \subseteq \mathcal{X}$ is a closed linear subspace, which is invariant with respect to R. There exists a subset $\mathbb{N}_1 \subseteq \mathbb{N}$ such that

$$\mathcal{F} = \bigoplus_{m\in\mathbb{N}_1} \mathcal{X}_m, \tag{7.30}$$

where $\mathcal{X}_m$ is as in (7.19).

PROOF   1. Since $R(t) = e^{-\alpha t}Z(t)$ and $\mathcal{F}$ is linear, $\mathcal{F}$ is also invariant with respect to Z. We now show that also the orthogonal complement $\mathcal{F}^\perp$ of $\mathcal{F}$ in $\mathcal{X}$ is invariant with respect to Z. Indeed, given $U_0 \in \mathcal{F}^\perp$, take any $V_0 \in \mathcal{F}$. Since $Z(t)$ is unitary for all $t \in \mathbb{R}$, its adjoint satisfies

$$Z(t)^* = Z(t)^{-1} = Z(-t);$$

therefore,

$$\langle Z(t)U_0, V_0\rangle_{\mathcal{X}} = \langle U_0, Z(t)^*V_0\rangle_{\mathcal{X}} = \langle U_0, Z(-t)V_0\rangle_{\mathcal{X}}.$$

Since $Z(-t)V_0 \in \mathcal{F}$ because $\mathcal{F}$ is invariant, it follows that $\langle Z(t)U_0, V_0\rangle_{\mathcal{X}} = 0$; hence, $Z(t)U_0 \in \mathcal{F}^{\perp}$, and $\mathcal{F}^{\perp}$ is invariant with respect to Z.

2. We next show that for each $m \in \mathbb{N}$,

$$\mathcal{F} \cap \mathcal{X}_m \neq \{0\} \quad \Longleftrightarrow \quad \mathcal{X}_m \subseteq \mathcal{F}, \tag{7.31}$$

$$\mathcal{F}^{\perp} \cap \mathcal{X}_m \neq \{0\} \quad \Longleftrightarrow \quad \mathcal{X}_m \subseteq \mathcal{F}^{\perp}. \tag{7.32}$$

Consider in fact (7.31). Since $\mathcal{X}_m$ does not reduce to the origin, if $\mathcal{X}_m \subseteq \mathcal{F}$ then obviously $\mathcal{F} \cap \mathcal{X}_m = \mathcal{X}_m$ also does not reduce to $\{0\}$. Conversely, note that $\mathcal{F} \cap \mathcal{X}_m$ is a subspace of both $\mathcal{F}$ and $\mathcal{X}_m$, which is invariant with respect to Z. However, proposition (7.5) implies that $\mathcal{X}_m$ does not contain any such proper subspace. Since $\mathcal{F} \cap \mathcal{X}_m$ does not reduce to the origin, it must be $\mathcal{F} \cap \mathcal{X}_m = \mathcal{X}_m$, and thus $\mathcal{X}_m \subseteq \mathcal{F}$. This proves (7.31); the proof of (7.32) is analogous.

3. Set now, for $t \in \mathbb{R}$,

$$\tilde{Z}(t) := \tfrac{1}{2}(Z(t) + Z(-t)).$$

We claim that for all $m \in \mathbb{N}$, $t \in \mathbb{R}$ and $U \in \mathcal{X}$,

$$U \in \mathcal{X}_m \Longleftrightarrow \tilde{Z}(t)U = \cos(\omega_m t)U, \tag{7.33}$$

where $\omega_m = \sqrt{m^2 + \alpha_1}$ as in proposition 7.5. Indeed, if $U \in \mathcal{X}_m$, the identity at the right side of (7.33) follows from (7.23). Conversely, suppose $U \in \mathcal{X}$ satisfies the right side of (7.33), but $U \notin \mathcal{X}_m$. Then, recalling the second of (7.20), there must be $r \in \mathbb{N}$, with $r \neq m$, such that $\langle U, U_r\rangle_{\mathcal{X}} \neq 0$ for some $U_r \in \mathcal{X}_r$, for otherwise, if $\langle U, U_r\rangle = 0$ for all $r \neq m$,

$$U = \sum_{r=0}^{\infty} \langle U, U_r\rangle_{\mathcal{X}} U_r = \langle U, U_m\rangle_{\mathcal{X}} U_m \in \mathcal{X}_m.$$

Since $U \neq 0$, U is either in $\mathcal{F}$ or in $\mathcal{F}^{\perp}$. Since these subspaces are both invariant with respect to Z, $\tilde{Z}(t)U$ is in the same subspace where U is. Hence, from (7.33) we obtain that

$$\langle \tilde{Z}(t)U, U_r\rangle_{\mathcal{X}} = \cos(\omega_m t)\langle U, U_r\rangle_{\mathcal{X}} \neq 0. \tag{7.34}$$

It is now easy to check that each operator $\tilde{Z}(t)$ is also unitary; therefore, since the map $t \mapsto \tilde{Z}(t)$ is symmetric,

$$\langle \tilde{Z}(t)U, U_r\rangle_{\mathcal{X}} = \langle U, \tilde{Z}(t)^{-1}U_r\rangle_{\mathcal{X}} = \langle U, \tilde{Z}(-t)U_r\rangle_{\mathcal{X}} = \langle U, \tilde{Z}(t)U_r\rangle_{\mathcal{X}}. \tag{7.35}$$

But from the part of (7.33) we have already proven, $\tilde{Z}(t)U_r = \cos(\omega_r t)U_r$, because $U_r \in \mathcal{X}_r$. From (7.35) and (7.34) we have then that

$$\cos(\omega_m t)\langle U, U_r\rangle_{\mathcal{X}} = \cos(\omega_r t)\langle U, U_r\rangle_{\mathcal{X}};$$

since $\langle U, U_r\rangle_{\mathcal{X}} \neq 0$, this implies that $\cos(\omega_m t) = \cos(\omega_r t)$. Since $t \in \mathbb{R}$ is arbitrary, this forces $\omega_m = \omega_r$; since this is not true, we conclude that $U \in \mathcal{X}_m$.

4. We now show that for all $m \in \mathbb{N}$, either $\mathcal{X}_m \subseteq \mathcal{F}$, or $\mathcal{X}_m \subseteq \mathcal{F}^\perp$. Indeed, given $U \in \mathcal{X}_m \setminus \{0\}$, we can uniquely decompose

$$U = U_\mathcal{F} \oplus U_{\mathcal{F}}^\perp, \quad U_\mathcal{F} \in \mathcal{F}, \quad U_{\mathcal{F}}^\perp \in \mathcal{F}^\perp.$$

Then, by (7.33), for all $t \in \mathbb{R}$

$$\tilde{Z}(t)U = \cos(\omega_m t)U = \cos(\omega_m t)U_\mathcal{F} + \cos(\omega_m t)U_{\mathcal{F}}^\perp.$$

Since also

$$\tilde{Z}(t)U = \tilde{Z}(t)U_\mathcal{F} + \tilde{Z}(t)U_{\mathcal{F}}^\perp,$$

and $\tilde{Z}(t)U_\mathcal{F} \in \mathcal{F}$, $\tilde{Z}(t)U_{\mathcal{F}}^\perp \in \mathcal{F}^\perp$, it follows that

$$\tilde{Z}(t)U_\mathcal{F} = \cos(\omega_m t)U_\mathcal{F}, \quad \tilde{Z}(t)U_{\mathcal{F}}^\perp = \cos(\omega_m t)U_{\mathcal{F}}^\perp. \tag{7.36}$$

Since (7.33) is an equivalence, (7.36) implies that $U_\mathcal{F}$ and $U_{\mathcal{F}}^\perp$ are both in $\mathcal{X}_m$. Hence, $U_\mathcal{F} \in \mathcal{F} \cap \mathcal{X}_m$, and $U_{\mathcal{F}}^\perp \in \mathcal{F}^\perp \cap \mathcal{X}_m$. Since at least one of $U_\mathcal{F}$, $U_{\mathcal{F}}^\perp$ is not zero, the conclusion follows from (7.31) and (7.32).

5. We are now ready to conclude the proof of theorem 7.9. Given $\mathcal{F}$ as above, set

$$\mathbb{N}_1 := \{m \in \mathbb{N}: \mathcal{X}_m \subseteq \mathcal{F}\}.$$

We claim that, then, (7.30) holds. Indeed, it is clear that

$$\bigoplus_{m \in \mathbb{N}_1} \mathcal{X}_m \subseteq \mathcal{F}. \tag{7.37}$$

To show the inverse inclusion, let $U \in \mathcal{F}$. Recalling the second of (7.20), we decompose

$$U = \sum_{m \in \mathbb{N}_1} \alpha_m U_m + \sum_{m \notin \mathbb{N}_1} \alpha_m U_m =: U' + U''.$$

By (7.37), $U' \in \mathcal{F}$; consequently, $U'' = 0$, and $U = U' \in \bigoplus_{m \in \mathbb{N}_1} \mathcal{X}_m$. Thus, (7.30) follows, and the proof of theorem 7.9 is complete. $\Box$

3. As a consequence of theorem 7.9, we deduce

THEOREM 7.10
If $\alpha^2 < 2k$, the flow $R = (R(t))_{t \in \mathbb{R}}$ generated by the linearized equation (7.14) admits at most a countable family of closed, finite dimensional, invariant manifolds which contain the stationary point 0 and are differentiable at this point.

PROOF   By proposition 7.8, any such manifold $\mathcal{M}$ is a closed, linear subspace of $\mathcal{X}$. By theorem 7.9, $\mathcal{M}$, being finite dimensional, is the direct sum of at most a finite number of subspaces of $\mathcal{X}_m$. Since the family of finite subsets of $\mathbb{N}$ is countable, the conclusion follows. $\Box$

7.5   C^1 **Linearization Equivalence**

Theorem 7.10 characterizes a certain class of finite dimensional, invariant manifolds of the flow R generated by the linearized problem (7.14), namely those which contain the stationary point $U = 0$, and are differentiable at this point. Our goal is now to extend this result to the flow S generated by the original nonlinear problem (7.11), of which (7.14) is the linearization at its stationary state $P_1 = (1,0)$ (regarded as a constant function in $\mathcal{X}$). We achieve this by generalizing to the infinite dimensional case (i.e. to general Banach spaces) the Hartman-Grobman theorem A.8 on the topological equivalence at the origin of the systems of ODEs

$$\dot{x} = f(x), \qquad \dot{x} = f'(0)x,$$

where $f \in C^1(\mathbb{R}^n; \mathbb{R}^n)$ and $f(0) = 0$.

We start with the following C^1 linearization result.

THEOREM 7.11

Let $\mathcal{X}$ be a Banach space, and $F: \mathcal{X} \to \mathcal{X}$ be a map, not necessarily linear. Let $x_0 \in \mathcal{X}$ be a fixed point of F, and assume that F is of class C^1 in a neighborhood $\mathcal{U}$ of x_0. Let $L := F'(x_0)$ (the Fréchet derivative of F at x_0), and assume that L has a bounded inverse. Assume further that

$$\|F'(x) - L\|_{\mathcal{L}(\mathcal{X},\mathcal{X})} = O(\|x - x_0\|_{\mathcal{X}}) \qquad \text{as } x \to x_0 \ (x \in \mathcal{U}), \tag{7.38}$$

$$\|L^{-1}\|_{\mathcal{L}(\mathcal{X},\mathcal{X})} \|L\|^2_{\mathcal{L}(\mathcal{X},\mathcal{X})} < 1. \tag{7.39}$$

There exist then a neighborhood $\mathcal{V} \subseteq \mathcal{U}$ of x_0, with $F(\mathcal{V}) \subseteq \mathcal{V}$, and a C^1 diffeomorphism $\Phi: \mathcal{V} \to \Phi(\mathcal{V}) \subseteq \mathcal{X}$, such that $\Phi(x_0) = 0$, $\Phi'(x_0) = I_{\mathcal{X}}$,

$$\Phi'(x) - I_{\mathcal{X}} = O(\|x - x_0\|_{\mathcal{X}}) \qquad \text{as } x \to x_0 \ (x \in \mathcal{V}),$$

and

$$\Phi F = L \Phi \qquad \text{on } \mathcal{V}.$$

Moreover, Φ is unique, in the sense that if $\mathcal{W} \subseteq \mathcal{U}$ and $\Psi: \mathcal{W} \to \mathcal{X}$ also satisfy the same conditions as $\mathcal{V}$ and Φ, then $\Phi \equiv \Psi$ on $\mathcal{V} \cap \mathcal{W}$.

We postpone the proof of this theorem to section 7.10.1, and proceed instead to extend it to a general flow S on $\mathcal{X}$, and its linearization R at a stationary point x_0 of S.

THEOREM 7.12

Let $(S(t))_{t \in \mathbb{R}}$ be a flow on a Banach space $\mathcal{X}$, such that for each $t \in \mathbb{R}$, $S(t)$ is a C^2 diffeomorphism on a neighborhood $\mathcal{U}$ of a common fixed point x_0 (i.e., $S(t)x_0 = x_0$

for all $t \in \mathbb{R}$). For each $t \in \mathbb{R}$, let $R(t) := (S(t))'(x_0)$ (the Fréchet derivative of $S(t)$ at x_0). Assume there is $\tau \in \mathbb{R}$ such that the operators $F := S(\tau)$ and $L := R(\tau)$ satisfy conditions (7.38) and (7.39) of theorem 7.11. There exist then a neighborhood $\mathcal{V} \subseteq \mathcal{U}$ of x_0, and a C^1 diffeomorphism $\Phi \colon \mathcal{V} \to \Phi(\mathcal{V}) \subseteq \mathcal{X}$, such that $\Phi(x_0) = 0$, $\Phi'(x_0) = I_{\mathcal{X}}$,

$$\|\Phi'(x) - I_{\mathcal{X}}\|_{\mathcal{L}(\mathcal{X},\mathcal{X})} = O(\|x - x_0\|_{\mathcal{X}}) \qquad \text{as } x \to x_0 \ (x \in \mathcal{V}),$$

and the identity

$$\Phi S(t) = R(t)\Phi$$

holds for all $t \in \mathbb{R}$ in a ball $B(x_0, r) \subseteq \mathcal{V}$. This ball can be chosen independently of t if t is bounded from below.

We prove this theorem in section 7.10.2. We now show that this theorem can be applied to the flows S and R generated, respectively, by problem (7.11) and its linearization (7.14) at the stationary state P_1.

Indeed, choosing any $\tau > 0$, we immediately check that $F = S(\tau)$ and $L = R(\tau)$ do satisfy the assumptions of theorem 7.12. In fact, (7.38) holds because $S(\tau)$ is of class C^2, as we know from theorem 7.1; as for (7.39), recall from proposition 7.4 that $R(\tau) = \mathrm{e}^{-\alpha\tau} Z(\tau)$, and that $Z(\tau)$ is an isometry. Thus,

$$\|L\|_{\mathcal{L}(\mathcal{X},\mathcal{X})} = \|R(\tau)\|_{\mathcal{L}(\mathcal{X},\mathcal{X})} = \mathrm{e}^{-\alpha\tau}\|Z(\tau)\|_{\mathcal{L}(\mathcal{X},\mathcal{X})} = \mathrm{e}^{-\alpha\tau},$$

$$\|L^{-1}\|_{\mathcal{L}(\mathcal{X},\mathcal{X})} = \|R(-\tau)\|_{\mathcal{L}(\mathcal{X},\mathcal{X})} = \mathrm{e}^{\alpha\tau}\|Z(-\tau)\|_{\mathcal{L}(\mathcal{X},\mathcal{X})} = \mathrm{e}^{\alpha\tau}.$$

It follows that

$$\|L^{-1}\|_{\mathcal{L}(\mathcal{X},\mathcal{X})} \|L\|^2_{\mathcal{L}(\mathcal{X},\mathcal{X})} = \mathrm{e}^{-\alpha\tau} < 1$$

as required. We summarize this conclusion in

THEOREM 7.13

If $\alpha^2 < k$, the flow S generated by problem (7.11) is, near its stationary state P_1, C^1-equivalent to its linearization R, i.e. to the flow generated by the linearized problem (7.14). Consequently, there is a neighborhood $\mathcal{V}$ of P_1 in $\mathcal{X}$, with the property that there is one at most countable family $\mathcal{F}$ of submanifolds of $\mathcal{X}$, whose intersection with $\mathcal{V}$ are closed, finite dimensional manifolds, each of which is invariant with respect to S, contains the intersection $\mathcal{A} \cap \mathcal{V}$ (where $\mathcal{A}$ is the global attractor of S in $\mathcal{X}$), and is differentiable at P_1.

7.6   Perturbations of the Nonlinear Flow

1.   In this section we perturb the flow S generated by the nonlinear problem (7.11) in such a way that, on one hand, the long-time behavior of the perturbed flow is the same as that of S in a neighborhood of its stationary state P_1, and, on the other, near P_1 the global attractor of the perturbed flow, which must coincide with the global attractor $\mathcal{A}$ of S, is not contained in any of the at most countably many manifolds containing P_1. From this we deduce that S cannot admit an inertial manifold $\mathcal{M}$, because otherwise $\mathcal{A} \subseteq \mathcal{M}$ and, near P_1, $\mathcal{M}$ must be one of the at most countably many manifolds containing P_1.

2.   Going back to the second order IBVP (H_N), we realize that since the nonlinearity g is independent of x, and the boundary conditions are of Neumann type, the flow S admits a 2-dimensional invariant linear subspace in $\mathcal{X}$, consisting of functions that are independent of the space variable x. More precisely, let

$$L_c^2(0, \pi) := \{u \in L^2(0, \pi) \colon u(x) \equiv \text{const. a.e. in }]0, \pi[\}\,.$$

Then obviously $L_c^2(0, \pi) \subseteq H^1(0, \pi)$, and $\mathcal{X}_c := L_c^2(0, \pi) \times L_c^2(0, \pi) \subseteq \mathcal{X}$. Note that if P_{-1}, P_0 and P_1 are as in (7.13), then P_{-1}, P_0 and $P_1 \in \mathcal{X}_c$. We claim:

PROPOSITION 7.14
$\mathcal{X}_c$ is a 2-dimensional linear subspace of $\mathcal{X}$, invariant with respect to S.

PROOF $\mathcal{X}_c$ is obviously a linear subspace of $\mathcal{X}$. Let $(u_0, u_1) \in \mathcal{X}_c$, and set

$$(u(t), \tilde{u}(t)) := S(t)(u_0, u_1)\,.$$

Then, u solves (H_N). On the other hand, since u_0 and u_1 are constant functions, which we can identify with two numbers, we can consider the solution y of the Cauchy problem

$$\begin{cases} y'' + 2\alpha y' = k(y - y^3) \\ y(0) = u_0\,, \quad y'(0) = u_1\,. \end{cases} \tag{7.40}$$

Set $v(x, t) := y(t)$. Then, v solves (H_N) with the same initial data; therefore, by uniqueness (proposition 7.2), $v = u$ and $\tilde{u} = v_t$. But $(v(\cdot, t), v_t(\cdot, t)) \in \mathcal{X}_c$ for all $t \in \mathbb{R}$; thus, $(u(t), \tilde{u}(t)) := S(t)(u_0, u_1) \in \mathcal{X}_c$. This means that $S(t)\mathcal{X}_c \subseteq \mathcal{X}_c$ for all $t \in \mathbb{R}$.

Conversely, given $(u_0, u_1) \in \mathcal{X}_c$ and $\bar{t} \in \mathbb{R}$, let y be the solution of the Cauchy problem (7.40), with the initial values replaced by

$$y(\bar{t}) = u_0\,, \quad y'(\bar{t}) = u_1\,.$$

Then $(y(0), y'(0)) \in \mathcal{X}_c$, and $(u_0, u_1) = S(\bar{t})(y(0), y'(0)) \in S(\bar{t})\mathcal{X}_c$. This means that $\mathcal{X}_c \subseteq S(\bar{t})\mathcal{X}_c$. Thus, $\mathcal{X}_c$ is invariant with respect to S. Finally, $\mathcal{X}_c$ is obviously 2-dimensional, since it is isomorphic to $\mathbb{R}^2$. $\square$

3.   Since the equation of (7.40) has exactly the same structure of Duffing's equation (1.52), with $\lambda = 0$, we know from section 2.3.3 that there is a heteroclinic orbit γ joining the stationary states P_0 and P_1. As in figure 2.3, γ has the shape shown in figure 7.1; note that condition (7.16), i.e. $\alpha^2 < 2k$, guarantees that P_1 is an hyperbolic, asymptotically stable stationary state.

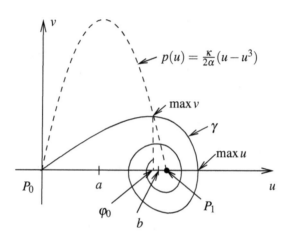

Figure 7.1: Heteroclinic orbit joining P_0 to P_1

Since γ is an orbit of the flow generated by (7.40) (and of S), there are functions $t \mapsto \varphi(t)$ and $t \mapsto \psi(t)$ such that

$$\gamma = \{(\varphi(t), \psi(t)) : t \in \mathbb{R}\}.$$

In particular, there is $t_0 \in \mathbb{R}$ such that

$$\varphi'(t) > 0, \quad \psi'(t) = \varphi''(t) > 0 \tag{7.41}$$

for $t < t_0$, while for $t \geq t_0$,

$$\varphi(t) \geq \varphi(t_0) =: \varphi_0, \tag{7.42}$$

with $0 < \varphi_0 < 1$. We fix then two numbers a and b, with

$$0 < a < \varphi_0 < b < 1, \tag{7.43}$$

and introduce the set of functions

$$\mathcal{J}(a,b) := \{f \in C^2([0,\pi] \times \mathbb{R}) : f(x,u) \equiv 0 \quad \text{if } u \leq a \text{ or } u \geq b\}.$$

This set is a Banach space with respect to the usual C^2 norm. For $f \in \mathcal{J}(a,b)$, we consider the "perturbed" IBVP

$$
\begin{cases}
u_{tt} + 2\alpha u_t - u_{xx} = g(u) + f(x,u) \\
u(0,x) = u_0(x), \quad u_t(0,x) = u_1(x) \\
u_x(t,0) = u_x(t,\pi) = 0,
\end{cases}
\tag{7.44}
$$

which we transform as usual into its first order formulation

$$
U_t + \alpha U = AU + G(U) + F(x,U),
\tag{7.45}
$$

with A and G as in (7.11), and $F(x,U) := (0, f(x,u))$, $U \in \mathcal{X}$. This problem, which is autonomous, generates a flow on $\mathcal{X}$; more precisely, as in theorem 7.1 we have

THEOREM 7.15
For each $f \in \mathcal{J}(a,b)$, problem (7.45) generates a continuous flow $S_f = (S_f(t))_{t \in \mathbb{R}}$ on $\mathcal{X}$, which admits a compact, global attractor $\mathcal{A}_f$ in $\mathcal{X}$. Moreover, for each compact interval $[t_0, t_1] \subset \mathbb{R}$, the map

$$
\mathcal{X} \times \mathcal{J}(a,b) \ni (U_0, f) \mapsto S_f(\cdot)U_0 \in C([t_0,t_1]; \mathcal{X})
\tag{7.46}
$$

is of class C^2.

7.7 Asymptotic Properties of the Perturbed Flow

1. Theorem 7.13 shows that there is a ball $B(P_1, r)$ in $\mathcal{X}$ on which the flow S is C^1-equivalent to its linearization R. If

$$
\eta \leq \eta_0 =: \min\{r, 1 - b\},
$$

then

$$
B(P_1, \eta) \cap (]a,b[\times \{0\}) = \emptyset \quad \text{in } \mathcal{X};
\tag{7.47}
$$

consequently, if $f \in \mathcal{J}(a,b)$ and $U = (u, \tilde{u}) \in B(P_1, \eta)$, $f(x,u) = 0$ for all $x \in [0, \pi]$. This implies that the restrictions of the flows S_f and S to $\mathcal{V}_\eta := B(P_1, \eta)$ coincide; therefore, the asymptotic behavior of these restrictions are the same. In particular, each S_f admits $\mathcal{A} \cap \mathcal{V}_\eta$ as a global attractor in $\mathcal{X}$, and, if S admitted an inertial manifold $\mathcal{M}$, $\mathcal{M} \cap \mathcal{V}_\eta$ is an inertial manifold (with boundary) for S_f. Moreover, since P_1 is a stable stationary state of both S and S_f, there is $\eta' \leq \eta$ such that for all $t \geq 0$,

$$
S(t)B(P_1, \eta') = S_f(t)B(P_1, \eta') \subseteq \mathcal{V}_\eta.
\tag{7.48}
$$

Since $\eta \leq r$, it follows that also the restriction of the flow S_f to $\mathcal{V}_\eta$ is C^1-equivalent to the restriction of the linearized flow R to $\mathcal{V}_\eta$. Hence, as in theorem 7.13, there are only at most countably many submanifolds of $\mathcal{X}$, whose intersection with $\mathcal{V}_\eta$ are closed, finite dimensional manifolds, which are invariant with respect to S_f, contain the intersection $\mathcal{A} \cap \mathcal{V}_\eta$, and are differentiable at P_1. Since S_f is independent of f on $\mathcal{V}_\eta$, this family is actually independent of f; in fact, it coincides with the family $\mathcal{F}_S$ of manifolds of the flow S, described in theorem 7.13.

2.   Our strategy is then the following. In proposition 7.17 below, we fix a choice of $\eta \in]0, \eta_0]$. In turn, this choice of η will determine another parameter δ. We show then that if the perturbation f is "sufficiently small", as measured by this parameter δ, then the corresponding flow S_f admits an heteroclinic orbit γ_f joining the stationary states P_0 to P_1. We know that $\gamma_f \subseteq \mathcal{A}$ (since $\mathcal{A}$ is the attractor of S_f); thus, if S admitted an inertial manifold $\mathcal{M}$ containing $\mathcal{A}$, then, since $\mathcal{M}$ is also an inertial manifold for S_f, $\gamma_f \subseteq \mathcal{M}$ as well. Since γ_f converges to P_1, it must enter $\mathcal{V}_\eta$; hence, γ_f must converge to P_1 along one of the manifolds of $\mathcal{F}_S$.

More precisely, if S admitted an inertial manifold $\mathcal{M}$, then there must exist a manifold $\mathcal{M}_m \in \mathcal{F}_S$, such that $\gamma_f \cap \mathcal{V}_\eta \subseteq \mathcal{M}_m$. We proceed then to show that there is at least one $f \in \mathcal{J}(a,b)$ such that this does not happen; i.e., that there is at least one point on the corresponding set $\gamma_f \cap \mathcal{V}_\eta$, which is not on any one of the manifolds of $\mathcal{F}_S$. This produces a contradiction, which shows that the flow S cannot admit a locally invariant inertial manifold $\mathcal{M}$, containing the attractor $\mathcal{A}$.

3.   We now show that when f is small, S_f admits an heteroclinic orbit γ_f joining P_0 to P_1. The choice of a in (7.43) uniquely determines a point $U_* = (a, \tilde{a}) \in \gamma$. For $f \in \mathcal{J}(a,b)$ and $t \in \mathbb{R}$, we let

$$U_f(t) := S_f U_* =: (u_f(t), \tilde{u}_f(t)) \tag{7.49}$$

(thus, $\tilde{u}_f = \alpha u_f + (u_f)_t$), and denote by $\gamma_f \subseteq \mathcal{X}$ the image of the curve $t \mapsto U_f(t)$. Finally, for $\rho > 0$, we denote by $\mathcal{J}_\rho$ the ball of $\mathcal{J}(a,b)$ with center 0 and radius ρ (in the C^2 norm). We claim:

PROPOSITION 7.16
For all $\eta \in]0, \eta_0]$, there is $\delta_1 \in]0, \eta]$ with the property that if $\delta \in]0, \delta_1]$ and $f \in \mathcal{J}_\delta$, then γ_f is an heteroclinic orbit for S_f, joining its stationary states P_0 and P_1.

PROOF   Since $f(\cdot, u) = 0$ for $u \leq a$, and φ is increasing for $t \leq t_0$ (recall (7.42)), it follows that $S_f(t)U_* = S(t)U_*$ for all $t \leq t_0$. Consequently,

$$\lim_{t \to -\infty} U_f(t) = (0,0) = P_0. \tag{7.50}$$

Since P_1 is an asymptotically stable stationary state of S, there is $T_\eta > 0$ such that for all $t \geq T_\eta$,

$$S(t)U_* \in B(P_1, \eta/2). \tag{7.51}$$

By the continuity of the map $(U_*, f) \mapsto S_f(\cdot)U_*$ (see (7.46)), (7.51) implies that there is $\delta_1 > 0$ such that if $f \in \mathcal{J}_{\delta_1}$ and $T_\eta \leq t \leq T_\eta + \delta_1$, $S_f(t)U_* \in \mathcal{V}_\eta$. From (7.48) we deduce then that $S_f(t)U_* = S(t)U_*$ for all $t \geq T_\eta$. Thus,

$$\lim_{t \to +\infty} U_f(t) = \lim_{t \to +\infty} S(t)U_* = P_1.$$

Together with (7.50), this implies that γ_f is an heteroclinic orbit for S_f, joining P_0 to P_1. □

7.8 The Nonexistence Result

For $\eta \leq \eta_0$, let $\delta \leq \delta_1 \leq \eta$, T_η be as in the proof of proposition 7.16, and $\mathcal{V}_\eta$ as in (7.47). Recall, from our previous discussion, that if S admits an inertial manifold $\mathcal{M}$ containing the global attractor $\mathcal{A}$, then there is $\mathcal{M}_m \in \mathcal{F}_S$ such that $\mathcal{V}_\eta \cap \mathcal{M} = B(P_1, \eta) \cap \mathcal{M}_m$.

We now define a map $\Phi \colon \mathcal{J}_\delta \to B(P_1, \eta) \subseteq \mathcal{X}$ by

$$\Phi(f) := S_f(T_\eta)U_*. \tag{7.52}$$

Then, $\Phi(f)$ belongs to $\mathcal{V}_\eta \cap \gamma_f \subseteq \mathcal{V}_\eta \cap \mathcal{A}$; hence,

$$\Phi(f) \in \mathcal{V}_\eta \cap \mathcal{M}_m, \tag{7.53}$$

for some $\mathcal{M}_m \in \mathcal{F}_S$.

Our goal is then to show that if δ is sufficiently small, there are many functions $f \in \mathcal{J}_\delta$ such that (7.53) does not hold, i.e. $\Phi(f)$ cannot belong to any of the at most countably many finite dimensional manifolds $\mathcal{M}_m$ of $\mathcal{F}_S$, which contain P_1 and are differentiable at P_1.

To this end, we prepare

PROPOSITION 7.17

For $\eta \in]0, \eta_0]$, let $\delta_1 \in]0, \eta]$ be as in proposition 7.16, and determine T_η, so that (7.51) holds. There exist numbers η and δ, with $0 < \delta \leq \eta \leq \eta_0$, and, correspondingly, a dense subset Λ_δ of $\mathcal{J}_\delta$, such that for all $f \in \Lambda_\delta$, the range of $\Phi'(f)$ is an infinite dimensional subspace of $\mathcal{X}$ (recall that $\Phi'(f) \in \mathcal{L}(\mathcal{J}_\delta; \mathcal{X})$).

We postpone the proof of this result to the next section. Proposition 7.17 determines our choice of η; in turn, this choice determines the values of δ (via δ_1 of proposition 7.16), and T_η, as well as the neighborhood $\mathcal{V}_\eta$ of P_1.

These choices allow us to finally arrive at Mora and Solà-Morales' nonexistence result:

THEOREM 7.18

Let $\mathcal{F}_S$ be the at most countable family of closed, finite dimensional C^1 manifolds of S, containing the stationary state P_1, and differentiable at P_1. Let δ be as in proposition 7.17, and $\Phi(f)$ be as in (7.52). There is a dense subset Σ of $\mathcal{J}_\delta$, with the property that if $f \in \Sigma$, there is no manifold $\mathcal{M} \in \mathcal{F}_S$ containing $\Phi(f)$.

PROOF   1. Let $\mathcal{F}_S = (\mathcal{M}_m)_{m \in \mathbb{N}}$ be the family of manifolds for S, described in theorem 7.13. For each $m \in \mathbb{N}$, define

$$\Sigma_m := \{f \in \mathcal{J}_\delta : \Phi(f) \notin \mathcal{M}_m\}.$$

We claim that each Σ_m is nonempty, open, and dense in $\mathcal{J}_\delta$. As a consequence of Baire's category theorem A.19, the set

$$\Sigma := \bigcap_{m \in \mathbb{N}} \Sigma_m$$

would then also be dense in $\mathcal{J}_\delta$, and the very definition of Σ_m would imply that Σ has the desired properties.

   2. If Σ_m were empty, then $\Phi(f) \in \mathcal{M}_m$ for all $f \in \mathcal{J}_\delta$, so that the range of $\Phi'(f)$ would be contained in the tangent space to $\mathcal{M}_m$ at $\Phi(f)$. Since $\mathcal{M}_m$ is finite dimensional, this tangent space is also finite dimensional; thus, if we take $f \in \Lambda_\delta$, we reach a contradiction with the conclusion of proposition 7.17. A similar argument shows that Σ_m is dense in $\mathcal{J}_\delta$. For, otherwise, there would be a $\tilde{g} \in \mathcal{J}_\delta$ and $\varepsilon_0 > 0$ such that $\|g - \tilde{g}\|_{\mathcal{J}_\delta} \geq \varepsilon_0$ for all $g \in \Sigma_m$. But this would imply that if $g \in \mathcal{J}_\delta$ and $\|g - \tilde{g}\|_{\mathcal{J}_\delta} < \varepsilon_0$, then $g \notin \Sigma_m$. Taking in particular $g \in \Lambda_\delta$ (which is possible since Λ_δ is dense in $\mathcal{J}_\delta$), we reach again a contradiction with proposition 7.17.

   3. Finally, we show that Σ_m is open, as a consequence of the continuity of the map $f \mapsto S_f$. Indeed, let $\tilde{f} \in \Sigma_m$. Then, recalling (7.52), $\Phi(\tilde{f}) = S_{\tilde{f}}(T)U_* \notin \mathcal{M}_m$. Since $\mathcal{M}_m$ is closed, there is an open neighborhood $\mathcal{W}$ of $\Phi(\tilde{f})$ in $\mathcal{X}$ such that $\mathcal{W} \subseteq \mathcal{V}_\eta$ and $\mathcal{W} \cap \mathcal{M}_m = \emptyset$. Then, $\Phi(f) \in \mathcal{W}$ for all f sufficiently close to $\tilde{f}$ and, for these f, $\Phi(f) \notin \mathcal{M}_m$. Hence, $f \in \Sigma_m$, and Σ_m is open. This concludes the proof of theorem 7.18, under the assumption that proposition 7.17 holds.   □

7.9   Proof of Proposition 7.17

1.   We first compute explicitly the Fréchet derivative $\Phi'(f)$ in $\mathcal{L}(\mathcal{J}(a,b);\mathcal{X})$, for $f \in \mathcal{J}(a,b)$. Thus, let $h \in \mathcal{J}(a,b)$. Recalling the definition (7.52) of $\Phi(f)$, we compute

$$\Phi'(f)h = \lim_{r \to 0} \frac{1}{r}(\Phi(f + rh) - \Phi(f)) = \lim_{r \to 0} \frac{1}{r}(S_{f+rh}(T_\eta)U_* - S_f(T_\eta)U_*). \quad (7.54)$$

Now, if u_{f+rh} and u_f are as in (7.49), they are, respectively, the solutions of the equations

$$u_{tt} + 2\alpha u_t - u_{xx} = g(u) + f(\cdot, u) + rh(\cdot, u),$$
$$v_{tt} + 2\alpha v_t - v_{xx} = g(v) + f(\cdot, v),$$

with common initial values $(u_*, \tilde{u}_*)$, and Neumann boundary conditions. Thus, writing $u_{f+rh} = u_f + rw$, we deduce that w satisfies the equation

$$r(w_{tt} + 2\alpha w_t - w_{xx}) = g(u_f + rw) - g(u_f) + f(x, u_f + rw) - f(x, u_f)$$
$$+ rh(x, u_f + rw).$$

Dividing by r and letting $r \to 0$, we obtain that w solves the linear, nonhomogeneous IBVP

$$\begin{cases} w_{tt} + 2\alpha w_t - w_{xx} = g'(u_f)w + f'(x, u_f) + h(x, u_f) \\ w(0, x) = 0, \quad w_t(0, x) = 0 \\ w_x(t, 0) = w_x(t, \pi) = 0. \end{cases} \tag{7.55}$$

Thus, recalling (7.54), we conclude that

$$\Phi'(f)h = (w(T_\eta, \cdot), w_t(T_\eta, \cdot) + \alpha w(T_\eta, \cdot)). \tag{7.56}$$

2. We can now proceed to prove proposition 7.17. We set

$$\Lambda := \bigcup_{a < a' < b' < b} \mathcal{J}(a', b'), \tag{7.57}$$

and, for $\delta > 0$,

$$\Lambda_\delta := \Lambda \cap \mathcal{J}_\delta.$$

We easily see that Λ is dense in $\mathcal{J}(a, b)$; as a consequence, Λ_δ is dense in $\mathcal{J}_\delta$. We now show that if δ is sufficiently small, Λ_δ satisfies the requirements of the proposition. Thus, we have to show that if $f \in \Lambda_\delta$, with δ sufficiently small, the range of $\Phi'(f)$ is an infinite dimensional subspace of $\mathcal{X}$. We will achieve this by constructing a sequence $(h_m)_{m \in \mathbb{N}}$ of linearly independent functions in $\mathcal{J}(a, b)$, such that the corresponding images $\Phi'(f)h_m$, defined as in (7.56), form a sequence of linearly independent functions in $\mathcal{X}$. To this end, we first adjust the parametrization $t \mapsto U(t) = (\varphi(t), \psi(t))$ of the heteroclinic orbit γ, so that

$$\varphi(0) = a, \quad \psi(0) = \tilde{a}. \tag{7.58}$$

Then, (7.41) and (7.42) remain valid on an interval $]-\infty, t_0[$, with $t_0 > 0$, since $\varphi_0 = \varphi(t_0) > a$. In fact, φ is strictly increasing on a larger interval $]-\infty, T_0[$, where $T_0 > t_0$ is the first value of t such that $\varphi(t) = 1$. In particular, φ is increasing on $[0, T_\eta]$, where T_η is as in the proof of proposition 7.16. Since $\psi = \varphi'$, we also have that φ is convex on $]-\infty, t_0[$. Next, we note that if $f \in \Lambda_\delta$, then $f \in \Lambda$, so by (7.57) there are numbers a_f, b_f, such that $a < a_f < b_f < b$, and $f \in \mathcal{J}(a_f, b_f)$.

3.   At the end of this section, we shall prove

LEMMA 7.19

Let η_0 be as in the beginning of section 7.7, and set $m_f := \min\{a_f, \varphi_0\}$. There are $\eta \leq \eta_0$ and, correspondingly, $\delta \leq \delta_1$ (recall that δ_1 depends on η), with the property that for all $f \in \Lambda_\delta$, there are numbers $c_f \in \,]a, m_f]$ and $t_f \in \,]0, t_0]$ such that $\varphi(t_f) \geq c_f$ and

$$\forall\, (t,x) \in \,]-\infty, t_f] \times [0,\pi]: \quad u_f(t,x) \leq \varphi(t) \leq a_f, \tag{7.59}$$
$$\forall\, (t,x) \in [t_f, +\infty[\, \times [0,\pi]: \quad u_f(t,x) \geq c_f. \tag{7.60}$$

Assuming this lemma to hold, we fix $f \in \Lambda_\delta$, and proceed with the construction of the desired sequence $(h_m)_{m \in \mathbb{N}}$. In addition to the properties described above, we further require that the IBV problems (7.55) corresponding to this sequence be as "simple" as possible; ideally, we would like to be able to solve these problems by simple separation of variables. To this end, we first determine the numbers a_f, b_f, c_f and t_f corresponding to f as per lemma 7.19, and remark that since $f(x,u) \equiv 0$ if $u \leq a_f$, by the uniqueness of the solutions to problems (7.44) and (7.11) it follows that

$$u_f(t,x) = \varphi(t) \tag{7.61}$$

for all $t \leq t_0$ and $x \in [0,\pi]$. Thus, if we restrict our attention to the interval $[0,t_0]$, and only consider functions supported in the interval $[a, a_f]$, the equation in (7.55) simplifies into

$$w_{tt} + 2\alpha w_t - w_{xx} - g'(\varphi(t))w = h(x, \varphi(t)). \tag{7.62}$$

As we have stated, we would like to solve this equation by separation of variables. Thus, it is natural to look for a sequence $(w_m)_{m \in \mathbb{N}}$ of solutions of the form

$$w_m(t,x) = \mu_m(t)\cos(mx). \tag{7.63}$$

It turns out that we can indeed do so, if we assume each function h_m to have the form

$$h_m(x,u) = \chi_m(u)\cos(mx), \tag{7.64}$$

with χ_m of class C^2, and supported in the interval $[a, a_f]$. With these choices, a direct replacement of (7.64) into (7.62) shows that for each $m \in \mathbb{N}$, the function μ_m in (7.63) should be determined as the solutions to the linear, nonhomogeneous Cauchy problem on $[0, t_0]$

$$\begin{cases} \mu'' + 2\alpha\mu' + m^2\mu - g'(\varphi(t))\mu = \chi_m(\varphi(t)) \\ \mu(0) = 0, \quad \mu'(0) = 0. \end{cases} \tag{7.65}$$

Moreover, to ensure that the corresponding functions $x \mapsto w_m(t,x)$ are linearly independent on $[0,\pi]$ for all $t \geq 0$ (in particular, for $t = T_\eta$, as desired), we require that there be at least one $\bar{t} \in [0, t_0]$ such that $\mu_m(\bar{t}) \neq 0$ for all $m \in \mathbb{N}$ (recall (7.63)).

4. It remains now to construct the functions $(\chi_m)_{m \in \mathbb{N}}$. To this end, we first define, for $m \in \mathbb{N}$, a function $t \mapsto \zeta_m(t)$ on $[0, t_0]$ as the solution of the homogeneous Cauchy problem

$$
\begin{cases}
\zeta'' + 2\alpha \zeta' + m^2 \zeta - g'(\varphi(t))\zeta = 0 \\
\zeta(t_f) = 0, \quad \zeta'(t_f) = 0.
\end{cases}
\tag{7.66}
$$

Then, recalling that, as we have previously remarked, the function $t \mapsto \varphi(t)$ is invertible on $]-\infty, t_0]$, and that $t_f \leq t_0$, for $m \in \mathbb{N}$ we set

$$
\chi_m(u) := \left. (\sigma'' \zeta_m + 2\sigma' \zeta_m' + 2\alpha\sigma' \zeta_m) \right|_{t = \varphi^{-1}(u)},
\tag{7.67}
$$

where $\sigma \colon \mathbb{R} \to \mathbb{R}$ is a C^2 function such that $\sigma(0) \equiv 0$ for $t \leq 0$ and $\sigma(t) \equiv 1$ for $t \geq \varphi^{-1}(c_f)$. With these choices, each function χ_m is indeed supported in $[a, c_f]$, because if $u \leq a$ (respectively, if $u \geq c_f$), then $t = \varphi^{-1}(u) \leq \varphi^{-1}(a) = 0$ (respectively, $t = \varphi^{-1}(u) \geq \varphi^{-1}(c_f)$), and, therefore, $\sigma(t) = 0$ (resp., $\sigma(t) = 1$). In either case, $\sigma'(t) = \sigma''(t) = 0$, and $\chi_m(u) = 0$. We can now verify that, for each $m \in \mathbb{N}$, the function $\mu_m(t) := \sigma(t)\zeta_m(t)$ does solve the Cauchy problem (7.65), with χ_m defined by (7.67). Indeed, since $\sigma(0) = \sigma'(0) = 0$, the initial conditions $\mu_m(0) = \mu_m'(0) = 0$ are taken. Next, recalling (7.66), we compute that the left side of the equation in (7.65) equals

$$
\sigma''(t)\zeta_m(t) + 2\sigma'(t)\zeta_m'(t) + 2\alpha\sigma'(t)\zeta_m(t),
\tag{7.68}
$$

while its right side equals, by (7.67),

$$
\chi_m(\varphi(t)) = \left. (\sigma'' \zeta_m + 2\sigma' \zeta_m' + 2\alpha\sigma' \zeta_m) \right|_{t = \varphi^{-1}(\varphi(t))},
$$

which is exactly (7.68). Finally, we note that

$$
\mu_m(t_f) = \sigma(t_f)\zeta_m(t_f) = 1.
$$

In fact, $\varphi(t_f) \geq c_f$ by lemma 7.19; therefore, $t_f \geq \varphi^{-1}(c_f)$, and $\sigma(t_f) = 1$.

5. In conclusion, proposition 7.17 is completely proven, as soon as we see that lemma 7.19 holds. To prove this lemma, we first show that for each $\eta \in]0, \eta_0]$ and $f \in \Lambda$ (the set defined in (7.57)), there is a constant $K_\eta > 0$ such that for all $t \in [0, T_\eta]$,

$$
\max_{0 \leq x \leq \pi} |u_f(t, x) - \varphi(t)| \leq K_\eta F(1 - e^{-t}),
\tag{7.69}
$$

where $F := \|f\|_{C^2([0, \pi])}$, and T_η is determined so that (7.51) holds. Note that K depends on η, at least via T_η. Because of the imbedding $H^1(0, \pi) \hookrightarrow C([0, \pi])$, to obtain (7.69) it is sufficient to estimate the norm of the difference $z(t, \cdot) := u_f(t, \cdot) -$

$\varphi(t)$ in $H^1(0,\pi)$. To this end, we resort to (by now familiar) energy estimates on z, which satisfies the IBVP

$$\begin{cases} z_{tt} + 2\alpha z_t - z_{xx} = g(u_f) - g(\varphi) + f(x, u_f) \\ z(0,x) = 0, \qquad z_t(0,x) = 0 \\ z_x(t,0) = z_x(t,\pi) = 0. \end{cases} \tag{7.70}$$

Set

$$E(t) := \|z_t(t,\cdot)\|^2 + \alpha\langle z(t,\cdot), z_t(t,\cdot)\rangle + \alpha^2\|z(t,\cdot)\|^2 + \|z_x(t,\cdot)\|^2.$$

Acting as in the proof of proposition 7.2, we easily obtain, after multiplication of the equation of (7.70) in $L^2(0,\pi)$ by $2z_t$ and αz, and addition of the two resulting identities, that E satisfies on $[0, T_\eta]$ the estimate

$$\frac{d}{dt}E(t) \leq 2C_1 F\sqrt{E(t)} + 2C_2 E(t), \tag{7.71}$$

where C_1 is independent of t, but C_2 depends on T_η, u_f and φ, via the quantities

$$\max_{0 \leq t \leq T_\eta} \|u_f(t,\cdot)\|_{H^1(0,\pi)}, \quad \max_{0 \leq t \leq T_\eta} \|\varphi(t)\|.$$

From (7.71) we obtain

$$\frac{d}{dt}\sqrt{E(t)} \leq C_1 F + C_2\sqrt{E(t)}; \tag{7.72}$$

since $E(0) = 0$, integration of (7.72) yields

$$\sqrt{E(t)} \leq \tfrac{C_1}{C_2} F(e^{C_2 t} - 1) \leq \tfrac{C_1}{C_2} F e^{C_2 T_\eta}(1 - e^{-C_2 t}). \tag{7.73}$$

By Sobolev's imbedding and Schwarz' inequalities, there is $C_3 > 0$ such that

$$\|z(t,\cdot)\|_{C^0([0,\pi])} \leq C_3\sqrt{E(t)}; \tag{7.74}$$

hence, (7.69) follows from (7.74) and (7.73), with $K_\eta := C_1 C_2^{-1} C_3 e^{C_2 T_\eta}$.

6.   Let now t_0 be as in (7.42). Without loss of generality, we can assume that $T_\eta \geq t_0$. Consider first the case $0 \leq t \leq t_0$. Then, since φ is convex on $[0, t_0]$, for $0 < s < t \leq t_0$ we have, recalling (7.58),

$$\varphi(t) \geq \varphi(s) + \varphi'(s)(t - s) \geq \varphi(s) + \varphi'(0)(t - s) = \varphi(s) + \tilde{a}(t - s). \tag{7.75}$$

Let now $f \in \Lambda_\delta$, so that $F \leq \delta$. By (7.69), and (7.75) with $s = \tfrac{1}{2}t$ we obtain that for all $x \in [0, \pi]$

$$u_f(t,x) \geq \varphi(t) - K_\eta F(1 - e^{-t}) \geq \varphi(t) - K_\eta \delta t \geq \varphi(\tfrac{1}{2}t) + \tfrac{1}{2}\tilde{a}t - K_\eta \delta t.$$

If $\tilde{a} \geq 2K_\eta \delta$, this implies that

$$u_f(t,x) \geq \varphi(\tfrac{1}{2}t), \qquad 0 \leq t \leq t_0.$$

Consider next the case that $t_0 \leq t \leq T_\eta$. Keeping in mind that φ is still increasing on $[0, T_\eta]$, we choose $t = t_0$ and $s = \tfrac{1}{2}t_0$ in (7.75), and obtain

$$u_f(t,x) \geq \varphi(t) - K_\eta F(1 - e^{-t}) \geq \varphi(t_0) - K_\eta \delta \geq \varphi(\tfrac{1}{2}t_0) + \tfrac{1}{2}\tilde{a}t_0 - K_\eta \delta.$$

If also $\tilde{a}t_0 \geq 2K_\eta \delta$, this implies that

$$u_f(t,x) \geq \varphi(\tfrac{1}{2}t_0). \tag{7.76}$$

Since

$$\varphi(\tfrac{1}{2}t_0) \leq \varphi(t_0) = \varphi_0,$$

if η is sufficiently small (7.76) also holds for all $t \geq T_\eta$, because $u_f(x,t) \in V_\eta$. In fact, it is sufficient that $\eta \leq \min\{\eta_0, 1 - \varphi_0\}$. In conclusion, we have established that if

$$\eta \leq \min\{R, 1 - b, 1 - \varphi_0\} \tag{7.77}$$

and

$$\delta \leq \frac{\tilde{a}}{2K_\eta} \min\{1, t_0\}, \tag{7.78}$$

then

$$u_f(t,x) \geq \begin{cases} \varphi(\tfrac{1}{2}t) & \text{if } 0 \leq t \leq t_0, \\ \varphi(\tfrac{1}{2}t_0) & \text{if } t_0 \leq t. \end{cases}$$

Thus, to conclude the proof of lemma 7.19 it is sufficient to take

$$t_f := \begin{cases} t_0 & \text{if } a_f \geq \varphi_0, \\ \varphi^{-1}(a_f) & \text{if } a_f \leq \varphi_0, \end{cases} \qquad c_f := \varphi(\tfrac{1}{2}t_f).$$

Indeed, the inequality $t_f \leq t_0$ holds trivially if $a_f \geq \varphi_0$, while if $a_f < \varphi_0$, it follows from

$$\varphi(t_f) = a_f \leq \varphi_0 = \varphi(t_0)$$

and the invertibility of φ on $]-\infty, t_0]$. The inequality $c_f \leq m_f$ follows from

$$c_f = \varphi(\tfrac{1}{2}t_f) \leq \varphi(t_f) = \begin{cases} \varphi(t_0) = \varphi_0 = m_f & \text{if } \varphi_0 \leq a_f, \\ a_f = m_f & \text{if } \varphi_0 \geq a_f. \end{cases}$$

Since $t_f \leq t_0$, (7.59) follows from (7.61) and

$$u_f(t,x) = \varphi(t) \leq \varphi(t_f) = \begin{cases} \varphi(t_0) = \varphi_0 \leq a_f & \text{if } \varphi_0 \leq a_f, \\ a_f & \text{if } \varphi_0 \geq a_f. \end{cases}$$

Finally, (7.60) follows from (7.76) and

$$u_f(t,x) \geq \varphi(\tfrac{1}{2}t_0) \geq \varphi(\tfrac{1}{2}t_f) = c_f.$$

With this, the proof of lemma 7.19 is complete. Note that we first define η by (7.77), and then δ by (7.78); thus, δ depends on η, not only through δ_1 of proposition 7.16, but also via K_η, which depends on T_η.

In conclusion, proposition 7.17 and, consequently, Mora and Solà-Morales' nonexistence theorem 7.18 are now completely proven.

7.10   The C^1 Linearization Equivalence Theorems

In this section we prove theorems 7.11 and 7.12.

7.10.1   Equivalence for a Single Operator

1.   We start with the case of a single operator. For convenience, we reproduce the statement of theorem 7.11, and abbreviate $\| \cdot \| := \| \cdot \|_{\mathcal{X}}$.

THEOREM 7.20
Let $\mathcal{X}$ be a Banach space, and $F: \mathcal{X} \to \mathcal{X}$ be a map, not necessarily linear. Let $x_0 \in \mathcal{X}$ be a fixed point of F, and assume that F is of class C^1 in a neighborhood $\mathcal{U}$ of x_0. Let $L := F'(x_0)$ (the Fréchet derivative of F at x_0), and assume that L has a bounded inverse. Assume further that

$$\|F'(x) - L\|_{\mathcal{L}(\mathcal{X};\mathcal{X})} = O(\|x - x_0\|) \qquad as\ x \to x_0\ (x \in \mathcal{U}), \tag{7.79}$$

$$\|L^{-1}\|_{\mathcal{L}(\mathcal{X};\mathcal{X})} \|L\|^2_{\mathcal{L}(\mathcal{X};\mathcal{X})} < 1. \tag{7.80}$$

There exist then a neighborhood $\mathcal{V} \subseteq \mathcal{U}$ of x_0, with $F(\mathcal{V}) \subseteq \mathcal{V}$, and a C^1 diffeomorphism $\Phi: \mathcal{V} \to \Phi(\mathcal{V}) \subseteq \mathcal{X}$, such that $\Phi(x_0) = 0$, $\Phi'(x_0) = I_{\mathcal{X}}$,

$$\|\Phi'(x) - I_{\mathcal{X}}\|_{\mathcal{L}(\mathcal{X};\mathcal{X})} = O(\|x - x_0\|) \qquad as\ x \to x_0\ (x \in \mathcal{V}), \tag{7.81}$$

and

$$\Phi F = L\Phi \qquad on\ \mathcal{V}. \tag{7.82}$$

Moreover, Φ is unique, in the sense that if $\mathcal{W} \subseteq \mathcal{U}$ and $\Psi: \mathcal{W} \to \mathcal{X}$ also satisfy the same conditions as $\mathcal{V}$ and Φ, then $\Phi \equiv \Psi$ on $\mathcal{V} \cap \mathcal{W}$.

PROOF   **1.** We first show that condition (7.80) implies that L is a strict contraction, i.e. that $\|L\|_{\mathcal{L}(\mathcal{X};\mathcal{X})} < 1$. Assuming otherwise, for x and $y \in \mathcal{X}$ let $u = L^{-1}x$ and

$v = L^{-1}y$. Then

$$\|x-y\|_{\mathcal{X}} \le \|L^{-1}\|_{\mathcal{L}(\mathcal{X};\mathcal{X})} \|L\|_{\mathcal{L}(\mathcal{X};\mathcal{X})} \|x-y\|_{\mathcal{X}} < \frac{1}{\|L\|_{\mathcal{L}(\mathcal{X};\mathcal{X})}} \|x-y\|_{\mathcal{X}} \le \|x-y\|_{\mathcal{X}},$$

which is a contradiction.

2.   We want to determine Φ as the solution of the operator equation (7.82), which we rewrite as

$$\Phi = L^{-1}\Phi F. \tag{7.83}$$

Without loss of generality, we can assume that $x_0 = 0$; thus, $F(0) = 0$. Near 0 we can then write $F = L + G$, and $\Phi = I + \Psi$, where I is the identity in $\mathcal{X}$, and (7.83) becomes

$$I + \Psi = L^{-1}(I+\Psi)(L+G) = L^{-1}(L+G+\Psi(L+G))$$
$$= I + L^{-1}G + L^{-1}\Psi(L+G),$$

that is,

$$\Psi = L^{-1}\Psi(L+G) + L^{-1}G = K_1(\Psi) + L^{-1}G =: K(\Psi). \tag{7.84}$$

In this equation, L and G are known, and we seek to determine the unknown Ψ as a fixed point of the map K defined by (7.84). To this end, for $\delta > 0$ we denote by $\mathcal{B}_\delta$ the ball of $\mathcal{X}$ of center 0 and radius δ, and introduce the space

$$\mathcal{H} := \{\psi \in C^1(\overline{B}(0,\delta); \mathcal{X}): \ \psi(0) = 0, \ \psi'(0) = 0, \ \|\psi'(x)\|_{\mathcal{L}} = O(\|x\|) \ (x \to 0)\}.$$

$\mathcal{H}$ is easily seen to be a Banach space with respect to the norm

$$\|\psi\|_{\mathcal{H}} := \sup_{0 < \|x\| \le \delta} \frac{\|\psi'(x)\|_{\mathcal{L}}}{\|x\|}. \tag{7.85}$$

We immediately have

PROPOSITION 7.21
Let $G := F - L$. For all $\delta > 0$, the restrictions of G and $L^{-1}G$ to $\overline{B}(0,\delta)$, which we still denote by G and $L^{-1}G$, are in $\mathcal{H}$.

PROOF   To show that $G \in \mathcal{H}$, note first that $G(0) = F(0) - L(0) = 0$, and $G'(0) = F'(0) - L = 0$, because $L = F'(x_0) = F'(0)$. By (7.79), with $x_0 = 0$,

$$\|G'(x)\|_{\mathcal{L}(\mathcal{X};\mathcal{X})} = \|F'(x) - L\|_{\mathcal{L}(\mathcal{X};\mathcal{X})} = O(\|x\|)$$

as $x \to 0$. Consequently, $(L^{-1}G)(0) = L^{-1}(G(0)) = 0$, $(L^{-1}G)'(0) = L^{-1}(G'(0)) = 0$, and

$$\|(L^{-1}G)'(x)\|_{\mathcal{L}(\mathcal{X};\mathcal{X})} = \|L^{-1}(G'(x))\|_{\mathcal{L}(\mathcal{X};\mathcal{X})}$$

$$\leq \|L^{-1}\|_{\mathcal{L}(\mathcal{X};\mathcal{X})} \|G'(x)\|_{\mathcal{L}(\mathcal{X};\mathcal{X})} = O(\|x\|)$$

as $x \to 0$. Thus, $L^{-1}G \in \mathcal{H}$ as well.   □

As a consequence of proposition 7.21, the maps K_1 and K introduced in (7.84) are well defined, because if δ is sufficiently small, $L+G$ maps $\overline{B}(0,\delta)$ into itself. In fact, since $G \in \mathcal{H}$, $\|G(x)\| = O(\|x\|)$ as $x \to 0$; thus, if $\|x\| \leq \delta$,

$$\|Lx + G(x)\| \leq \|L\|_{\mathcal{L}(\mathcal{X};\mathcal{X})} \|x\| + c\|x\|^2 \leq (\|L\|_{\mathcal{L}(\mathcal{X};\mathcal{X})} + c\delta)\|x\|. \qquad (7.86)$$

Since $\|L\|_{\mathcal{L}(\mathcal{X};\mathcal{X})} < 1$, there is $\delta_0 > 0$ such that $\|L\|_{\mathcal{L}(\mathcal{X};\mathcal{X})} + c\delta \leq 1$ for all $\delta \in \ ]0, \delta_0]$. Hence, (7.86) implies that $(L+G)(x) \in \overline{B}(0,\delta)$ if $x \in \overline{B}(0,\delta)$ and $\delta \leq \delta_0$. Consequently, the composition of functions $\psi \in \mathcal{H}$ with $L+G$ is defined.

3. We now show that if δ is sufficiently small, K is a strict contraction of $\mathcal{H}$ into itself.

PROPOSITION 7.22
Let $\delta \leq \delta_0$. For all $\psi \in \mathcal{H}$, $K_1(\psi) \in \mathcal{H}$. If also $\delta \|L^{-1}\|_{\mathcal{L}(\mathcal{X};\mathcal{X})} \leq 1$, then K is a strict contraction on $\mathcal{H}$.

PROOF   1. Since both ψ and $G \in \mathcal{H}$,

$$[K_1(\psi)](0) = [L^{-1}\psi(L+G)](0) = L^{-1}(\psi(0)) = 0.$$

Next, we compute the Fréchet derivative of $K_1(\psi)$ at a generic point $x \in \mathcal{X}$. Using the symbol $\circ$ to denote the composition of linear operators, by the chain rule we compute that

$$(K_1(\psi))'(x) = L^{-1} \circ \psi'(Lx + G(x)) \circ (L + G'(x)). \qquad (7.87)$$

Thus,

$$(K_1(\psi))'(0) = L^{-1} \circ \psi'(0) \circ (L + G'(0)) = 0.$$

As in the proof of proposition 7.21,

$$\|L + G'(x)\|_{\mathcal{L}(\mathcal{X};\mathcal{X})} \leq \|L\|_{\mathcal{L}(\mathcal{X};\mathcal{X})} + O(\|x\|) \leq \|L\|_{\mathcal{L}(\mathcal{X};\mathcal{X})} + c_1\delta_0; \qquad (7.88)$$

since G is continuous and $G(0) = 0$, $Lx + G(x) \to 0$ as $x \to 0$. Thus, since $\psi \in \mathcal{H}$, we obtain from (7.86), (7.87) and (7.88)

$$\begin{aligned}
\|(K_1(\psi))'(x)\|_{\mathcal{L}} &\leq \|L^{-1}\|_{\mathcal{L}(\mathcal{X};\mathcal{X})} \|\psi'(Lx + G(x))\|_{\mathcal{L}(\mathcal{X};\mathcal{X})} \|L + G'(x)\|_{\mathcal{L}(\mathcal{X};\mathcal{X})} \\
&\leq \|L^{-1}\|_{\mathcal{L}(\mathcal{X};\mathcal{X})} c_2 \|Lx + G(x)\|_{\mathcal{L}(\mathcal{X};\mathcal{X})} (\|L\|_{\mathcal{L}(\mathcal{X};\mathcal{X})} + c_1\delta_0) \\
&\leq \|L^{-1}\|_{\mathcal{L}(\mathcal{X};\mathcal{X})} c_2 (\|L\|_{\mathcal{L}(\mathcal{X};\mathcal{X})} + c\delta)\|x\| (\|L\|_{\mathcal{L}(\mathcal{X};\mathcal{X})} + c_1\delta_0) \\
&=: c_3 \|x\|.
\end{aligned}$$

This completes the proof that $K_1(\psi) \in \mathcal{H}$. Since $L^{-1}G \in \mathcal{H}$ by proposition 7.21, it follows that also $K(\psi) \in \mathcal{H}$.

2.   To show that K is a contraction, let $\psi_1, \psi_2 \in \mathcal{H}$. Then $K(\psi_1) - K(\psi_2) = K_1(\psi_1) - K_1(\psi_2)$, and, recalling (7.85),

$$\|K_1(\psi_1) - K_1(\psi_2)\|_{\mathcal{H}} = \sup_{\substack{x \in \overline{B}_\delta \\ x \neq 0}} \frac{1}{\|x\|} \|(K_1(\psi_1))'(x) - (K_1(\psi_2))'(x)\|_{\mathcal{L}(\mathcal{X};\mathcal{X})} . \quad (7.89)$$

Let $x \in \overline{B}(0, \delta)$. From (7.87) we have

$$(K_1(\psi_1))'(x) - (K_1(\psi_2))'(x) = L^{-1} \circ (\psi_1'(Lx + G(x)) - \psi_2'(Lx + G(x))) \circ (L + G'(x));$$

thus, recalling (7.86) and (7.89), and that $Lx + G(x) \in \overline{B}(0, \delta)$ if $x \in \overline{B}(0, \delta)$,

$$\frac{1}{\|x\|} \|(K_1(\psi_1))'(x) - (K_1(\psi_2))'(x)\|_{\mathcal{L}(\mathcal{X};\mathcal{X})}$$

$$\leq \|L^{-1}\|_{\mathcal{L}(\mathcal{X};\mathcal{X})} \frac{\|\psi_1'(Lx + G(x)) - \psi_2'(Lx + G(x))\|_{\mathcal{L}(\mathcal{X};\mathcal{X})}}{\|Lx + G(x)\|}$$

$$\cdot \frac{\|Lx + G(x)\|}{\|x\|} \|L + G'(x)\|_{\mathcal{L}(\mathcal{X};\mathcal{X})} \quad (7.90)$$

$$\leq \|L^{-1}\|_{\mathcal{L}(\mathcal{X};\mathcal{X})} \|\psi_1' - \psi_2'\|_{\mathcal{H}} (\|L\|_{\mathcal{L}(\mathcal{X};\mathcal{X})} + c\|x\|)(\|L\|_{\mathcal{L}(\mathcal{X};\mathcal{X})} + c_1\|x\|)$$

$$\leq \|\psi_1' - \psi_2'\|_{\mathcal{H}} \Big(\|L^{-1}\|_{\mathcal{L}(\mathcal{X};\mathcal{X})} \|L\|_{\mathcal{L}(\mathcal{X};\mathcal{X})}^2 + (c + c_1) \|L^{-1}\|_{\mathcal{L}(\mathcal{X};\mathcal{X})} \|L\|_{\mathcal{L}(\mathcal{X};\mathcal{X})}$$

$$+ cc_1 \|L^{-1}\|_{\mathcal{L}(\mathcal{X};\mathcal{X})} \|x\|^2 \Big)$$

$$\leq \|\psi_1' - \psi_2'\|_{\mathcal{H}} (\|L^{-1}\|_{\mathcal{L}(\mathcal{X};\mathcal{X})} \|L\|_{\mathcal{L}(\mathcal{X};\mathcal{X})}^2 + c_L \|x\|), \quad (7.91)$$

where

$$c_L := (c + c_1) \|L^{-1}\|_{\mathcal{L}(\mathcal{X};\mathcal{X})} \|L\|_{\mathcal{L}(\mathcal{X};\mathcal{X})} + cc_1 \|L^{-1}\| \delta_0 .$$

Because of (7.80), we can find $\delta_1 \in]0, \delta_0]$ such that

$$\|L^{-1}\|_{\mathcal{L}(\mathcal{X};\mathcal{X})} \|L\|_{\mathcal{L}(\mathcal{X};\mathcal{X})}^2 + c_L \delta_1 = \tfrac{1}{2}(1 + \|L^{-1}\|_{\mathcal{L}(\mathcal{X};\mathcal{X})} \|L\|_{\mathcal{L}(\mathcal{X};\mathcal{X})}^2) =: \rho < 1.$$

With this choice of δ_1, we conclude from (7.88) and (7.91) that if $\delta \leq \delta_1$,

$$\|K_1(\psi_1) - K_1(\psi_2)\|_{\mathcal{H}} \leq \rho \|\psi_1 - \psi_2\|_{\mathcal{H}} .$$

Thus, K is a strict contraction in $\mathcal{H}$, as claimed.   □

4.   From proposition 7.22 it follows that equation (7.84) has a unique solution $\Psi \in \mathcal{H}$; setting $\Phi = I + \Psi$, we conclude that (7.83), and therefore (7.82), hold. Set then $\mathcal{V} := B(0, \delta)$. Since $\Psi \in C^1(\overline{B}(0, \delta); \mathcal{X})$, by theorem A.45 we conclude that Φ is a diffeomorphism between $\mathcal{V}$ and its image. Consequently, the proof of theorem 7.11 is complete.   □

7.10.2   Equivalence for Groups of Operators

We can now apply theorem 7.11 to extend the C^1-equivalence theorem to a group of operators. For convenience, we reproduce the statement of theorem 7.12:

THEOREM 7.23
Let $(S(t))_{t\in\mathbb{R}}$ be a flow on a Banach space $\mathcal{X}$, such that for each $t \in \mathbb{R}$, $S(t)$ is a C^2 diffeomorphism on a neighborhood $\mathcal{U}$ of a common fixed point x_0 (i.e., $S(t)x_0 = x_0$ for all $t \in \mathbb{R}$). For each $t \in \mathbb{R}$, let $R(t) := (S(t))'(x_0)$ (the Fréchet derivative of $S(t)$ at x_0). Assume there is $\tau \in \mathbb{R}$ such that the operators $F := S(\tau)$ and $L := R(\tau)$ satisfy conditions (7.38) and (7.39) of theorem 7.11. There exist then: a neighborhood $\mathcal{V} \subseteq \mathcal{U}$ of x_0, and a C^1 diffeomorphism $\Phi \colon \mathcal{V} \to \Phi(\mathcal{V}) \subseteq \mathcal{X}$, such that $\Phi(x_0) = 0$, $\Phi'(x_0) = I_{\mathcal{X}}$,

$$\|\Phi'(x) - I_{\mathcal{X}}\|_{\mathcal{L}(\mathcal{X};\mathcal{X})} = O(\|x - x_0\|_{\mathcal{X}}) \qquad \text{as } x \to x_0 \ (x \in \mathcal{V}), \tag{7.92}$$

and the identity

$$\Phi S(t) = R(t)\Phi \tag{7.93}$$

holds for all $t \in \mathbb{R}$ in a ball $B(x_0, r) \subseteq \mathcal{V}$. This ball can be chosen independently of t if t is bounded from below.

PROOF   1. By theorem 7.20, there are a neighborhood $\mathcal{V}$ of x_0 and a C^1 diffeomorphism $\Phi \colon \mathcal{V} \to \Phi(\mathcal{V})$ such that $\Phi(x_0) = 0$, $\Phi'(x_0) = I$ and (7.81), (7.82) hold. $\mathcal{V}$ and Φ depend on τ, via the positions $F = S(\tau)$ and $L = R(\tau)$. Fix $t \in \mathbb{R}$, $t \neq \tau$, and set $\mathcal{V}_t := S(-t)\mathcal{V}$, $\Phi_t := R(-t)\Phi S(t)$. We claim that Φ_t satisfies, on $\mathcal{V}_t$, the same conditions satisfied by Φ on $\mathcal{V}$. In fact: at first,

$$\Phi_t(x_0) = R(-t)\Phi S(t)x_0 = 0,$$

because $S(t)x_0 = 0$. Next, recalling that $R(\theta)$ is a linear operator for each $\theta \in \mathbb{R}$, that $(S(t))'(x_0) = R(t)$, and that $\Phi'(x_0) = I$, we compute that

$$\Phi_t'(x_0) = R(-t) \circ \Phi'(S(t)x_0) \circ (S(t))'(x_0)$$
$$= R(-t) \circ \Phi'(x_0) \circ R(t) = R(-t)R(t) = I. \tag{7.94}$$

2. To show that Φ_t satisfies (7.92), we write again $\Phi = I + \Psi$, $\Psi \in \mathcal{H}$, as in the proof of theorem 7.11. Then, from (7.94), for all $z \in \mathcal{V}_t$:

$$\Phi_t'(z) - I = R(-t) \circ (I + \Psi'(S(t)z)) \circ (S(t))'(z) - I$$
$$= [R(-t) \circ (S(t))'(z) - I] + R(-t) \circ \Psi'(S(t)z) \circ (S(t))'(z)$$
$$=: \Lambda_1(z) + \Lambda_2(z). \tag{7.95}$$

Since $S(t) \in C^2(\mathcal{X}; \mathcal{X})$, we can estimate

$$\|\Lambda_1(z)\|_{\mathcal{L}(\mathcal{X};\mathcal{X})} = \|R(-t)((S(t))'(z) - R(t))\|_{\mathcal{L}}$$

$$\le \|R(-t)\|_{\mathcal{L}(\mathcal{X};\mathcal{X})}\|(S(t))'(z) - (S(t))'(x_0)\|_{\mathcal{L}(\mathcal{X};\mathcal{X})}$$
$$\le c_1(t)\|z - x_0\|. \tag{7.96}$$

As for Λ_2, we note that since $z \in V_t = S(-t)V$, the point $x := S(t)z$ is in V; thus, by (7.92), as in (7.96),

$$\|\Lambda_2(z)\|_{\mathcal{L}(\mathcal{X};\mathcal{X})} = \|R(-t)\|_{\mathcal{L}(\mathcal{X};\mathcal{X})}\|\Psi'(x)\|_{\mathcal{L}(\mathcal{X};\mathcal{X})}\|(S(t))'(z)\|_{\mathcal{L}(\mathcal{X};\mathcal{X})}$$
$$\le c\|R(-t)\|_{\mathcal{L}(\mathcal{X};\mathcal{X})}\|x - x_0\|(\|z - x_0\| + \|R(t)\|_{\mathcal{L}(\mathcal{X};\mathcal{X})}). \tag{7.97}$$

By Taylor's formula,

$$\|x - x_0\| = \|S(t)z - S(t)x_0\| = \|R(t)(z - x_0)\| + O(\|z - x_0\|^2);$$

thus, if $\|z - x_0\| \le \delta$, we obtain from (7.97)

$$\|\Lambda_2(z)\|_{\mathcal{L}(\mathcal{X};\mathcal{X})} = c\|R(-t)\|_{\mathcal{L}(\mathcal{X};\mathcal{X})}(\|R(t)\|_{\mathcal{L}(\mathcal{X};\mathcal{X})} + c_2(t)\delta)\|z - x_0\| \le c_3(t)\|z - x_0\|.$$

Putting this and (7.96) back into (7.95), we deduce that

$$\|\Phi'_t(z) - I\|_{\mathcal{L}(\mathcal{X};\mathcal{X})} \le c_4(t)\|z - x_0\|, \tag{7.98}$$

i.e. that (7.92) holds for Φ_t in V_t (recall that $x_0 = S(-t)x_0 \in V_t$).

3. Finally, we show that (7.93) also holds for Φ_t. In fact, we prove more, that is, that for all $t \in \mathbb{R}$,

$$\Phi_t = R(-t)\Phi_t S(t) = R(-\tau)\Phi_t S(\tau). \tag{7.99}$$

Indeed, recalling that $\Phi_t := R(-t)\Phi S(t)$, $L = R(\tau)$ and $F = S(\tau)$, from (7.83) we compute

$$R(-\tau)\Phi_t S(\tau) = R(-\tau)R(-t)\Phi S(t)S(\tau)$$
$$= R(-t)R(-\tau)\Phi S(\tau)S(t) = R(-t)\Phi S(t) = \Phi_t.$$

Thus, by uniqueness, $\Phi_t = \Phi$ on $V \cap V_t$, and we can take this common operator as the required Ψ; in particular, (7.93) follows from (7.99).

4. The dependence of c_4 on t in (7.98) shows that, in general, identity (7.93) holds on a neighborhood of x_0 whose diameter may depend on t. To show that this neighborhood can be determined independently of t if t is bounded from below, fix $t_0 \in \mathbb{R}$, and consider only $t \ge t_0$. It is then sufficient to show that the set

$$\mathcal{W} := \bigcap_{t \ge t_0} S(-t)\mathcal{V}$$

is a neighborhood of x_0. To this end, we split

$$\mathcal{W} = \mathcal{W}_1 \cap \mathcal{W}_2 := \left(\bigcap_{t_0 \le t \le t_0 + \tau} S(-t)\mathcal{V}\right) \cap \left(\bigcap_{t \ge t_0 + \tau} S(-t)\mathcal{V}\right),$$

and show separately that $\mathcal{W}_1$ and $\mathcal{W}_2$ are neighborhoods of x_0.

Assume for the moment that it is possible to find $r > 0$ such that for all $t \in [t_0, t_0 + \tau]$,

$$S(t)B(x_0, r) \subseteq \mathcal{V}. \tag{7.100}$$

Then $B(x_0, r) \subseteq S(-t)\mathcal{V}$ for all $t \in [t_0, t_0 + \tau]$; hence, $\mathcal{W}_1 \supseteq B(x_0, r)$ and, therefore, $\mathcal{W}_1$ is a neighborhood of x_0. As for $\mathcal{W}_2$, we first recall that, by construction, $S(\tau)\mathcal{V} \subseteq \mathcal{V}$. Arguing by induction, it follows that $S(m\tau)\mathcal{V} \subseteq \mathcal{V}$ for all integer $m \geq 1$. Given then $t \geq t_0 + \tau$, we write $t = t_0 + \theta + m\tau$, with $m \geq 1$ and $0 \leq \theta < \tau$, so that, by (7.100),

$$S(t)B(x_0, r) = S(m\tau)S(t_0 + \theta)B(x_0, r) \subseteq S(m\tau)\mathcal{V} \subseteq \mathcal{V}.$$

Consequently, $\mathcal{W}_2 \supseteq B(x_0, r)$, and $\mathcal{W}_2$ is a neighborhood of x_0.

5. To conclude the proof of theorem 7.23 it is therefore sufficient to find $r > 0$ such that (7.100) holds. To achieve this, let $\eta > 0$ be such that $B(x_0, \eta) \subseteq \mathcal{V}$. Since $S(t)$ is continuous and $S(t)x_0 = x_0$ for all $t \in \mathbb{R}$, there is $\tilde{\rho}(t) > 0$ such that

$$S(t)B(x_0, \tilde{\rho}(t)) \subseteq B(x_0, \eta). \tag{7.101}$$

Set

$$\rho(t) := \sup\{\tilde{\rho}(t) : (7.101) \text{ holds}\}, \tag{7.102}$$
$$r := \inf\{\rho(t) : t_0 \leq t \leq t_0 + \tau\}. \tag{7.103}$$

We claim that $r > 0$. Assuming this, then for all $t \in [t_0, t_0 + \tau]$,

$$S(t)B(x_0, r) \subseteq B(x_0, \eta) \subseteq \mathcal{V},$$

and (7.100) holds. Clearly, $r > 0$ if ρ were a continuous function, because ρ is positive over the compact interval $[t_0, t_0 + \tau]$. Otherwise, assume that $r = 0$. By (7.103), there is a minimizing sequence $(\theta_m)_{m \in \mathbb{N}}$, such that $\rho(\theta_m) \to 0$. Since $(\theta_m)_{m \in \mathbb{N}} \subseteq [t_0, t_0 + \tau]$, there exists a subsequence, which we still denote by $(\theta_m)_{m \in \mathbb{N}}$, converging to some $\theta \in [t_0, t_0 + \tau]$. Since the map $(x, t) \mapsto S(t)x$ is continuous, and $S(\theta)x_0 = x_0$, given η as above there is $\beta > 0$ such that

$$\|x - x_0\| + |t - \theta| < \beta \quad \Longrightarrow \quad \|S(t)x - x_0\| < \eta. \tag{7.104}$$

We can choose $m_0 \in \mathbb{N}$ such that $\|\theta_m - \theta\| < \frac{1}{2}\beta$ and $\rho(\theta_m) \leq \frac{1}{4}\beta$ for all $m \geq m_0$. By (7.102), there is $y_m \in S(\theta_m)B(x_0, \frac{1}{2}\beta)$ such that

$$\|y_m - x_0\| \geq \eta. \tag{7.105}$$

Let $x_m \in B(x_0, \frac{1}{2}\beta)$ be such that $y_m = S(\theta_m)x_m$. Then, $x = x_m$ and $t = \theta_m$ satisfy the left side of (7.104). Consequently,

$$\|S(\theta_m)x_m - x_0\| = \|y_m - x_0\| < \eta,$$

contradicting (7.105). Thus, $r > 0$, and (7.100) follows. The proof of theorem 7.23 is now complete. $\Box$

Appendix: Selected Results from Analysis

In this appendix we put together a number of definitions and results on various topics in Analysis, that we have referred to in the previous chapters. We present these results mostly without proof, but always with at least one reference, indicating where a proof can be found. In these citations, a format like "sct. 1.2.3" refers to subsection 3 of section 2 of chapter 1.

A.1 Ordinary Differential Equations

In this section we report some well known results on the well-posedness and stability of classical and generalized solutions of ODEs in $\mathbb{R}^N$.

A.1.1 Classical Solutions

We consider the system of ODEs for a vector valued unknown function $\mathbb{R} \ni t \mapsto x(t) \in \mathbb{R}^N$

$$\dot{x} = f(t,x), \tag{A.1}$$

with f defined at least on the product $I \times \mathcal{U}$, $I \subseteq \mathbb{R}$ an interval, and $\mathcal{U} \subseteq \mathbb{R}^N$ an open domain. We also assign to (A.1) the initial condition

$$x(t_0) = x_0, \qquad (t_0, x_0) \in I \times \mathcal{U}. \tag{A.2}$$

Sufficient conditions for the local well-posedness of classical solutions to the initial value problem (IVP in short) (A.1)+(A.2) are well known:

THEOREM A.1
Let $f \in \mathrm{C}(I \times \mathcal{U}; \mathbb{R}^N)$. There exists a closed neighborhood

$$\{(t,x) \in I \times \mathcal{U}: |t - t_0| \le \alpha, \ \|x - x_0\| \le \beta\} =: I_0 \times \mathcal{B}_0$$

of (t_0, x_0), and a function $x \in \mathrm{C}^1(I_0; \mathcal{B}_0)$, solution of the IVP (A.1)+(A.2).
If in addition f satisfies the t-uniform Lipschitz condition in x

$$\exists L > 0 \, \forall (t,x), (t,\bar{x}) \in I \times \mathcal{U}: \quad \|f(t,x) - f(t,\bar{x})\| \le L\|x - \bar{x}\|,$$

then the solution x is uniquely determined by the initial values (t_0, x_0). *In fact, x depends continuously on* (t_0, x_0), *in the sense that if* $t \mapsto x(t)$ *and* $t \mapsto \bar{x}(t)$ *are solutions of (A.1), corresponding to initial values* (t_0, x_0) *and* $(\bar{t}_0, \bar{x}_0)$, *and defined on neighborhoods* I_0 *and* $\bar{I}_0$ *of* t_0 *and* $\bar{t}_0$ *respectively, then there exists* $C > 0$, *independent of* (t_0, x_0) *and* $(\bar{t}_0, \bar{x}_0)$, *such that for all* $t \in I_0 \cap \bar{I}_0$,

$$\|x(t) - \bar{x}(t)\| \le C \left(|t_0 - \bar{t}_0| + \|x_0 - \bar{x}_0\| \right).$$

PROOF   See e.g. Coddington-Levinson, [CL55, ch. 1].   ☐

By repeated applications of the local existence theorem A.1, local solutions can be extended to a so-called MAXIMAL SOLUTION, that is, a solution of the IVP (A.1)+(A.2) defined on a maximal interval $I' \subseteq I$, with $t_0 \in I'$. If $I' = I$, we say that the IVP (A.1)+(A.2) can be solved GLOBALLY in I. As the following result shows, extension of a local solution to a global one can be achieved if a time-independent bound on the local solution is available. Such bounds are usually established by means of so-called A PRIORI estimates on the local solutions.

THEOREM A.2
Assume there is $M > 0$ *with the property that if* $I' \subseteq I$ *is any interval containing* t_0, *and* $x \in C^1(I', \mathbb{R}^N)$ *is a solution of the IVP (A.1)+(A.2) on* I', *then*

$$\sup_{t \in I'} \|x(t)\| \le M. \tag{A.3}$$

Then, the IVP (A.1) + (A.2) can be solved globally in I.

PROOF   See e.g. Coddington-Levinson, [CL55, sct. 2.1].   ☐

The bound in estimate (A.3) is called *a priori*, because M has to be found independently of any particular interval I' on which a local solution is determined. As we have seen in section 2.9 of chapter 2, a priori estimates can be established by means of integral or differential inequalities, such as the Gronwall or the exponential inequalities of propositions 2.62 and 2.64.

A.1.2   Generalized Solutions

More generally, we can consider GENERALIZED SOLUTIONS to the IVP (A.1) + (A.2), in the sense that we require the ODEs (A.1) to be satisfied only almost everywhere in t. In this context, it is natural to look for solutions of (A.1) that are absolutely continuous.

DEFINITION A.3   *Let* $I \subseteq \mathbb{R}$ *be an interval, and* $\mathcal{X} = \mathbb{R}^N$ *(or, more generally, a reflexive Banach space). A function* $f: I \to \mathcal{X}$ *is said to be* ABSOLUTELY CONTIN-

UOUS *if for all* $\varepsilon > 0$ *there is* $\delta > 0$ *such that for any finite choice of nonoverlapping subintervals* $]a_i, b_i[\subset I, i = 1, \ldots, r$,

$$\sum_{i=1}^{r} |a_i - b_i| \leq \delta \implies \sum_{i=1}^{r} \|f(a_i) - f(b_i)\|_{\mathcal{X}} \leq \varepsilon.$$

We denote by $\mathrm{AC}(I; \mathcal{X})$ *the space of absolutely continuous functions from* I *into* $\mathcal{X}$.

Recalling then from sections A.4 and A.6 below the definitions of the Lebesgue spaces $\mathrm{L}^1(I)$ and $\mathrm{L}^1(I; \mathcal{X})$, we have the following

THEOREM A.4
Let $I \subseteq \mathbb{R}$ *be an interval. A function* $f : I \to \mathcal{X}$ *is absolutely continuous if and only if* f *is differentiable almost everywhere in* I, *with* $f' \in \mathrm{L}^1(I; \mathcal{X})$.

PROOF   If $\mathcal{X} = \mathbb{R}^N$, see e.g. Rudin, [Rud74, ch. 8]. For a generalization to functions valued into a reflexive Banach space, see Komura, [Kom70].   ⬜

We can then state a result, usually known as CARATHÉODORY's theorem, which describes sufficient conditions for the existence of generalized solutions to the IVP (A.1)+(A.2).

THEOREM A.5
Assume that $f(\cdot, x)$ *is measurable in* I *for each* $x \in \mathcal{U}$, *and that* $f(t, \cdot)$ *is continuous on* $\mathcal{U}$ *for each* $t \in I$. *Assume further that there is a function* $\varphi \in \mathrm{L}^1(I)$, *such that for almost all* $t \in I$,

$$\sup_{x \in \mathcal{U}} \|f(x, t)\| \leq \varphi(t).$$

Then, there exists $I' \subseteq I$, *containing* t_0, *and a function* $x \in \mathrm{AC}(I'; \mathcal{U})$, *which is a generalized solution of the IVP (A.1) + (A.2) on* I'.

PROOF   See e.g. Coddington-Levinson, [CL55, sct. 2.1].   ⬜

A.1.3   Stability for Autonomous Systems

We now restrict our attention to systems of ODEs that are AUTONOMOUS, i.e. systems (A.1) in which f is independent of t; that is, of the form

$$\dot{x} = f(x). \tag{A.4}$$

We take $t_0 = 0$ in the initial condition (A.2), and assume that the IVP (A.4)+(A.2) has a global solution on the right interval $[0, +\infty[$. To emphasize the dependence of this solution on its initial value, we write it as $x(\cdot, x_0)$. Finally, we call the image of the map $t \mapsto x(t, x_0)$ in $\mathbb{R}^N$ the ORBIT starting at x_0.

DEFINITION A.6   *A point $x \in \mathbb{R}^N$ is a* STATIONARY POINT *(or, an* EQUILIBRIUM POINT*), for the autonomous system (A.4) if $f(x) = 0$. A stationary point x_0 of S is said to be:*

1. STABLE, *if for any neighborhood $\mathcal{U}$ of x_0 there is a neighborhood $\mathcal{V} \subset \mathcal{U}$ of x such that any solution $t \mapsto x(t, x_1)$ with $x_1 \in \mathcal{V}$ is such that $x(t, x_1) \in \mathcal{U}$ for all $t \geq 0$;*

2. UNSTABLE, *if it is not stable;*

3. ASYMPTOTICALLY STABLE, *if x_0 is stable and there is a neighborhood $\mathcal{V}$ of x such that any orbit starting in $\mathcal{V}$ converges to x_0, i.e. if for all $x_1 \in \mathcal{V}$,*

$$\lim_{t \to +\infty} x(t, x_1) = x_0.$$

This definition corresponds to the so-called stability in the sense of Lyapunov. The following is a well known criterion for the stability of the stationary points of autonomous systems.

THEOREM A.7
Assume that $f \in C^1$, and let x_0 be a stationary point of the ODE (A.4). Consider the matrix $A := f'(x_0)$, and assume that all the eigenvalues of A have nonzero real part. Then:

1. *If the real part of all the eigenvalues of A are negative, x_0 is asymptotically stable;*

2. *If A has at least one eigenvalue with positive real part, x_0 is unstable.*

PROOF   See e.g. Coddington-Levinson, [CL55, sct. 13.1].   □

The proof of theorem A.7 is based on a result, known as the HARTMAN-GROB-MAN EQUIVALENCE THEOREM, on the topological equivalence, near the stationary point x_0, of system (A.4) and its linearization at x_0, i.e. the system

$$\dot{y} = A(y - x_0), \qquad A := f'(x_0), \tag{A.5}$$

of which x_0 is also a stationary point.

THEOREM A.8
Assume that $f \in C^1$, and that $f(x_0) = 0$. Let $A := f'(x_0)$, and assume that all the eigenvalues of A have nonzero real part. There exist then two neighborhoods $\mathcal{U}$ and

$\mathcal{V}$ of x_0 in $\mathbb{R}^N$, and a homeomorphism $\Phi\colon \mathcal{U} \to \mathcal{V}$, such that for all $x_1 \in \mathcal{U}$, there is a neighborhood I of 0 in $\mathbb{R}$, such that for all $t \in I$,

$$\Phi(x(t,x_1)) = x_0 + e^{tA}(\Phi(x_1) - x_0).$$

That is, Φ maps orbits of the "original" system (A.4), which are near x_0, into orbits of the linearized system (A.5), which also are near x_0.

PROOF   See e.g. Perko, [Per91, sct. 2.8]. $\Box$

We remark that, in the proof of the Hartman-Grobman theorem, it is often assumed that $x_0 = 0$. This is without loss of generality; for, otherwise, we can consider the change of unknown $u(t) := x(t) - x_0$. Then, u satisfies the ODE

$$\dot{u} = f(u + x_0) =: g(u),$$

and the function $t \mapsto u(t) \equiv 0$ is a solution of this ODE, since $g(0) = f(x_0) = 0$; finally, $g'(0) = f'(x_0) = A$.

In the scalar case, i.e. when $N = 1$, theorem A.7 just means that if x_0 is a stationary point, then x_0 is asymptotically stable if $f'(x_0) < 0$, while it is unstable if $f'(x_0) > 0$. For example, consider the ODE

$$\dot{x} = x(1 - x). \tag{A.6}$$

The stationary points are $x_0 = 0$ and $x_1 = 1$, with $f'(0) = 1$ and $f'(1) = -1$. Thus, x_0 is unstable, while x_1 is asymptotically stable, as it is easy to see by direct analysis of the ODE (A.6).

We conclude this section with the celebrated POINCARÉ-BENDIXON THEOREM, which gives a complete description of the ω-limits sets of the orbits of the autonomous system (A.4) in the planar case $N = 2$.

THEOREM A.9
Let $\Omega \subseteq \mathbb{R}^2$ be an open domain, and $f \in C^1(\Omega; \mathbb{R}^2)$. Assume that the orbit $\gamma_+ = (x(t))_{t \geq 0}$ of a solution x of (A.4) is contained in a compact subset of Ω, and that its ω-limit set $\omega(\gamma_+)$ contains no stationary points. Then, either γ_+ or $\omega(\gamma_+)$ is a periodic orbit (called LIMIT CYCLE).

PROOF   See e.g. Coddington-Levinson, [CL55, ch. 16]. $\Box$

Roughly speaking, theorem A.9 states that either γ_+ is already a periodic orbit (in which case $\omega(\gamma_+) = \gamma_+$), or γ_+ converges to $\omega(\gamma_+)$. Typically, γ_+ "spirals" around $\omega(\gamma_+)$.

A.2   Linear Spaces and their Duals

In this section, we assume as known the definitions of Banach and Hilbert spaces, and that of linear maps between linear spaces. Unless otherwise stated, the linear structure of all the spaces we consider refers to the scalar field $\mathbb{R}$, and all Banach spaces we consider are assumed to be separable. If $\mathcal{X}$ is a Banach space, we denote by $\|\cdot\|_{\mathcal{X}}$ the norm in $\mathcal{X}$ and, if in addition $\mathcal{X}$ is a Hilbert space, we denote by $\langle\cdot,\cdot\rangle_{\mathcal{X}}$ the inner product in $\mathcal{X}$. Most of the material we report can be found in Dautray and Lions, [DL88, ch. VI], or Kato, [Kat95, chs. 3 and 5], to which we refer for additional details and for the proof of those results we occasionally state without giving an explicit justification.

A.2.1   Orthonormal Bases in Hilbert spaces

DEFINITION A.10   *Let $\mathcal{X}$ be a Hilbert space, and $e := (e_m)_{m\in\mathbb{N}}$ a sequence in $\mathcal{X}$.*

1. e is an ORTHONORMAL SYSTEM *in $\mathcal{X}$ if for each $j, k \in \mathbb{N}$,*

$$\langle e_j, e_k\rangle_{\mathcal{X}} = \begin{cases} 1 & \text{if } j = k \\ 0 & \text{if } j \neq k \end{cases}.$$

2. e is TOTAL *in $\mathcal{X}$, if the only $x \in \mathcal{X}$ such that $\langle x, e_j\rangle_{\mathcal{X}} = 0$ for all $j \in \mathbb{N}$ is $x = 0$.*

3. A total orthonormal system in $\mathcal{X}$ is called a TOTAL BASIS *of $\mathcal{X}$.*

PROPOSITION A.11
The elements of an orthonormal system are linearly independent. The span of a total system is dense in $\mathcal{X}$, and every nontrivial Hilbert space contains a total basis $(e_j)_{j\in\mathbb{N}}$. Relative to this basis, every $x \in \mathcal{X}$ admits the FOURIER SERIES EXPANSION

$$x = \sum_{j=0}^{\infty} \langle x, e_j\rangle_{\mathcal{X}} e_j, \tag{A.7}$$

with the series converging in $\mathcal{X}$. Moreover, the following PARSEVAL IDENTITY *holds:*

$$\|x\|_{\mathcal{X}}^2 = \sum_{j=0}^{\infty} |\langle x, e_j\rangle_{\mathcal{X}}|^2.$$

PROOF   See e.g. Dautray-Lions, [DL88, sct. VI.1.6.3]. □

A.2.2 Dual Spaces and the Hahn-Banach Theorem

1. Dual spaces.

DEFINITION A.12 *The* TOPOLOGICAL DUAL $\mathcal{X}'$ *of* $\mathcal{X}$ *is the set of all linear and continuous functions from* $\mathcal{X}$ *into* $\mathbb{R}$. *We denote by*

$$\mathcal{X}' \times \mathcal{X} \ni (x',x) \mapsto \langle x',x \rangle_{\mathcal{X}' \times \mathcal{X}} \in \mathbb{R}$$

(or, when the space $\mathcal{X}$ *is fixed and there is no danger of confusion, simply by* $\langle x',x \rangle$*) the so-called* DUALITY PAIRING *between* $\mathcal{X}$ *and its dual. That is,* $\langle x',x \rangle$ *is the value of the real number which is the image of* $x \in \mathcal{X}$ *under the linear function* $x' \in \mathcal{X}'$.

The dual space $\mathcal{X}'$ is itself a Banach space, with respect to the norm

$$\|x'\|_{\mathcal{X}'} := \sup_{\substack{x \in \mathcal{X} \\ x \neq 0}} \frac{|\langle x',x \rangle_{\mathcal{X}' \times \mathcal{X}}|}{\|x\|_{\mathcal{X}}} = \sup_{\substack{x \in \mathcal{X} \\ \|x\|_{\mathcal{X}}=1}} |\langle x',x \rangle_{\mathcal{X}' \times \mathcal{X}}| \tag{A.8}$$

(see e.g. Yosida, [Yos80, sct. IV.7]).

The following result, known as the HAHN-BANACH THEOREM, guarantees the existence of nontrivial linear functionals.

THEOREM A.13
Let $\mathcal{X}$ *be a normed linear space on* $\mathbb{R}$, $\mathcal{X}_0$ *a subspace of* $\mathcal{X}$, *and* $x_0' \in \mathcal{X}_0'$. *There exists* $x' \in \mathcal{X}'$, *such that for all* $x \in \mathcal{X}_0$,

$$\langle x',x \rangle_{\mathcal{X}' \times \mathcal{X}} = \langle x_0',x \rangle_{\mathcal{X}_0' \times \mathcal{X}_0}$$

(i.e., x' *is an extension of* x_0'*), and* $\|x'\|_{\mathcal{X}'} = \|x_0'\|_{\mathcal{X}_0'}$.

PROOF See e.g. Yosida, [Yos80, sct. IV.5]. ◻

2. Biduals. Since the dual $\mathcal{X}'$ of a Banach space $\mathcal{X}$ is itself a Banach space, we can in turn consider its dual $(\mathcal{X}')'$, which is called the BIDUAL space of $\mathcal{X}$, and denoted by $\mathcal{X}''$. The so-called "CANONICAL" INJECTION $j \colon \mathcal{X} \to \mathcal{X}''$ is defined by

$$\mathcal{X} \ni x \mapsto x'' = j(x) \quad \overset{\text{def.}}{\Longleftrightarrow} \quad \forall x' \in \mathcal{X}', \quad \langle x'',x' \rangle_{\mathcal{X}'' \times \mathcal{X}'} = \langle x',x \rangle_{\mathcal{X}' \times \mathcal{X}}. \tag{A.9}$$

The image $j(\mathcal{X})$ of $\mathcal{X}$ into $\mathcal{X}''$ is a linear subspace, in general proper, of $\mathcal{X}''$.

DEFINITION A.14 *Let* $\mathcal{X}$ *be a Banach space, and* $\mathcal{X}''$ *its bidual, defined by (A.9).* $\mathcal{X}$ *is said to be* REFLEXIVE *if* $j(\mathcal{X}) = \mathcal{X}''$. *In this case,* j *is an isomorphism.*

Thus, if $\mathcal{X}$ is reflexive, it can be identified to its bidual $\mathcal{X}''$ by the canonical isomorphism j defined in (A.9).

3. Weak and weak* convergence.

DEFINITION A.15 *Let $\mathcal{X}$ be a normed linear space on $\mathbb{R}$, and $\mathcal{X}'$ its topological dual.*

1. *A sequence $(x_m)_{m \in \mathbb{N}} \subset \mathcal{X}$ is said to* CONVERGE WEAKLY *to $x \in \mathcal{X}$ if for all $x' \in \mathcal{X}'$,*

$$\langle x', x_m \rangle \to \langle x', x \rangle \quad \text{in } \mathbb{R}.$$

2. *A sequence $(x'_m)_{m \in \mathbb{N}} \subset \mathcal{X}'$ is said to* CONVERGE WEAKLY* *to $x' \in \mathcal{X}'$ if for all $x \in \mathcal{X}$,*

$$\langle x'_m, x \rangle \to \langle x', x \rangle \quad \text{in } \mathbb{R}.$$

THEOREM A.16
Let $\mathcal{X}$ be a normed linear space on $\mathbb{R}$, and $\mathcal{X}'$ its topological dual. Then:

1. *Every weakly convergent sequence in $\mathcal{X}$ (respectively, every weakly* convergent sequence in $\mathcal{X}'$) is bounded.*

2. *If in addition $\mathcal{X}$ is reflexive, every bounded sequence in $\mathcal{X}$ (respectively, in $\mathcal{X}'$) contains a weakly (respectively, a weakly*) convergent subsequence.*

PROOF   See e.g. Yosida, [Yos80, scts. V.1, V.2].   ⬜

A.2.3 Linear Operators in Banach Spaces

In this section we review the definitions and most important results concerning various types of bounded and unbounded operators. In particular, we recall the spectral theory of linear operators with compact inverse. Again, we assume that all spaces $\mathcal{X}$, $\mathcal{Y}$, etc. we consider are at least separable Banach spaces.

1. Bounded, unbounded, and closed operators.

DEFINITION A.17   *Let $\mathcal{X}$ and $\mathcal{Y}$ be two Banach spaces, and $A \colon \mathcal{X} \to \mathcal{Y}$ be a linear operator.*

1. *A is said to be* BOUNDED, *if there is a constant K such that for all $x \in \mathcal{X}$,*

$$\|Ax\|_{\mathcal{Y}} \le K \|x\|_{\mathcal{X}}. \tag{A.10}$$

Otherwise, A is said to be UNBOUNDED.

2. *We denote the set of all linear, bounded operators $A \colon \mathcal{X} \to \mathcal{Y}$ by $\mathcal{L}(\mathcal{X}, \mathcal{Y})$.*

3. *If* $\mathcal{Y} = \mathcal{X}$, *and* $A: \mathcal{X} \to \mathcal{X}$ *is a linear operator, but not necessarily bounded, the* DOMAIN *of A is the set*

$$\mathrm{dom}(A) := \{x \in \mathcal{X} : Ax \in \mathcal{X}\}.$$

4. *An unbounded operator* $A: \mathcal{X} \to \mathcal{Y}$ *is said to be* CLOSED *if whenever* $(x_m)_{m \in \mathbb{N}} \subset \mathrm{dom}(A)$ *is a sequence such that*

$$x_m \to x \ \text{ in } \mathcal{X} \quad \text{and} \quad Ax_m \to y \ \text{ in } \mathcal{Y}$$

as $m \to +\infty$, *then* $x \in \mathrm{dom}(A)$ *and* $y = Ax$.

It is clear from (A.10) that a bounded linear operator is continuous; moreover, $\mathcal{L}(\mathcal{X}, \mathcal{Y})$ is itself a Banach space, endowed with the norm

$$\|A\|_{\mathcal{L}(\mathcal{X},\mathcal{Y})} := \sup_{\substack{x \in \mathcal{X} \\ x \neq 0}} \frac{\|Ax\|_{\mathcal{Y}}}{\|x\|_{\mathcal{X}}} = \sup_{\substack{x \in \mathcal{X} \\ \|x\|_{\mathcal{X}} = 1}} \|Ax\|_{\mathcal{Y}}. \tag{A.11}$$

Likewise, if $A: \mathcal{X} \to \mathcal{X}$ is closed, $\mathrm{dom}(A)$ is a Banach space, with respect to the graph norm defined by

$$\|u\|_{\mathrm{dom}(A)}^2 := \|u\|_{\mathcal{X}}^2 + \|Au\|_{\mathcal{X}}^2, \qquad u \in \mathrm{dom}(A).$$

(For a proof, see e.g. Dautray-Lions, [DL88, ch. VI].)

In particular, recalling definition A.12, we have that $\mathcal{X}' = \mathcal{L}(\mathcal{X}, \mathbb{R})$, and (A.8) is in accord with (A.11).

PROPOSITION A.18
Assume $A \in \mathcal{L}(\mathcal{X}, \mathcal{Y})$ *is bijective. Then* $A^{-1} \in \mathcal{L}(\mathcal{Y}, \mathcal{X})$.

PROOF   See e.g. Dautray-Lions, [DL88, sct. VI.1.2].   ☐

Finally, we recall that many results on linear operators are based on the following result, known as BAIRE'S CATEGORY LEMMA:

THEOREM A.19
Let $\mathcal{X}$ *be a complete metric space, and assume that* $(A_n)_{n \in \mathbb{N}}$ *is a sequence of nonempty, open subsets of* $\mathcal{X}$, *each dense in* $\mathcal{X}$. *Then, the set*

$$A := \bigcap_{n=0}^{\infty} A_n$$

is also dense in $\mathcal{X}$.

PROOF   See e.g. Dautray-Lions, [DL88, sct. VI.1].   ☐

2. Isometries and unitary operators.

DEFINITION A.20   *Let $\mathcal{X}$ and $\mathcal{Y}$ be Hilbert spaces, and $A: \mathcal{X} \to \mathcal{Y}$ be a linear operator. A is said to be:*

   1. An ISOMETRY, *if for all $x \in \mathcal{X}$, $\|Ax\|_{\mathcal{Y}} = \|x\|_{\mathcal{X}}$.*

   2. A UNITARY OPERATOR *if it is a surjective isometry.*

Note that an isometry is obviously injective, and the inverse of a unitary operator is unitary.

3. Resolvent, spectrum, eigenvalues and eigenvectors.

DEFINITION A.21   *Let $\mathcal{X}$ be an Hilbert space, and $A: \mathcal{X} \to \mathcal{X}$ be a linear operator, not necessarily bounded. Let I denote the identity in $\mathcal{X}$. Then:*

   1. The RESOLVENT SET *of A is the set*

$$\rho(A) := \{\lambda \in \mathbb{C}: \lambda I - A \ \text{is bijective in } \mathcal{X}\}.$$

   2. For $\lambda \in \rho(A)$, the linear operator

$$R(A, \lambda) := (\lambda I - A)^{-1}: \mathcal{X} \to \mathcal{X}$$

   is called the RESOLVENT *of A.*

   3. The POINT SPECTRUM *of A is the set*

$$\sigma_{\mathrm{p}}(A) := \{\lambda \in \mathbb{C}: \lambda I - A \ \text{is not injective in } \mathcal{X}\}.$$

   4. Each $\lambda \in \sigma_{\mathrm{p}}(A)$ is called an EIGENVALUE *of A, and each $x \in \mathrm{dom}(A) \setminus \{0\}$ such that*

$$(\lambda I - A)x = 0$$

   is called an EIGENVECTOR *of A, corresponding to the eigenvalue λ.*

4. Compact operators.

DEFINITION A.22   *Let $\mathcal{X}$ and $\mathcal{Y}$ be two Banach spaces. A bounded operator $A \in \mathcal{L}(\mathcal{X}, \mathcal{Y})$ is said to be* COMPACT, *if the image $(Ax_m)_{m \in \mathbb{N}}$ of any sequence $(x_m)_{m \in \mathbb{N}}$ bounded in $\mathcal{X}$ contains a subsequence $(Ax_{m_k})_{k \in \mathbb{N}}$, converging in $\mathcal{Y}$. We denote the set of compact operators in $\mathcal{L}(\mathcal{X}, \mathcal{Y})$ by $\mathcal{K}(\mathcal{X}, \mathcal{Y})$.*

It is easy to see that $\mathcal{K}(\mathcal{X}, \mathcal{Y})$ is a closed subspace of $\mathcal{L}(\mathcal{X}, \mathcal{Y})$, with respect to the topology of the uniform convergence of operators (i.e., with respect to the norm (A.11)).

Of particular importance are operators whose inverse are compact. More generally:

DEFINITION A.23   *Let $\mathcal{X}$ be a Hilbert space, and $A\colon \mathcal{X} \to \mathcal{X}$ be a linear operator, with nonempty resolvent set. A is said to have* COMPACT RESOLVENT, *if its resolvent $R(A,\lambda)$ is compact for all $\lambda \in \rho(A)$.*

THEOREM A.24
Let $\mathcal{X}$ be a Hilbert space, and $A\colon \mathcal{X} \to \mathcal{X}$ be a linear operator, with nonempty resolvent set. Then, A has compact resolvent if and only if the injection $j\colon \mathrm{dom}(A) \to \mathcal{X}$ is compact. Moreover, the existence of just one $\lambda \in \rho(A)$ such that $R(A,\lambda)$ is compact is sufficient to guarantee that A has compact resolvent.

PROOF   See e.g. Engel-Nagel, [EN00, sct. 2.4.d].   ▯

A.2.4   Adjoint of a Bounded Operator

1.   The notion of the ADJOINT of a bounded operator is a straightforward generalization of that of the transpose of a matrix in $\mathbb{R}^N$.

DEFINITION A.25   *Let $\mathcal{X}$ and $\mathcal{Y}$ be Banach spaces, with duals $\mathcal{X}'$ and $\mathcal{Y}'$, and let $A \in \mathcal{L}(\mathcal{X},\mathcal{Y})$. The* ADJOINT *(or* TRANSPOSE*) of A is the operator $A'\colon \mathcal{Y}' \to \mathcal{X}'$ defined by*

$$\mathcal{Y}' \ni y' \mapsto x' = A'y' \quad \overset{\mathrm{def.}}{\Longleftrightarrow} \quad \forall x \in \mathcal{X}\colon \quad \langle x',x\rangle_{\mathcal{X}'\times\mathcal{X}} = \langle y',Ax\rangle_{\mathcal{Y}'\times\mathcal{Y}}. \tag{A.12}$$

It is immediate to see that $A' \in \mathcal{L}(\mathcal{Y}',\mathcal{X}')$. Moreover, the adjoint of a compact operator is compact (see e.g. Kato, [Kat95, sct. III.4.2]).

2.   If $\mathcal{X}$ and $\mathcal{Y}$ are Hilbert spaces, and $A \in \mathcal{L}(\mathcal{X},\mathcal{Y})$, we can identify $\mathcal{X}$ and $\mathcal{Y}$ with their duals $\mathcal{X}'$ and $\mathcal{Y}'$, by means of Riesz' representation theorem (see e.g. Yosida, [Yos80, sct. III.6]). Then, we can define a linear operator $A^*\colon \mathcal{Y} \to \mathcal{X}$, by the composition

$$A^* := \rho_{\mathcal{X}}^{-1} \circ A' \circ \rho_{\mathcal{Y}}, \tag{A.13}$$

where $\rho_{\mathcal{X}}\colon \mathcal{X} \to \mathcal{X}'$ and $\rho_{\mathcal{Y}}\colon \mathcal{Y} \to \mathcal{Y}'$ are, respectively, the Riesz' isomorphisms identifying $\mathcal{X}$ and $\mathcal{Y}$ with their duals. The operator A^* defined in (A.13) is also called the ADJOINT of A. Clearly, $A^* \in \mathcal{L}(\mathcal{Y},\mathcal{X})$, and identity (A.12) reads

$$\langle A^*y,x\rangle_{\mathcal{X}} = \langle y,Ax\rangle_{\mathcal{Y}}, \tag{A.14}$$

for all $x \in \mathcal{X}$ and $y \in \mathcal{Y}$. In fact, recalling that $\rho_{\mathcal{X}}$ is defined by the identity

$$\langle \rho_{\mathcal{X}}x,y\rangle_{\mathcal{X}'\times\mathcal{X}} = \langle x,y\rangle_{\mathcal{X}}, \qquad x,y \in \mathcal{X},$$

and analogously for ρ_y, recalling (A.12) we have

$$\langle A^*y, x \rangle_{\mathcal{X}} = \langle \rho_{\mathcal{X}}^{-1} A' \rho_y y, x \rangle_{\mathcal{X}} = \langle A' \rho_y y, x \rangle_{\mathcal{X}' \times \mathcal{X}} = \langle \rho_y y, Ax \rangle_{\mathcal{Y}' \times \mathcal{Y}} = \langle y, Ax \rangle_{\mathcal{Y}}.$$

Thus, when $\mathcal{X}$ is a Hilbert space, and $\mathcal{Y} = \mathcal{X}$, it makes sense to give the following

DEFINITION A.26   *Let $\mathcal{X}$ be a Hilbert space, and $A \in \mathcal{L}(\mathcal{X}, \mathcal{X})$. The operator A is said to be* SELF-ADJOINT *if its adjoint, defined in (A.13), is such that $A^* = A$.*

Recalling (A.14), a self-adjoint operator $A \in \mathcal{L}(\mathcal{X}, \mathcal{X})$ satisfies the identity

$$\forall x, y \in \mathcal{X}: \quad \langle Ax, y \rangle_{\mathcal{X}} = \langle x, Ay \rangle_{\mathcal{X}}. \tag{A.15}$$

THEOREM A.27
Let $\mathcal{X}$ and $\mathcal{Y}$ be Hilbert spaces, and $A \in \mathcal{L}(\mathcal{X}, \mathcal{Y})$ be a unitary operator. Then its adjoint A^* satisfies the equivalent identities

$$A^*A = I_{\mathcal{X}}, \quad AA^* = I_{\mathcal{Y}}, \quad A^{-1} = A^*,$$

where $I_{\mathcal{X}}$ and $I_{\mathcal{Y}}$ are, respectively, the identity operators in $\mathcal{X}$ and $\mathcal{Y}$.

PROOF   See e.g. Kato, [Kat95, sct. 5.2]. ◻

3.   More generally, let $\mathcal{X}$ be a reflexive Banach space, $\mathcal{Y} = \mathcal{X}'$ and $A \in \mathcal{L}(\mathcal{X}, \mathcal{X}')$. Then, the transpose operator A' is in $\mathcal{L}(\mathcal{X}'', \mathcal{X}')$, and we can define an operator $A^*: \mathcal{X} \to \mathcal{X}'$, by the composition

$$A^* := A' \circ j, \tag{A.16}$$

where $j: \mathcal{X} \to \mathcal{X}''$ is the canonical injection of $\mathcal{X}$ into its bidual, defined in (A.9). This operator is also called the ADJOINT of A; recalling (A.9), this terminology is justified by the identities

$$\langle A^*x, y \rangle_{\mathcal{X}' \times \mathcal{X}} = \langle A'j(x), y \rangle_{\mathcal{X}' \times \mathcal{X}} = \langle j(x), Ay \rangle_{\mathcal{X}'' \times \mathcal{X}'} = \langle Ay, x \rangle_{\mathcal{X}' \times \mathcal{X}},$$

for all $x, y \in \mathcal{X}$. Consequently, we can give

DEFINITION A.28   *Let $\mathcal{X}$ be a reflexive Banach space, and $A \in \mathcal{L}(\mathcal{X}, \mathcal{X}')$. The operator A is said to be* SELF-ADJOINT *if its adjoint, defined in (A.16), is such that $A^* = A$.*

Recalling (A.14), a self-adjoint operator $A \in \mathcal{L}(\mathcal{X}, \mathcal{X}')$ satisfies the identity

$$\langle Ax, y \rangle_{\mathcal{X}' \times \mathcal{X}} = \langle Ay, x \rangle_{\mathcal{X}' \times \mathcal{X}}, \quad \forall x, y \in \mathcal{X}.$$

A.2.5   Adjoint of an Unbounded Operator

Let now $\mathcal{X}$ be a Hilbert space, and assume that $A\colon \mathcal{X} \to \mathcal{X}$ is a linear operator, but not necessarily bounded. To define the adjoint A^* of A, also as an unbounded operator on $\mathcal{X}$, we proceed as follows. First, we introduce the set

$$\mathcal{X}_0 := \{ v \in \mathcal{X} \colon (\exists w \in \mathcal{X} \, \forall u \in \mathrm{dom}(A) \colon \langle Au, v \rangle_{\mathcal{X}} = \langle u, w \rangle_{\mathcal{X}}) \}, \tag{A.17}$$

and note that if $v \in \mathcal{X}_0$, the map

$$\mathrm{dom}(A) \ni u \mapsto \langle Au, v \rangle_{\mathcal{X}}$$

is continuous on $\mathrm{dom}(A)$ with respect to the norm induced by $\mathcal{X}$. Indeed, this follows from the estimate

$$|\langle Au, v \rangle_{\mathcal{X}}| = |\langle u, w \rangle_{\mathcal{X}}| \leq \|w\|_{\mathcal{X}} \|u\|_{\mathcal{X}} =: C_w \|u\|_{\mathcal{X}} \, .$$

If we assume that $\mathrm{dom}(A)$ is dense in $\mathcal{X}$, the element w in (A.17) is uniquely determined by v. This justifies the following

DEFINITION A.29   *Let $\mathcal{X}$ be a Hilbert space, and A a linear operator on $\mathcal{X}$, with domain $\mathrm{dom}(A)$ dense in $\mathcal{X}$. Define $\mathcal{X}_0$ as in (A.17). The linear operator $A^*\colon \mathcal{X} \to \mathcal{X}$, with domain $\mathrm{dom}(A^*) = \mathcal{X}_0$, defined by*

$$A^* v := w, \quad v \in \mathcal{X}_0,$$

is called the ADJOINT *of A in $\mathcal{X}$.*

Note that, in general, A^* is an unbounded operator if so is A.

DEFINITION A.30   *Let $\mathcal{X}$ be a Hilbert space, and A a linear operator on $\mathcal{X}$, with domain $\mathrm{dom}(A)$ dense in $\mathcal{X}$. A is said to be:*
1) SYMMETRIC, *if $\mathrm{dom}(A) \subseteq \mathrm{dom}(A^*)$, and for all $u, v \in \mathrm{dom}(A)$,*

$$\langle Au, v \rangle_{\mathcal{X}} = \langle Av, u \rangle_{\mathcal{X}} \, . \tag{A.18}$$

(Note that the left side of (A.18) equals $\langle u, A^* v \rangle_{\mathcal{X}}$.)
2) SELF-ADJOINT *if $\mathrm{dom}(A) = \mathrm{dom}(A^*)$, and (A.18) holds for all $u, v \in \mathrm{dom}(A)$.*

A.2.6   Gelfand Triples of Hilbert Spaces

1.   We consider two Hilbert spaces $\mathcal{V}$ and $\mathcal{H}$, with $\mathcal{V} \hookrightarrow \mathcal{H}$ densely and continuously. In accord with (A.10), the continuity of the injection means that there is a constant C such that for all $u \in \mathcal{V}$,

$$\|u\|_{\mathcal{H}} \leq C \|u\|_{\mathcal{V}} \, . \tag{A.19}$$

2. Let $\mathcal{H}'$ and $\mathcal{V}'$ denote, respectively, the topological duals of $\mathcal{H}$ and $\mathcal{V}$. The adjoint j^* of the injection $j: \mathcal{V} \to \mathcal{H}$ is then a natural injection of $\mathcal{H}'$ into $\mathcal{V}'$, defined, for $h' \in \mathcal{H}'$, by the identity

$$\langle j^*h', u\rangle_{\mathcal{V}' \times \mathcal{V}} := \langle h', ju\rangle_{\mathcal{H}' \times \mathcal{H}}, \qquad u \in \mathcal{V}.$$

It follows then that j^* is injective, and the image $j^*(\mathcal{H}')$ is dense in $\mathcal{V}'$. Hence, $\mathcal{H}'$ can be identified to a dense subspace of $\mathcal{V}'$. Since $\mathcal{H}'$ can also be identified to $\mathcal{H}$, by means of Riesz' representation theorem, we finally arrive at the sequence of continuous injections

$$\mathcal{V} \hookrightarrow \mathcal{H} \equiv \mathcal{H}' \hookrightarrow \mathcal{V}'.$$

In this case, we call the three spaces $\mathcal{V}$, $\mathcal{H}$ and $\mathcal{V}'$ a GELFAND TRIPLE. Finally, we often assume that the injection $\mathcal{V} \hookrightarrow \mathcal{H}$ is not only continuous, but also compact, in the sense of definition (A.23); this means that every sequence in $\mathcal{V}$ which is bounded with respect to the norm of $\mathcal{V}$ must contain a subsequence, converging in $\mathcal{H}$.

A.2.7 Linear Operators in Gelfand Triples

1. Let now $\mathcal{V} \hookrightarrow \mathcal{H} \hookrightarrow \mathcal{V}'$ be a Gelfand triple, and $A \in \mathcal{L}(\mathcal{V}, \mathcal{V}')$. Since $\mathcal{V}$ is dense in $\mathcal{H}$, we can consider A as an unbounded operator in $\mathcal{H}$, with domain

$$\mathrm{dom}(A) := \{u \in \mathcal{V}: Au \in \mathcal{H}\}.$$

Then, the identity

$$\langle Au, v\rangle_{\mathcal{V}' \times \mathcal{V}} = \langle Au, v\rangle_{\mathcal{H}} \tag{A.20}$$

holds for all $u \in \mathrm{dom}(A)$ and $v \in \mathcal{V}$. This implies the following

PROPOSITION A.31

Let A be as described. Then, A is self-adjoint as an operator from $\mathcal{V}$ into $\mathcal{V}'$ (i.e., in the sense of definition A.28), if and only if A is self-adjoint as an operator in $\mathcal{H}$ (i.e., in the sense of definition A.30).

PROOF This is a consequence of (A.15) and (A.20), from which it follows that for all $u, v \in \mathrm{dom}(A)$,

$$\langle Au, v\rangle_{\mathcal{H}} = \langle Au, v\rangle_{\mathcal{V}' \times \mathcal{V}} = \langle Av, u\rangle_{\mathcal{V}' \times \mathcal{V}} = \langle Av, u\rangle_{\mathcal{H}}.$$

⬜

2. In the same conditions, we also have

THEOREM A.32
Assume that A is a bijection from $\mathcal{V}$ into $\mathcal{V}'$. Then A, as an operator in $\mathcal{H}$, is invertible on all of $\mathcal{H}$, and its inverse A^{-1} is linear and continuous (i.e., $A^{-1} \in \mathcal{L}(\mathcal{H},\mathcal{H})$), and self-adjoint. If in addition the injection $\mathcal{V} \hookrightarrow \mathcal{H}$ is compact, A^{-1} is a compact operator.

PROOF   We recall that A^{-1} is defined, as an operator in $\mathcal{L}(\mathcal{V}',\mathcal{V})$. Thus, for any $h \in \mathcal{H} \hookrightarrow \mathcal{V}'$, $A^{-1}h$ is defined in $\mathcal{V}$, hence in $\mathcal{H}$. The linearity of A is clear; its continuity follows from the estimate

$$\|A^{-1}h\|_{\mathcal{H}} \leq C_1 \|A^{-1}h\|_{\mathcal{V}} \leq C_1 \|A^{-1}\|_{\mathcal{L}} \|h\|_{\mathcal{V}'} \leq C_1 \|A^{-1}\|_{\mathcal{L}} C_2 \|h\|_{\mathcal{H}},$$

where C_1 and C_2 are determined, respectively, by the continuity of the injections $\mathcal{V} \hookrightarrow \mathcal{H}$ and $\mathcal{H} \hookrightarrow \mathcal{V}'$ (as in (A.19)). To show that A^{-1} is self-adjoint, let $u, v \in \mathcal{H}$. Then, recalling that A is self-adjoint,

$$\langle u, A^{-1}v \rangle_{\mathcal{H}} = \langle u, A^{-1}v \rangle_{\mathcal{V}' \times \mathcal{V}} = \langle AA^{-1}u, A^{-1}v \rangle_{\mathcal{V}' \times \mathcal{V}} = \langle A^*A^{-1}v, A^{-1}u \rangle_{\mathcal{V}' \times \mathcal{V}}$$
$$= \langle AA^{-1}v, A^{-1}u \rangle_{\mathcal{V}' \times \mathcal{V}} = \langle v, A^{-1}u \rangle_{\mathcal{V}' \times \mathcal{V}} = \langle v, A^{-1}u \rangle_{\mathcal{H}}.$$

Finally, let $(u_n)_{n\in\mathbb{N}}$ be a bounded sequence in $\mathcal{H}$. For each $n \in \mathbb{N}$, let $v_n := A^{-1}u_n$. Since the sequence $(u_n)_{n\in\mathbb{N}}$ is also bounded in $\mathcal{V}'$, the sequence $(v_n)_{n\in\mathbb{N}}$ is bounded in $\mathcal{V}$. Thus, if $\mathcal{V} \hookrightarrow \mathcal{H}$ compactly, there is a subsequence $(v_{n_k})_{k\in\mathbb{N}}$ converging in $\mathcal{H}$. Since $v_{n_k} = A^{-1}u_{n_k}$, this means that A^{-1} is compact. $\square$

3.   We now come to the question of the solvability of an abstract equation of the form

$$Au = f, \tag{A.21}$$

where $A \in \mathcal{L}(\mathcal{V},\mathcal{V}')$, and $f \in \mathcal{V}'$.

DEFINITION A.33   *Let V be a Banach space. An operator $A \in \mathcal{L}(\mathcal{V},\mathcal{V}')$ is said to be:*

1. POSITIVE, *if for all $u \in \mathcal{V}$,*

$$\langle Au, u \rangle_{\mathcal{V}' \times \mathcal{V}} \geq 0.$$

2. STRICTLY POSITIVE, *if for all $u \in \mathcal{V} \setminus \{0\}$,*

$$\langle Au, u \rangle_{\mathcal{V}' \times \mathcal{V}} > 0. \tag{A.22}$$

3. COERCIVE, *if there is a constant α such that for all $u \in \mathcal{V}$,*

$$\langle Au, u \rangle_{\mathcal{V}' \times \mathcal{V}} \geq \alpha \|u\|_{\mathcal{V}}^2. \tag{A.23}$$

Note that a coercive operator is strictly positive. Moreover, we have the following result, known as the LAX-MILGRAM LEMMA:

THEOREM A.34
Let V be a Hilbert space, and $A \in \mathcal{L}(V, V')$. If A is self-adjoint and coercive, A is an isomorphism between V and V'. Consequently, for all $f \in V'$ problem (A.21) has a unique solution $u \in V$.

PROOF   See e.g. Dautray-Lions, [DL88, sct. VI.3.2.5]. $\square$

Note that the map
$$V \times V \ni (u, v) \mapsto \langle Au, v \rangle_{V' \times V}$$
is bilinear and continuous. Thus, we can more generally consider a bilinear continuous map $a\colon V \times V \to \mathbb{R}$, and the associated problem of finding, for given $f \in V'$, a solution $u \in V$ of the problem

$$\forall v \in V: \quad a(u, v) = \langle f, v \rangle_{V' \times V}. \tag{A.24}$$

This problem, which is called the VARIATIONAL FORMULATION of problem (A.21), can be solved by means of the following version of the Lax-Milgram lemma.

THEOREM A.35
Let a be a bilinear continuous map on $V \times V$ as above, and assume that a is coercive, in the sense that (compare to (A.23)) there is $\alpha > 0$ such that for all $u \in V$,

$$a(u, u) \geq \alpha \|u\|_V^2.$$

Then for all $f \in V'$, problem (A.24) is uniquely solvable in V.

PROOF   See e.g. Dautray-Lions, [DL88, sct. VII.1]. $\square$

A.2.8 Eigenvalues of Compact Operators

The following theorem describes the structure of the point spectrum of a compact operator.

THEOREM A.36
Let X be a Hilbert space, and $L \in \mathcal{L}(X, X)$ be a self-adjoint, strictly positive, compact operator. Then:

1. *The point spectrum $\sigma_{\mathrm{p}}(L)$ of L is not empty; in fact, either $\|L\|_{\mathcal{L}(X;X)} \in \sigma_{\mathrm{p}}(L)$ or $-\|L\|_{\mathcal{L}(X;X)} \in \sigma_{\mathrm{p}}(L)$.*

2. *All eigenvalues of L are real and strictly positive, and have finite multiplicity.*

3. *These eigenvalues can be ordered into a nonincreasing sequence* $(\mu_k)_{k\in\mathbb{N}} \subset \mathbb{R}_{>0}$, *such that*

$$\lim_{k\to+\infty} \mu_k = 0.\qquad\qquad(A.25)$$

4. *Eigenvectors of L corresponding to distinct eigenvalues are orthogonal.*

5. *In fact, L admits a complete orthonormal system of eigenvectors.*

PROOF   A proof of part of this theorem can be found e.g. in Zeidler, [Zei95, sct. 4.2], except for the fact that the eigenvalues of L are at most countable and strictly positive. The countability of the eigenvalues is a consequence of the compactness of L; see e.g. Kato, [Kat95, sct. 3.6.7]. The positivity of the eigenvalues is a consequence of the strict positivity of L. Indeed, if $u \neq 0$ is an eigenvector of L corresponding to the real eigenvalue μ, from (A.22) we deduce that

$$0 < \langle Lu, u\rangle_{\mathcal{X}} = \langle \mu u, u\rangle_{\mathcal{X}} = \mu\|u\|_{\mathcal{X}}^2.$$

$\square$

From theorem A.36 we immediately deduce

THEOREM A.37
Let $\mathcal{V} \hookrightarrow \mathcal{H} \hookrightarrow \mathcal{V}'$ *be a Gelfand triple of Hilbert spaces, with the injection* $\mathcal{V} \hookrightarrow \mathcal{H}$ *compact. Let* $A \in \mathcal{L}(\mathcal{V},\mathcal{V}')$ *be a self-adjoint, strictly positive operator. Then, A admits a sequence* $(\lambda_k)_{k\in\mathbb{N}}$ *of real, strictly positive eigenvalues, each having finite multiplicity. The eigenvalues can be ordered into a nondecreasing sequence, such that*

$$\lim_{k\to+\infty} \lambda_k = +\infty,\qquad\qquad(A.26)$$

and the corresponding eigenvectors form a complete orthonormal system in $\mathcal{H}$.

PROOF   By theorem A.32, A^{-1} is, as an operator in $\mathcal{H}$, linear, continuous, self-adjoint and compact. A^{-1} is also strictly positive, since so is A. Indeed, let $u \in \mathcal{H}$, and set $v := A^{-1}u \in \mathcal{H}$. Since in fact $v \in \mathcal{V}$, and $Av = u \in \mathcal{H}$, it follows that $v \in \mathrm{dom}(A)$. Then,

$$\langle A^{-1}u, u\rangle_{\mathcal{H}} = \langle v, Av\rangle_{\mathcal{H}} > 0.$$

By theorem A.36, A^{-1} admits a system of eigenvectors $(w_j)_{j\in\mathbb{N}}$ corresponding to eigenvalues $(\mu_j)_{j\in\mathbb{N}} \in \,]0,+\infty[$, with $\mu_j \to 0$ as $j \to +\infty$. From the identities

$$A^{-1}w_j = \mu_j w_j, \qquad j \in \mathbb{N},$$

setting $\lambda_j := \frac{1}{\mu_j}$ we deduce that

$$Aw_j = \lambda_j w_j, \qquad j \in \mathbb{N}; \tag{A.27}$$

that is, each λ_j is an eigenvalue of A, relative to the same eigenvector w_j. Clearly, (A.26) follows from (A.25). The rest of the proof is immediate.   □

We remark that if in addition A is coercive, we can endow $\mathcal{V}$ with the equivalent inner product defined by

$$\mathcal{V} \times \mathcal{V} \ni (u,v) \mapsto ((u,v))_{\mathcal{V}} := \langle Au, v \rangle_{\mathcal{V}' \times \mathcal{V}}.$$

Indeed, (A.23) implies that for all $u \in \mathcal{V}$,

$$\alpha \|u\|_{\mathcal{V}}^2 \leq \langle Au, u \rangle_{\mathcal{V}' \times \mathcal{V}} \leq \|A\|_{\mathcal{L}} \|u\|_{\mathcal{V}}^2.$$

With respect to this choice of inner product, the system of eigenvectors $(w_j)_{j \in \mathbb{N}}$ is orthogonal also in $\mathcal{V}$. Indeed, for each $j, k \in \mathbb{N}$ we have

$$((w_j, w_k))_{\mathcal{V}} = \langle Aw_j, w_k \rangle_{\mathcal{V}' \times \mathcal{V}} = \lambda_j \langle w_j, w_k \rangle_{\mathcal{V}' \times \mathcal{V}} = \lambda_j \langle w_j, w_k \rangle_{\mathcal{H}}.$$

A.2.9 Fractional Powers of Positive Operators.

In this section we define the fractional powers of strictly positive, self-adjoint, compact operators in the context of a Gelfand triple $\mathcal{V} \hookrightarrow \mathcal{H} \hookrightarrow \mathcal{V}'$ of Hilbert spaces. In this case, the existence of an orthonormal system of eigenvectors allows us to define the fractional powers of the operator by means of Fourier series expansions, generalizing (A.7).

Let $A \in \mathcal{L}(\mathcal{V}, \mathcal{V}')$, and assume that all assumptions of theorem A.37 are satisfied. In particular, the injection $\mathcal{V} \hookrightarrow \mathcal{H}$ is compact. Consider the orthonormal system of eigenvectors $(w_j)_{j \in \mathbb{N}}$ of A, corresponding to the sequence of eigenvalues $(\lambda_j)_{j \in \mathbb{N}}$. For $s \in \mathbb{R}$, we can define a linear, self-adjoint, unbounded operator $A^s \colon \mathcal{H} \to \mathcal{H}$, in the following way. When $s \geq 0$, we assign as the domain of A^s the set

$$\mathrm{dom}(A^s) := \left\{ u \in \mathcal{H} \colon \sum_{j=0}^{\infty} \lambda_j^{2s} |\langle u, w_j \rangle_{\mathcal{H}}|^2 < +\infty \right\}; \tag{A.28}$$

this set is clearly a linear subspace of $\mathcal{H}$. When $s < 0$, we define $\mathrm{dom}(A^s)$ as the completion of $\mathcal{H}$ with respect to the norm

$$u \mapsto \left(\sum_{j=0}^{\infty} \lambda_j^{2s} |\langle u, w_j \rangle_{\mathcal{H}}|^2 \right)^{1/2}. \tag{A.29}$$

For $u \in \mathrm{dom}(A^s)$, $s \in \mathbb{R}$, we define

$$A^s u := \sum_{j=0}^{\infty} \lambda_j^s \langle u, w_j \rangle_{\mathcal{H}} w_j, \tag{A.30}$$

the series converging in $\mathcal{H}$.

THEOREM A.38
In the above described assumptions, for each $s \in \mathbb{R}$ the space $\mathrm{dom}(A^s)$ is a Hilbert space, with respect to the inner product defined by

$$(u,v) \mapsto \langle u,v \rangle_s := \sum_{j=0}^{\infty} \lambda_j^{2s} \langle u, w_j \rangle_{\mathcal{H}} \langle v, w_j \rangle_{\mathcal{H}}.$$

For each $s_1, s_2 \in \mathbb{R}$, with $s_1 \geq s_2$,

$$\mathrm{dom}(A^{s_1}) \hookrightarrow \mathrm{dom}(A^{s_2}) \tag{A.31}$$

densely, and with compact injection. The operator $A^{s_1-s_2}$ is an isomorphism from $\mathrm{dom}(A^{s_1})$ into $\mathrm{dom}(A^{s_2})$. In particular, if $s \geq 0$, each space $\mathrm{dom}(A^s)$ is dense in $\mathcal{H}$; $\mathrm{dom}(A^{-s}) := (\mathrm{dom}(A^s))'$ (that is, the topological dual of the domain of A^s, defined in (A.28)), and $A^{-s} = (A^s)^*$ (that is, A^{-s} is the adjoint of A^s, as introduced in definition A.29).

PROOF   Most of the claims are immediate; see also Zeidler, [Zei95, sct. 5.8]. To show the compactness of the injection (A.31), consider a sequence $(u_m)_{m \in \mathbb{N}}$, bounded in $\mathrm{dom}(A^{s_1})$. Then, there is a subsequence $(u_{m_k})_{k \in \mathbb{N}}$, converging weakly to some element u in $\mathrm{dom}(A^{s_1})$. By replacing u_{m_k} with $u_{m_k} - u$, we can assume that $u = 0$. We have to show that $u_{m_k} \to 0$ in $\mathrm{dom}(A^{s_2})$ strongly, that is (recalling the definition (A.29) of the norm in $\mathrm{dom}(A^s)$), that for all $\eta > 0$ there is $K > 0$ such that for all $k \geq K$,

$$\sum_{j=0}^{\infty} \lambda_j^{2s_2} \left| \langle u_{m_k}, w_j \rangle_{\mathcal{H}} \right|^2 \leq \eta. \tag{A.32}$$

For $j_0 \in \mathbb{N}_{>0}$ to be determined, split

$$\sum_{j=0}^{\infty} \lambda_j^{2s_2} \left| \langle u_{m_k}, w_j \rangle_{\mathcal{H}} \right|^2 = \underbrace{\sum_{j=0}^{j_0-1} \lambda_j^{2s_2} \left| \langle u_{m_k}, w_j \rangle_{\mathcal{H}} \right|^2}_{=:S_{0k}} + \underbrace{\sum_{j=j_0}^{\infty} \lambda_j^{2s_2} \left| \langle u_{m_k}, w_j \rangle_{\mathcal{H}} \right|^2}_{=:R_{0k}}. \tag{A.33}$$

Since the sequence $(u_{m_k})_{k \in \mathbb{N}}$ is bounded in $\mathrm{dom}(A^{s_1})$, we can estimate

$$R_{0k} = \sum_{j=j_0}^{\infty} \lambda_j^{-2(s_1-s_2)} \lambda_j^{2s_1} \left| \langle u_{m_k}, w_j \rangle_{\mathcal{H}} \right|^2 \leq \lambda_{j_0}^{-2(s_1-s_2)} \sum_{j=j_0}^{\infty} \lambda_j^{2s_1} \left| \langle u_{m_k}, w_j \rangle_{\mathcal{H}} \right|^2$$

$$\leq \lambda_{j_0}^{-2(s_1-s_2)} \|u_{m_k}\|_{\mathrm{dom}(A^{s_1})}^2 \leq M \lambda_{j_0}^{-2(s_1-s_2)}, \tag{A.34}$$

for suitable constant M independent of k and j. Since $\lambda_j \to +\infty$ as $j \to +\infty$, it follows from (A.34) that we can fix $j_0 \in \mathbb{N}_{>0}$ such that for all $k \in \mathbb{N}$,

$$R_{0k} \leq \tfrac{1}{2}\eta. \tag{A.35}$$

Since the subsequence $(u_{m_k})_{k \in \mathbb{N}}$ converges weakly to 0 in $\mathrm{dom}(A^{s_1})$, we have that for each $j = 0, \ldots, j_0 - 1$,

$$\langle u_{m_k}, w_j \rangle_{\mathrm{dom}(A^{s_1})} = \sum_{i=0}^{\infty} \lambda_i^{2s_1} \langle u_{m_k}, w_i \rangle_{\mathcal{H}} \langle w_i, w_j \rangle_{\mathcal{H}} = \lambda_j^{2s_1} \langle u_{m_k}, w_j \rangle_{\mathcal{H}} \to 0$$

as $k \to +\infty$. Consequently, also

$$\left| \langle u_{m_k}, w_j \rangle_{\mathcal{H}} \right|^2 \to 0,$$

so that we can determine $K > 0$ such that for all $k \geq K$,

$$\left| \langle u_{m_k}, w_j \rangle_{\mathcal{H}} \right|^2 \leq \frac{\eta}{2 j_0 \lambda_{j_0}^{s_2}}.$$

Then, for $k \geq K$ we have that

$$S_{0k} \leq \lambda_{j_0}^{2s_2} \sum_{j=0}^{j_0-1} \left| \langle u_{m_k}, w_j \rangle_{\mathcal{H}} \right|^2 \leq \lambda_{j_0}^{2s_2} \frac{\eta}{2 j_0 \lambda_{j_0}^{2s_2}} j_0 = \frac{1}{2} \eta.$$

From this and (A.35), (A.33), we deduce that (A.32) holds. This completes the proof of the compactness of the injection (A.31). $\quad\square$

In particular, for $s = 0$, $s = \frac{1}{2}$ and $s = 1$, we have that

$$\mathrm{dom}(A^0) = \mathcal{H}, \quad \mathrm{dom}(A^{1/2}) = \mathcal{V}, \quad \mathrm{dom}(A^1) = \mathrm{dom}(A), \tag{A.36}$$

as sets and as Hilbert spaces. In particular, the first of (A.36) follows from (A.7).

REMARK A.39   A definition of fractional powers of positive operators can also be given in the more general case of a self-adjoint linear operator A in a separable Hilbert space $\mathcal{H}$, with dense domain $\mathrm{dom}(A)$, under the assumption that A is COERCIVE in $\mathcal{H}$, i.e. such that (compare to definition A.33) there is $\alpha > 0$ such that for all $x \in \mathrm{dom}(A)$,

$$\langle Au, u \rangle_{\mathcal{H}} \geq \alpha \|u\|_{\mathcal{X}}^2.$$

We refer e.g. to Dautray-Lions, [DL90, sct. VIII.6]. $\quad\square$

A.2.10   Interpolation Spaces

The theory of interpolation describes the construction, starting from two normed spaces $\mathcal{X}$ and $\mathcal{Y}$, of a family of "intermediate" normed spaces $\mathcal{Z}_\theta$, parametrized by $\theta \in [0, 1]$, so that

$$\mathcal{Z}_0 := \mathcal{X} \cap \mathcal{Y} \hookrightarrow \mathcal{Z}_\theta \hookrightarrow \mathcal{Y} =: \mathcal{Z}_1. \tag{A.37}$$

These spaces have, in particular, the property that for all $x \in \mathcal{X} \cap \mathcal{Y}$,

$$\|x\|_{\mathcal{Z}_\theta} \leq C \|x\|_{\mathcal{X}}^{1-\theta} \|x\|_{\mathcal{Y}}^{\theta}, \tag{A.38}$$

with C independent of x. The spaces $\mathcal{Z}_\theta$ are called INTERPOLATION SPACES between $\mathcal{X} \cap \mathcal{Y}$ and $\mathcal{Y}$, and are denoted by

$$\mathcal{Z}_\theta = [\mathcal{X} \cap \mathcal{Y}, \mathcal{Y}]_\theta \, ;$$

inequality (A.38) is called the corresponding INTERPOLATION INEQUALITY.

For an overview of the general theory of the interpolation spaces, we refer e.g. to Bergh-Löfström, [BL76]; here, we limit ourselves to follow Lions-Magenes, [LM72, sct. 1.2], and recall one possible construction of interpolation spaces, starting from two separable Hilbert spaces $\mathcal{X}$ and $\mathcal{Y}$. Since this construction rests heavily on the notion of the fractional powers of a positive operator, we assume that $\mathcal{X} \hookrightarrow \mathcal{Y}$ densely, with continuous and compact injection; note, however, that, by remark A.39, the assumption of compactness is not required.

Given $\mathcal{X}$ and $\mathcal{Y}$ as above, we consider $\mathcal{X} \hookrightarrow \mathcal{Y} \hookrightarrow \mathcal{X}'$ as a Gelfand triple, and define a linear, unbounded operator $A \colon \mathcal{X} \to \mathcal{X}'$ as follows. First, we assign as its domain the set of all $u \in \mathcal{X}$ such that the linear map

$$\mathcal{X} \ni v \mapsto \langle u, v \rangle_{\mathcal{X}} \in \mathbb{R} \tag{A.39}$$

is continuous on $\mathcal{X}$ with respect to the (weaker) topology induced by $\mathcal{Y}$. Thus, if $u \in \mathrm{dom}(A)$, (A.39) defines a linear operator $A \colon \mathcal{X} \to \mathcal{X}'$, by

$$\langle Au, v \rangle_{\mathcal{X}' \times \mathcal{X}} = \langle u, v \rangle_{\mathcal{X}}, \qquad v \in \mathcal{X}. \tag{A.40}$$

As in section A.2.7, we can then consider A as an unbounded linear operator in $\mathcal{Y}$, with domain $\mathrm{dom}(A)$ dense in $\mathcal{Y}$. Moreover, (A.40) implies that A is self-adjoint and strictly positive; in fact, for all $u \in \mathrm{dom}(A)$,

$$\langle Au, u \rangle_{\mathcal{Y}} = \langle Au, u \rangle_{\mathcal{X}' \times \mathcal{X}} = \langle u, u \rangle_{\mathcal{X}} = \|u\|_{\mathcal{X}}^2 .$$

Consequently, for $s \geq 0$ we can define, as in section A.2.9, the fractional powers A^s of A, as linear, unbounded operators in $\mathcal{Y}$. In particular, the operator $\Lambda := A^{1/2}$ is also positive and self-adjoint; moreover, recalling (A.36), $\mathrm{dom}(\Lambda) = \mathcal{X}$ and, by (A.40), for all $u, v \in \mathcal{X}$,

$$\langle u, v \rangle_{\mathcal{X}} = \langle \Lambda u, \Lambda v \rangle_{\mathcal{Y}}. \tag{A.41}$$

We define then, for $\theta \in [0,1]$, the interpolation spaces

$$[\mathcal{X}, \mathcal{Y}]_\theta := \mathrm{dom}(\Lambda^{1-\theta}). \tag{A.42}$$

These spaces are Banach spaces, with respect to the graph norm defined by

$$\|u\|_\theta^2 := \|u\|_{\mathcal{Y}}^2 + \|\Lambda^{1-\theta} u\|_{\mathcal{Y}}^2 .$$

In particular, note that

$$[\mathcal{X}, \mathcal{Y}]_0 = \mathrm{dom}(\Lambda) = \mathcal{X}, \quad [\mathcal{X}, \mathcal{Y}]_1 = \mathrm{dom}(I_{\mathcal{Y}}) = \mathcal{Y}, \tag{A.43}$$

in accord with (A.37).

A.2.11 Differential Calculus in Banach Spaces

In this section we recall some basic results concerning homeomorphisms and diffeomorphisms between Banach spaces.

DEFINITION A.40   *Let $\mathcal{X}$ and $\mathcal{Y}$ be Banach spaces, and $\mathcal{U} \subseteq \mathcal{X}$ be open. Consider a map $F: \mathcal{U} \to \mathcal{Y}$, and let $x \in \mathcal{U}$. F is said to be (Fréchet)* DIFFERENTIABLE *at x, if there exists an operator $A \in \mathcal{L}(\mathcal{X}, \mathcal{Y})$ such that*

$$\lim_{\substack{h \in \mathcal{X} \\ h \to 0}} \frac{\|F(x+h) - F(x) - Ah\|_{\mathcal{Y}}}{\|h\|_{\mathcal{X}}} = 0.$$

It is easily seen that A is uniquely determined by x (see e.g. Ambrosetti-Prodi, [AP93, sct. 1.1]); we call A the FRÉCHET derivative of F at x, and write $A =: F'(x)$. In particular, note that if F is linear and continuous (i.e., $F \in \mathcal{L}(\mathcal{X}, \mathcal{Y})$), then F is differentiable at each $x \in \mathcal{X}$, with $F'(x) = F$ (this is because $F(x+h) - F(x) = F(h)$).

DEFINITION A.41   *In the same conditions of definition A.40, assume that F is differentiable at each $x \in \mathcal{U}$. The map*

$$\mathcal{U} \ni x \mapsto F'(x) \in \mathcal{L}(\mathcal{X}, \mathcal{Y})$$

is called the DERIVATIVE *of F. If F' is continuous, we say that F is* CONTINUOUSLY DIFFERENTIABLE *in $\mathcal{U}$, and write $F \in \mathrm{C}^1(\mathcal{U}, \mathcal{Y})$.*

DEFINITION A.42   *Let $\mathcal{X}$ and $\mathcal{Y}$ be normed linear spaces, and $\mathcal{U} \subseteq \mathcal{X}$, $\mathcal{V} \subseteq \mathcal{Y}$. A map $F: \mathcal{U} \to \mathcal{V}$ is called a* HOMEOMORPHISM, *if F is a continuous bijection, and its inverse $F^{-1}: \mathcal{V} \to \mathcal{U}$ is also continuous. If in addition both $\mathcal{X}$ and $\mathcal{Y}$ are Banach spaces and both F and F^{-1} are continuously differentiable, F is called a* DIFFEOMORPHISM.

DEFINITION A.43   *A mapping $F: \mathcal{D} \subseteq \mathcal{V} \to \mathcal{V}$ in a normed, linear space $\mathcal{V}$ is called* NORM-COERCIVE *if for any $\gamma \geq 0$ there exists a closed, bounded set $\mathcal{D}_\gamma \subseteq \mathcal{D}$ such that $\|Fy\| > \gamma$ for all $y \in \mathcal{D} \setminus \mathcal{D}_\gamma$. (Compare e.g. with Rheinboldt, [Rhe69, def. 3.6].)*

THEOREM A.44
Let $\mathcal{V}$ be a normed linear space and $G: \mathcal{D} \subseteq \mathcal{V} \to \mathcal{V}$ a sequential compact mapping on the $\mathcal{P}(\mathcal{V})$-path-connected[1] set $\mathcal{D} \subseteq \mathcal{V}$. Suppose further that $F = I - G$ is a norm-coercive local homeomorphism and that $F(\mathcal{D})$ is open. Then G is a bijection, and hence a homeomorphism, from $\mathcal{D}$ onto $\mathcal{V}$.

---

[1] In [Rhe69], $\mathcal{P}(\mathcal{V})$ denotes the set of all continuous paths in $\mathcal{V}$.

PROOF See e.g. [Rhe69, thm. 3.7] ☐

The following theorem describes a sufficient condition for a differentiable map to be a local diffeomorphism.

THEOREM A.45
Let $\mathcal{X}$ and $\mathcal{Y}$ be Banach spaces, $F \in C^1(\mathcal{X}, \mathcal{Y})$, and $x \in \mathcal{X}$ be such that $F'(x)$ is invertible (as a linear map from $\mathcal{X}$ into $\mathcal{Y}$). Then F is locally invertible at x, with a C^1 inverse. More precisely, there are neighborhoods $\mathcal{U}$ of x in $\mathcal{X}$, and $\mathcal{V}$ of $F(x)$ in $\mathcal{Y}$, such that F is a diffeomorphism from $\mathcal{U}$ into $\mathcal{V}$. Moreover, for all $y \in \mathcal{V}$, the inverse differentiation formula

$$\left(F^{-1}\right)'(y) = \left(F'(F^{-1}(y))\right)^{-1}$$

holds.

PROOF See e.g. Ambrosetti-Prodi, [AP93, sct. 2.1, thm. 1.2]. ☐

A.3 Semigroups of Linear Operators

In this section we report the most fundamental results on semigroup theory, as relevant to their applications to semilinear evolution equations of the form (4.4). Most of the material we present is taken from Pazy, [Paz83]. As before, we assume that the spaces we consider are at least separable Banach spaces.

A.3.1 General Results

1. We start by introducing the definition of semigroups depending on a real parameter.

DEFINITION A.46 *Let $S = (S(t))_{t \geq 0}$ be a family of linear, continuous operators in $\mathcal{X}$ (i.e., $S(t) \in \mathcal{L}(\mathcal{X}, \mathcal{X})$ for all $t \geq 0$).*

1. *S is said to be a* SEMIGROUP *if*

$$S(0) = I_{\mathcal{X}} \tag{A.44}$$

(the identity in $\mathcal{X}$), and for all $t, \theta \geq 0$,

$$S(t + \theta) = S(t)S(\theta) = S(\theta)S(t). \tag{A.45}$$

2. *A semigroup S is said to be* CONTINUOUS *if for each $x \in \mathcal{X}$, the map $t \mapsto S(t)x$ is continuous from $[0, +\infty[$ to $\mathcal{X}$. In this case, S is also called a C^0-semigroup.*

Given a semigroup S on $\mathcal{X}$, we can define a linear operator A on $\mathcal{X}$, in general unbounded, in the following way. First, we assign as its domain the subspace

$$\text{dom}(A) := \{x \in \mathcal{X}: \lim_{t \to 0^+} \frac{S(t)x - x}{t} =: S'_+(0)x \quad \text{exists in } \mathcal{X}\};$$

then, we define $A: \text{dom}(A) \to \mathcal{X}$ by

$$A := S'_+(0).$$

It is clear that A is a linear operator; A is called the INFINITESIMAL GENERATOR (or, more simply, the GENERATOR) of the semigroup S.

The following fundamental result, known as the HILLE-YOSIDA THEOREM, relates the properties of a C^0-semigroup and its infinitesimal generator.

THEOREM A.47

Let $A: \mathcal{X} \to \mathcal{X}$ be a linear operator, not necessarily bounded. A is the infinitesimal generator of a C^0-semigroup S, if and only if its domain $\text{dom}(A)$ is dense in $\mathcal{X}$, A is closed, the resolvent set $\rho(A)$ (recall definition A.21) contains the interval $[0, +\infty[$, and for all $\lambda > 0$,

$$\|R(A, \lambda)\|_{\mathcal{L}} \leq \frac{1}{\lambda}.$$

PROOF See e.g. Pazy, [Paz83, sct. 1.3]. $\square$

2. The definition of semigroup can be extended to families of linear continuous operators on $\mathcal{X}$, depending on a complex parameter, which necessarily varies in an additive semigroup of the complex plane $\mathbb{C}$. In particular, we will consider parameters varying in an open sector

$$\Sigma := \{z \in \mathbb{C}: \varphi_1 < \arg z < \varphi_2, \ \varphi_1 < 0 < \varphi_2\}, \tag{A.46}$$

containing the nonnegative real axis $[0, +\infty[$.

DEFINITION A.48 *Let Σ be as in (A.46), and $S = (S(z))_{z \in \Sigma}$ be a family of linear, continuous operators in $\mathcal{X}$. S is said to be an ANALYTIC SEMIGROUP if the following conditions are satisfied:*

1. *The map $z \mapsto S(z)$ is analytic from Σ to $\mathcal{L}(\mathcal{X}, \mathcal{X})$;*

2. *The semigroup properties (A.44) and (A.45) hold, i.e. if $S(0) = I_{\mathcal{X}}$ and for all $z, \zeta \in \Sigma$,*

$$S(z + \zeta) = S(z)S(\zeta) = S(\zeta)S(z);$$

3. For each $x \in \mathcal{X}$, the map $z \mapsto S(z)x$ is continuous from Σ to $\mathcal{X}$.

The following result extends to analytic semigroups the characterization of C^0-semigroups given by the Hille-Yosida theorem A.47.

THEOREM A.49

Let $A: \mathcal{X} \to \mathcal{X}$ be the infinitesimal generator of a semigroup S. Then, S is analytic (that is, S can be extended to an analytic semigroup defined on a sector Σ as in (A.46)) if and only if there are positive constants C and Λ such that for all $n \in \mathbb{N}_{>0}$ and all $\lambda > n\Lambda$,

$$\|AR(A,\lambda)^{n+1}\|_{\mathcal{L}} \leq \frac{C}{n\lambda^n} \,.$$

PROOF   See e.g. Pazy, [Paz83, sct. 2.5].   ⬜

3.   Definition A.46 can be naturally extended to that of a GROUP $S = (S(t))_{t \in \mathbb{R}}$ of linear, continuous operators on $\mathcal{X}$, by requiring that all statements in the definition be valid for all $t, \theta \in \mathbb{R}$. In particular, we have the following characterization of unitary groups (see definition A.20) in Hilbert spaces, known as STONE'S THEOREM.

THEOREM A.50

Let $\mathcal{X}$ be a Hilbert space, and $A: \mathcal{X} \to \mathcal{X}$ be a linear operator, with domain $\mathrm{dom}(A)$ dense in $\mathcal{X}$. Then, A is the generator of a unitary group S on $\mathcal{X}$ if and only if $A^* = -A$.

PROOF   See e.g. Engel-Nagel, [EN00, sct. 2.3].   ⬜

We recall that if $\mathcal{X}$ is a linear space on $\mathbb{C}$, with imaginary unit i, the condition $A^* = -A$ is equivalent to the requirement that iA be self-adjoint (because $\bar{\mathrm{i}} = -\mathrm{i}$).

A.3.2   Applications to PDEs

The theory of semigroups allows us to solve the initial value problem for evolution equations of the form (3.14), i.e.

$$U_t + AU = F(U), \tag{A.47}$$

by interpreting them as abstract ODEs in a Banach space $\mathcal{X}$.

DEFINITION A.51   *Let $A: \mathcal{X} \to \mathcal{X}$ be a linear operator, not necessarily bounded. Let $U_0 \in \mathcal{X}$, and $T > 0$.*

1. *A function $U \in C([0,T]; \mathcal{X})$ is a* MILD SOLUTION *of the initial value problem for (A.47), with initial value $U(0) = U_0$, if U satisfies the integral equation*

$$U(t) = e^{-tA}U_0 + \int_0^t e^{-(t-\theta)A} F(U(\theta)) \, d\theta, \qquad 0 \le t \le T.$$

2. *A function $U \in AC([0,T]; \mathcal{X})$ is a* STRONG SOLUTION *of the same IVP, if equation (A.47) is satisfied for almost all $t \in [0,T]$. (Recall that, by theorem A.4, U is differentiable almost everywhere in $[0,T]$.)*

THEOREM A.52

Assume that A is the generator of a C^0-semigroup on $\mathcal{X}$, and that $F: \mathcal{X} \to \mathcal{X}$ is globally Lipschitz continuous. Then for all $U_0 \in \mathcal{X}$, the initial value problem for (A.47), with initial value $U(0) = U_0$, has a unique mild solution U. Moreover, for all $T > 0$ the map

$$\mathcal{X} \ni U_0 \mapsto U \in C([0,T]; \mathcal{X})$$

is Lipschitz continuous. If in addition $U_0 \in \mathrm{dom}(A)$, then U is a strong solution.

PROOF See e.g. Pazy, [Paz83, sct. 6.1]. ⬜

Theorem A.52 can be applied directly to evolution equations of "parabolic" type, such as the heat equation (3.1), where $A = -\Delta$, which we know to generate an analytic semigroup on $L^2(\Omega)$ (see theorem A.79 below). This is not the case for the wave equation (3.4), which has first to be converted into an equivalent first order system of the form (3.14). Recalling (5.113) and (5.114), the operator A has then the matrix form

$$A = \begin{pmatrix} 0 & -\frac{1}{\varepsilon} \\ -\Delta & \frac{1}{\varepsilon} \end{pmatrix}, \tag{A.48}$$

and $\mathcal{X}$ is the product space $H_0^1(\Omega) \times L^2(\Omega)$. In this case, we resort instead to the following result:

THEOREM A.53

Let $\varepsilon > 0$, and A be as in (A.48). Then, A generates a C^0-semigroup on $\mathcal{X} = H_0^1(\Omega) \times L^2(\Omega)$.

PROOF See e.g. Pazy, [Paz83, sct. 7.2]. ⬜

A.4 Lebesgue Spaces

In this section we review the main properties of the Lebesgue spaces $L^p(\Omega)$, $1 \leq p \leq +\infty$. Here, Ω denotes an arbitrary domain of $\mathbb{R}^N$; we allow $\Omega = \mathbb{R}^N$. We consider in Ω the standard Lebesgue measure, and integration in Ω is meant in the Lebesgue sense. When we do not provide a reference for the proof of a result, such proof can be found e.g. in Adams, [Ada78, ch. 2].

A.4.1 The Spaces $L^p(\Omega)$

DEFINITION A.54   *Let $p \in [1, +\infty]$.*

1. *If $p \neq +\infty$, $L^p(\Omega)$ is the space of all equivalence classes, with respect to the equivalence relation*

$$f \sim g \iff \operatorname{meas}\{x \in \Omega : f(x) \neq g(x)\} = 0,$$

of the measurable functions $f : \Omega \to \mathbb{R}$, such that

$$\int_\Omega |f(x)|^p \, dx < +\infty.$$

2. *If $p = +\infty$, $L^\infty(\Omega)$ is the space of all equivalence classes of the measurable functions $f : \Omega \to \mathbb{R}$ which are* ESSENTIALLY BOUNDED; *that is,*

$$f \in L^\infty(\Omega) \overset{\text{def.}}{\iff} \exists M > 0 : |f(x)| \leq M \quad \text{a.e. in } \Omega. \tag{A.49}$$

As usual, with abuse of notation we identify an equivalence class with any one of its representatives, which we still call a "function".

3. *Finally, we define the space $L^p_{\text{loc}}(\Omega)$ of locally p-integrable (or, if $p = +\infty$, locally bounded) functions by*

$$f \in L^p_{\text{loc}}(\Omega) \overset{\text{def.}}{\iff} (\forall \mathcal{K} \subset \Omega, \ \mathcal{K} \text{ compact} : f|_K \in L^p(\mathcal{K})).$$

Each $L^p(\Omega)$ is a linear space, which can be endowed with the norms

$$|u|_p := \left(\int_\Omega |u(x)|^p \, dx \right)^{1/p} \qquad \text{if } 1 \leq p < +\infty, \tag{A.50}$$

$$|u|_\infty := \operatorname*{supess}_{x \in \Omega} |u(x)| := \inf\{M > 0 : (A.49) \text{ holds}\}. \tag{A.51}$$

We list the major properties of these spaces in the following theorem.

THEOREM A.55
Let $1 \leq p \leq +\infty$, and consider in $L^p(\Omega)$ the norms defined in (A.50) and (A.51). Then:

1. $L^p(\Omega)$ *is a Banach space;*

2. $L^2(\Omega)$ *is a Hilbert space, with respect to the scalar product*

$$\langle f,g \rangle := \int_\Omega f(x)g(x)\,dx \quad \text{or} \quad \langle f,g \rangle := \int_\Omega f(x)\overline{g(x)}\,dx, \qquad (A.52)$$

   depending respectively on whether the underlying scalar field is $\mathbb{R}$, *or* $\mathbb{C}$;

3. *If* $1 \le p < +\infty$, *the space* $C_0^\infty(\Omega)$ *of the infinitely differentiable functions with support compact in* Ω *is dense in* $L^p(\Omega)$;

4. *If* $1 \le p < +\infty$, $L^p(\Omega)$ *is separable, while* $L^\infty(\Omega)$ *is not;*

5. *If* $1 < p < +\infty$, $L^p(\Omega)$ *is reflexive; neither* $L^1(\Omega)$ *nor* $L^\infty(\Omega)$ *are reflexive.*

We can also give a definition of the spaces $L^p(\Gamma)$, where Γ is a sufficiently smooth $(N-1)$-dimensional submanifold of $\mathbb{R}^N$ (see e.g. definition 5.1 of chapter 5). This is done in a natural way, by means of charts of local coordinates, which are required to have an image in $L^p(\mathbb{R}^N)$.

A.4.2 Inequalities

We denote by p, q etc. generic numbers in $[1,+\infty]$, set $\frac{1}{\infty} := 0$, and call p and q CONJUGATE INDICES if

$$\frac{1}{p} + \frac{1}{q} = 1.$$

In particular, the pairs $(p,q) = (2,2)$, $(p,q) = (1,+\infty)$, and $(p,q) = (+\infty,1)$ are conjugate indices.

PROPOSITION A.56

Let p, $q \in\,]1,+\infty[$ *be conjugate indices. Then,* YOUNG'S INEQUALITY

$$ab \le \frac{1}{p}a^p + \frac{1}{q}b^q \qquad (A.53)$$

holds, for all a, $b \ge 0$. *More generally, for all* $\eta > 0$ *there is* $C_\eta > 0$ *such that*

$$ab \le \eta a^p + C_\eta b^q. \qquad (A.54)$$

In particular, for $p = q = 2$, *(A.53) and (A.54) read*

$$ab \le \frac{1}{2}a^2 + \frac{1}{2}b^2, \quad ab \le \eta a^2 + \frac{1}{4\eta}b^2.$$

THEOREM A.57

Let p, $q \in [1,+\infty]$ *be conjugate indices. Then for all* $u \in L^p(\Omega)$ *and* $v \in L^q(\Omega)$, *the product* $uv \in L^1(\Omega)$, *and* HÖLDERS'S INEQUALITY

$$|uv|_1 \le |u|_p |v|_q \qquad (A.55)$$

holds. More generally, for all $\eta > 0$ there is $C_\eta > 0$ such that

$$|uv|_1 \leq \eta |u|_p^p + C_\eta |v|_q^q .$$

As a consequence, MINKOWSKI'S INEQUALITY *holds, for $u,\ v \in L^p(\Omega)$:*

$$|u+v|_p \leq |u|_p + |v|_p .$$

For functions in the Hilbert space $L^2(\Omega)$, Hölder's inequality (A.55) reads

$$\langle f,g \rangle \leq |f|_2 |g|_2 ,$$

and is usually known as SCHWARZ' INEQUALITY.

Hölder's inequality (A.55) can be generalized to the product of any finite number of functions.

THEOREM A.58
Let r and $p_1,\dots,p_k \in [1,+\infty]$ be such that

$$\sum_{j=1}^{k} \frac{1}{p_j} = \frac{1}{r}. \tag{A.56}$$

Assume $f_j \in L^{p_j}(\Omega)$, for $1 \leq j \leq k$. Then the product $f_1 \cdots f_k$ is in $L^r(\Omega)$, and satisfies the estimate

$$|f_1 \cdots f_k|_r \leq |f_1|_{p_1} \cdots |f_k|_{p_k}. \tag{A.57}$$

PROOF   The result can be easily proven by induction on k.   ◻

A.4.3   Other Properties of the Spaces $L^p(\Omega)$

1. The Dual of $L^p(\Omega)$.

THEOREM A.59
Let $p \in [1,+\infty[$, and q be its conjugate index. Then, the topological dual of $L^p(\Omega)$ is isometrically isomorphic to the space $L^q(\Omega)$. Consequently, if $p \in\,]1,+\infty[$ and $(f_m)_{m\in\mathbb{N}} \subset L^p(\Omega)$, then $f_m \to f$ weakly in $L^p(\Omega)$ if and only if for all $g \in L^q(\Omega)$,

$$\int_\Omega f_m(x)g(x)\,\mathrm{d}x \;\to\; \int_\Omega f(x)g(x)\,\mathrm{d}x .$$

Analogously, $f_m \to f$ weakly* in $L^\infty(\Omega)$ if and only if for all $g \in L^1(\Omega)$,

$$\int_\Omega f_m(x)g(x)\,\mathrm{d}x \;\to\; \int_\Omega f(x)g(x)\,\mathrm{d}x .$$

2. $(L^p(\Omega))_{p \geq 1}$ **as a Family of Interpolation Spaces.** If $p < q < r$, it is natural to expect the space $L^q(\Omega)$ to be an "intermediate" space between $L^r(\Omega)$ and $L^p(\Omega)$. This is indeed the case if Ω has finite measure, in which case actually $L^r(\Omega) \hookrightarrow L^q(\Omega) \hookrightarrow L^p(\Omega)$.

PROPOSITION A.60

Let Ω have finite measure $|\Omega|$. If $1 \leq p \leq q \leq +\infty$, then $L^q(\Omega) \hookrightarrow L^p(\Omega)$, with continuous imbedding. More precisely, for all $u \in L^q(\Omega)$,

$$|u|_p \leq |\Omega|^{\frac{1}{p} - \frac{1}{q}} |u|_q .$$

More generally, we can characterize the spaces $L^p(\Omega)$ as a family of interpolation spaces (see section A.2.10).

THEOREM A.61

Let $p, r \in [1, +\infty]$, with $p \leq r$. For $0 \leq \lambda \leq 1$, define q by

$$\frac{1}{q} = \frac{\lambda}{p} + \frac{1 - \lambda}{r} . \tag{A.58}$$

Then $q \in [p, r]$, with $q = p$ if $\lambda = 1$ and $q = r$ if $\lambda = 0$, and

$$L^q(\Omega) = [L^r(\Omega) \cap L^p(\Omega), L^p(\Omega)]_\lambda .$$

That is, if $u \in L^p(\Omega) \cap L^r(\Omega)$, then $u \in L^q(\Omega)$, and satisfies the interpolation inequality

$$|u|_q \leq |u|_p^\lambda |u|_r^{1-\lambda} . \tag{A.59}$$

PROOF See Bergh-Löfström, [BL76, ch. 5]. Note that (A.59) is a consequence of Hölder's inequality: indeed, letting $\alpha = \lambda q$ and $\beta = (1 - \lambda)q$, we easily verify that (A.58) implies that $a = \frac{p}{\alpha}$ and $b = \frac{r}{\beta}$ are conjugate indices. $\qquad \Box$

A.5   Sobolev Spaces of Scalar Valued Functions

In this section we review the main properties of the Sobolev spaces $H^s(\Omega)$, $s \in \mathbb{R}$. Here, we assume that either $\Omega = \mathbb{R}^N$, or that Ω is bounded. In the latter case, we assume that its boundary $\partial\Omega$ is sufficiently smooth; more precisely, that $\partial\Omega$ is an $(N - 1)$-dimensional submanifold of $\mathbb{R}^N$, of class C^m, for suitable $m \geq 0$. As in section A.4 we consider in Ω the Lebesgue measure, and integration is meant in the Lebesgue sense. Proof of the results for which we do not provide a reference can be found e.g. in Adams, [Ada78, chs. 3-6], or in Lions-Magenes, [LM72, ch. 1].

A.5.1 Distributions in Ω

The linear space $C_0^\infty(\Omega)$ of the infinitely differentiable functions with compact support in Ω can be endowed with a locally convex (but not metrizable) topology (see e.g. Rudin, [Rud73, ch. 6]). The corresponding topological space is denoted by $D(\Omega)$, and called the space of TEST FUNCTIONS. The space of linear continuous functionals on $D(\Omega)$, called DISTRIBUTIONS, is denoted by $D'(\Omega)$. We also denote by $\langle \cdot, \cdot \rangle_{D(\Omega)}$ the duality product between $D'(\Omega)$ and $D(\Omega)$.

Given $\alpha := (\alpha_1, \ldots, \alpha_N) \in \mathbb{N}^N$, we call α a MULTIINDEX, define its LENGTH as $|\alpha| := \alpha_1 + \cdots + \alpha_N$, and set

$$D^\alpha := \frac{\partial^{|\alpha|}}{\partial x_1^{\alpha_1} \cdots \partial x_N^{\alpha_N}}.$$

Any function $f \in L^2(\Omega)$ defines a distribution F, by

$$\langle F, \varphi \rangle_{D(\Omega)} := \int_\Omega f(x)\varphi(x)\,dx; \qquad (A.60)$$

to denote the dependence of F on f, we write $F := T_f$. The converse, however, is not true in general; for example, for the so-called DIRAC δ distribution, defined by

$$\langle \delta, \varphi \rangle_{D(\Omega)} := \varphi(0),$$

there is no $f \in L^2(\Omega)$ such that $\delta = T_f$. This fact motivates the following definition.

DEFINITION A.62 *A distribution F is* REGULAR, *if there is $f \in L^2(\Omega)$ such that $F = T_f$.*

The importance of distributions in the theory of PDEs resides in the fact that any distribution $F \in D'(\Omega)$ can be differentiated in distributional sense. That is, given any multiindex α, we can define a new distribution $D^\alpha F$, called the DISTRIBUTIONAL DERIVATIVE of F of order α, by the identity

$$\langle D^\alpha F, \varphi \rangle_{D(\Omega)} := (-1)^{|\alpha|} \langle F, D^\alpha \varphi \rangle_{D(\Omega)}, \qquad \varphi \in D(\Omega). \qquad (A.61)$$

This definition is natural, in the sense that if a distribution F is regular, i.e. $F = T_f$, and $f \in C^m(\overline{\Omega})$, then we can integrate the right side of (A.61) by parts, and deduce that $D^\alpha F = T_{D^\alpha f}$ for all α with $|\alpha| \le m$.

A.5.2 The Spaces $H^m(\Omega)$, $m \in \mathbb{N}$

The spaces $H^m(\Omega)$, $m \in \mathbb{N}$, consist of subspaces of $L^2(\Omega)$, whose functions possess weak derivatives of all orders up to m. More precisely, note that, given any function $f \in L^2(\Omega)$, the distribution T_f can be differentiated in distributional sense, as in (A.61). Consequently, we can give

DEFINITION A.63   *Let $f \in L^2(\Omega)$, and $\alpha \in \mathbb{N}^N$. We say that f has a WEAK DERIVATIVE $f_\alpha \in L^2(\Omega)$ if the distribution $D^\alpha T_f$ is regular; that is, if there is $f_\alpha \in L^2(\Omega)$ such that $D^\alpha T_f = T_{f_\alpha}$. In this case, with abuse of notation we write that $D^\alpha T_f \in L^2(\Omega)$.*

This definition A.63 makes sense, since it is easily seen that f_α is uniquely determined (up to a set of measure zero in Ω) by the distribution $D^\alpha T_f$. Recalling (A.61) and (A.60), definition A.63 means that $f_\alpha \in L^2(\Omega)$ is the weak derivative of order α of a function $f \in L^2(\Omega)$ if for all $\varphi \in \mathrm{dom}(\Omega)$,

$$\int_\Omega f_\alpha(x)\varphi(x)\,dx = (-1)^{|\alpha|} \int_\Omega f(x)(D^\alpha \varphi)(x)\,dx. \tag{A.62}$$

Note that (A.62) generalizes the classical integration by parts formula, which would hold if $f \in C^{|\alpha|}(\Omega)$; indeed, in this case $f_\alpha = D^\alpha f$, and the vanishing of φ in a neighborhood of $\partial\Omega$ causes the absence of the boundary terms in (A.62).

DEFINITION A.64   *Let $m \in \mathbb{N}$. We set*

$$\mathrm{H}^m(\Omega) := \{ u \in L^2(\Omega) \colon (\forall \alpha \in \mathbb{N}^N, |\alpha| \le m \colon D^\alpha T_u \in L^2(\Omega)) \}.$$

With abuse of notation, we set $D^\alpha T_u =: D^\alpha u$.

Thus, if $m \ge 1$, functions in $\mathrm{H}^m(\Omega)$ have weak derivatives of order up to m in $L^2(\Omega)$. This notion of differentiability, however, obviously does not coincide with the classical one. In particular, neither of the spaces $\mathrm{H}^m(\Omega)$ and $C^m(\Omega)$ is in general contained in the other.

Each $\mathrm{H}^m(\Omega)$ is a linear space, which can be endowed with the norm

$$\|u\|_m := \left(\sum_{|\alpha| \le m} \int_\Omega |D^\alpha u(x)|^2 \, dx \right)^{1/2}.$$

In particular, $\mathrm{H}^0(\Omega) = L^2(\Omega)$ (as a set and as a normed space).

We list the major properties of these spaces in the following theorem.

THEOREM A.65
For each $m \in \mathbb{N}$, $\mathrm{H}^m(\Omega)$ is a separable Banach space, in which the subspace $C^\infty(\Omega) \cap \mathrm{H}^m(\Omega)$ is dense. In fact, $\mathrm{H}^m(\Omega)$ is a Hilbert space, with respect to the scalar product

$$\langle f, g \rangle_m := \sum_{|\alpha| \le m} \int_\Omega (D^\alpha f)(x)(D^\alpha g)(x)\,dx$$

(with modification analogous to the second of (A.52) if the underlying scalar field is $\mathbb{C}$).

REMARK A.66   In general, it is not true that $C_0^\infty(\Omega)$ is dense in $\mathrm{H}^s(\Omega)$; this does hold, however, if $m = 0$, or if $\Omega = \mathbb{R}^N$. $\square$

A.5.3   The Spaces $H^s(\Omega)$, $s \in \mathbb{R}_{\geq 0}$

The spaces $H^s(\Omega)$, for general $s \in [0, +\infty[$, can be defined by interpolation. More precisely, given $s \geq 0$, we choose $m \in \mathbb{N}$ such that $0 \leq s \leq m$, and set

$$H^s(\Omega) := [H^m(\Omega), L^2(\Omega)]_{1-s/m}. \tag{A.63}$$

The following result shows that this definition is consistent.

THEOREM A.67
The space $H^s(\Omega)$ defined in (A.63) is independent, up to norm equivalence, of the choice of $m \geq s$. Moreover, $H^s(\Omega)$ can be defined by interpolation between consecutive integers; that is, setting $m := \lfloor s \rfloor$ (the integer part of s), again up to norm equivalence we have that

$$H^s(\Omega) := [H^{m+1}(\Omega), H^m(\Omega)]_{m+1-s}. \tag{A.64}$$

PROOF   See e.g. Lions-Magenes, [LM72, sct. 1.9].   ⬜

REMARK A.68   When $s \in \mathbb{N}$, definition (A.64) is in accord with the definition given in A.64. Indeed, if $s \in \mathbb{N}$, then $m = s$, so that, by (A.43),

$$H^s(\Omega) = [H^{m+1}(\Omega), H^m(\Omega)]_1 = H^m(\Omega).$$

⬜

As a consequence of general results of interpolation theory, the spaces $H^s(\Omega)$ inherit the same properties listed in theorem A.65. The following theorem gives some additional properties of the spaces $H^s(\Omega)$.

THEOREM A.69
For $s \geq 0$, let $H^s(\Omega)$ be defined as in (A.63). Then:

1. $H^s(\Omega)$ **as interpolation spaces.**   *For s_1 and s_2 such that $0 \leq s_2 \leq s_1$, and $\theta \in [0, 1]$, we have*

$$[H^{s_1}(\Omega), H^{s_2}(\Omega)]_\theta = H^{(1-\theta)s_1 + \theta s_2}.$$

2. **Imbeddings of $H^s(\Omega)$.**   *The following imbeddings are continuous:*

   2.1.   If $2s < N$, $H^s(\Omega) \hookrightarrow L^q(\Omega)$, for $2 \leq q \leq \frac{2N}{N-2s}$.

   2.2.   If $2s = N$, $H^s(\Omega) \hookrightarrow L^q(\Omega)$, for $2 \leq q < +\infty$.

   2.3.   If $2s > N$, $H^s(\Omega) \hookrightarrow C_b(\overline{\Omega})$, where $C_b(\overline{\Omega}) := C(\overline{\Omega})$ if Ω is bounded, while $C_b(\mathbb{R}^N)$ denotes the subspace of the functions in $C(\mathbb{R}^N)$ which are bounded.

If in addition Ω is bounded and $s > 0$, the imbedding $H^s(\Omega) \hookrightarrow H^{s-\varepsilon}(\Omega)$ is compact for all $\varepsilon \in]0, s]$.

3. **$H^s(\Omega)$ as an algebra.** *If $2s > N$, $H^s(\Omega)$ is an algebra; that is, if f and $g \in H^s(\Omega)$, their pointwise product $f \cdot g$ (which makes sense, because f and g are continuous, by part 2.2 above) is also in $H^s(\Omega)$.*

PROOF   In addition to Adams, [Ada78, chs. 3-6], or in Lions-Magenes, [LM72, ch. 1]. For the imbeddings when s is not an integer, see also Triebel, [Tri95, scts. 2.3, 2.6], or Peetre, [Pee66].   ▯

The imbeddings described in theorem A.69 are a consequence of the following more general result, due to Gagliardo and Nirenberg.

THEOREM A.70
Let $m \in \mathbb{N}$, $p, r \in [1, +\infty]$, and $u \in H^m(\Omega) \cap L^r(\Omega)$. For integer $j \le m$, and $\theta \in [\frac{j}{m}, 1]$ (with the exception $\theta \ne 1$ if $m - j - \frac{N}{2} \in \mathbb{N}$), define q by

$$\frac{1}{q} = \frac{j}{N} + \theta \left(\frac{1}{2} - \frac{m}{N} \right) + \frac{1}{r}(1 - \theta). \tag{A.65}$$

Then for any $\gamma \in \mathbb{N}^N$, with $|\gamma| = j$, $D^\gamma u \in L^q(\Omega)$, and satisfies the GAGLIARDO-NIRENBERG *inequality*

$$\|D^\gamma u\|_{L^q(\Omega)} \le C \sum_{|\alpha|=m} \|D^\alpha u\|_{L^2(\Omega)}^\theta \|u\|_{L^r(\Omega)}^{1-\theta} + C_1 \|u\|_{L^s(\Omega)}, \tag{A.66}$$

where $s := \max(2, r)$, and $C > 0$, $C_1 \ge 0$ are independent of u. The choice $C_1 = 0$ is admissible if $\Omega = \mathbb{R}^N$.

PROOF   See e.g. Racke, [Rac92, ch. 4].   ▯

REMARK A.71   1. The choice $j = 0$, $\theta = 1$ in (A.65) yields the original Sobolev imbedding $H^m(\Omega) \hookrightarrow L^q(\Omega)$, with $q = \frac{2N}{N-2m}$, described in part (2.1) of theorem A.69 for integer $s = m$. The choice of the lowest possible value $\theta = \frac{j}{m}$ in (A.65) yields the dimensionless version of estimate (A.66)

$$\|D^\gamma u\|_{L^q(\Omega)} \le C \sum_{|\alpha|=m} \|D^\alpha u\|_{L^2(\Omega)}^{j/m} \|u\|_{L^r(\Omega)}^{1-j/m} + C_1 \|u\|_{L^s(\Omega)},$$

with $|\gamma| = j$ and $q \ge 1$ defined by

$$\frac{1}{q} = \frac{j}{2m} + \frac{1}{r} \left(1 - \frac{j}{m} \right).$$

2. Unless u satisfies some extra conditions, such as having vanishing "trace" at $\partial\Omega$ (see section A.5.4 below), it is not in general possible to choose $C_1 = 0$ in (A.66) if $\Omega \neq \mathbb{R}^N$. To see this, it is sufficient to consider the example $u(t) = t$ in $\Omega =]0,1[\subset \mathbb{R}$, with $m = 2$, $j = 1$, $\theta = \frac{1}{2}$ and $r = q = 2$: if (A.66) held with $C_1 = 0$, we would deduce the contradiction $u'(t) \equiv 0$.   ☐

We recall that when $\Omega = \mathbb{R}^N$, the spaces $H^s(\mathbb{R}^N)$ can be defined in an alternative way, by means of the Fourier transform. The two definitions coincide, up to isomorphisms. In particular, we can then define, by means of charts of local coordinates, the Sobolev spaces $H^s(\Gamma)$, where Γ is a $(N-1)$-dimensional submanifold of $\mathbb{R}^N$ of class C^m, $m \geq s$.

A.5.4   The Spaces $H_0^s(\Omega)$, $s \in \mathbb{R}_{\geq 0}$, and $H^s(\Omega)$, $s \in \mathbb{R}_{<0}$

We now assume that Ω is bounded, and that its boundary is, as already stated, a $(N-1)$-dimensional submanifold of $\mathbb{R}^N$ of class C^m, $m \geq 0$. Given a function $f \in C^m(\overline{\Omega})$, we can then define its TRACES at $\partial\Omega$

$$\gamma_j f := \frac{\partial^j f}{\partial \nu^j}, \qquad 0 \leq j \leq m, \tag{A.67}$$

where ν denotes the unit outward normal vector to $\partial\Omega$. If $m \geq 1$ and $s \in]\frac{1}{2}, m]$, we can extend the definition of these traces to functions in $H^s(\Omega)$, by means of a density argument.

THEOREM A.72
Let Ω be as stated above, and $0 \leq s \leq m$. The space $C^m(\overline{\Omega})$ is dense in $H^s(\Omega)$.

THEOREM A.73
Let Ω be as stated above, and assume that $m \geq s > \frac{1}{2}$. Let m_0 be the largest integer such that $0 \leq m_0 < s - \frac{1}{2}$. Then for $0 \leq j \leq m_0$, the TRACE OPERATOR γ_j defined in (A.67) extends by continuity to a linear, continuous map from $H^s(\Omega)$ to $H^{s-j-1/2}(\partial\Omega)$. The map

$$\tilde{\gamma} : H^s(\Omega) \mapsto \prod_{j=0}^{m_0} H^{s-j-1/2}(\partial\Omega) \tag{A.68}$$

is surjective, and has a continuous inverse.

PROOF   See e.g. Lions - Magenes, [LM72, sct. 1.9].   ☐

In particular, functions in $H^1(\Omega)$ have a trace in $H^{1/2}(\partial\Omega)$.

For $s \geq 0$ we define the subspace $H_0^s(\Omega)$ of $H^s(\Omega)$ as the closure of $C_0^\infty(\Omega)$ in $H^s(\Omega)$.

Since the map $\tilde{\gamma}$ defined in (A.68) is surjective, and vanishes on $C_0^\infty(\Omega)$, it follows that this space cannot be dense in $H^s(\Omega)$ if $s > \frac{1}{2}$. Hence, in this case, $H_0^s(\Omega)$ is a proper subspace of $H^s(\Omega)$ (unless $\Omega = \mathbb{R}^N$).

THEOREM A.74
The space $C_0^\infty(\Omega)$ is dense in $H^s(\Omega)$ if and only if $s \leq \frac{1}{2}$, in which case, $H_0^s(\Omega) = H^s(\Omega)$. If $s > \frac{1}{2}$,

$$H_0^s(\Omega) = \{u \in H^s(\Omega): \gamma_j(u) = 0,\ 1 \leq j < s - \tfrac{1}{2}\}.$$

Furthermore, if $s - \frac{1}{2} \notin \mathbb{N}$, and $m \in \mathbb{N}$ is such that $m \geq s$,

$$H_0^s(\Omega) = [H_0^m(\Omega), L^2(\Omega)]_{1-s/m}. \tag{A.69}$$

PROOF   See e.g. Lions-Magenes, [LM72, ch. 11]; in particular, (A.69) is contained in their theorem 11.6, since $H_0^0(\Omega) = L^2(\Omega)$ by the first claim of theorem A.74.   □

Finally, for $s < 0$ we set

$$H^s(\Omega) := \left(H_0^{-s}(\Omega)\right)'$$

(topological dual). Thus, in particular, $H^{-1}(\Omega)$ is defined to be the dual of $H_0^1(\Omega)$.
We conclude with the following interpolation theorem.

THEOREM A.75
Let $s_1, s_2 \geq 0$, with $s_2 \neq m + \frac{1}{2}$, $m \in \mathbb{N}$. Then, for $\theta \in [0,1]$,

$$[H^{s_1}(\Omega), H^{-s_2}(\Omega)]_\theta = H^{(1-\theta)s_1 - \theta s_2}(\Omega),$$

provided that $(1-\theta)s_1 - \theta s_2 \neq n + \frac{1}{2}$, $n \in \mathbb{N}$.

PROOF   See e.g. Lions-Magenes, [LM72, sct. 1.12.4].   □

A.5.5 The Laplace Operator

In chapter 3, we have considered the standard example of the Gelfand triple consisting of the Sobolev spaces

$$\mathcal{V} := H_0^1(\Omega) \hookrightarrow \mathcal{H} := L^2(\Omega) \hookrightarrow \mathcal{V}' = H^{-1}(\Omega),$$

with $\Omega \subset \mathbb{R}^n$ a bounded domain. In particular, the boundedness of Ω implies that the injection $H_0^1 \hookrightarrow L^2(\Omega)$ is compact (by part 2 of theorem A.69). In the sequel, we denote as usual by $\|\cdot\|_m$ the norm in $H^m(\Omega)$, and abbreviate $\|\cdot\|_0 = \|\cdot\|$.

1. The operator $A := -\Delta$ is clearly in $\mathcal{L}(\mathcal{V}, \mathcal{V}')$; as such, A is self-adjoint and strictly positive, as we see from the identities

$$\langle -\Delta u, v \rangle_{\mathcal{V}' \times \mathcal{V}} = \langle \nabla u, \nabla v \rangle_{\mathcal{H}} = \langle -\Delta v, u \rangle_{\mathcal{V}' \times \mathcal{V}},$$
$$\langle -\Delta u, u \rangle_{\mathcal{V}' \times \mathcal{V}} = \langle \nabla u, \nabla u \rangle_{\mathcal{H}} = \|\nabla u\|^2. \tag{A.70}$$

Moreover, by theorem A.32, A has compact inverse. Consequently, from theorem A.37 we deduce

THEOREM A.76
The operator $A = -\Delta$ admits an unbounded sequence of positive eigenvalues $(\lambda_j)_{j \in \mathbb{N}}$. This sequence can be ordered so that

$$0 < \lambda_1 < \lambda_2 \leq \cdots \leq \lambda_j \leq \cdots, \qquad \lambda_j \to +\infty. \tag{A.71}$$

For each $j \in \mathbb{N}$, the corresponding eigenvector w_j is in $C^\infty(\Omega) \cap C(\overline{\Omega})$, and the sequence $(w_j)_{j \in \mathbb{N}}$ is a complete orthogonal system in $L^2(\Omega)$.

2. We now show that and that we can choose in $\mathcal{V} = H_0^1(\Omega)$ the norm

$$\|u\|_{\mathcal{V}} := \|\nabla u\|, \qquad u \in H_0^1(\Omega). \tag{A.72}$$

Formally, this is a consequence of (A.27). Indeed, given $u \in \mathcal{V}$, if we consider its Fourier series expansion (A.7), in terms of the total basis of the eigenvectors of $-\Delta$, recalling (A.71) and that these eigenvectors are orthonormal in $L^2(\Omega)$, we have that

$$\|\nabla u\|^2 = \langle -\Delta u, u \rangle = \left\langle -\Delta \sum_{j=0}^{\infty} \alpha_j w_j, \sum_{k=0}^{\infty} \alpha_k w_k \right\rangle = \left\langle \sum_{j=0}^{\infty} \lambda_j \alpha_j w_j, \sum_{k=0}^{\infty} \alpha_k w_k \right\rangle$$
$$= \sum_{j=0}^{\infty} \lambda_j \alpha_j^2 \geq \lambda_0 \sum_{j=0}^{\infty} \alpha_j^2 = \lambda_0 \|u\|^2.$$

That is, for all $u \in H_0^1(\Omega)$,

$$\|\nabla u\|^2 \geq \lambda_0 \|u\|^2, \tag{A.73}$$

which is known as POINCARÉ'S INEQUALITY.

It follows that $-\Delta$ is well defined and bijective from $\mathcal{V}$ into $\mathcal{V}'$. This is a consequence of Lax-Milgram's theorem A.34, since A is self-adjoint, positive and coercive, as we see from (A.70).

3. As an operator in $L^2(\Omega)$, $-\Delta$ has domain

$$\text{dom}(A) = H^2(\Omega) \cap H_0^1(\Omega);$$

to see this, we need to recall the following elliptic regularity result.

THEOREM A.77

Consider the elliptic boundary value problem

$$\begin{cases} -\Delta u = f & \text{in } \Omega, \\ \quad u = 0 & \text{on } \partial\Omega. \end{cases} \tag{A.74}$$

Let $f \in H^{-1}(\Omega)$, and $u \in H_0^1(\Omega)$ be the corresponding unique solution of (A.74), whose existence is assured by the Lax-Milgram theorem. If in addition $f \in L^2(\Omega)$, then $u \in H^2(\Omega)$, and there is $C > 0$, independent of u, such that

$$\|u\|_2 \leq C(\|\Delta u\| + \|u\|).$$

PROOF See e.g. Gilbarg-Trudinger, [GT83, sct. 8.4]. $\quad\square$

In conclusion, theorems A.37 and A.38 are applicable, and we can thus define the fractional powers $(-\Delta)^s$, for all $s \in \mathbb{R}$.

4. We now show that, if for $s \geq 0$ we set

$$\tilde{H}^s(\Omega) := \operatorname{dom}((-\Delta)^{s/2}), \tag{A.75}$$

then

THEOREM A.78

For all $s \geq 0$, such that $s - \frac{1}{2} \notin \mathbb{N}$, $\tilde{H}^s(\Omega) = H_0^s(\Omega)$.

PROOF Given $s \geq 0$, fix $m \in \mathbb{N}$, with $m \geq s$, and consider the operator

$$A_m := (-\Delta)^m \colon H_0^m(\Omega) \to L^2(\Omega),$$

as an unbounded operator with domain $H^{2m}(\Omega) \cap H_0^m(\Omega)$. We do this in the same way as in theorem A.77 (which in fact yields the case $m = 1$); that is, we consider the elliptic boundary value problem

$$\begin{cases} (-\Delta)^m u = f & \text{in } \Omega, \\ \dfrac{\partial^j u}{\partial v^j} = 0 & \text{on } \partial\Omega, \; 0 \leq j \leq m - 1. \end{cases} \tag{A.76}$$

Since A_m is an elliptic operator which is also uniformly strongly elliptic (see e.g. Dautray-Lions, [DL88, sct. VII.5]), given $f \in H^{-m}(\Omega)$, by Lax-Milgram's theorem there is a unique $u \in H_0^m(\Omega)$, solution of (A.76). If in addition $f \in L^2(\Omega)$, then $u \in H^{2m}(\Omega)$, as follows from theorem 5.3 of chapter 2, sct. 5.3, and remark 1.3 of chapter 2, sct. 1.4, of Lions-Magenes, [LM72] (see also Dautray-Lions, [DL88, sct. VII.1.7]). Consequently, we can consider the operator

$$\Lambda = A_m^{1/2} = (-\Delta)^{m/2}$$

as an operator that defines the interpolation spaces $[H_0^m(\Omega), L^2(\Omega)]_\theta$, in accord with (A.42). Recalling then (A.69), we have that

$$H_0^s(\Omega) = \left[H_0^m(\Omega), L^2(\Omega)\right]_{1-s/m} = \mathrm{dom}((-\Delta)^{\frac{m}{2}\frac{s}{m}}) = \mathrm{dom}((-\Delta)^{s/2}) = \tilde{H}^s(\Omega),$$

as per definition (A.75). ▯

In particular, for $m = 1$ we are reduced to the case $\mathcal{X} = H_0^1(\Omega)$, $\mathcal{Y} = L^2(\Omega)$, with $A_1 = -\Delta$ and

$$[\mathcal{X}, \mathcal{Y}]_\theta = \mathrm{dom}((-\Delta)^{(1-\theta)/2}).$$

Taking $\theta = 0$ yields then that

$$\mathrm{dom}((-\Delta)^{1/2}) = \mathcal{X} = H_0^1(\Omega),$$

and (A.41) reduces to the identity

$$\langle u, v \rangle_\mathcal{X} = \langle \nabla u, \nabla v \rangle,$$

which corresponds to the choice in $H_0^1(\Omega)$ of the scalar product inducing the norm (A.72). In this way, the operator $(-\Delta)^{1/2}$ can be formally related (but not identified) with the operator ∇.

5. We conclude this section by mentioning the following result on the operator $-\Delta$.

THEOREM A.79
Let Ω be a bounded domain of $\mathbb{R}^N$. The operator $-\Delta$ generates an analytic semigroup in $L^2(\Omega)$.

PROOF This is a consequence of a more general result, concerning strongly elliptic operators (even with variable coefficients). See e.g. Pazy, [Paz83, sct. 7.2, thm. 2.7].
▯

A.6 Sobolev Spaces of Vector Valued Functions

A.6.1 Lebesgue and Sobolev Spaces

In this section we introduce the Lebesgue and Sobolev spaces $L^p(a, b; \mathcal{X})$ and $H^m(a, b; \mathcal{X})$, where $]a, b[\subseteq \mathbb{R}$ is an interval, and $\mathcal{X}$ is a separable Hilbert space.

For $1 \leq p \leq +\infty$, we define $L^p(a, b; \mathcal{X})$ to be the space of the (equivalence classes of) functions $u:]a, b[\to \mathcal{X}$ which are strongly measurable, and such that

$$\int_a^b \|u(t)\|_\mathcal{X}^p \, dt < +\infty \tag{A.77}$$

if $1 \leq p < +\infty$, or

$$\text{sup ess}\{\|u(t)\|_{\mathcal{X}} : a < t < b\} < +\infty.$$

In (A.77), the integral is meant in the sense of Bochner (see e.g. Yosida, [Yos80, sct. V.5]),

It can be shown that $L^p(a,b;\mathcal{X})$ is a Banach space, with respect to the norms

$$\|u\|_p := \left(\int_a^b \|u(t)\|_{\mathcal{X}}^p \, dt \right)^{1/p}, \qquad 1 \leq p < +\infty,$$

$$\|u\|_\infty := \text{sup ess}\{\|u(t)\|_{\mathcal{X}} : a < t < b\}.$$

If $p = 2$, $L^2(a,b;\mathcal{X})$ is a Hilbert space, with respect to the scalar product

$$\langle u, v \rangle_0 := \int_a^b \langle u(t), v(t) \rangle_{\mathcal{X}} \, dt.$$

We can introduce in $L^p(a,b;\mathcal{X})$ the notion of weak (i.e., distributional) derivatives with respect to t, in a manner totally analogous to definition A.63. For $k \in \mathbb{N}$ and $u \in L^p(a,b;\mathcal{X})$, we denote by $u^{(k)}$ its strong derivative of order k. For $m \in \mathbb{N}$ we define then

$$H^m(a,b;\mathcal{X}) := \{u \in L^2(a,b;\mathcal{X}) : u^{(j)} \in L^2(a,b;\mathcal{X}), \, 0 \leq j \leq m\}.$$

This is a Hilbert space, with respect to the scalar product

$$\langle u, v \rangle_m := \sum_{j=0}^m \int_a^b \langle u^{(j)}(t), v^{(j)}(t) \rangle_{\mathcal{X}} \, dt.$$

A.6.2   The Intermediate Derivatives Theorem

In the sequel, we consider two Hilbert spaces $\mathcal{X}$ and $\mathcal{Y}$, with $\mathcal{X} \hookrightarrow \mathcal{Y}$, densely and continuous. For integer $m \geq 0$, we set

$$W(a,b;\mathcal{X},\mathcal{Y}) := \{u \in L^2(a,b;\mathcal{X}) : u^{(m)} \in L^2(a,b;\mathcal{Y})\};$$

when there is no possibility of confusion, we abbreviate $W(a,b;\mathcal{X},\mathcal{Y}) = \mathcal{W}$. Then, recalling section A.2.10, we have the following result:

THEOREM A.80

Let $u \in \mathcal{W}$. Then, for $0 \leq j \leq m$,

$$u^{(j)} \in L^2(a,b;[\mathcal{X},\mathcal{Y}]_{j/m}).$$

Moreover, if $0 \leq j \leq m-1$ and $\theta_j := \frac{1}{m}\left(j + \frac{1}{2}\right)$,

$$u^{(j)} \in C([a,b];[\mathcal{X},\mathcal{Y}]_{\theta_j}).$$

The linear maps $u \mapsto u^{(j)}$ of $\mathcal{W}$ into $L^2(a,b;[\mathcal{X},\mathcal{Y}]_{j/m})$, $0 \le j \le m$ (respectively, into $C([a,b];[\mathcal{X},\mathcal{Y}]_{\theta_j})$, $0 \le j \le m-1$), are continuous. In particular, there are positive constants C_1 and C_2, with C_1 independent of the length of $]a,b[$, such that for all $u \in \mathcal{W}$ and $1 \le j \le m-1$,

$$\max_{a \le t \le b} \|u^{(j)}(t)\|_{[\mathcal{X},\mathcal{Y}]_{\theta_j}} \le C_1 \|u\|_{L^2(a,b;\mathcal{X})}^{1-\theta_j} \|u^{(m)}\|_{L^2(a,b;\mathcal{Y})}^{\theta_j} + C_2 \|u\|_{L^2(a,b;\mathcal{X})}.$$

PROOF   See e.g. Lions-Magenes, [LM72, scts. 1.2, 1.3].   ⬚

As a consequence, recalling the definition (A.67) of $H^s(\Omega)$, as well as theorem A.75, we deduce from theorem A.80 the following imbeddings results.

THEOREM A.81
Let $m \ge r \ge -1$, and $0 \le k \le m$. Assume that

$$u \in \{u \in L^2(a,b;H^m(\Omega)): u^{(k)} \in L^2(a,b;H^r(\Omega))\}.$$

Let $p := m - \frac{j}{k}(m-r)$ and $q := p - \frac{m-r}{2k}$. Then, for $0 \le j \le k-1$,

$$u^{(j)} \in L^2(a,b;H^p(\Omega)) \cap C([a,b];H^q(\Omega)).$$

Moreover, there exist positive constants C_1 and C_2, with C_1 independent of the length of $]a,b[$, such that

$$\max_{a \le t \le b} \|u^{(j)}(t)\|_q \le C_1 \|u\|_{L^2(a,b;H^m(\Omega))}^{1-(j+1/2)/k} \|u^{(k)}\|_{L^2(a,b;H^r(\Omega))}^{(j+1/2)/k} + C_2 \|u\|_{L^2(a,b;H^m(\Omega))}.$$

We conclude with the following compact imbedding result.

THEOREM A.82
Let $\mathcal{X}$, $\mathcal{Y}$ and $\mathcal{Z}$ be reflexive Banach spaces, with $\mathcal{X} \hookrightarrow \mathcal{Y} \hookrightarrow \mathcal{Z}$, the injection $\mathcal{X} \hookrightarrow \mathcal{Y}$ being compact. Let $p,q \in]1,+\infty[$. Then the injection

$$\{u \in L^p(a,b;\mathcal{X}): u' \in L^q(a,b;\mathcal{Z})\} \hookrightarrow L^p(a,b;\mathcal{Y}),$$

which is continuous, is also compact.

PROOF   See e.g. Lions, [Lio69, sct. 1.5].   ⬚

A.7   The Spaces H(div, Ω) and H(curl, Ω)

In this section we consider certain subspaces of the space $L^2(\Omega)$, which are particularly suited to the mathematical study of various systems of PDEs which arise in fluid and electromagnetic dynamics. For a proof of the results we mention without justification, we refer e.g. to Duvaut-Lions, [DL69, sct. 5.7], or to Foias-Temam, [FT78].

A.7.1 Notations

We assume that Ω is a bounded open domain of $\mathbb{R}^N$, simply connected, and that its boundary $\partial\Omega$ is at least of Lipschitz class; we denote by $\vec{v}$ the outward unit normal to $\partial\Omega$. As usual, we denote by $\|\cdot\|$ and $\langle\cdot,\cdot\rangle$ the norm and scalar product in $L^2(\Omega)$. In addition, we adopt the following

CONVENTION A.83 *If $\vec{u} = (u_1,\dots,u_N)\colon \Omega \to \mathbb{R}^N$, $N \geq 2$, is a vector valued function, and $\mathcal{X}$ is a space of scalar functions on Ω, such as $L^2(\Omega)$ or $C(\Omega)$, then, with abuse of notation, we write $\vec{u} \in \mathcal{X}$ to mean that all the components of $\vec{u}$ are in $\mathcal{X}$. In particular, if the context is clear and there is no danger of confusion, we use the same notation $\mathcal{X}$ to denote the space $\mathcal{X}^N$ as well.*

We consider the differential operators "div" and "curl", formally defined as follows. Given $\vec{u} = (u_1,\dots,u_N)\colon \mathbb{R}^N \to \mathbb{R}^N$, for $j = 1,\dots,N$ we set $\partial_j = \partial/\partial x_j$, and

$$\operatorname{div}\vec{u} := \sum_{j=0}^{N} \partial_j u_j\,;$$

thus, $x \mapsto (\operatorname{div}\vec{u})(x)$ is a scalar function. Then, if $N = 3$ we define

$$\operatorname{curl}\vec{u} := (\partial_2 u_3 - \partial_3 u_2,\, \partial_3 u_1 - \partial_1 u_3,\, \partial_1 u_2 - \partial_2 u_1)\,,$$

while if $N = 2$ we define

$$\operatorname{curl}\vec{u} := \partial_1 u_2 - \partial_2 u_1\,;$$

thus, $x \mapsto (\operatorname{curl}\vec{u})(x)$ is a scalar function if $N = 2$, and a vector function if $N = 3$.

A.7.2 The Space $H(\operatorname{div},\Omega)$

1.   We set

$$H(\operatorname{div},\Omega) := \{\vec{u} \in L^2(\Omega)\colon \operatorname{div}\vec{u} \in L^2(\Omega)\}\,.$$

This linear subspace of $L^2(\Omega)$ is a Hilbert space with respect to the norm defined by

$$\|\vec{u}\|_{\operatorname{div}}^2 := \|\vec{u}\|^2 + \|\operatorname{div}\vec{u}\|^2\,.$$

Clearly, $H^1(\Omega) \hookrightarrow H(\operatorname{div},\Omega)$; in fact, we have

THEOREM A.84
The space $C^\infty(\overline{\Omega})$ is dense in $H(\operatorname{div},\Omega)$.

As a consequence, the NORMAL COMPONENT trace operator $\vec{u} \mapsto \vec{v} \cdot \vec{u}$ can be extended by continuity from a linear continuous map of $C^\infty(\overline{\Omega})$ into $C^\infty(\partial\Omega)$, to a linear continuous map, still denoted by $\vec{u} \mapsto \vec{v} \cdot \vec{u}$, of H(div, Ω) into $H^{-1/2}(\partial\Omega)$. More precisely, denoting by $\langle \cdot, \cdot \rangle_{1/2}$ the duality pairing between $H^{-1/2}(\partial\Omega)$ and $H^{1/2}(\partial\Omega)$, the trace $\vec{v} \cdot \vec{u}$ is defined as follows. Given any $\psi \in H^{1/2}(\partial\Omega)$, we choose $\Psi \in H^1(\Omega)$ such that $\gamma_0(\Psi) = \psi$ (this is possible by the surjectivity of γ_0; see theorem A.73). Then, we define

$$\langle \vec{v} \cdot \vec{u}, \psi \rangle_{1/2} := \langle \operatorname{div} \vec{u}, \Psi \rangle + \langle \vec{u}, \nabla \Psi \rangle. \tag{A.78}$$

Indeed, we easily see that the right side of (A.78) is independent of the choice of Ψ (as long as $\gamma_0(\Psi) = \psi$, of course), and that, if $\vec{u} \in H(\operatorname{div}, \Omega)$, it is continuous in Ψ with respect to the H^1 norm. Since the dependence of Ψ on ψ is also continuous, the right side of (A.78) depends continuously on ψ; hence, it defines an element of $H^{-1/2}(\partial\Omega)$. This element is precisely the desired normal component of $\vec{u}$. Note that (A.78) generalizes the well known integration by parts formula

$$\int_{\partial\Omega} (\vec{v} \cdot \vec{u}) v \, ds = \int_\Omega (\operatorname{div} \vec{u}) v \, ds + \int_\Omega \vec{u} \cdot \nabla v \, ds,$$

which is valid for all $\vec{u}$ and v in the space $C^1(\overline{\Omega})$, which is dense in H(div, Ω).

2. We can then introduce the following subspaces of H(div, Ω).

$$H_0(\operatorname{div}, \Omega) := \{ \vec{u} \in H(\operatorname{div}, \Omega) : \vec{v} \cdot \vec{u} = 0 \},$$
$$H^0(\operatorname{div}, \Omega) := \{ \vec{u} \in H(\operatorname{div}, \Omega) : \operatorname{div} \vec{u} = 0 \},$$
$$H^0_0(\operatorname{div}, \Omega) := H_0(\operatorname{div}, \Omega) \cap H^0(\operatorname{div}, \Omega).$$

THEOREM A.85
The space

$$C^\infty_{0,\operatorname{div}}(\Omega) := \{ \vec{u} \in C^\infty_0(\Omega) : \operatorname{div} \vec{u} = 0 \}$$

is dense in $H^0_0(\operatorname{div}, \Omega)$. *Moreover, setting*

$$H^1_v(\Omega) := \{ \vec{u} \in H^1(\Omega) : \vec{v} \cdot \vec{u} = 0 \}, \tag{A.79}$$

the FRIEDRICHS' INEQUALITY

$$\|\vec{u}\| \leq C \left(\| \operatorname{div} \vec{u} \| + \| \operatorname{curl} \vec{u} \| \right) \tag{A.80}$$

holds for all $\vec{u} \in H^1_v(\Omega)$, *with C independent of* $\vec{u}$.

A.7.3   The Space H(curl, Ω)

1. We set

$$H(\operatorname{curl}, \Omega) := \{ \vec{u} \in L^2(\Omega) : \operatorname{curl} \vec{u} \in L^2(\Omega) \}.$$

This linear subspace of $L^2(\Omega)$ is a Hilbert space with respect to the norm defined by

$$\|\vec{u}\|^2_{curl} := \|\vec{u}\|^2 + \|\operatorname{curl}\vec{u}\|^2.$$

Clearly, $H^1(\Omega) \hookrightarrow H(curl, \Omega)$; in fact, we have

THEOREM A.86
The space $C^\infty(\overline{\Omega})$ is dense in $H(curl, \Omega)$.

As a consequence, the TANGENTIAL COMPONENT trace operator $\vec{u} \mapsto \vec{v} \times \vec{u}$ can be extended by continuity from a linear continuous map of $C^\infty(\overline{\Omega})$ into $C^\infty(\partial\Omega)$, to a linear continuous map, still denoted by $\vec{u} \mapsto \vec{v} \times \vec{u}$, of $H(curl, \Omega)$ into $H^{-1/2}(\partial\Omega)$. This is done exactly as for the definition of the normal component $\vec{v} \cdot \vec{u}$; that is, we define $\vec{v} \times \vec{u} \in H^{-1/2}(\partial\Omega)$ by the identity

$$\langle \vec{v} \times \vec{u}, \psi \rangle_{1/2} := \langle \operatorname{curl}\vec{u}, \Psi \rangle - \langle \vec{u}, \operatorname{curl}\Psi \rangle, \tag{A.81}$$

for $\vec{u} \in H(curl, \Omega)$, $\psi \in H^{1/2}(\partial\Omega)$, and $\Psi \in H^1(\Omega)$ such that $\gamma_0(\Psi) = \psi$. Note that (A.81) generalizes the well known integration by parts formula

$$\int_{\partial\Omega} \vec{v} \times \vec{u} \cdot \vec{v}\,\mathrm{d}s = \int_\Omega \operatorname{curl}\vec{u} \times \vec{v}\,\mathrm{d}s - \int_\Omega \vec{u} \times \operatorname{curl}\vec{v}\,\mathrm{d}s,$$

which is valid for all $\vec{u}$ and $\vec{v}$ in the space $C^1(\overline{\Omega})$, which is dense in $H(curl, \Omega)$.

2.   We can then introduce the following subspaces of $H(curl, \Omega)$.

$$H_0(curl, \Omega) := \{\vec{u} \in H(curl, \Omega) : \vec{v} \times \vec{u} = \vec{0}\},$$
$$H^0(curl, \Omega) := \{\vec{u} \in H(curl, \Omega) : \operatorname{curl}\vec{u} = \vec{0}\},$$
$$H^0_0(curl, \Omega) := H_0(curl, \Omega) \cap H^0(curl, \Omega).$$

THEOREM A.87
The space

$$C^\infty_{0,curl}(\Omega) := \{\vec{u} \in C^\infty_0(\Omega) : \operatorname{curl}\vec{u} = \vec{0}\}$$

is dense in $H^0_0(curl, \Omega)$. Moreover, setting

$$H^1_\tau(\Omega) := \{\vec{u} \in H^1(\Omega) : \vec{v} \times \vec{u} = \vec{0}\}, \tag{A.82}$$

the Friedrichs' inequality (A.80) also holds for all $\vec{u} \in H^1_\tau(\Omega)$.

A.7.4   Relations between H(div, Ω) and H(curl, Ω)

In this section we report some results on spaces that are constructed from both H(div, Ω) and H(curl, Ω). In particular, we present a well known decomposition theorem of L$^2(\Omega)$.

THEOREM A.88
Let H$_\nu^1(\Omega)$ *and* H$_\tau^1(\Omega)$ *be the spaces defined in (A.79) and (A.82). Then,*

$$H_0(\text{curl}, \Omega) \cap H(\text{div}, \Omega) = H_\tau^1(\Omega),$$
$$H(\text{curl}, \Omega) \cap H_0(\text{div}, \Omega) = H_\nu^1(\Omega),$$

and the norm defined by

$$\|\vec{u}\|_{\mathcal{X}}^2 := \|\text{curl}\,\vec{u}\|^2 + \|\text{div}\,\vec{u}\|^2,$$

where $\mathcal{X}$ *denotes either of the spaces* H$_\nu^1(\Omega)$ *or* H$_\tau^1(\Omega)$, *is equivalent to the* H^1 *norm in* $\mathcal{X}$. *Consequently, if* $\vec{u} \in \mathcal{X}$ *is such that* div $\vec{u} = 0$ *and* curl $\vec{u} = \vec{0}$, *then* $\vec{u} = \vec{0}$.

PROPOSITION A.89
For all $\vec{a} \in$ H$_0(\text{curl}, \Omega)$, curl $\vec{a} \in$ H$_0^0(\text{div}, \Omega)$.

PROOF   Since div(curl $\vec{a}$) = $\vec{0}$, curl $\vec{a} \in$ H$^0(\text{div}, \Omega)$, so that its normal component is defined, with $\vec{v} \cdot \text{curl}\,\vec{a} \in$ H$^{-1/2}(\partial\Omega)$. Given then arbitrary $\varphi \in$ H$^{1/2}(\partial\Omega)$, let $\Phi \in$ H$^1(\Omega)$ be such that $\gamma_0(\Phi) = \varphi$. By density, we can without loss of generality assume that $\Phi \in$ H$^2(\Omega)$, so that $\gamma_0(\nabla\Phi) \in$ H$^{1/2}(\partial\Omega)$. Then, recalling (A.78) and (A.81),

$$\langle \vec{v} \cdot \text{curl}\,\vec{a}, \varphi \rangle_{1/2} = \langle \text{curl}\,\vec{a}, \nabla\Phi \rangle = \langle \vec{v} \times \vec{a}, \gamma_0(\nabla\Phi) \rangle_{1/2} = 0. \qquad (A.83)$$

$\square$

PROPOSITION A.90
For all $\psi \in$ H$_0^1(\Omega)$, $\nabla\psi \in$ H$_0^0(\text{curl}, \Omega)$.

PROOF   Since curl($\nabla\psi$) = $\vec{0}$, $\nabla\psi \in$ H$^0(\text{curl}, \Omega)$, so that its tangential component is defined, with $\vec{v} \times \nabla\psi \in$ H$^{-1/2}(\partial\Omega)$. Given then arbitrary $\varphi \in$ H$^{1/2}(\partial\Omega)$, let $\Phi \in$ H$^2(\Omega)$ be such that $\gamma_0(\Phi) = \varphi$. Then, recalling (A.81) and (A.78),

$$\langle \vec{v} \times \nabla\psi, \varphi \rangle_{1/2} = -\langle \nabla\psi, \text{curl}\,\Phi \rangle = \langle \vec{v} \cdot \text{curl}\,\Phi, \psi \rangle_{1/2} = 0. \qquad (A.84)$$

$\square$

We now set

$$\mathcal{V}_0 := H_0(\text{curl}, \Omega) \cap H^0(\text{div}, \Omega),$$

and note that, by Friedrichs' inequality (A.80), the norm defined by

$$\|\vec{u}\|_{\mathcal{V}_0} := \|\text{curl}\,\vec{u}\|$$

is a norm in $\mathcal{V}_0$ equivalent to the one induced by $H^1(\Omega)$.

PROPOSITION A.91
For all $\vec{f} \in H_0^0(\text{div}, \Omega)$, there exists a unique $\vec{u} \in \mathcal{V}_0$ such that $\text{curl}\,\vec{u} = \vec{f}$.

PROOF   Define a bilinear form a on $\mathcal{V}_0 \times \mathcal{V}_0$ by

$$a(u, v) := \langle \text{curl}\,\vec{u}, \text{curl}\,\vec{v} \rangle.$$

Then, a is bilinear and coercive on $\mathcal{V}_0$; indeed, for all $\vec{u} \in \mathcal{V}_0$,

$$a(u, u) = \|\text{curl}\,\vec{u}\|^2 = \|\vec{u}\|_{\mathcal{V}_0}^2.$$

Let $f \in \mathcal{V}_0'$ be defined by

$$\langle f, \vec{v} \rangle_{\mathcal{V}_0' \times \mathcal{V}_0} := \langle \vec{f}, \text{curl}\,\vec{v} \rangle, \qquad \vec{v} \in \mathcal{V}_0.$$

By the Lax-Milgram theorem A.35, there is a unique $\vec{u} \in \mathcal{V}_0$, such that for all $\vec{v} \in \mathcal{V}_0$,

$$\langle \text{curl}\,\vec{u}, \text{curl}\,\vec{v} \rangle = \langle \vec{f}, \text{curl}\,\vec{v} \rangle. \tag{A.85}$$

We now show that $\vec{w} := \text{curl}\,\vec{u} - \vec{f} = \vec{0}$, as desired. Indeed, we can compute $\text{curl}\,\vec{w}$ at least in distributional sense; more precisely, since $\vec{w} \in L^2(\Omega)$, recalling (A.61) and (A.60) we have that, for all $\vec{\varphi} \in D(\Omega)$,

$$\langle \text{curl}\,\vec{w}, \vec{\varphi} \rangle_D = \langle \vec{w}, \text{curl}\,\vec{\varphi} \rangle_D = \langle \vec{w}, \text{curl}\,\vec{\varphi} \rangle. \tag{A.86}$$

Given $\vec{\varphi} \in D(\Omega)$, let $\beta \in C^\infty(\Omega) \cap C^0(\overline{\Omega})$ be defined as the solution to the elliptic BVP

$$\begin{cases} -\Delta\beta = \text{div}\,\vec{\varphi}, \\ \beta_{|\partial\Omega} = 0. \end{cases}$$

Then, by proposition A.90, $\vec{\varphi} + \nabla\beta \in \mathcal{V}_0$; therefore, by (A.85),

$$\langle \vec{w}, \text{curl}\,\vec{\varphi} \rangle = \langle \vec{w}, \text{curl}(\vec{\varphi} + \nabla\beta) \rangle = 0. \tag{A.87}$$

Together with (A.86), (A.87) implies that $\text{curl}\,\vec{w} = \vec{0}$ in $D'(\Omega)$. But then, this also means that $\text{curl}\,\vec{w} \in L^2(\Omega)$, so that $\vec{w} \in H^0(\text{curl}, \Omega)$. Next, we note that $\text{div}\,\vec{w} = \text{div}\,\vec{f} = 0$, so that $\vec{w} \in H^0(\text{div}, \Omega)$ as well. In particular, the normal component $\vec{\nu} \cdot \vec{w}$

is defined in $\mathrm{H}^{-1/2}(\partial\Omega)$; since $\vec{v} \times \vec{u} = \vec{0}$ and $\vec{v} \cdot \vec{f} = 0$, recalling (A.83) we deduce that $\vec{v} \cdot \vec{w} = 0$. In conclusion, we have that $\vec{w} \in \mathrm{H}_v^1(\Omega)$, and curl $\vec{w} = \vec{0}$, div $\vec{w} = 0$. By the last part of theorem A.88, we conclude that $\vec{w} = \vec{0}$, as claimed. Finally, to prove the uniqueness of $\vec{u}$ it is sufficient to note that if $\vec{v} \in V_0$ is also such that curl $\vec{v} = \vec{f}$, then the difference $\vec{z} := \vec{u} - \vec{v}$ is such that

$$\mathrm{curl}\,\vec{z} = \vec{0}, \qquad \mathrm{div}\,\vec{z} = 0, \qquad \vec{v} \times \vec{z} = \vec{0}.$$

Hence, arguing as we did above for $\vec{w}$, we deduce that $\vec{z} = \vec{0}$. $\quad\square$

REMARK A.92   We can also determine $\vec{u}$ by resorting to general results in the classical theory of elliptic systems (see e.g. Agmon-Douglis-Nirenberg, [ADN64]). Indeed, it is sufficient to note that $\vec{u}$ can be determined as the solution of the system

$$\begin{cases} \mathrm{curl}\,\vec{u} + \nabla p = \vec{f}, \\ \mathrm{div}\,\vec{u} \qquad\;\; = 0, \\ \vec{v} \times \vec{u} \qquad\; = \vec{0}. \end{cases} \tag{A.88}$$

In fact, it turns out that this system is strongly elliptic, that its boundary conditions are complementing, and that the source data $\{\vec{f}, 0\}$ are orthogonal to the range of the adjoint of the operator $L := (\mathrm{curl} + \nabla, \mathrm{div})$ (which is $L^* = (\mathrm{curl} - \nabla, -\mathrm{div})$). Hence, (A.88) has a unique solution $(\vec{u}, p)$. However, $\nabla p = \vec{0}$, because p is also a solution of the elliptic Neumann BVP

$$\begin{cases} -\Delta p = -\mathrm{div}\,\vec{f} = 0, \\ \frac{\partial p}{\partial v} = \vec{v} \cdot \nabla p = \vec{v} \cdot (\vec{f} - \mathrm{curl}\,\vec{u}) = 0 \end{cases}$$

(the last step follows from (A.83)). Hence, curl $\vec{u} = \vec{f}$. $\quad\square$

In fact, the converse of proposition A.91 is also true.

THEOREM A.93
The following identities hold, as sets and as Hilbert spaces:

$$\mathrm{H}_0^0(\mathrm{div}, \Omega) = \{\vec{u} \in \mathrm{L}^2(\Omega) : (\exists\,\vec{a} \in \mathrm{H}(\mathrm{curl}, \Omega) : \vec{u} = \mathrm{curl}\,\vec{a})\}, \tag{A.89}$$
$$\mathrm{H}_0^0(\mathrm{curl}, \Omega) = \{\vec{u} \in \mathrm{L}^2(\Omega) : (\exists\,\psi \in \mathrm{H}_0^1(\Omega) : \vec{u} = -\nabla\psi)\}. \tag{A.90}$$

PROOF   The inclusions $\supseteq$ are a consequence of propositions A.89 and A.90. For the reverse inclusion in (A.89), use proposition A.91, with $\vec{f} = \vec{u}$. For the inclusion $\subseteq$ in (A.90), given $\vec{u} \in \mathrm{H}_0^0(\mathrm{curl}, \Omega)$ define ψ as the solution of the elliptic Dirichlet BVP

$$\begin{cases} -\Delta\psi = \mathrm{div}\,\vec{u}, \\ \psi|_{\partial\Omega} = 0. \end{cases}$$

Since $\operatorname{div} \vec{u} \in \mathrm{H}^{-1}(\Omega)$, by theorem A.77 we deduce that ψ is uniquely determined in $\mathrm{H}_0^1(\Omega)$. Let $\vec{z} := \vec{u} + \nabla \psi$. Then,

$$\begin{cases} \operatorname{curl} \vec{z} = \operatorname{curl} \vec{u} = \vec{0}, \\ \operatorname{div} \vec{z} = \operatorname{div} \vec{u} + \Delta \psi = 0, \\ \vec{v} \times \vec{z} = \vec{v} \times \nabla \psi = \vec{0} \end{cases}$$

(the last step following as in (A.84)). Hence, $\vec{z} = \vec{0}$, by the last part of theorem A.88; that is, $\vec{u} = -\nabla \psi$, as desired. $\qquad \square$

We can then finally state the following orthogonality result.

THEOREM A.94
Let $\Omega \subset \mathbb{R}^3$ be a bounded, simply connected open domain, with a Lipschitz boundary $\partial \Omega$. Then,

$$\mathrm{L}^2(\Omega) = \mathrm{H}_0^0(\operatorname{div}, \Omega) \oplus \mathrm{H}^0(\operatorname{curl}, \Omega).$$

In other words, for all $\vec{u} \in \mathrm{L}^2(\Omega)$, there exist $\vec{a} \in \mathrm{H}_0(\operatorname{curl}, \Omega)$ and $\psi \in \mathrm{H}^1(\Omega)$, such that

$$\vec{u} = \operatorname{curl} \vec{a} - \nabla \psi. \qquad (A.91)$$

The functions ψ and $\vec{a}$ of theorem A.93 are usually called the SCALAR and VECTOR POTENTIALS of $\vec{u}$; in general, they are not uniquely determined.

PROOF The inclusion $\supseteq$ is obvious. For the converse, we introduce the space

$$\mathcal{V}_1 := \mathrm{H}^1(\Omega) \big| \mathbb{R}$$

of the equivalence classes $\Phi = [\varphi]$ of functions in $\mathrm{H}^1(\Omega)$, with respect to the equivalence relation defined by

$$f \sim g \quad \overset{\mathrm{def.}}{\Longleftrightarrow} \quad \exists c \in \mathbb{R}: f(x) - g(x) = c \text{ a.e.}$$

We denote such equivalence classes by capital letters like Ψ, and any of their representatives by a lower case letter like ψ. The space $\mathcal{V}_1$ is a Hilbert space, with respect to the norm

$$\|\Psi\|_{\mathcal{V}_1} = \|\nabla \psi\|, \quad \psi \in \Psi \qquad (A.92)$$

(see e.g. Nečas, [Ne£67, sct. 2.7.2]; note that the right side of (A.92) is independent of the particular representative ψ of Ψ). Consider then the bilinear form on $\mathcal{V}_1$ defined by

$$a(\Phi, \Psi) := \langle \nabla \varphi, \nabla \psi \rangle, \quad \varphi \in \Phi, \ \psi \in \Psi.$$

This form is continuous and coercive on $\mathcal{V}_1$; in particular,

$$a(\Phi, \Phi) = \|\nabla \varphi\|^2 = \|\Phi\|^2_{\mathcal{V}_1}.$$

The map

$$\Phi \ni \Phi \mapsto -\langle \vec{u}, \nabla \varphi \rangle, \qquad \varphi \in \Phi,$$

is linear continuous on $\mathcal{V}_1$; therefore, by the Lax-Milgram theorem A.35, there is a unique $\Psi \in \mathcal{V}_1$ such that for all $\Phi \in \mathcal{V}_1$,

$$a(\Psi, \Phi) = \langle \nabla \psi, \nabla \varphi \rangle = \langle \vec{u}, \nabla \varphi \rangle. \tag{A.93}$$

Fix now any $\psi \in \Psi$ (thus, $\psi \in H^1(\Omega)$). Given any $\zeta \in H^1_0(\Omega)$, we compute that, because of (A.93),

$$\langle -\operatorname{div}(\nabla \psi + \vec{u}), \zeta \rangle_{H^{-1}(\Omega) \times H^1_0(\Omega)} = \langle \nabla \psi + \vec{u}, \nabla \zeta \rangle = 0.$$

Consequently, $\nabla \psi + \vec{u} \in H^0(\operatorname{div}, \Omega)$; therefore, its normal component is defined in $H^{-1/2}(\partial \Omega)$. Given any $\alpha \in H^{1/2}(\partial \Omega)$, let $A \in H^1(\Omega)$ be such that $\gamma_0(A) = \alpha$. Then, again by (A.93),

$$\langle \vec{v} \cdot (\nabla \psi + \vec{u}), \alpha \rangle_{1/2} = (\nabla \psi + \vec{u}, \nabla A) = 0.$$

Consequently, $\nabla \psi + \vec{u} \in H^0_0(\operatorname{div}, \Omega)$, so that, by (A.89) of theorem A.93, there is $\vec{a} \in H_0(\operatorname{curl}, \Omega)$ such that $\nabla \psi + \vec{u} = \operatorname{curl} \vec{a}$, as desired in (A.91). $\qquad \Box$

A.8   Almost Periodic Functions

In this section we briefly recall some basic facts concerning almost periodic functions, valued in a separable Banach space $\mathcal{X}$. All the material in this section is taken from Amerio-Prouse, [AP71].

DEFINITION A.95   *A subset $\mathcal{J} \subseteq \mathbb{R}$ is said to be* RELATIVELY DENSE, *if there is $\ell > 0$ such that any interval of $\mathbb{R}$ of length ℓ contains at least one point of $\mathcal{J}$.*

DEFINITION A.96   *Let $f \colon \mathbb{R} \to \mathcal{X}$ be continuous and, for $\tau \in \mathbb{R}$, define the translated function $t \mapsto f_\tau(t) := f(t + \tau)$. The function f is said to be* ALMOST PERIODIC *if for all $\varepsilon > 0$ there is a relatively dense set $\mathcal{T} \subseteq \mathbb{R}$ such that for all $\tau \in \mathcal{T}$,*

$$\sup_{t \in \mathbb{R}} \|f(t + \tau) - f(t)\|_{\mathcal{X}} < \varepsilon.$$

Each $\tau \in \mathcal{T}$ is called an ε-ALMOST PERIOD of f.

The following characterization of almost periodic functions is generally known as
BOCHNER'S THEOREM.

THEOREM A.97
The following conditions are equivalent:

1. *The function $f \colon \mathbb{R} \to \mathcal{X}$ is almost periodic;*

2. *The set of the translations $(f_\tau)_{\tau \in \mathbb{R}}$ is relatively compact in the space $C_b(\mathbb{R})$, endowed with the sup norm;*

3. *There exists a relatively dense sequence $(\tau_m)_{m \in \mathbb{N}}$ such that the sequence of translated functions $(f(\cdot + \tau_m))_{m \in \mathbb{N}}$ is relatively compact in $C_b(\mathbb{R})$.*

PROOF   See e.g. Amerio-Prouse, [AP71, ch. 1]. ⬜

In particular, since for all $x \in \mathcal{X}$ and $\lambda \in \mathbb{R}$, the function $t \mapsto e^{i\lambda t}x$ is periodic, we deduce that all trigonometric polynomials of the form

$$\sum_{k=1}^{N} e^{i\lambda_k t} a_k,$$

with $\lambda_k \in \mathbb{R}$ and $a_k \in \mathcal{X}$, are almost periodic. The following theorem characterizes almost periodic functions as the sum of a uniformly convergent trigonometric series.

THEOREM A.98
A function $f \colon \mathbb{R} \to \mathbb{R}^N$ is almost periodic if and only if there are sequences $(a_k)_{k \in \mathbb{N}} \subseteq \mathbb{R}^N$ and $(\lambda_k)_{k \in \mathbb{N}} \subset \mathbb{R}$ such that for all $t \in \mathbb{R}$,

$$f(t) = \sum_{k=1}^{\infty} e^{i\lambda_k t} a_k.$$

PROOF   See e.g. Amerio-Prouse,[AP71, ch. 2]. ⬜

Bibliography

[Ada78]   R.A. Adams. *Sobolev spaces*. Academic Press, Inc., New York, 1978.

[ADN64]   S. Agmon, A. Douglis, and L. Nirenberg. Estimates near the boundary for the solution of elliptic differential equations satisfying general boundary values, II. *Comm. Pure Appl. Math.*, 17:35–92, 1964.

[Ama90]   H. Amann. *Ordinary Differential Equations: An Introduction to Nonlinear Analysis*. De Gruyter, Berlin, 1990.

[AP71]   L. Amerio and G. Prouse. *Almost-periodic functions and functional equations. (The University Series in Higher Mathematics.)*. Van Nostrand Reinhold Company VIII, New York, 1971.

[AP93]   A. Ambrosetti and G. Prodi. *A Primer of Nonlinear Analysis*. Cambridge University Press, Cambridge, 1993.

[ASY96]   K.T. Alligood, T.D. Sauer, and J.A. Yorke. *Chaos. An Introduction to Dynamical Systems*. Springer-Verlag, New York, 1996.

[Bal73]   J.M. Ball. Stability theory for an extensible beam. *J. Diff. Eqs.*, 14:399–418, 1973.

[BdMCR98]   L. Boutet de Monvel, I.D. Chueshov, and A.V. Rezounenko. Inertial manifolds for retarded semilinear parabolic equations. *Nonlinear Analysis*, 34:907–925, 1998.

[BF95]   A. Bensoussan and F. Flandoli. Stochastic inertial manifold. *Stochastics Rep.*, 53(1-2):13–39, 1995.

[BJ89]   P.W. Bates and C.K.R.T. Jones. Invariant manifolds for semilinear partial differential equations. In U. Kirchgrabner and H.O. Walther, editors, *Dynamics Reported, Vol. 2*, pages 1–38. John Wiley & Sons, New York, 1989.

[BL76]   J. Bergh and J. Löfström. *Interpolation Spaces. An Introduction*, volume 223 of *Grundlehren der mathematischen Wissenschaften*. Springer-Verlag, Berlin, 1976.

[Bre73]   H. Brezis. *Opérateurs Maximaux Monotones*. North-Holland, Amsterdam, 1973.

[BV92]   A. V. Babin and M. I. Vishik. *Attractors of Evolution Equations*. North Holland, Amsterdam, 1992.

[CF88] P. Constantin and C. Foias. *Navier-Stokes Equations.* The University of Chicago Press, Chicago, 1988.

[CFNT86] P. Constantin, C. Foias, B. Nicolaenko, and R. Temam. Nouveaux resultats sur les variétés inertielles pour les équations différentielles dissipatives. *C. R. Acad. Sci., Paris, Ser. I,* 302:375–378, 1986.

[CFNT89] P. Constantin, C. Foias, B. Nicolaenko, and R. Temam. *Integral Manifolds and Inertial Manifolds for Dissipative Partial Differential Equations,* volume 70 of *Applied Mathematical Sciences.* Springer-Verlag, Berlin, 1989.

[Chu95] I.D. Chueshov. Approximate inertial manifolds of exponential order for semilinear parabolic equations subjected to additive white noise. *J. Dyn. Differ. Equations,* 7(4):549–566, 1995.

[CI74] N. Chaffee and E. Infante. A bifurcation problem for a nonlinear parabolic equation. *Appl. Anal.,* 4:17–37, 1974.

[CL55] E.A. Coddington and N. Levinson. *Theory of Ordinary Differential Equations.* McGill-Hill Book Company, Inc., 1955.

[CL99] T. Caraballo and J.A. Langa. Tracking properties of trajectories on random attracting sets. *Stochastic Anal. Appl.,* 17(3):339–358, 1999.

[CS01] I.D. Chueshov and M. Scheutzow. Inertial manifolds and forms for stochastically perturbed retarded semilinear parabolic equations. *J. Dyn. Differ. Equations,* 13(2):355–380, 2001.

[DG91] F. Demengel and J.M. Ghidaglia. Some remarks on the smoothness of inertial manifolds. *Nonlinear Anal., Theory Methods Appl.,* 16(1):79–87, 1991.

[DL69] G. Duvaut and J.L. Lions. *Les Inéquations en Mécanique et en Physique.* Dunod-Gauthier-Villars, Paris, 1969.

[DL88] R. Dautray and J.-L. Lions. *Mathematical Analysis and Numerical Methods for Science and Technology. Volume 2: Functional and Variational Methods.* Springer-Verlag, Berlin, 1988.

[DL90] R. Dautray and J.-L. Lions. *Mathematical Analysis and Numerical Methods for Sciences and Technology. Volume 3: Spectral Theory and Applications.* Springer-Verlag, Berlin, 1990.

[DLS03] J. Duan, K. Lu, and B. Schmalfuss. Invariant manifolds for stochastic partial differential equations. *Ann. Probab.,* 31(4):2109–2135, 2003.

[EFK98] A. Eden, C. Foias, and V. Kalantarov. A remark on two constructions of exponential attractors for α-contractions. *J. Dyn. Differ. Equations,* 10(1):37–45, 1998.

[EFNT94] A. Eden, C. Foias, B. Nicolaenko, and R. Temam. *Exponential Attractors for Dissipative Evolution Equations.* Research in Applied Mathematics. Wiley and Masson, 1994.

[EM93] A. Eden and A. Milani. Exponential attractors for extensible beam equations. *Nonlinearity*, 6:457–479, 1993.

[EM95] A. Eden and A. Milani. On the convergence of attractors and exponential attractors for singularly perturbed hyperbolic equations. *Turkish Math. J.*, 19:102–117, 1995.

[EMN92] A. Eden, A. Milani, and B. Nicolaenko. Finite dimensional exponential attractors for semilinear wave equations with damping. *J. Math. An. Appl.*, 169(2):408–419, 1992.

[EN00] K.-J. Engel and R. Nagel. *One-Parameter Semigroups for Linear Evolution Equations.*, volume 194 of *Graduate Texts in Mathematics*. Springer-Verlag, Berlin, 2000.

[Fal85] K.J. Falconer. *The Geometry of Fractal Sets*, volume 85 of *Cambridge Tracts in Mathematics*. Cambridge University Press, Cambridge, 1985.

[Fal90] K. Falconer. *Fractal Geometry: Mathematical Foundations and Applications*. John Wiley & Sons, Chichester, 1990.

[Fei78] M. J. Feigenbaum. Qualitative universality for a class of nonlinear transformations. *J. Stat. Phys.*, 19(1):25–52, 1978.

[FGMZ03] P. Fabrie, C. Galusinski, A. Miranville, and S. Zelik. Uniform Exponential Attractors for a Singularly Perturbed Damped Wave Equation. Preprint, 2003.

[FNST85] C. Foias, B. Nicolaenko, G.R. Sell, and R. Temam. Variétés inertielles pour l'équation de Kuramoto-Sivashinsky. *C. R. Acad. Sci., Paris, Ser. I*, 301:285–288, 1985.

[FNST88] C. Foias, B. Nicolaenko, G.R. Sell, and R. Temam. Inertial manifolds for the Kuramoto-Sivashinsky equation and an estimate of their lowest dimension. *J. Math. Pures Appl., IX.*, 67(3):197–226, 1988.

[FST85] C. Foias, G.R. Sell, and R. Temam. Variétés inertielles des équations différentielles dissipatives. (Inertial manifolds for dissipative differential equations). *C. R. Acad. Sci., Paris, Ser. I*, 301:139–141, 1985.

[FST88] C. Foias, G.R. Sell, and R. Temam. Inertial manifolds for nonlinear evolutionary equations. *J. Differ. Equations*, 73:309–353, 1988.

[FST89] C. Foias, G.R. Sell, and E.S. Titi. Exponential tracking and approximation of inertial manifolds for dissipative nonlinear equations. *J. Dyn. Differ. Equations*, 1(2):199–244, 1989.

[FT78]   C. Foias and R. Temam. Remarques sur les équations de Navier-Stokes stationnaires et les phénomènes successifs de bifurcation. *Ann. Sc. Norm. Sup. Pisa, Ser. IV*, 5:29–63, 1978.

[GGMP03]   S. Gatti, M. Grasselli, A. Miranville, and V. Pata. A Construction of a Robust Family of Exponential Attractors. Preprint, 2003.

[GH83]   J. Guckenheimer and P. Holmes. *Nonlinear Oscillations, Dynamical Systems, and Bifurcations of Vector Fields*. Springer-Verlag, New York, 1983.

[GP02]   M. Grasselli and V. Pata. On the damped semilinear wave equation with critical exponent. In *Proc. of IV Int. Conf. on Dyn. Systs. and PDEs, May 24-27*, Wilmington, NC, USA, 2002.

[GT83]   D. Gilbarg and N.S. Trudinger. *Elliptic Partial Differential Equations of Second Order*, volume 224 of *Grundlehren der mathematischen Wissenschaften*. Springer-Verlag, Berlin, 1983.

[GV97]   A.Y. Goritskij and M.I. Vishik. Local integral manifolds for a nonautonomous parabolic equation. *J. Math. Sci., New York*, 85(6):2428–2439, 1997.

[Had01]   J. Hadamard. Sur l'iteration et les solutions asymptotiques des equations différentielles. *Bull. Soc. Math. France*, 29:224–228, 1901.

[Hal88]   J. Hale. *Asymptotic Behavior of Dissipative Systems*, volume 25 of *AMS Math. Surveys and Monographs*. Providence, Rhode Island, 1988.

[Har91]   A. Haraux. *Systemes Dynamiques Dissipatifs et Applications*. Masson, Paris, 1991.

[Hen81]   D. Henry. *Geometric Theory of Semilinear Parabolic Equations*, volume 850 of *Lecture Notes in Mathematics*. Springer-Verlag, 1981.

[Hen85]   D. Henry. Some infinite-dimensional Morse-Smale systems defined by parabolic partial differential equations. *J. Differ. Equations*, 59:165–205, 1985.

[HP76]   M. Hénon and Y. Pomeau. Two strange attractors with a simple structure. In R. Temam, editor, *Turbulence and Navier-Stokes Equations*, volume 565 of *Lecture Notes Math.*, pages 29–68. Springer-Verlag, Berlin, 1976.

[HR88]   J. K. Hale and G. Raugel. Upper semicontinuity of the attractor for a singularly perturbed hyperbolic equation. *J. Diff. Eqs.*, 73(2):197–214, 1988.

[HR90]   J. K. Hale and G. Raugel. Lower semicontinuity of the attractor for a singularly perturbed hyperbolic equation. *J. Dyn. Diff. Eqs.*, 2(1):19–67, 1990.

[HS93]   M. Hirsch and S. Smale. *Differential Equations, Dynamical Systems, and Linear Algebra*. Academic Press, Boston, 1993.

[Jol89]   M.S. Jolly. Explicit construction of an inertial manifold for a reaction diffusion equation. *J. Differ. Equations*, 78(2):220–261, 1989.

[JS87]   D. W. Jordan and P. Smith. *Nonlinear Ordinary Differential Equations*. Clarendon Press, Oxford, 1987.

[JT96]   D.A. Jones and E.S. Titi. C^1 approximations of inertial manifolds for dissipative nonlinear equations. *J. Differ. Equations*, 127(1):54–86, 1996.

[Kat95]   T. Kato. *Perturbation Theory for Linear Operators*. Classics in Mathematics. Springer-Verlag, Berlin, 1995.

[Kom70]   Y. Komura. Nonlinear semigroups in Hilbert spaces. In *Proc. Int. Conf. Funct. Anal. Rel. Topics, Tokyo 1969*, pages 260–268. 1970.

[KS02a]   N. Koksch and S. Siegmund. Inertial manifolds for nonautonomous dynamical systems and for nonautonomous dynamical systems. *CD ROM Proceedings of the Equadiff 10, Prague 2001*, pages 221–266, 2002.

[KS02b]   N. Koksch and S. Siegmund. Pullback attracting inertial manifolds for nonautonomous dynamical systems. *J. Dyn. Differ. Equations*, 14(4):889–941, 2002.

[KS03]   N. Koksch and S. Siegmund. Cone invariance and squeezing properties for inertial manifolds of nonautonomous evolution equations. *Banach Center Publications, vol. 60, Warzawa*, pages 27–48, 2003.

[Kur66]   C. Kuratowski. *Topology*, volume I. Academic Press, New York, 1966.

[Lio69]   J. L. Lions. *Quelques Méthodes de Résolution des Problémes aux Limites non Linéaires*. Dunod, Paris, 1969.

[Lio73]   J. L. Lions. *Perturbations Singulières dans les Problémes aux Limites et en Contrôle Optimal*, volume 323 of *Lect. Notes Math.* Springer-Verlag, Berlin, 1973.

[LL99]   Y. Latushkin and B. Layton. The optimal gap condition for invariant manifolds. *Discrete and Continuous Dynamical Systems*, 5(2):233–268, 1999.

[LM72]   J. L. Lions and E. Magenes. *Non-Homogeneous Boundary Value Problems*, volume I. Springer-Verlag, New York, 1972.

[Lor63]   E.N. Lorenz. Deterministic nonperiodic flow. *J. Atmospherical Sciences*, 20(2):130–141, 1963.

[LS89] M. Luskin and G.R. Sell. Approximation theories for inertial manifolds. *RAIRO, Modelisation Math. Anal. Numer.*, 23(3):445–461, 1989.

[Lya47] A.M. Lyapunov. *Problème Gènèral de la Stabilité de Mouvement*, volume 17 of *Annals of Math. Studies*. Princeton University Press, Princeton, 1947.

[Lya92] A.M. Lyapunov. *The General Problem of Stability of Motion. With a biography of Lyapunov by V.I. Smirnov and a bibliography of Lyapunov's work by J.F. Barret.* Transl. and ed. by A.T. Fuller. Taylor & Francis, London, 1992.

[Mañ81] B. Mañé. *On the Dimension of the Compact Invariant Sets of Certain Nonlinear Maps*, volume 898 of *Lecture Notes Math.*, pages 230–242. Springer-Verlag, Berlin, 1981.

[MM76] J. Marsden and M. McCracken. *Hopf Bifurcation and its Applications.* Springer-Verlag, New York, 1976.

[Moo92] F. C. Moon. *Chaotic and Fractal Dynamics.* Wiley, New York, 1992.

[Mor87] X. Mora. Finite-dimensional attracting invariant manifolds for damped semilinear wave equations. In *Contributions to Nonlinear Partial Differential Equations, Vol. II, Proc. 2nd Franco-Span. Colloq., Paris 1985, Pitman Res. Notes Math. Ser. 155*, pages 172–183. 1987.

[MPS88] J. Mallet-Paret and G.R. Sell. Inertial manifolds for reaction diffusion equations in higher space dimensions. *J. Am. Math. Soc.*, 1(4):804–866, 1988.

[MSM87] X. Mora and J. Solà-Morales. *Existence and Nonexistence of Finite Dimensional Globally Attracting Invariant Manifolds in Semilinear Damped Wave Equations*, pages 187–210. Springer-Verlag, New York, 1987.

[Neč67] J. Nečas. *Les Méthodes Directes en Théorie des Équations Elliptiques.* Academia, Prague, 1967.

[Nin93] H. Ninomiya. Inertial manifolds and squeezing properties. *Lect. Notes Numer. Appl. Anal.*, 12:175–202, 1993.

[Ole96] O. Oleinik. *Some Asymptotic Problems in the Theory of Partial Differential Equations.* Lezioni Fermiane. Cambridge Univ. Press, Cambridge, 1996.

[Paz83] A. Pazy. *Semigroups of Linear Operators and Applications to Partial Differential Equations.* Springer-Verlag, New York, 1983.

[Pee66] J. Peetre. Espaces d'interpolation et théorème de Sobolev. *Ann. Inst. Fourier*, 16:279 – 317, 1966.

[Per28] O. Perron. Über Stabilität und asymptotisches Verhalten der Integrale von Differentialgleichungssystemen. *Math. Zeitschrift*, 29:129–160, 1928.

[Per29] O. Perron. Über Stabilität und asymptotisches Verhalten der Lösungen eines Systems endlicher Differenzengleichungen. *J. Reine Angew. Math.*, 161:41–64, 1929.

[Per30] O. Perron. Die Stabilitätsfrage bei Differentialgleichungen. *Math. Zeitschrift*, 32:703–728, 1930.

[Per91] L. Perko. *Differential Equations and Dynamical Systems*. Texts in Applied Mathematics 7. Springer-Verlag, Berlin, 1991.

[Rac92] R. Racke. *Lectures on Nonlinear Evolution Equations*. Vieweg, Braunschweig, 1992.

[Rhe69] W.C. Rheinboldt. Local mapping relations and global implicit function theorems. *Trans. Am. Math. Soc.*, 138:183–198, 1969.

[Rob93] J.C. Robinson. Inertial manifolds and the cone condition. *Dyn. Syst. Appl.*, 2(3):311–330, 1993.

[Rob96] J.C. Robinson. The asymptotic completeness of inertial manifolds. *Nonlinearity*, 9:1325–1340, 1996.

[Rob99] J.C. Robinson. Global attractors: Topology and finite-dimensional dynamics. *J. Dyn. Differ. Equations*, 11(3):557–581, 1999.

[Rob01] J.C. Robinson. *Infinite-Dimensional Dynamical Systems. An Introduction to Dissipative Parabolic PDEs and the Theory of Global Attractors*. Cambridge Texts in Applied Mathematics. Cambridge University Press, Cambridge, 2001.

[Rud73] W. Rudin. *Functional Analysis*. McGraw-Hill Series in Higher Mathematics. McGraw-Hill Book Comp., New York, 1973.

[Rud74] W. Rudin. *Real and Complex Analysis. 2nd ed.* McGraw-Hill Series in Higher Mathematics. McGraw-Hill Book Comp., New York, 1974.

[Spa82] C. Sparrow. *The Lorenz Equations: Bifurcations, Chaos and Strange Attractors*, volume 41 of *Lecture Notes in Applied Mathematics*. Springer-Verlag, Berlin, 1982.

[SY02] G.R. Sell and Y. You. *Dynamics of Evolutionary Equations*, volume 143 of *Applied Mathematical Sciences*. Springer-Verlag, New York, 2002.

[Tan79] H. Tanabe. *Equations of Evolution*. Pitman, London, 1979.

[Tem83] R. Temam. *Navier-Stokes Equations and Nonlinear Functional Analysis*. CBMS-NSF Reg. Conf. Series in Appl. Math., SIAM, Philadelphia, 1983.

[Tem88] R. Temam. *Infinite-Dimensional Dynamical Systems in Mechanics and Physics*, volume 68 of *Applied Mathematical Sciences*. Springer-Verlag, New York, 1988.

[Tri95] H. Triebel. *Interpolation Theory, Function Spaces, Differential Operators*. J.A. Barth Verlag, Leipzig, 1995.

[TY94] M. Taboado and Y. You. Invariant manifolds for retarded semilinear wave equations. *J. Differ. Equations*, 114(2):337–369, 1994.

[Ver90] F. Verhulst. *Nonlinear Ordinary Differential Equations and Dynamical Systems*. Springer-Verlag, Berlin, 1990.

[WF97] Z. Wang and X. Fan. Inertial manifolds for nonautonomous evolution equations. *Adv. Math., Beijing*, 26(2):185–186, 1997.

[Yos80] K. Yosida. *Functional Analysis. 6th ed.*, volume 123 of *Grundlehren der mathematischen Wissenschaften*. Springer-Verlag, Berlin, 1980.

[Zei95] E. Zeidler. *Applied Functional Analysis. Applications to Mathematical Physics. Vol. 1.*, volume 108 of *Applied Mathematical Sciences*. Springer-Verlag, Berlin, 1995.

[ZM03] S. Zheng and A. Milani. Global attractors for singular perturbations of the Cahn-Hilliard equations. Preprint, 2003.

[ZM05] S. Zheng and A. Milani. Exponential attractors and inertial manifolds for singular perturbations of the Cahn-Hilliard equations. *To appear in Nonlinear Analysis - TMA*, 2005.

Index

Nomenclature

$\mathrm{AC}(\mathcal{A},\mathcal{B})$ the space of the absolutely continuous functions from $\mathcal{A}$ to $\mathcal{B}$, 325

$\alpha(\mathcal{Y})$ the α-limit set of $\mathcal{Y}$, 53

$\alpha(x)$ the α-limit set of the point x, 55

$\alpha(\mathcal{A})$ the compactness measure of $\mathcal{A}$, 74

A' the adjoint (or transpose) of the operator As, 333

A^s a fractional power of A, 340

A^* the adjoint of the operator A, 333

$B(a,r)$ the ball with center a and radius r in X, 105

$\overline{B}(a,r)$ the closed ball with center a and radius r in X, 155

$B_\delta(a,r)$ the ball with center a and radius r with respect to the pseudometric δ, 80

$\mathrm{C}(\mathcal{A};\mathcal{B})$ the space of continuous functions from $\mathcal{A}$ into $\mathcal{B}$, 5

$\mathrm{C}_0^\infty(\Omega)$ the space of the infinitely differentiable functions with compact support in Ω, 350

$\mathrm{C}^1(\mathcal{A};\mathcal{B})$ the space of continuously differentiable functions from $\mathcal{A}$ into $\mathcal{B}$, 5

$\mathcal{C}_L$ the cone in X with respect to the projections π_1, π_2 and the parameter L, 189

$\mathcal{C}_P$ the cone in X with respect to the projection P, 138

D^α derivative to the multiindex α, 353

$\mathrm{diam}(\mathcal{M})$ the diameter of $\mathcal{M}$, 74

$\dim_F(\mathcal{K})$ the fractal dimension of $\mathcal{K}$, 81

$d(a,b)$ the distance of the points a and b, 41

$d(a,\mathcal{B})$ the distance of the point a to the set $\mathcal{B}$, 41

$\partial(\mathcal{A},\mathcal{B})$ the semidistance from the $\mathcal{A}$ to the set $\mathcal{B}$, 41

$\mathrm{dist}(\mathcal{A},\mathcal{B})$ the distance of the sets $\mathcal{A}$ and $\mathcal{B}$, 41

$\mathrm{D}(\Omega)$ the space of the test functions on Ω, 353

$\mathrm{D}'(\Omega)$ the space of the distributions on Ω, 353

$\mathcal{E}_{>a}$ the set $\{x \in \mathcal{E}: x > a\}$, 41

$\mathcal{E}_{\geq a}$ the set $\{x \in \mathcal{E}: x \geq a\}$, 41

$\mathcal{E}_{<a}$ the set $\{x \in \mathcal{E}: x < a\}$, 41

$\mathcal{E}_{\leq a}$ the set $\{x \in \mathcal{E}: x \leq a\}$, 41

F' the Fréchet derivative of F, 344

$\gamma(x)$ the complete orbit through x, 50

$\gamma_-(x)$ the backward orbit starting at x, 50

$\gamma_+(x)$ the forward orbit starting at x, 50

$H^m(\Omega)$ the Sobolev space of the functions on Ω having weak derivatives up to the order m in $L^2(\Omega)$, 354

$H^m(a,b;\mathcal{X})$ the Sobolev space of the functions from $]a,b[$ into $\mathcal{X}$ having weak derivatives up to the order m in $L^2(a,b;\mathcal{X})$, 362

$H^s(\Omega)$ the interpolation space $[H^m(\Omega),L^2(\Omega)]_{1-s/m}$ with $m \geq s$, 355

$H_0^s(\Omega)$ the closure of $C_0^\infty(\Omega)$ in $H^s(\Omega)$, 357

$\hookrightarrow$ the continuous injection, 336

$\lfloor x \rfloor$ the largest integer less than or equal to x, 24

$[a,b[$ the left closed, right open interval from a to b, 5

$[a,b]$ the closed interval from a to b, 19

$]a,b[$ the open interval from a to b, 5

$]a,b]$ the left open, right closed interval from a to b, 21

$\mathcal{K}(\mathcal{X},\mathcal{Y})$ the set of all linear, compact operators from $\mathcal{X}$ into $\mathcal{Y}$, 332

$\mathcal{L}(\mathcal{X},\mathcal{Y})$ the set of all linear, bounded operators from $\mathcal{X}$ into $\mathcal{Y}$, 330

$L^\infty(\Omega)$ the space of the equivalence classes of the measurable, essentially bounded functions on Ω, 349

$L^p(\Omega)$ the space of the equivalence classes of Lebesgue p-integrable functions on Ω, 349

$L_{\text{loc}}^p(\Omega)$ the space of the equivalence classes of locally Lebesgue p-integrable functions on Ω, 349

$L^p(a,b;\mathcal{X})$ the space of the p-integrable functions on $]a,b[$ with values in $\mathcal{X}$, 361

$L^\infty(a,b;\mathcal{X})$ the space of measurable, essentially bounded functions on $]a,b[$ with values in $\mathcal{X}$, 361

$\mathcal{M}^s(x)$ the stable manifold through x, 57

$\mathcal{M}^u(x)$ the unstable manifold through x, 57

$\mathbb{N}$ the set of natural numbers (including 0), 2

$\omega(\mathcal{Y})$ the ω-limit set of $\mathcal{Y}$, 53

$\omega(x)$ the ω-limit set of the point x, 55

$\mathbb{R}$ the set of real numbers, xiv

$\rho(A)$ the resolvent set of A, 332

$R(A,\lambda)$ the resolvent of A, 332

$\sigma_p(A)$ the point spectrum of A, 332

$\mathcal{T}_*^2$ the set $\{(t,\tau) \in \mathcal{T} \times \mathcal{T} : t \geq \tau\}$, 2

$\mathcal{W}(a,b;\mathcal{X},\mathcal{Y})$ for fixed m, the intermediate space of all functions in $L^2(a,b;\mathcal{X})$ whose m-th. derivative belongs to $L^2(a,b;\mathcal{Y})$, 362

$\mathcal{X}'$ the (topological) dual of $\mathcal{X}$, 329

$[\mathcal{X},\mathcal{Y}]_\theta$ the interpolation space between $\mathcal{X}$ and $\mathcal{Y}$ to the parameter θ, 343

$\mathbb{Z}$ the set of integer numbers, 2